COMBINATORIAL SYNTHESIS OF NATURAL PRODUCT-BASED LIBRARIES

CRITICAL REVIEWS IN COMBINATORIAL CHEMISTRY

Series Editors

BING YAN
School of Pharmaceutical Sciences
Shandong University, China

ANTHONY W. CZARNIK
Department of Chemistry
University of Nevada–Reno, U.S.A.

A series of monographs in molecular diversity and combinatorial chemistry, high-throughput discovery, and associated technologies.

Combinatorial and High-Throughput Discovery and Optimization of Catalysts and Materials
Edited by Radislav A. Potyrailo and Wilhelm F. Maier

Combinatorial Synthesis of Natural Product-Based Libraries
Edited by Armen M. Boldi

COMBINATORIAL SYNTHESIS OF NATURAL PRODUCT-BASED LIBRARIES

Edited by
Armen M. Boldi

CRC Press
Taylor & Francis Group
Boca Raton London New York

CRC Press is an imprint of the
Taylor & Francis Group, an **informa** business

First published 2006 by Taylor & Francis

Published 2019 by CRC Press
Taylor & Francis Group
6000 Broken Sound Parkway NW, Suite 300
Boca Raton, FL 33487-2742

First issued in paperback 2019

No claim to original U.S. Government works

ISBN 13: 978-0-367-45362-6 (pbk)
ISBN 13: 978-0-8493-4000-0 (hbk)

Library of Congress Card Number 2005046686

Library of Congress Cataloging-in-Publication Data

Combinatorial synthesis of natural product-based libraries / edited by Armen M. Boldi.
p. cm. -- (Critical reviews in combinatorial chemistry)
Includes bibliographical references and index.
ISBN 0-8493-4000-4 (alk. paper)
1. Combinatorial chemistry. 2. Natural products--Synthesis. I. Boldi, Armen M. II. Series.

RS419.C68 2006
615'19--dc22   2005046686

Visit the Taylor & Francis Web site at
http://www.taylorandfrancis.com

and the CRC Press Web site at
http://www.crcpress.com

Preface

As we view the world we live in, we become inspired by the beauty, variety, complexity, and grandeur of the natural realm. Many disciplines obtain inspiration for their own worthy pursuits from that which exists in nature. As synthetic organic chemists, medicinal chemists, pharmaceutical scientists, and natural product researchers, we are inspired by the molecules found in nature to synthesize what nature synthesizes, to make variations of what we discover in nature, and to look to nature for inspiration in new directions for our profession. What motivates us is not only a desire to better understand biological processes and how molecules interact and control these pathways, but we are looking for new ways to modulate biological responses in order to treat disease.

From the beginning of history, we have utilized the resources available to us to treat diseases and disorders. From the rudimentary beginnings of selecting plants to treat common ailments, we have entered the age in human history in which we can take what is available in nature and modify it, adapt it, or mimic it. The goal is to produce sophisticated and targeted therapeutic treatments. Biologically, the approaches range from the isolation and use of natural products as therapeutics to the manipulation of organisms through genetic engineering. Chemically, natural products are functionalized, degraded, and transformed or, alternatively, serve as the starting point for creative and varied approaches to generate a range of new molecular structures that interact with biological targets.

Modern drug discovery has relied upon many different tools and approaches. The discovery of small-molecule therapeutics by synthesis and from the isolation of natural products — from plants, microorganisms, and marine organisms — continues to be an important source for new drugs. In principle, combinatorial chemistry is high-throughput organic synthesis. New terms have appeared for this approach (i.e., diversity-oriented synthesis, target-oriented synthesis, parallel synthesis), but common to all of these modern counterparts to traditional synthesis is the accelerated synthesis of a wide range of diverse molecules. This development, coupled with the advances in molecular biology, high-throughput screening, and modern drug discovery, are fulfilling a demand for more compounds with better properties for treating disease. Furthermore, the NIH Roadmap Initiative will certainly play an important role in the development of modern drug discovery.

So it comes as no surprise that today natural products and combinatorial chemistry are converging into a common approach: the combinatorial synthesis of natural product-based libraries. This book will by no means attempt to provide a comprehensive treatment of every publication and every perspective that falls within the scope of the topic. However, the goal is to introduce the reader to the area through the work of several leading scientists and to point the reader to the key literature. In the last few years, a number of excellent reviews have been written, and we have attempted to summarize the work in the field with this volume. This book will serve as a useful textbook in an advanced undergraduate or graduate course in organic synthesis, medicinal chemistry, natural products, or pharmaceutical science by focusing upon the interplay of drug discovery, natural products, and organic synthesis.

In Chapter 1, Boldi and Dragoli provide an overview of the fields of natural products research and combinatorial chemistry. The latter half of the chapter examines several areas of research in the synthesis of natural-product-based libraries.

In Chapter 2, Ganesan gives a historical overview of the development of natural products research and shows the importance of combinatorial chemistry approaches to such compounds in the drug discovery process. Examples from his research help to illustrate the value and future of such an integrated approach to drug discovery using natural-product-like libraries.

In Chapter 3, Stahura and Bajorath delve into the computational analysis of natural molecules and discuss strategies for the design of compound libraries based upon natural products. The growing bioinformatics approach to library design indicates that natural products will play a critical role in future drug discovery programs.

In Chapter 4, Dong and Myles outline the exciting area of engineered biosynthesis of natural products. A range of topics including polyketide biosynthesis, nonribosomal peptide biosynthesis, and glycosylation and tailoring are described.

In Chapter 5, Eckard, Abel, Rasser, Simon, Sontag, and Hansske describe a program of producing scaffolds from a proprietary microorganism library. Yield optimization has furnished adequate quantities of some interesting scaffolds that have been screened or used for the generation of analog libraries.

In Chapter 6, Ley, Baxendale, and Myers shift our focus to the use of polymer-supported reagents and scavengers for the synthesis of natural products and natural product analogs. The methods are widely applicable and powerful for exploring the structural diversity of compounds found in nature.

In Chapter 7, Boldi outlines a number of small-molecule libraries derived from carbohydrates.

In Chapter 8, Sofia describes two approaches for the synthesis of novel antibiotics using moenomycin and anisomycin as synthetic templates.

In Chapter 9, Pirrung, Li, and Liu describe the library synthesis of asterriquinones and illudins, analogs of known fungal natural products.

In Chapter 10, Doi and Takahashi describe several classes of natural product-based libraries prepared in their group including vitamin D_3 analogs, trisaccharides, macrosphelide analogs, and cyclic depsipeptides.

In Chapter 11, Abreu, Branco, and Matthew provide a rather comprehensive, highlighted overview of natural-product-like combinatorial libraries representing a range of secondary metabolites including carbohydrates, fatty acid derivatives, polyketides, peptides, terpenoids, steroids, alkaloids, and flavonoids.

I do hope that, as you read through the chapters, you will grow to appreciate the body of work done in this area and grasp the importance and significance of this emerging field. Consequently, it is my hope that this work will stimulate further thought, discussion, and research in this area.

Armen M. Boldi

Acknowledgments

The preparation of this book began with an idea germinated from the work done by the researchers in the area of natural product-based libraries. This led me to study and to pursue work in the area that I have presented as early as 2001. In early 2004, Dr. Bing Yan, now at Shandong University in China, in light of my interest in both natural products and combinatorial chemistry and recognizing the need to summarize the growing body of literature in the area, suggested that I prepare this book. I thank him for the help and suggestions he has provided throughout the process. I thank all the researchers in the area and especially those who contributed chapters to this volume. The desire to pursue synthetic chemistry was cultivated and inspired by my past advisors: Don Deardorff at Occidental College in Los Angeles, California, Francois Diederich at the Eigenössische Technische Hochschule in Zürich, Switzerland, and Amos B. Smith, III, at the University of Pennsylvania, in Philadelphia, Pennsylvania. I have had the privilege of working with many excellent chemists over the years at Discovery Partners International and previously at Arris Pharmaceutical (now Applera Corporation-Celera Genomics Group). In particular, I thank Jeff Dener, Hisham Eissa, Thutam Hopkins, Cheng Hu, and Elaine Krueger for reducing to practice my inspiration for natural product-based libraries into some of the chemistries that I describe in portions of this book. I am also indebted to my colleagues Dean Dragoli, Chuck Johnson, and Dan Harvey at Discovery Partners International for reading parts of this manuscript and providing helpful suggestions. I thank my wife, Rubina, for her loving support as I assembled this volume, and I thank my son, Andrew, for being an inspiration to prepare a future generation of researchers.

Armen M. Boldi

Editor

Armen M. Boldi received his B.A. degree in chemistry from Occidental College in Los Angeles, California. He earned his Ph.D. in chemistry from the University of California at Los Angeles (UCLA), working with François Diederich. He went on to the University of Pennsylvania to work as a National Institutes of Health (NIH) postdoctoral fellow with Amos B. Smith, III, on the total synthesis of the spongistatins. In 1997, he joined the drug discovery efforts of Arris Pharmaceutical (now Applera Corporation–Celera Genomics Group) in the combinatorial chemistry group. Dr. Boldi was actively involved with the development and growth of this group, which developed first into a division and subsequently into a wholly owned subsidiary. The combinatorial chemistry business unit located in South San Francisco was subsequently acquired by Discovery Partners International in 2001. He continued to work at Discovery Partners International in the chemistry division on numerous library programs with pharmaceutical and biotech companies in the United States, Europe, and Japan. Acting as director of project management, he managed various synthetic chemistry collaborations. He recently moved to Codexis, Inc., Redwood City, California, to serve as manager of research and development operations.

For the past nine years, his group has focused on the development and the application of synthetic methodologies to the generation of discovery libraries and targeted libraries for lead identification. His research interests span the various disciplines in the drug discovery process including medicinal, combinatorial, and synthetic chemistry. He has a special interest in integrating small-molecule natural product-based libraries into lead discovery programs. Dr. Boldi has published over 20 technical papers in the areas of synthetic organic chemistry and combinatorial chemistry.

Contributors

Ulrich Abel
Santhera Pharmaceuticals
Heidelberg, Germany

Pedro M. Abreu
REQUIMTE, Departamento de Química
FCT-UNL
Caparica, Portugal

Jürgen Bajorath
Department of Life Science Informatics
B-IT International Center for Information
 Technology
Rheinische Friedrich-Wilhelms-University
Bonn, Germany

Ian R. Baxendale
Department of Chemistry
University of Cambridge
Cambridge, United Kingdom

Armen M. Boldi
Codexis, Inc.
Redwood City, California
(formerly of Discovery Partners
 International
South San Francisco, California)

Paula S. Branco
REQUIMTE, Departamento de Química
FCT-UNL
Caparica, Portugal

Takayuki Doi
Department of Applied Chemistry
Tokyo Institute of Technology
Tokyo, Japan

Steven D. Dong
Kosan Biosciences, Inc.
Hayward, California

Dean R. Dragoli
ChemoCentryx, Inc.
Mountain View, California

Peter Eckard
Discovery Partners International GmbH
Heidelberg, Germany

A. Ganesan
School of Chemistry
University of Southampton
Southampton, United Kingdom

Friedrich G. Hansske
Discovery Partners International GmbH
Heidelberg, Germany

Steven V. Ley
Department of Chemistry
University of Cambridge
Cambridge, United Kingdom

Zhitao Li
Department of Chemistry
Duke University
Durham, North Carolina

Hao Liu
Department of Chemistry
Duke University
Durham, North Carolina

Susan Matthew
REQUIMTE, Departamento de Química
FCT-UNL
Caparica, Portugal

Rebecca M. Myers
Department of Chemistry
University of Cambridge
Cambridge, United Kingdom

David C. Myles
Kosan Biosciences, Inc.
Hayward, California

Michael C. Pirrung
Department of Chemistry
University of California
Riverside, California

Hans-Falk Rasser
Discovery Partners International GmbH
Heidelberg, Germany

Werner Simon
Discovery Partners International GmbH
Heidelberg, Germany

Michael J. Sofia
Pharmasset, Inc.
Princeton, New Jersey

Bernd Sontag
Discovery Partners International GmbH
Heidelberg, Germany

Florence L. Stahura
Institute for Chemical Genomics
Seattle, Washington

Takashi Takahashi
Department of Applied Chemistry
Tokyo Institute of Technology
Tokyo, Japan

Table of Contents

1 Chemistry on the Interface of Natural Products and Combinatorial Chemistry

Armen M. Boldi and Dean R. Dragoli

CONTENTS

1.1 NATURAL PRODUCTS: A RICH SOURCE OF DRUG DISCOVERY LEADS

Stereochemically and functionally rich natural products are abundantly present in all organisms and are essential ingredients for all life. Plants, microorganisms, and marine organisms are three fertile sources of natural products that exhibit a range of biological activities.[1] The *Dictionary of Natural Products* catalogs the myriad of compounds isolated and characterized from nature; representative natural products are shown in Table 1.1.[2]

Natural products and natural product-based molecules command a pivotal role in the current era of drug discovery. In fact, natural products have played an integral role in the treatment of human disease long before the development of the modern pharmaceutical industry (see Chapter 2, Section 2.2).[3,4] For thousands of years, Egyptians, Chinese, and Greeks have used plant extracts to treat various ailments.[5] It was not until the 1800s with the isolation of natural products such as strychnine, morphine, atropine, and colchicine that the modern era of drug discovery began to unfold. With the first commercially pure natural product, morphine in 1826, and the first semisynthetic natural product, aspirin in 1899, the manner in which human ailments are treated began to change. The discovery of antibiotics such as penicillin in the 20th century helped propel the pharmaceutical industry into the modern era of drug discovery, which has included the development of natural products as commercial drugs.[6]

The influence of natural products in the discovery of new marketed therapeutics continues to be significant in various therapeutic areas. *Burger's Medicinal Chemistry and Drug Discovery* reviews natural products as leads for new pharmaceutical products for the central nervous system, neuromuscular disease, cancer, bacterial infections, cardiovascular disease, asthma, and parasites.[7] Drugs such as morphine, penicillin, cyclosporine A, lovastatin, acarbose, FK506 (tacrolimus), and paclitaxel (Taxol®)

TABLE 1.1
Dictionary of Natural Products **Classification of Natural Products**

Classification	Representative Natural Product	Name
Aliphatics		Prostaglandin D_3
Polyketides		Spongistatin 1
Carbohydrates		α-D-Glucose
Oxygen heterocycles		Kojic acid
Simple aromatics		Griseofulvin

TABLE 1.1 (CONTINUED)
Dictionary of Natural Products Classification of Natural Products

Classification	Representative Natural Product	Name
Benzofuranoids		Angeolide
Benzopyranoids		Myrsinoic acid C
Flavonoids		Crotafuran B
Tannins		Thonningianin B
Lignans		Gomisin A

TABLE 1.1 (CONTINUED)
Dictionary of Natural Products **Classification of Natural Products**

Classification	Representative Natural Product	Name
Polycyclic aromatics		β-Rubromycin
Terpenoids		Dysidiolide
Steroids		Digitoxigenin
Alkaloids		Mappicine
Amino acids		L-4-Hydroxyproline

TABLE 1.1 (CONTINUED)
Dictionary of Natural Products Classification of Natural Products

Classification	Representative Natural Product	Name
Polypyrroles	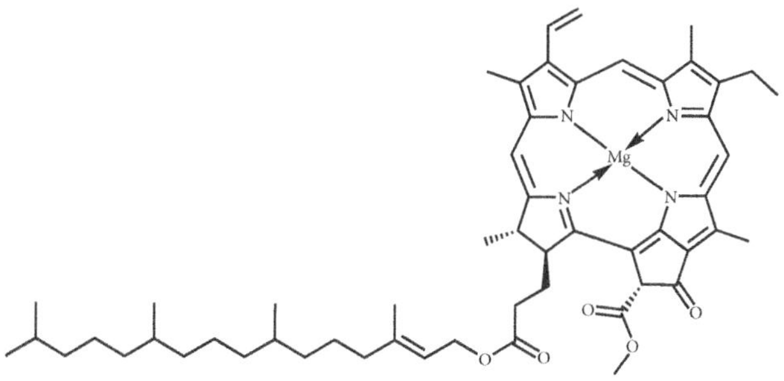	Chlorophyll A

are just a few natural compounds that have made a significant impact on the treatment of human disease. Not only are many drugs natural products, but many drugs are inspired by or derived from natural compounds.[8] A number of semisynthetic derivatives have made it to market. Compounds such as simvastatin (derived from lovastatin and an analog of mevastatin), topotecan and irinotecan (semisynthetic derivatives of camptothecin), and miglitol (an analog of 1-deoxynojirimycin) are some of the natural product-like drugs that have been recently approved (Figure 1.1). In the top 35 drugs sold worldwide, natural product-derived drugs are well-represented.[9]

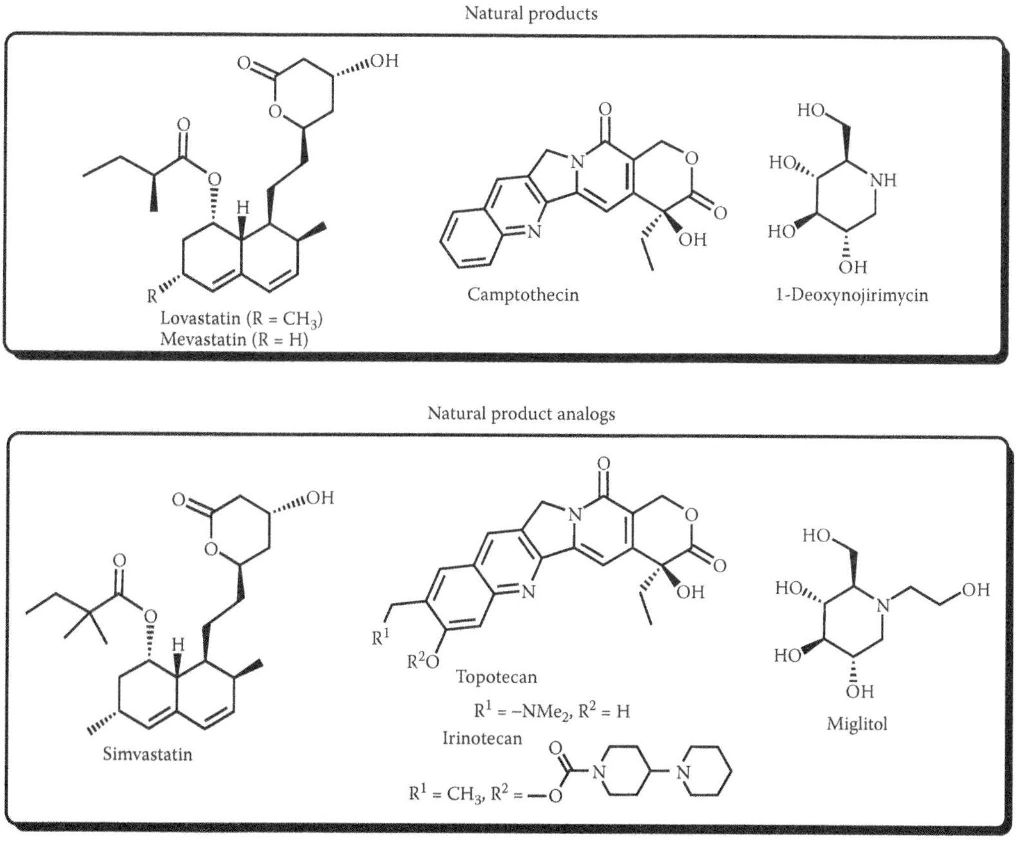

FIGURE 1.1 Marketed drugs derived from natural products.

In 2003 and 2004 alone, six additional natural product-based drugs were launched.[10] In addition to the several recently approved low-molecular-weight therapeutics shown in Table 1.2, a number of derivatives are currently being evaluated as drug candidates, primarily in the oncology and antiinfective therapeutic areas. Interestingly, marine natural products have seen the least commercial development[11]; development of molecules such as discodermolide show promise for this emerging source of drugs. Furthermore, a survey of drugs approved in the U.S. from 1981 to 2002 described the place of natural products in nonsynthetic new chemical entities (NCE).[12] Of the 877 small-molecule NCEs from all diseases, countries, and sources during this time period, 49% came from nonsynthetic origins. In the cancer area, 62% of the small-molecule NCEs were of natural origin. Forty-eight out of the 74 antihypertensive drugs are derived from natural product structures or mimics. In the area of antimigraine therapeutics, seven of the ten drugs are based upon serotonin, a low-molecular-weight natural product. Furthermore, many infectious disease drugs are derived from natural products.

Despite the prevalence of natural products as marketed drugs, the pharmaceutical industry began to look elsewhere for drugs. Several factors drove this trend: (1) high-throughput screening of molecular targets encouraged the use of chemical libraries instead of natural product extract libraries, (2) combinatorial chemistry promised greater chemical diversity than natural product libraries, and (3) the increase of molecular targets led to short timelines and made natural product-driven discovery impractical.[13] Yet, advances in the utilization of natural product extract libraries, the slow pace at which combinatorial chemistry has yielded new clinical candidates, and the appeal of using natural products as probes of biological pathways has led to a renewed interest in natural products as a strategic component of the drug discovery process. This has been driven in part by the favorable properties, high chemical diversity, and biochemical specificity that natural products have as lead compounds. Although synthetic small molecules continue to hold certain advantages (i.e., physicochemical properties such as Lipinski's "Rule of Five"),[14] natural products are privileged structures for modulating the activity of cellular pathways. Furthermore, advances in screening technologies, and molecular biology have made it more practical to incorporate natural products into the drug discovery process.

With the exception of several important low-molecular-weight natural products such as amine neurotransmitters (i.e., noradrenaline, adrenaline, serotonin, and melatonin), most natural products are different from synthetic drugs or drug candidates in several ways.[15] They have more stereogenic centers, are more architecturally complex with greater conformational biases and constraints, and contain more oxygen and less nitrogen. Other differences include molecular weight; natural products typically violate Lipinski's Rule of Five[14] by generally having molecular weights greater than 500. Synthetic molecules designed by medicinal chemists, on the other hand, tend to have a higher proportion of aromatic and heteroaromatic rings, fewer stereocenters, and lower molecular weights (complying with Lipinski's Rule of Five). Figure 1.2 illustrates these differences between natural and synthetic drugs by comparing marketed anticancer (Taxol® and Gleevec®) and hypercholester-olaemia (Mevacor® and Lipitor®) drugs. These differences suggest the necessity of exploring both natural products and synthetic molecules as therapeutic agents.

1.2 NATURAL PRODUCT-BASED COMBINATORIAL SYNTHESIS FOR LEAD DISCOVERY

Combinatorial chemistry grew in the 1990s as a technology-based solution to the demand for compounds in high-throughput screening campaigns against various therapeutic targets.[16] Small molecules generated via high-throughput synthesis began to dominate preclinical drug discovery programs. Many approaches to combinatorial chemistry, ranging from the synthesis of mixtures using chemical and radiofrequency tags[48] to discrete compounds on solid support or in solution, were successfully developed and utilized.[17] Furthermore, various high-throughput methods were

TABLE 1.2
Representative Examples of Recently Approved Natural Product, Natural Product-Derived, or Semisynthetic Natural Product Small-Molecule Drugs

Generic Name (Brand Name)	Structure	Disease Area	Company
Miglitol (Glyset®)		Diabetes	Bayer
Miglustat (Zavesca®)		Type 1 Gaucher's disease (metabolic disorder)	Pfizer, Actelion
Mycophenolate sodium (Myfortic®)		Immuno-suppression	Novartis
Oseltamivir (Tamiflu®)		Antiviral	Hoffmann-La Roche, Gilead
Pitavastatin (Livalo®)		Dypslipidemia	Sankyo, Kowa, Nissan

TABLE 1.2 (CONTINUED)
Representative Examples of Recently Approved Natural Product, Natural Product-Derived, or Semisynthetic Natural Product Small-Molecule Drugs

Generic Name (Brand Name)	Structure	Disease Area	Company
Rosuvastatin (Crestor®)		Dypslipidemia	Astra-Zeneca, Shionogi
Topiramate (Topamax®)		Anticonvulsant, antiepileptic	Ortho-McNeil, Johnson & Johnson
Voglibose (Basen, Glustat®)		Diabetes	Takeda, Abbott
Zanamivir (Relenza®)		Antiviral	GlaxoSmithKline

developed for solution-phase array syntheses including polymer-supported reagents, polymer-supported scavengers, and fluorous chemistry (see Chapter 6).[18] Methods for producing a range of molecular structures have been extensively reviewed and described.[19] Furthermore, engineered biosynthesis and biotransformations to generate compounds, as indicated in the preceding text, are yet other methods for compound synthesis (see Chapter 4 and Chapter 5).[20]

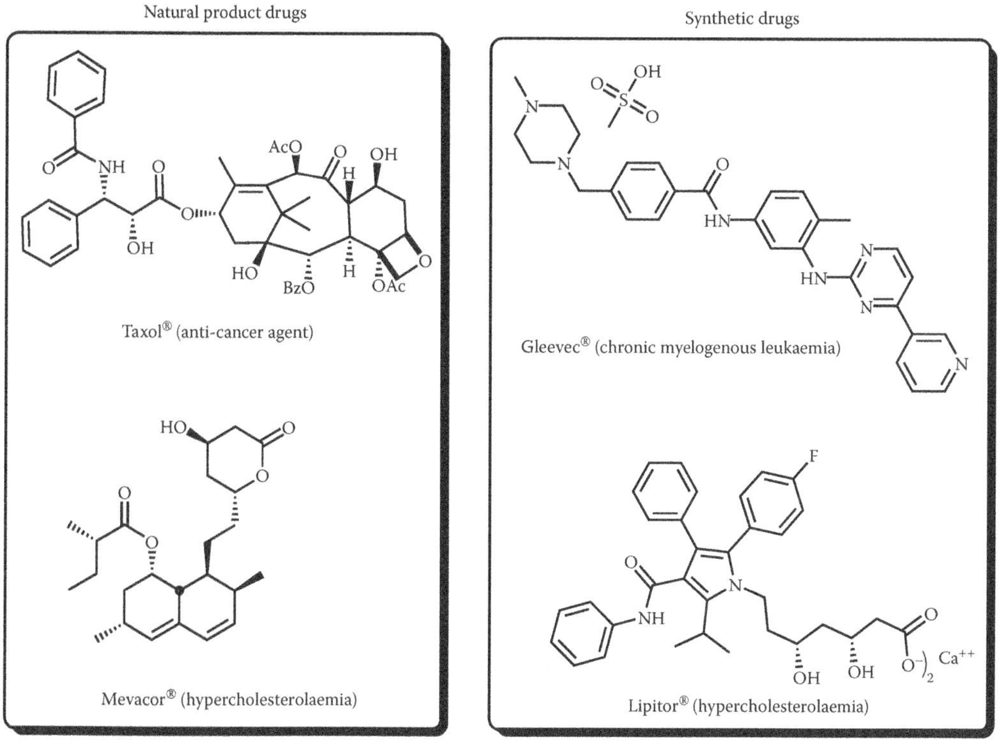

FIGURE 1.2 Differences between natural product drugs and synthetic drugs.

Combinatorial chemistry is equivalent to high-throughput synthesis of compound arrays in which side-chain, core structure, and stereochemical diversity are varied. At the heart of combinatorial chemistry is the parallel synthesis of compounds that may be lead-like,[21] drug-like,[15] or natural product-like (Figure 1.3). Two terms, recently introduced by Schreiber, define directionality of such libraries — *target-oriented synthesis* (TOS) and *diversity-oriented synthesis* (DOS).[22,23] In the strictest sense, these two types of libraries fall within the scope of combinatorial chemistry yet possess unique characteristics. Targeted libraries generated by TOS aim to elicit a specific biological response based on a gene family or a therapeutic area. DOS libraries, on the other hand, seek to generate more diversity than what has historically been the case for combinatorial libraries, by varying the skeletal and stereochemical elements of the core library structures.[24] Tan has described several categories of such DOS libraries: (1) core scaffolds of individual natural products, (2) specific substructures from classes of natural products, and (3) general structural characteristics of natural products.[25]

Although a significant number of biologically active compounds have been generated by combinatorial chemistry, the field continues to be criticized for its inability to generate leads and drugs.[26] This could not be farther from the truth. For example, Golebiowski and coworkers at Procter & Gamble described leads, with "sufficient potential (as measured by potency, selectivity, pharmacokinetics, physicochemical properties, novelty, and absence of toxicity) to progress to a full drug development program," discovered from libraries.[27] These leads originated from diversity libraries, thematic libraries (natural products, privileged scaffolds, and protein surface motifs), or focused libraries. Breitenbucher and Lee, emphasizing the impact of combinatorial chemistry on target-focused libraries, also illustrated the usefulness of libraries for analyzing structure–activity relationships (SARs).[28]

In the last 5 to 10 years, there has been a renaissance in natural products research[29] and a movement to combine combinatorial chemistry with natural products.[30–34] A cursory examination of the literature reveals many articles and reviews written in the area.[35,36] In 2001, Hall provided one of the earlier surveys of solution- and solid-phase strategies for libraries based on natural

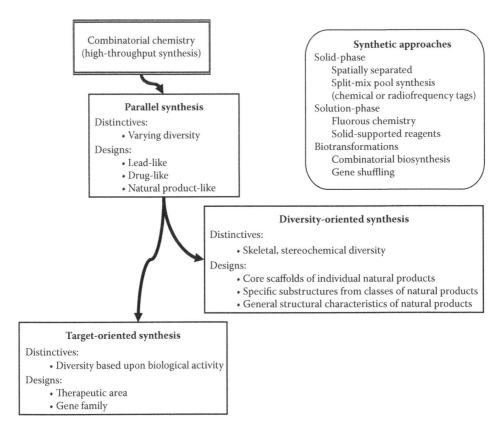

FIGURE 1.3 Combinatorial chemistry approaches.

product templates.[35c] Because natural products already possess known biological activity, they are good starting points for the design and synthesis of combinatorial libraries (see Chapter 2, Section 2.3).[37] A number of computational design studies validated the premise that natural product-like arrays improve biological relevance of combinatorial libraries.[38] For example, Feher and Schmidt, upon examining natural products, combinatorial compounds, and drug molecules, found that integration of natural product distribution properties make arrays more valuable in exploring cellular pathways.[39] Comparison of natural products, synthetic compounds, and marketed drugs have also been made by Henkel,[40] Schneider,[41] and Bajorath (see Chapter 3).[42] In the protein structure and bioinformatics design approach of Waldmann and coworkers described later in this chapter (Section 1.6), a powerful new approach to designing natural product-like libraries has been validated.

The molecular complexity present in nature is, in fact, diversity generated by combinatorial processes.[43] The immune system is a classic example of shuffling gene segments in order to assemble different antibodies for recognizing foreign antigens. Following carefully choreographed combinatorial synthetic steps, biological macromolecules, such as polypeptides, oligonucleotides, and polysaccharides, are assembled biosynthetically. Researchers have harnessed this biosynthetic machinery with techniques such as phage display[44] and gene shuffling.[45] Furthermore, combinatorial biosynthesis of natural products such as macrolide antibiotics is achieved by the assembly and shuffling of polyketide synthases (see Chapter 4).[46]

Although the biological machinery in living organisms generates natural products (see Chapter 4 and Chapter 5), synthetic chemistry is the primary method used by the pharmaceutical industry for generating natural product-based molecules. Several synthetic organic approaches have been used to increase the diversity of compounds related to natural products; all chemical approaches either start from a natural product or a synthetic starting material (Figure 1.4).[47] Using natural

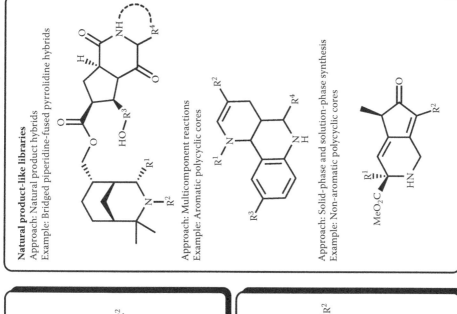

FIGURE 1.4 Natural product-based library synthetic approaches.

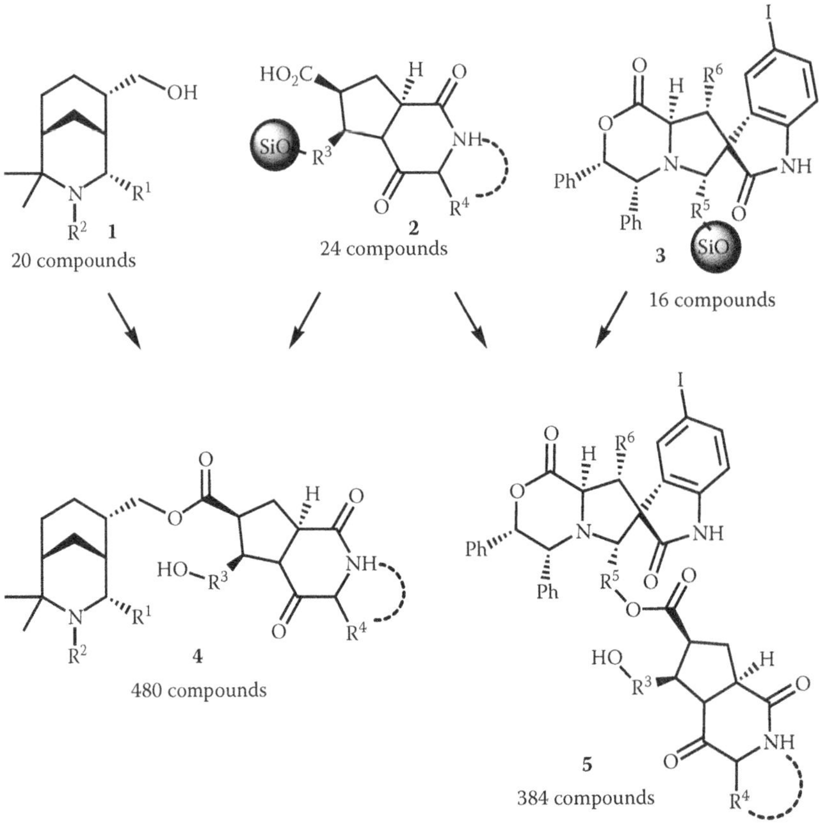

SCHEME 1.1 Natural product hybrid libraries.

products as starting materials, libraries of natural-product analogs are prepared by derivatization or decoration of natural products with diversity. A library of Taxol® analogs, prepared by Nicoloau and coworkers, exemplifies this traditional approach in natural products drug discovery programs.[48] Alternatively, natural products can be partially degraded and the core functionalized.[49] Meroquinene-derived piperidines, synthesized from quinine by Johnson and Zhang, illustrate this approach.[50,51]

A number of different approaches to natural product-like libraries have also been developed (Figure 1.4). One interesting new approach recently reviewed by Tietze and coworkers is the concept of natural product hybrids.[52] The idea is to combine portions of two different natural products into one molecule with the goal of discovering new or attenuated biological activity. A recent example described by Schreiber and coworkers illustrates this approach by using three natural product subunits, bridged piperidines **1**, fused pyrrolidines **2**, and spirocyclic oxindoles **3**, to prepare two natural product hybrid libraries **4** and **5** (Scheme 1.1).[53] Libraries of the three subunits were prepared and assembled by the formation of ester linkages.

Another approach is the use of multicomponent reactions (MCRs) to rapidly and efficiently construct structurally complex and varied polycyclic natural product-like compounds (Figure 1.5).[54] A number of synthetic transformations played a key role in the rapid assembly of such molecules including isocyanide-based reactions, aza- and non-aza [4+2] cycloadditions, [3+2] cycloadditions, and transition-metal-catalyzed reactions. Using isocyanide-based MCRs, pyrrolopyridines exemplified by mappicine represent an attractive library target for their biological activity.[55] Furthermore, the furoquinoline alkaloid tecleabine represents a common quinoline alkaloid core similar to structures found in a polycyclic library.[56]

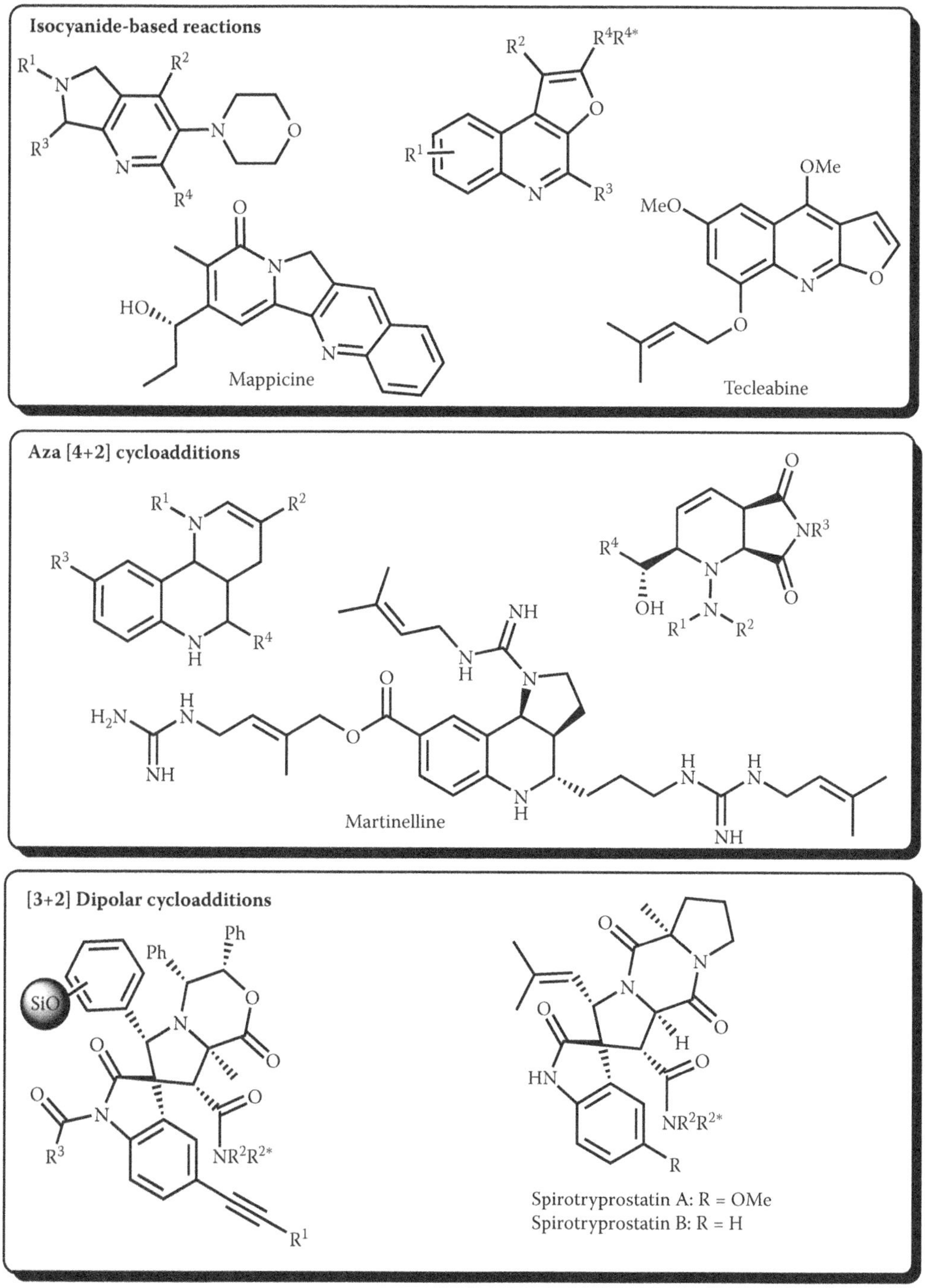

FIGURE 1.5 Aromatic polycyclic cores constructed using multicomponent reactions.

Cycloadditions are well-established reactions that allow rapid, efficient assembly of structurally and stereochemically complex molecules. Tetrahydroquinolines, a bicyclic ring system known in natural products such as martinelline, can be synthesized by the aza [4+2] cycloaddition, a particularly useful method for generating polycyclic nitrogen heterocycles.[57] Hall and coworkers

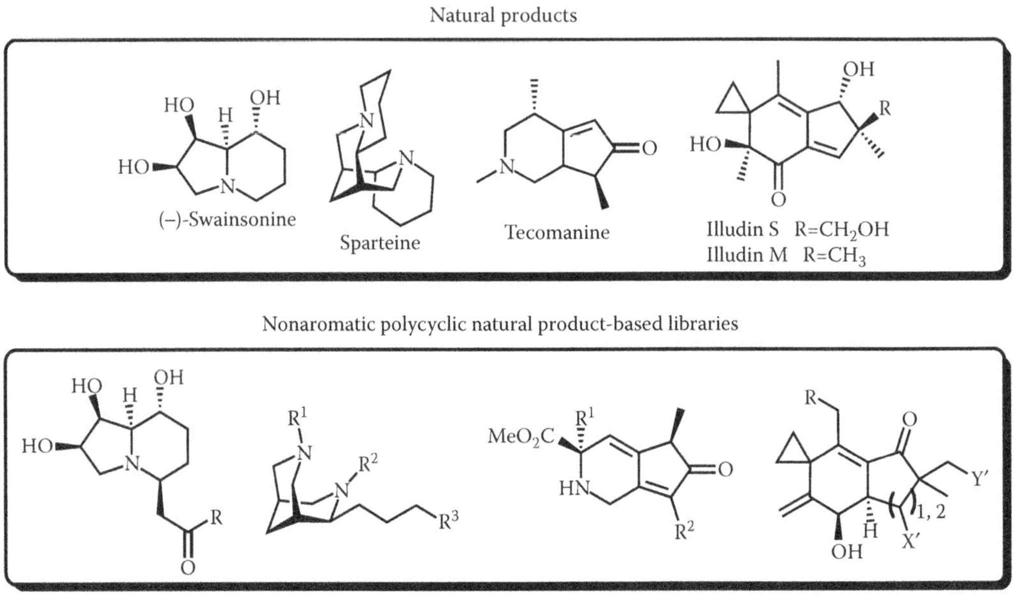

Natural products

(–)-Swainsonine

Sparteine

Tecomanine

Illudin S   R=CH$_2$OH
Illudin M   R=CH$_3$

Nonaromatic polycyclic natural product-based libraries

FIGURE 1.6 Nonaromatic polycyclic cores constructed using solution-phase and solid-phase methods.

prepared α-hydroxyalkyl piperidines, a class of molecules represented by natural products such as quinine, the palustrines, and various azasugars.[58] Finally, a third representative class of MCRs are [3+2] cycloadditions. The synthesis of an array of spirooxindoles similar to the spirotryprostatins was achieved using the Williams' three-component reaction.[59] This synthetic approach to natural product-based libraries unquestionably is a powerful method for supplementing the diversity found in natural products.

A number of nonaromatic polycyclic core libraries resembling natural products have also been reported (Figure 1.6).[60] The antitumor and immunomodulator (–)-swainsonine was the basis for a library of 5-substituted swainsonine analogs.[61] Substituted bispidines, the core structure of lupanine alkaloids such as sparteine, are of considerable interest for the activity in the cardiovascular therapeutic area.[62] Using a fluorous mixture synthesis approach, Curran and coworkers synthesized 4-alkylidene cyclopentenones analogous to the alkaloid, tecomanine.[63] Based on the illudin sesquiterpenes, Pirrung and Liu synthesized illudinoids using the Padwa–Kinder cycloaddition approach (see Chapter 9). Some of the library compounds exhibited similar anticancer activity to the naturally occurring illudins.[64] Later in this chapter, libraries based on dysidiolide (Section 1.6) and L-hydroxyproline (Section 1.7) illustrate some other examples of nonaromatic polycyclic compounds synthesized on solid phase.

Natural product-like arrays, synthesized by high-throughput synthesis, are playing a critical role in the chemogenomic exploration of cellular pathways.[23] As biologically validated starting points that bind to protein receptor surfaces, natural products are privileged structures and, consequently, natural-product-guided library synthesis increases the likelihood of finding lead compounds. Multifaceted approaches[65] utilize natural products including modification of natural product core structures, total synthesis of molecular skeletons, and libraries around privileged structures identified from natural products.[66] Many of these libraries, represented by terpenoids, alkaloids, and various heteocyclic scaffolds, exhibit a wide range of interesting biological activity. In the following sections of this chapter, research of several key contributors to this field will illustrate the work done in this area. In subsequent chapters, more detailed discussions of various aspects of combinatorial synthesis of natural product-based libraries will follow.

1.3 SOLID-PHASE SYNTHESIS OF PROSTAGLANDINS AND VANCOMYCIN

In the earliest of papers in the field of combinatorial chemistry,[67] Jonathan Ellman indicated that there was a need to establish methods for accessing larger collections of medicinally relevant organic molecules, or "privileged structures."[68] In that article, he described the first solid-phase synthesis of an array of 1,4-benzodiazepines, described therein as "one of the most important classes of bioavailable therapeutic agents." The benzodiazepines include natural products such as the benzomalvins[69] and sclerotegenin,[70] in addition to the large number of synthetic derivatives that have been investigated.[71] At that time, Ellman's research group focused on the synthesis of several classes of natural products, including β-turn mimetics,[72] prostaglandins, and shortly thereafter, vancomycin analogs. One of the goals was to establish the importance of combinatorial methodology in accessing natural product analogs, providing ready access to these highly important derivatives with more diversity and in larger numbers. The solid-phase synthesis of prostaglandins and vancomycin analogs illustrate the utility of combinatorial chemistry.

The attractiveness of prostaglandins as highly active therapeutics[73] has been tempered by the difficulty of their synthesis. Many methods have been described to make single derivatives using chemistry that addresses the delicate nature of the intermediates and final products, yet no common, general methodology to produce sets of derivatives could be described. The sensitive nature of these natural products forced many research groups to abandon prostaglandins as viable targets, and companies such as Pharmacia (now Pfizer) largely eliminated significant efforts to pursue prostaglandins as therapeutic agents.

The first report of a solid-phase prostaglandin synthesis was published by Janda and Chen in 1997 (see Chapter 11, Subsection 11.3.1).[74] In this work, the synthesis of PGE_2 methyl ester used a soluble-polymer approach with the three-component coupling methodology first popularized by Noyori.[75] Subsequently, Janda and coworkers prepared 16 derivatives using split-and-mix methodology.[76] But to access both the 1- and 2-series in addition to both the E and F derivatives, the general route established by Ellman[77] is highly preferable. Utilizing a single route, Ellman demonstrated that it is possible to develop combinatorial methodology to produce prostaglandins in a parallel format, while avoiding many limitations that had previously affected other established routes to these compounds.

In subsequent work,[78] Ellman and coworkers demonstrated the rapid synthesis of diverse prostaglandin analogs using parallel synthesis. As more than just a demonstration that prostanoid compounds could be produced on a bead, the goal of the project was to provide prostaglandin derivatives toward an actual biological target in collaboration with other researchers. Knowing that the α- and ω-side chains are critical in determining receptor affinity and specificity, a synthetic route allowing for variation of these components was considered a high priority. However, the main obstacles to completing a successful synthesis and incorporating desired diversity using the existing methodology included β-elimination of the hydroxyl moiety and enolate migration during the α-chain addition through alkylation.

To this end, a synthetic route was developed to include mild transformations that were highly compatible with multiple functionalities (Scheme 1.2). Using the method developed by Johnson and Braun[79] as a starting point, the Ellman route involved the incorporation of α-chains through Suzuki coupling and ω-side chains through cuprate additions. Additionally, to avoid the issue of β-elimination, the carbonyl functionality was reduced to the alcohol under Luche conditions prior to elaboration. The use of Suzuki conditions and a masked carboxylic acid allowed the researchers to incorporate a wide range of functionality in the upper side chain, including carboxylic acids, amides, and aliphatics from a single core. To access amides, N-acylsulfonamides were incorporated in the α-chain and then activated by alkylation with bromoacetonitrile immediately prior to amine displacement. It was also found that using higher-order cuprates allowed for the addition of both aryl and vinyl cuprates, giving access to two different classes of ω-chains for the ensuing biological

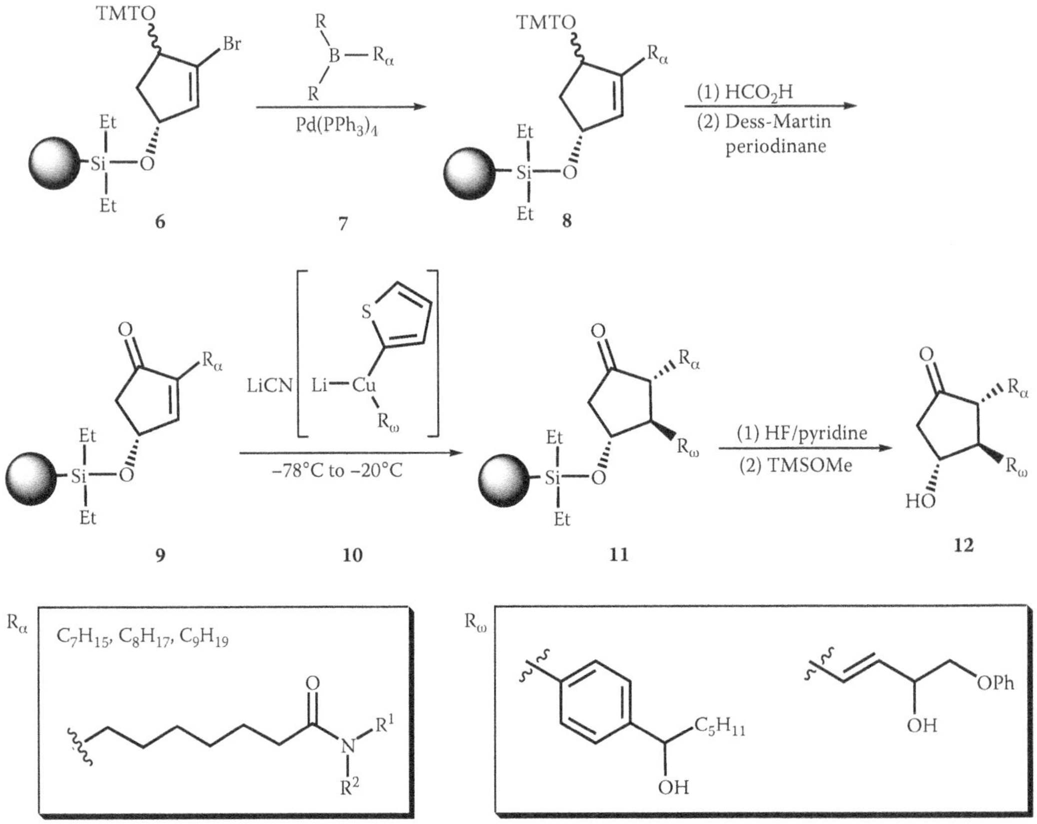

SCHEME 1.2 Solid-phase synthesis of prostaglandins.

studies, again using a common template. Specifically, thiophene was found to be the ideal "dummy ligand" in the preparation of the cuprate reagents as it provides for preferential transfer of both vinyl and aryl derivatives. Final cleavage from the solid support with HF-pyridine cleanly provided 26 final derivatives **12** with no elimination. Above all, this work demonstrated that modification of known total syntheses can provide access to highly desirable and diverse natural product analogs in a parallel format.

Concerns over the increase of resistance to vancomycin, the so-called "drug of last resort" in the treatment of aggressive, methicillin-resistant bacterial infections, has had many researchers searching for new analogs of this natural product.[80] In an effort to obtain derivatives active in aqueous media through combinatorial methodology, Ellman and coworkers chose to maintain the majority of the right-hand carboxylate-binding pocket of vancomycin and replace the left-hand side with a tripeptide that could be diversified through split synthesis (Figure 1.7).[81] The benefit of this strategy is that although many of the key binding elements are retained (hydrogen bonds and hydrophobic interactions), the repulsive interaction between the amide carbonyl and the oxygen of the lactate should be eliminated owing to free rotation of the peptidic side chain.

In designing the library, the researchers broke the synthesis into two parts: the essential, invariant core, and the tripeptide (Figure 1.8). Split-and-mix methodology was employed in preparing a library using 34 amino acids, providing 39,304 tripeptides on solid support, which were then coupled with the single core component. The key step in producing the biaryl ether core was a thallium(III) trinitrate-mediated oxidative cyclization, which was performed in moderate yields. After completing nine-step synthesis of the core and the preparation of the tripeptide precursors, the two components were coupled with PyBOP and HOAt. The final vancomycin analogs were then screened against

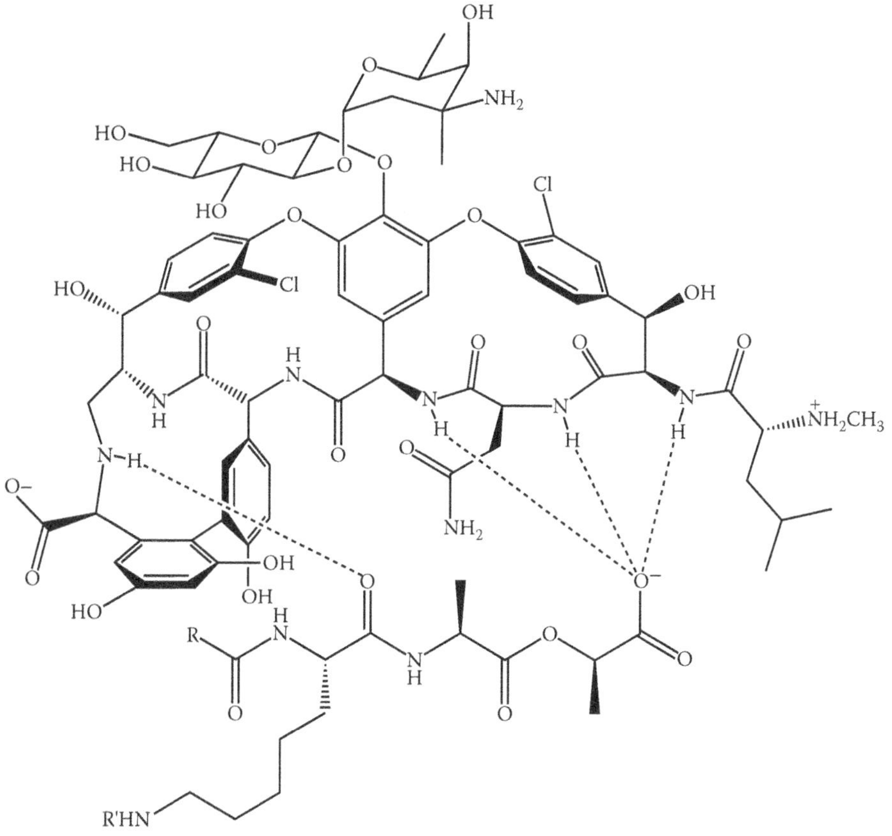

FIGURE 1.7 Vancomycin and tripeptide substrates.

both the L-Lys-D-Ala-D-Ala and L-Lys-D-Ala-D-Lac substrates with surprisingly high selectivity observed. From the on-bead assay, 59 beads active against the L-Lys-D-Ala-D-Lac substrate and 106 bead active against L-Lys-D-Ala-D-Ala were identified. Of those selected for L-Lys-D-Ala-D-Lac, 66% had a conserved L-Tpi in the first position with 49% overall having L-Tpi-L-His-X conserved. The tripeptide L-Tpi-L-His-L-Dapa exhibited better than fivefold binding over vancomycin. More impressively, these unique derivatives maintained significant activity against L-Lys-D-Ala-D-Ala, losing only sixfold affinity when compared with vancomycin, which is significant considering the simplicity of these compounds compared to the natural product.

The lesson learned from this synthesis is in the simplicity of the approach to obtaining active compounds using the natural receptor as merely a guideline for exploration. Rather than attempting to complete a total synthesis of complex natural product analogs, merely maintaining the key binding elements and exploring diversity space in the nonessential regions provided new lead compounds. However, there are other approaches that have proved fruitful, specifically peripheral modification of the vancomycin core[82] and covalent dimerization.[83] In the cited peripheral modification example, Nicolaou and coworkers demonstrated that derivatization of the sugar moiety on support-bound vancomycin can lead to enhanced activity against resistant strains of the bacteria. Although they found that a single sugar generally resulted in lower binding affinity, elaboration of the free amine of the vancosamine led to several compounds with high activity against several strains of both vancomycin-susceptible and vancomycin-resistant bacteria.

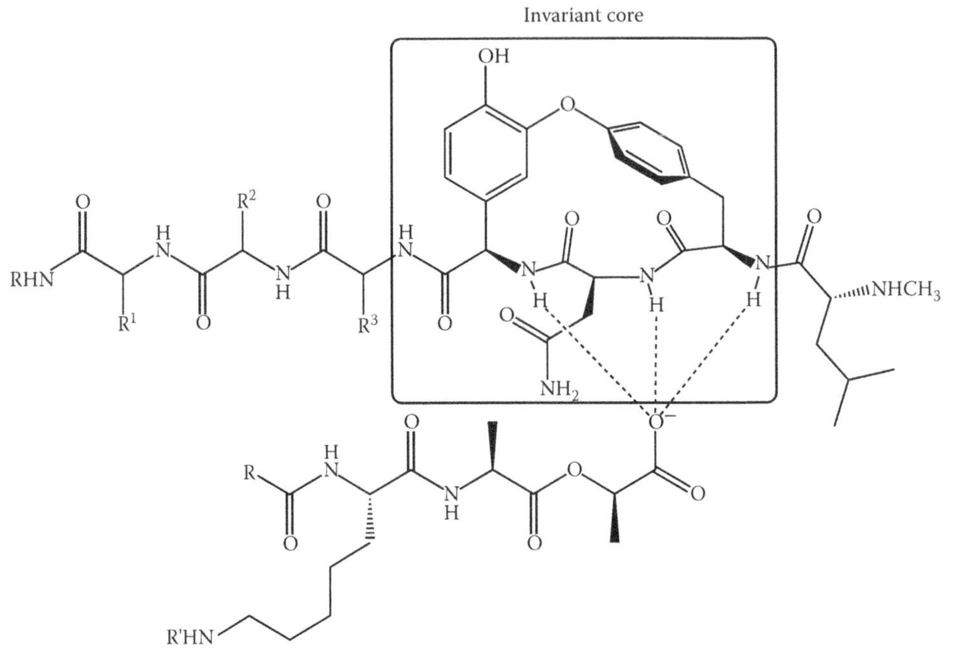

FIGURE 1.8 Vancomycin library and tripeptide substrates.

1.4 BENZOPYRANS: A PRIVILEGED PLATFORM FOR DRUG DISCOVERY

One of the earliest researchers in the area of natural product-like libraries is K.C. Nicolaou.[84] Inspired by the total synthetic efforts of his group on many important natural product targets and coupled with his interest in synthetic methodology and chemical biology, Nicolaou reported a number of elegant examples (see Chapter 11) that include natural product-like libraries based upon Taxol®,[48] sarcodictyins,[85] epothiolones,[86] vancomycin,[87] muscone,[88] and psammaplin a.[89]

As the chemical community began to adopt a combinatorial approach to natural product-like libraries in the late 1990s, Nicolaou applied the concept of "privileged structures" to the synthesis of natural product-like libraries with a series of papers on the solid-phase synthesis of benzopyrans.[90,91] The term "privileged structure" was first introduced by Evans and coworkers to describe benzodiazepines and benzazepines as structures that bind to multiple classes of proteins as high-affinity ligands.[68] According to the *Dictionary of Natural Products*,[2] there are 4636 benzopyranoid natural products, and of these, 461 contain the 2,2,-dimethyl-2*H*-benzopyran moiety. Benzopyrans exhibit a wide range of biological activities.[64a]

A unique linker strategy was utilized to generate benzopyrans. *o*-Prenylated phenols **13** were reacted with solid-supported selenyl bromide **14** to give resin-bound benzopyrans **15** (Scheme 1.3). Subsequent to elaboration on solid-support, the benzopyrans were oxidatively cleaved off of solid-support accompanied by *syn*-elimination to give olefinic benzopyrans **16**. Using IRORI radio-frequency encoding tags, MacroKan technologies, and the NanoKan system, all developed by Discovery Partners International, over 10,000 compounds were synthesized and used in biological assays. Representative types of benzopyrans synthesized include chalcones, pyranocoumarins, chromene glycosides, tetrazoles, and analogs of an aldosterone biosynthesis inhibitor and phosphodiesterase inhibitor. Further extension of the library of compounds prepared on solid phase was also achieved by derivatization of the olefinic bond.[64c]

Three examples demonstrate the power of this approach of natural product-based discovery of new biologically active compounds. First, inhibitors of NADH:ubiquinone oxidoreductase (complex

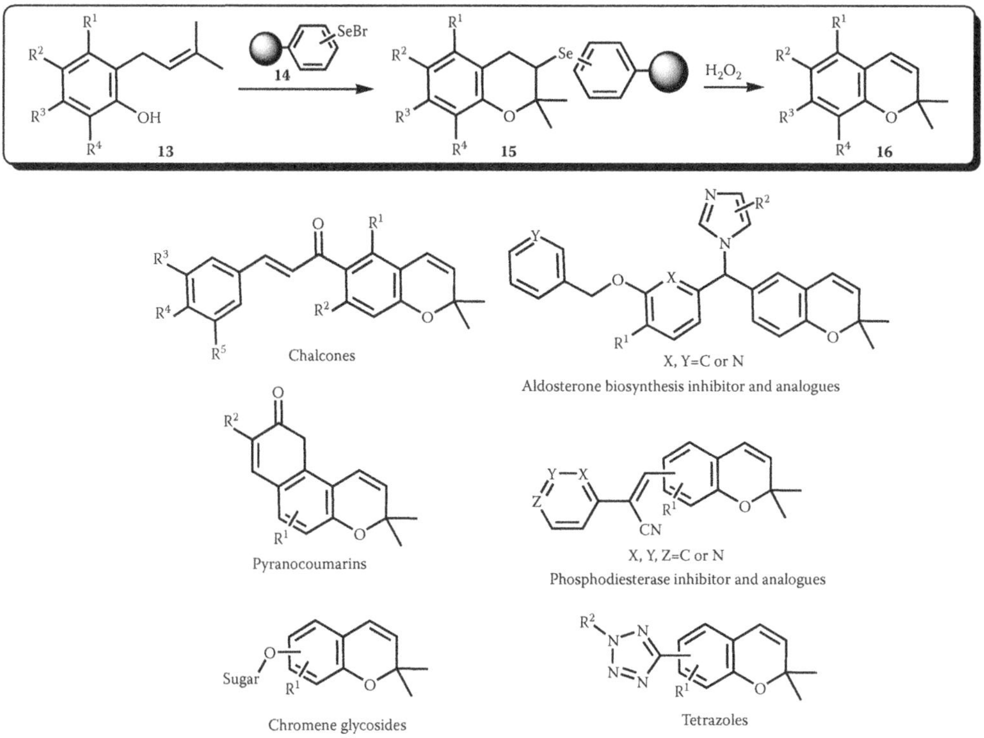

SCHEME 1.3 Benzopyrans synthesized on solid support utilizing a novel selenium linker and IRORI radio-frequency tags, MacroKan technologies, and the NanoKan system.

I), an enzyme involved in oxidative phosphorylation and a target of interest for insecticides and anticancer agents, were identified in screening these benzopyrans.[92] A family of naturally occurring inhibitors found in Cubé resin, used as a botanical insecticide for many years, was used as the starting point. One of the constituents, deguelin, is a 6.9-nM inhibitor of NADH:ubiquinone oxidoreductase (Figure 1.9). Generation of a discovery library of benzopyrans furnished nanomolar lead compounds. Of the benzopyran-aromatic moiety bridges evaluated, the ester and ether linkages gave the most potent compounds. Several focused libraries were designed to explore the SAR and to optimize for potency. Several nanomolar lead compounds were evaluated in cell-based assays and showed significant activity against various cancer cell lines.

With the emergence of drug-resistant strains of bacteria such as methicillin-resistant *Staphylococcus aureus* (MRSA) and the heavy reliance on antibiotics such as vancomycin (see Section 1.3), there is a serious need today for new antibiotics. As we will see in Sofia's chapter later in this book (see Chapter 8) and elsewhere,[93] parallel synthetic methodologies help in the development of antiinfective natural

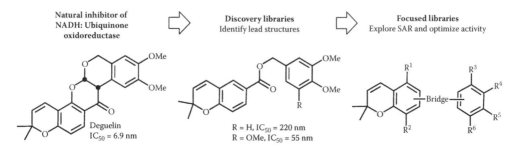

FIGURE 1.9 Discovery of benzopyran-based anticancer compounds from deguelin.

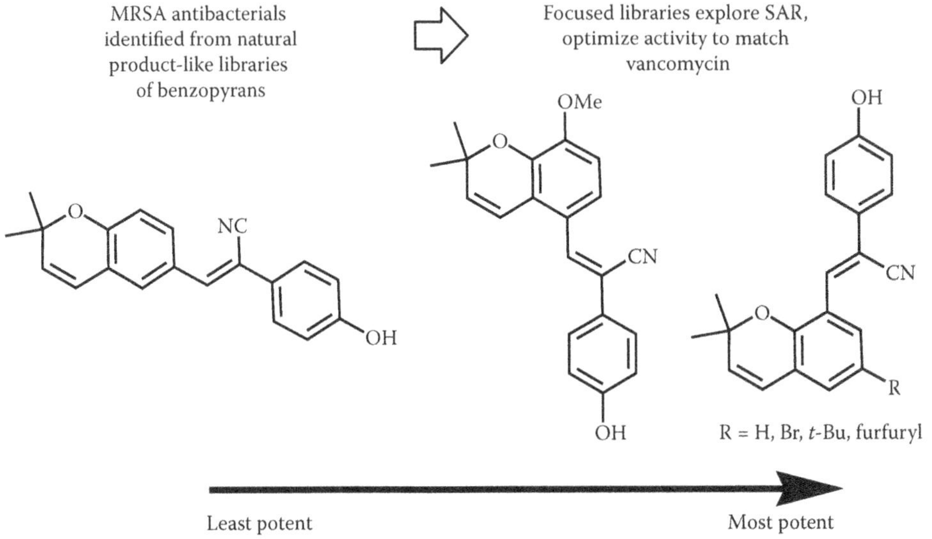

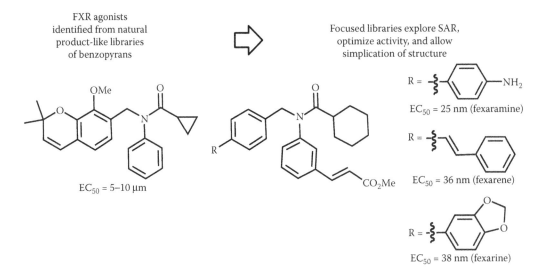

FIGURE 1.10 Discovery of benzopyran-based antibacterial agents.

products. Using the privileged scaffold screening approach, benzopyrans, specifically cyanostilbenes, with activity against several MRSA strains were discovered from combinatorial libraries (Figure 1.10).[94] Follow-up focused libraries explored SARs and identified a number of active antibiotics with *in vitro* potencies comparable to vancomycin. Interestingly, activity was correlated to the orientation of the stilbene moiety on the benzopyran. Furthermore, whereas the phenol was found to be important for activity, the nitrile was shown to be unimportant.

Finally, the farnesoid X receptor is a nuclear hormone receptor that functions as a bile acid sensor that coordinates cholesterol metabolism, lipid homeostatis, and absorption of dietary fat and vitamins.[95] Modulation of FXR may be useful for the treatment of cholestatis and diseases associated with bile acids. With only one known high-affinity nonsteroidal FXR agonist, a screening campaign with the benzopyran screening library furnished lead compounds for SAR development and design of new chemotypes for this target (Figure 1.11).[96] After several follow-up focused libraries, FXR

FIGURE 1.11 Discovery of FXR agonists from screening a benzopyran natural product-like library.

agonists, fexaramine, fexarine, and fexarene were identified. The fexaramine ligand has made possible the structure elucidation of FXR.[95]

1.5 DIVERSITY-ORIENTED SYNTHESIS FOR CHEMICAL BIOLOGY

Schreiber and coworkers recently described a synthetic strategy using σ elements, functionality that encodes skeletal information that can be transformed into products with different skeletons.[22,97] The diversity-oriented synthetic approach, aimed at understanding the function of proteins[98] has taken root in a number of laboratories around the world. DOS develops on what combinatorial chemistry initially sought to achieve. As previously indicated, DOS as a combinatorial approach to parallel synthesis expands the array diversity by increasing the number of stereoisomers per array and varying the number of core scaffolds per array.

A number of examples such as 1,3-dioxanes,[99] macrolactones,[100] ring-containing biaryls,[101,102] spirooxindoles, alkaloid-like compounds,[103] and polycyclic compounds[104,105] from the Schreiber group illustrate this approach to natural product-like libraries (see Chapter 11). An early example converted shikimic acid into intermediate tetracyclic γ-butyrolactones,[106] which were then functionalized around the core structure (see Chapter 11, Subsection 11.10.2). γ-Butyrolactones, found in about 10% of all natural products and which exhibit a broad range of biological activities, are a key element in a number of recent natural product-like compounds.[107] A more recent example, inspired by the rich skeletal diversity of indole alkaloids, utilized the rhodium(II)-catalyzed consecutive cyclization-cycloaddition reactions developed by Padwa and coworkers (Scheme 1.4).[108] A stereocontrolled tandem reaction utilizing the versatile scaffold allowed for multiple modes of intramolecular reactions.

Another approach involved transforming a commercially available steroid into three structurally unique libraries through a skeletal transformation strategy (Scheme 1.5).[109] Starting from dehydroisoandrosterone 3-acetate (20), an epoxide was prepared and attached to macrobeads through a silyl ether linkage. Epoxide 21 was treated with amines and subsequently derivatized to give the first set of compounds 22. Using diethylaluminum chloride as a Lewis acid catalyst to accelerate the Diels–Alder reaction over the retro-Diels–Alder reaction, pentacyclic core structures 24 were obtained as a single regio- and diastereomer. Upon heating to 110°C, the retro-Diels–Alder reaction occurred to give the paracyclophanes 25. Using an encoded, split-pool synthesis, a total of 4275 different compounds were generated.

In a series of reviews that encompass his recent work, Arya and coworkers describe several natural product-like libraries that describe how a total synthetic approach can be applied to the synthesis of such libraries.[110] Using the IRORI split-and-mix approach for solid-phase synthesis, enantiomerically pure tetrahydroquinoline derivatives were prepared (Figure 1.12)[111] Tetrahydroquinoline-based[112] tricyclic derivatives were generated from an enantiomerically pure bicyclic scaffold.[113] Other tetrahydroquinoline solid-supported intermediates amenable to high-throughput synthesis were also described. Using a ring-closing metathesis approach, the polycyclic compound containing a ten-member ring was synthesized.[114] From a common enantiopure intermediate on solid support, three polycyclic tetrahydroquinoline-based scaffolds were synthesized including an eight-membered ring.[115]

Moreover, two indoline-like libraries were prepared (Scheme 1.6). A hydroxyindoline-derived scaffold 26 was used to prepare a tricyclic indoline-based library 27 by an IRORI split-and-mix approach on solid-support.[116] A later indoline-alkaloid-like polycyclic library incorporated additional diversity, and the stereochemistry at the ring junction was controlled.[117]

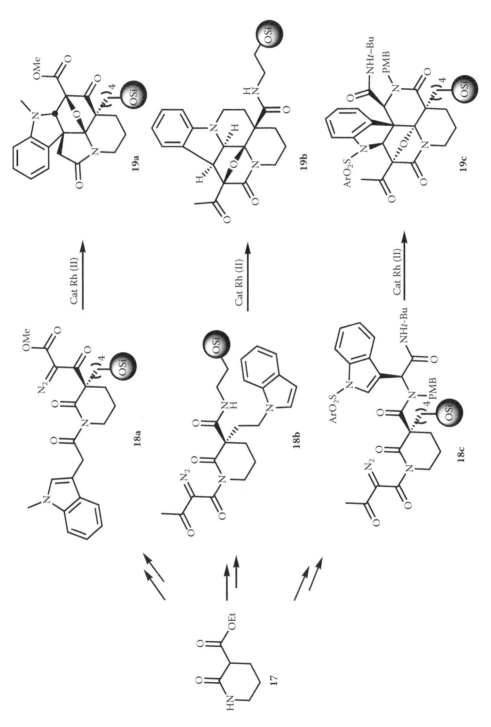

SCHEME 1.4 Tandem intramolecular cyclization-cycloaddition synthesis of indole alkaloid-like skeletons.

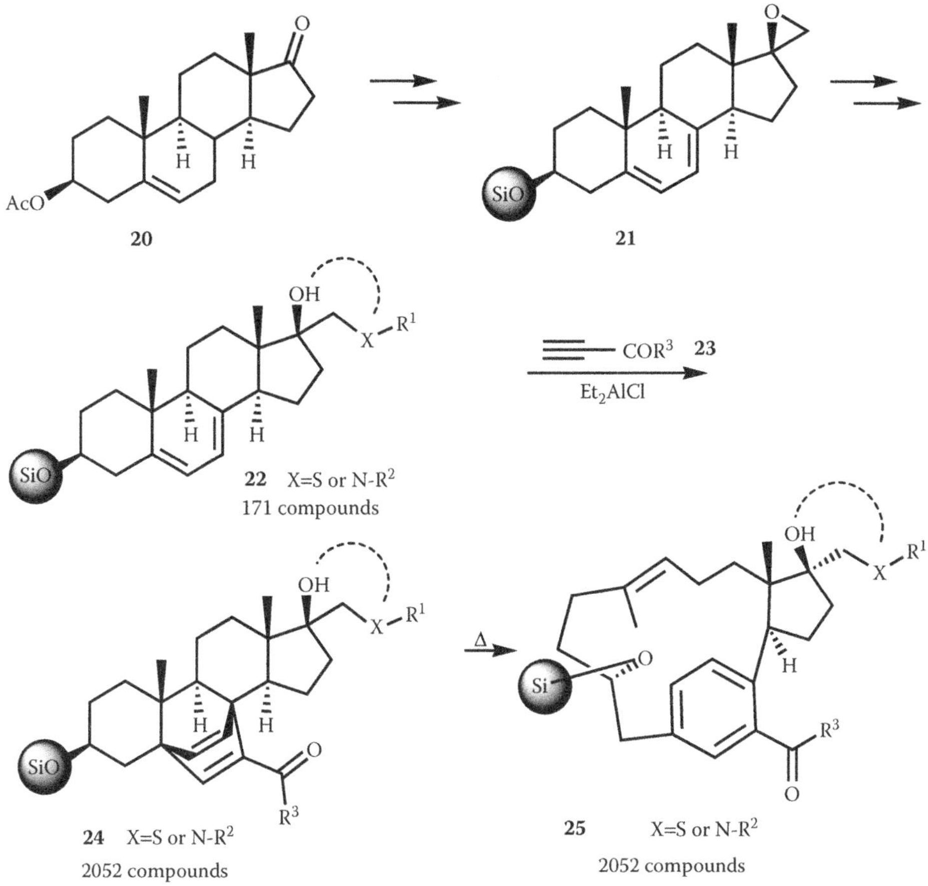

SCHEME 1.5 Skeletal transformations of steroidal compounds.

1.6 PROTEIN STRUCTURE SIMILARITY CLUSTERING FOR LIBRARY DESIGN

Whereas DOS seeks to achieve maximum diversity through structural complexity, natural product-guided synthesis is rooted more deeply in structural biology and starts from "a biologically validated starting point in structural space."[118] Waldmann and coworkers have described a successful approach of clustering protein three-dimensional structures with similar domains. The approach, described as "protein structure similarity clustering (PSSC)," groups proteins based on structure rather than gene family or sequence homology. Because there appear to be only about 1000 common protein folds that account for the observed protein domains, this structural conservation should serve as a guiding principle for identifying small-molecule modulators. The known ligands for one protein member of a given cluster should then serve as guides to library design for other proteins in the same cluster. Natural products as biologically "prevalidated ligands" serve as good starting points for the design of biologically active compounds.[119]

Waldmann analyzed several literature examples using the PSSC concept to demonstrate the validity of this approach. More importantly, he has successfully applied this approach to design a library of compounds based on dysidiolide, an inhibitor of phosphatase Cdc25A (Figure 1.13).[120] Cdc25A shares a structure that resembles acetylcholinesterase (AChE) and 11β-hydroxysteroid dehydrogenase (11βHSD type 1 and 11βHSD2).[121] After hypothesizing that the γ-hydroxybutenolide moiety[107] of dysidiolide was central to the phosphatase activity, a library of 147 γ-hydroxybutenolides and

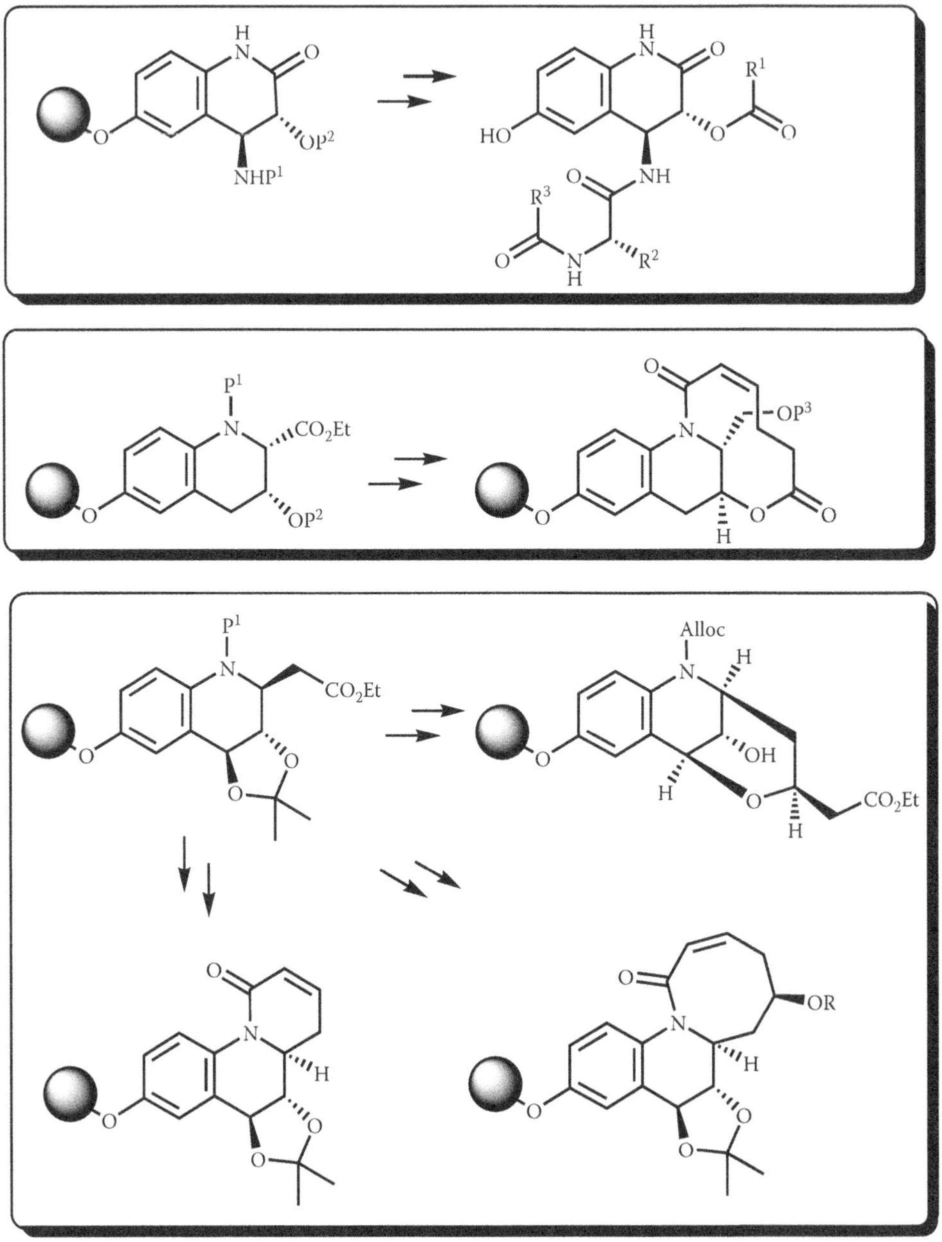

FIGURE 1.12 Solid-phase synthesis of tetrahydroquinolines.

α,β-unsaturated five-membered lactones was synthesized and screened for inhibition of the four targets. Inhibitors for all four targets were identified.

1.7 SMALL-MOLECULE LIBRARIES DERIVED FROM QUININE AND L-HYDROXYPROLINE

In addition to libraries derived from carbohydrates (see Chapter 7), a scaffold-based library approach using amino acids and alkaloids was implemented at Discovery Partners International. Quinine

Alloc
N
OBz
26
HO
N
N
R1
R2
O
O
H
27
SiO
Fmoc
N
O
CO2Et
NHAlloc
28
HO
N
N
R1
R2
O
O
CO2Et
NH
R3
O
29

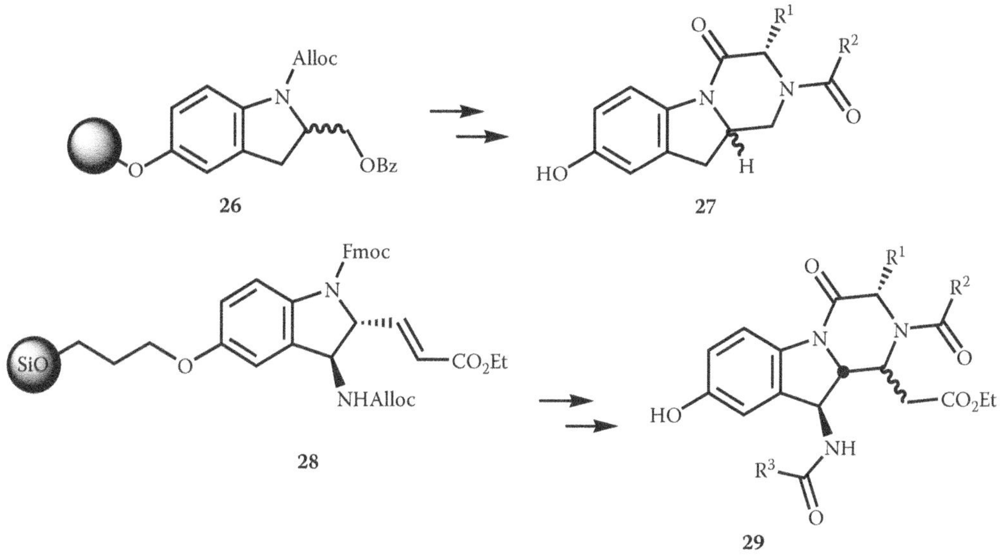

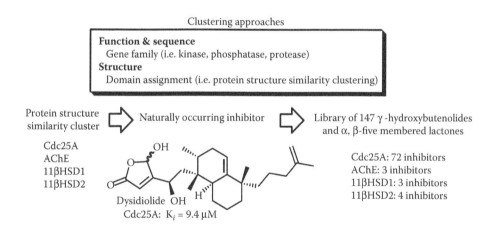

SCHEME 1.6 Solid-phase synthesis of indoline-like alkaloids.

Clustering approaches
Function & sequence
Gene family (i.e. kinase, phosphatase, protease)
Structure
Domain assignment (i.e. protein structure similarity clustering)
Protein structure similarity cluster
Naturally occurring inhibitor
Library of 147 γ-hydroxybutenolides and α, β-five membered lactones
Cdc25A
AChE
11βHSD1
11βHSD2
OH
O
O
H
Dysidiolide OH
Cdc25A: Ki = 9.4 μM
Cdc25A: 72 inhibitors
AChE: 3 inhibitors
11βHSD1: 3 inhibitors
11βHSD2: 4 inhibitors

FIGURE 1.13 Protein structure similarity clustering and natural product-driven library design.

(**30**) was degraded to a meroquinene scaffold **33** for the synthesis of substituted piperidines (Scheme 1.7).[50,51] L-4-Hydroxyproline was used to synthesize alkoxyproline[122] and pyrrolidinohydantoin libraries[123] through robust and general synthetic transformations highlighting a range of chemistries, methods, and high-throughput techniques in synthesis, analysis, and purification.[124]

Quinine is a white powder obtained from the bark of the cinchona tree found in the Andes mountains in Peru and Ecuador, and is used as a tonic in drinks, a treatment for muscle cramps, and an antimalarial drug. Meroquinene *t*-butyl ester **32** was synthesized following a literature procedure[125a,b] and scaled up to 1 kg (Scheme 1.7). Hydrolysis of *tert*-butyl ester **32** in 5 N hydrochloric acid at reflux for 2 h afforded the corresponding acid. Basification with sodium hydroxide and addition of BOC anhydride gave BOC-meroquinene acid **33**.

Marshall linker **34**[126] was esterified with the BOC-meroquinene scaffold **33** (Scheme 1.8). Cycloaddition of nitrile oxides **36** with the solid-supported terminal alkene **35** gave isoxazolines **37**. The regiochemistry of addition was expected to give predominantly the structure in which the

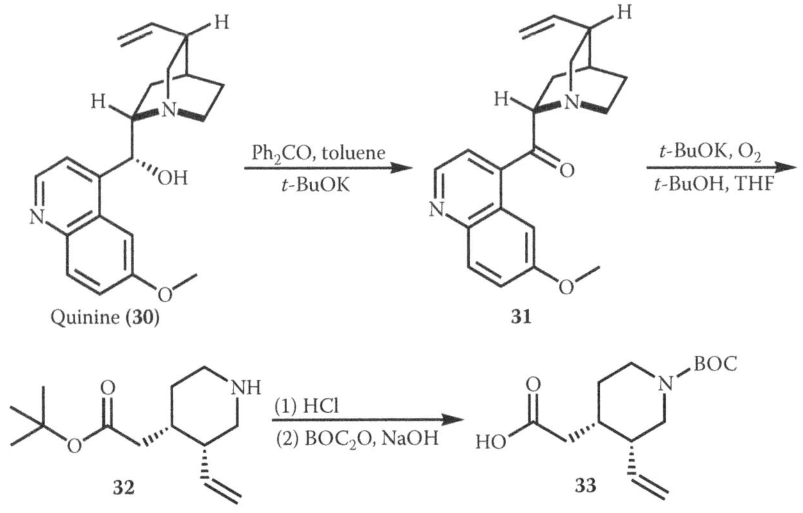

SCHEME 1.7 Degradation of quinine to BOC-protected meroquinene piperidine.

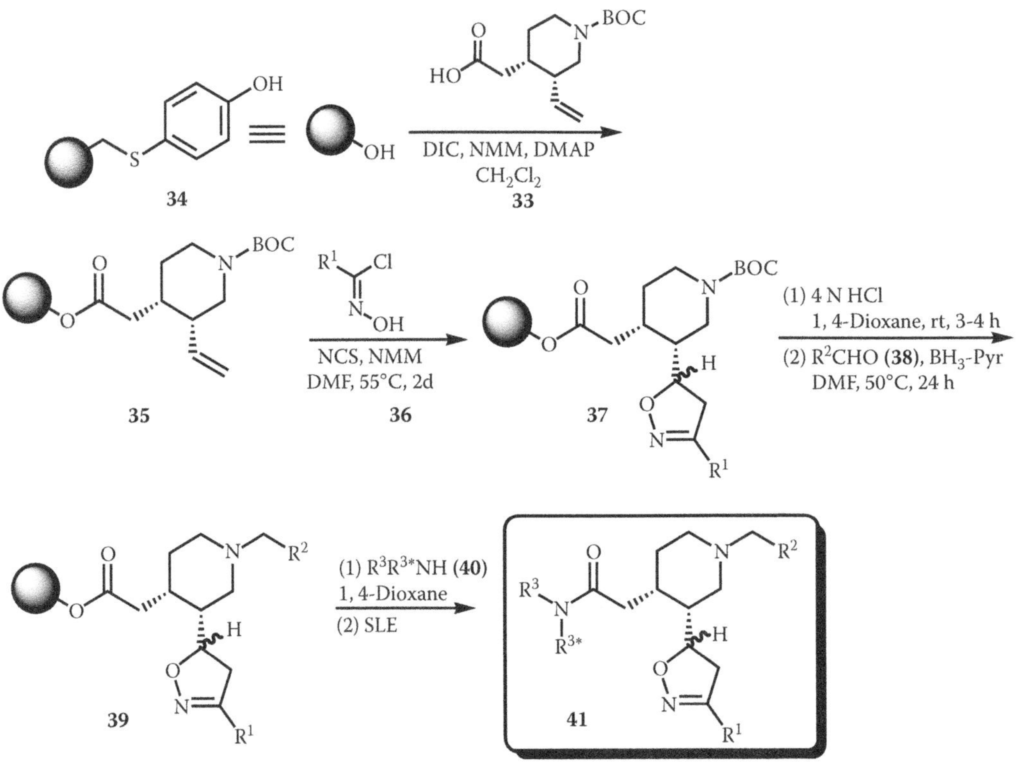

SCHEME 1.8 Solid-phase synthesis of meroquinene-piperidines.

oxygen of the nitrile oxide is attached to the more substituted carbon of the double bond.[127] Removal of the BOC protecting group with hydrochloric acid, followed by reductive amination of the amine hydrochloride salt proceeded smoothly using borane•pyridine complex.[128] No additional acid catalyst (e.g., acetic acid) was needed. Cleavage off solid-support of **39** using nucleophilic amines and SLE[129] to remove excess amine gave isoxazolines **41**.

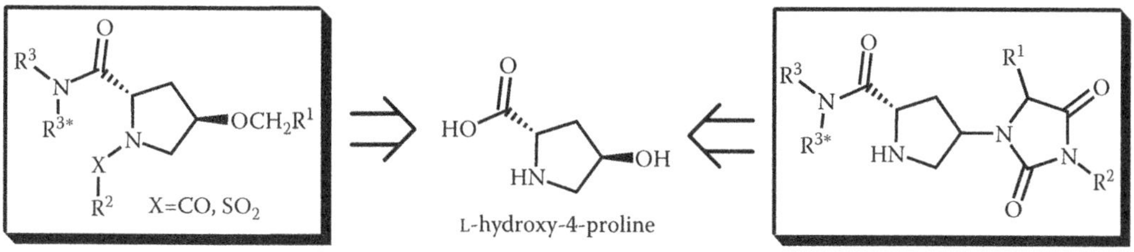

FIGURE 1.14 Alkoxyproline and pyrrolidinohydantoin libraries synthesized from L-hydroxy-4-proline.

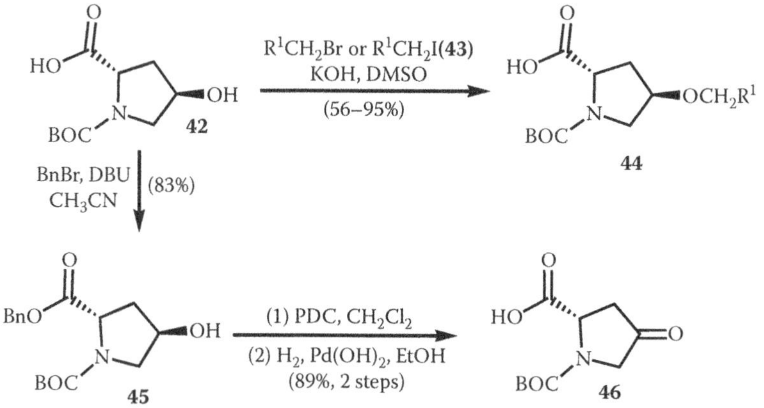

SCHEME 1.9 Preparation of library scaffolds from BOC-protected L-hydroxy-4-proline.

L-4-Hydroxyproline is a critical amino acid for the left-handed triple helix structure of collagen (three amino acid residues per turn — Gly-x-Hyp or Gly-x-Pro) and is produced by hydroxylation of proline in the rough endoplasmic reticulum.[130] From this commercially available natural product, two natural product-like libraries were synthesized (Figure 1.14). First, the BOC-protected derivative of L-4-hydroxy-4-proline **42** was converted into BOC-protected ethers **44** and the corresponding BOC-protected ketone **46** (Scheme 1.9). A practical multigram synthesis of alkoxyproline carboxylic acids **44** was achieved by alkylation of BOC-hydroxyproline **42,** with alkyl halides employing KOH as base in DMSO.[131] No chromatography was required in the syntheses of six carboxylic acids (56 to 95% yields). The synthesis of the keto-acid scaffold was based on a reported synthesis of ester **45**.[132] The benzyl ester was chosen because it could be removed in the last step by hydrogenation, thus avoiding basic conditions that might lead to side products. The oxidation of **45** with PDC in CH$_2$Cl$_2$ was complete in several hours.[133] On the largest scale attempted (157 g of **45**), careful monitoring of the temperature was essential when adding the portions of PDC in order to control the slightly exothermic reaction. The benzyl ester was removed using standard catalytic hydrogenation conditions. Purification of ketone **46** was achieved by recrystallization from acetonitrile.

The library synthesis of alkoxyprolines was achieved using an acid-stable, nucleophile-cleavable solid-support (Scheme 1.10).[122] Hydroxythiophenol (Marshall) linker **34** was esterified with the corresponding ethers of BOC-hydroxyproline.[126] Removal of the BOC protecting group with trifluoroacetic acid, followed by acylation gave solid-supported alkoxyproline derivatives **49**. Cleavage from the solid-support with excess primary amines or excess secondary amines, followed by purification of the crude products from the excess amine by supported liquid–liquid extraction (SLE)[129] gave alkoxyprolines **50** in high purity. Kinetic studies[134] on the cleavage of substrates off of this linker, and resin recycling studies have also been performed.[135] This solid-supported linker

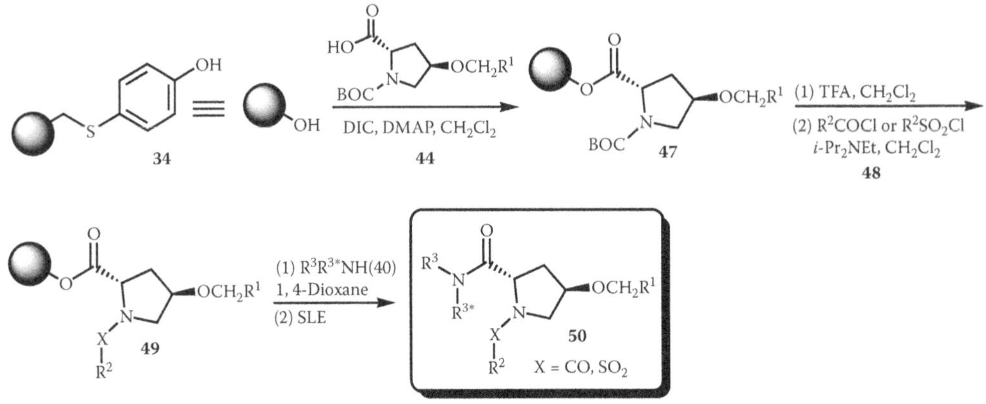

SCHEME 1.10 Solid-phase synthesis of alkoxyprolines.

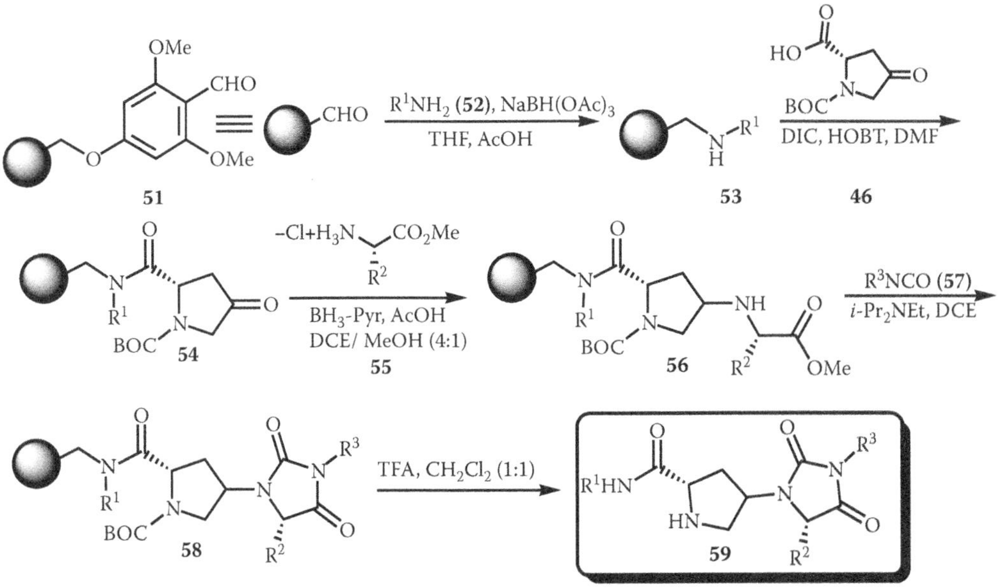

SCHEME 1.11 Solid-phase synthesis of pyrrolidinohydantoins.

is generally suitable for the synthesis of various small-molecule libraries, including the meroquinene-derived piperidines previously described.

Pyrrolidinohydantoins were prepared by first converting BOC-hydroxyproline **42** to the corresponding BOC-ketoproline **46** (Scheme 1.9). An acid-labile solid support **51**, the 4-formyl-3,5-dimethoxyphenoxymethyl linker, was functionalized with primary amines by reductive amination (Scheme 1.11). BOC-ketoproline **46** was coupled to solid-supported secondary amines **53**. The ketone was subsequently transformed into a 1:1 diastereomeric mixture of secondary amines **56** upon reductive amination with amino acid hydrochloride salts **55** using borane•pyridine complex. Upon treatment with isocyanates **57** in the presence of base, i-Pr$_2$NEt, solid-supported hydantoins **58** were generated. Using a 1:1 TFA/CH$_2$Cl$_2$ cleavage solution to ensure complete removal of the t-butyl and the BOC protecting groups on amino acid side chains, the final products **59** were isolated.

1.8 THE FUTURE OF NATURAL PRODUCT-BASED COMBINATORIAL LIBRARIES

In addition to their utility in drug discovery, natural product-based libraries serve as molecular probes of biologically activity. Chemogenomics seeks to identify a small molecule that perturbs the expression of each gene.[136] Consequently, small molecules have emerged as important tools for understanding and probing the expression of genes, and elucidating biological function. With a plethora of drug discovery targets emerging, there is a continued demand for new chemotypes in chemogenomic profiling. Natural product-like libraries will help us better understand how small molecules and proteins interact and regulate cellular processes important for health and disease.[137] Together, naturally occurring small molecules and small-molecule modulators of protein function are critical to understanding these SARs and to identifying new therapeutics.[138] The melding of high-throughput synthetic approaches to natural products research is opening many new and exciting possibilities for the future of pharmaceutical research.

REFERENCES

1. (a) Baker, D.D. and Alvi, K.A., Small-molecule natural products: new structures, new activities, *Curr. Opin. Biotechnol.*, 15, 576, 2004. (b) Myles, D.C., Novel biologically active natural and unnatural products, *Curr. Opin. Biotechnol.*, 14, 627, 2003.
2. Buckingham, J., *Dictionary of Natural Products*, Taylor and Francis/CRC Press, London, 2005 (CD-ROM).
3. Cragg, G.M. and Newman, D.J., Medicinals for the millennia: the historical record, *Ann. N.Y. Acad. Sci.*, 953, 3, 2001.
4. Newman, D.J., Cragg, G.M., and Snader, K.M., The influence of natural products upon drug discovery, *Nat. Prod. Rep.*, 17, 215, 2000.
5. (a) DeSmet, P.A.G.M., The role of plant-derived drugs and herbal medicines in healthcare, *Drugs*, 54, 801, 1997. (b) Hamburger, M. and Hostettmann, K., *Phytochemistry*, 30, 3864, 1991.
6. Drews, J., Drug discovery: a historical perspective, *Science*, 287, 1960, 2000.
7. Buss, A.D., Cox, B., and Waigh, R.D., in *Burger's Medicinal Chemistry and Drug Discovery*, 6th ed., Abraham, D.J., Ed., Vol. 1: Drug Discovery, John Wiley & Sons, New York, 2003, pp. 847–900.
8. Grabley, S. and Sattler, I., Natural products for lead identification: nature is a valuable resource for providing tools, in *Modern Methods of Drug Discovery*, Hillisch, A. and Hilgenfeld, R., Eds., 2003, pp. 87–107.
9. Butler, M.S., The role of natural product chemistry in drug discovery, *J. Nat. Prod.*, 67, 2141, 2004.
10. Butler, M.S., Natural products to drugs: natural product derived compounds in clinical trials, *Nat. Prod. Rep.*, 22, 162, 2005.
11. Newman, D.J. and Cragg, G.M., Marine natural products and related compounds in clinical and advanced preclinical trials, *J. Nat. Prod.*, 67, 1216, 2004.
12. Newman, D.J., Cragg, G.M., and Snader, K.M., Natural products as sources of new drugs over the period 1981–2002, *J. Nat. Prod.*, 66, 1022, 2003.
13. Koehn, F.E. and Carter, G.T., The evolving role of natural products in drug discovery, *Nat. Rev. Drug Discovery*, 4, 206, 2005.
14. Lipinski, C.A. et al., Experimental and computational approaches to estimate solubility and permeability in drug discovery and development settings, *Adv. Drug Delivery Rev.*, 23, 3, 1997.
15. Clardy, J. and Walsh, C., Lessons from natural molecules, *Nature*, 432, 829, 2004.
16. (a) Thompson, L.A. and Ellman, J.A., Synthesis and applications of small molecule libraries, *Chem. Rev.*, 96, 555, 1996. (b) Gallop, M.A. et al., Application of combinatorial technologies to drug discovery. 1. Background and peptide combinatorial libraries, *J. Med. Chem.*, 37, 1233, 1994. (c) Gordon, E.M. et al., Applications of combinatorial technologies to drug discovery. 2. Combinatorial organic synthesis, library screening strategies, and future directions, *J. Med. Chem.*, 37, 1385, 1994.

17. (a) Boldi, A.M. and Johnson, C.R., A system for the development and production of small-molecule libraries, in *Optimization of Solid-Phase Combinatorial Synthesis*, Yan, B. and Czarnik, A.W., Eds., Marcel Dekker: New York, 2001, pp. 1–27. (b) Seneci, P., *Solid-Phase Synthesis and Combinatorial Technologies*, John Wiley & Sons, New York, 2000. (c) Bunin, B.A., Dener, J.M., and Livingston, D.A., Application of combinatorial and parallel synthesis to medicinal chemistry, in *Annual Reports in Medicinal Chemistry*, Doherty, A., Ed., Vol. 34, Academic Press, San Diego, CA, 1999, pp. 267–286, chap. 27. (d) Bunin, B.A., *The Combinatorial Index*, Academic Press, San Diego, CA, 1998. (e) Thompson, L.A. and Ellman, J.A., *Chem. Rev.*, 96, 555, 1996.

18. (a) Zhang, W. et al., Solution-phase preparation of a 560-compound library of individual pure mappicine analogues by fluorous mixture synthesis, *J. Am. Chem. Soc.*, 124, 10443, 2002. (b) de Frutos, O. and Curran, D.P., Solution phase synthesis of libraries of polycyclic natural product analogues by cascade radical annulation: synthesis of a 64-member library of mappicine analogues and a 48-member library of mappicine ketone analogues, *J. Comb. Chem.*, 2, 639, 2000.

19. (a) Dolle, R.E., Comprehensive survey of combinatorial library synthesis: 2003, *J. Comb. Chem.*, 6, 623, 2004. (b) Dolle, R.E., Comprehensive survey of combinatorial library synthesis: 2002, *J. Comb. Chem.*, 5, 693, 2003. (c) Dolle, R.E., Comprehensive survey of combinatorial library synthesis: 2001, *J. Comb. Chem.*, 4, 369, 2002. (d) Dolle, R.E., Comprehensive survey of combinatorial library synthesis: 2000, *J. Comb. Chem.*, 3, 477, 2001. (e) Dolle, R.E., Comprehensive survey of combinatorial library synthesis: 1999, *J. Comb. Chem.*, 2, 383, 2000. (f) Dolle, R.E., Comprehensive survey of combinatorial library synthesis: 1998, *J. Comb. Chem.*, 1, 235, 1999.

20. Faber, K., *Biotransformations in Organic Chemistry: A Textbook*, 5th ed., Springer-Verlag, Berlin, 2004.

21. Teague, S.J. et al., The design of lead-like combinatorial libraries, *Angew. Chem., Int. Ed. Engl.*, 38, 3743, 1999.

22. (a) Burke, M.D., Berger, E.M., and Schreiber, S.L., A synthesis strategy yielding skeletally diverse small molecules combinatorially, *J. Am. Chem. Soc.*, 126, 14095, 2004. (b) Burke, M.D. and Schreiber, S.L., A planning strategy for diversity-oriented synthesis, *Angew. Chem., Int. Ed. Engl.*, 43, 46, 2004. (c) Schreiber, S.L., Target-oriented and diversity-oriented organic synthesis in drug discovery, *Science*, 287, 1964, 2000.

23. Arya, P., Chou, D.T.H., and Baek, M.-G., Diversity-based organic synthesis in the era of genomics and proteomics, *Angew. Chem., Int. Ed. Engl.*, 40, 339, 2001.

24. Burke, M.D., Berger, E.M., and Schreiber, S.L., Generating diverse skeletons of small molecules combinatorially, *Science*, 302, 613, 2003.

25. (a) Tan, D.S., Diversity-oriented synthesis: exploring the intersections between chemistry and biology, *Nat. Chem. Biol.*, 1, 74, 2005. (b) Shang, S. and Tan, D.S., Advanced chemistry and biology through diversity-oriented synthesis of natural product-like libraries, *Curr. Opin. Chem. Biol.*, 9, 248, 2005. (c) Tan, D.S., Current progress in natural product-like libraries for discovery screening, *Comb. Chem. High Throughput Screen.*, 7, 631, 2004.

26. Borman, S., Rescuing combichem, *Chem. Eng. News*, 82, 32, 2004.

27. Golebiowski, A., Klopfenstein, S.R., and Portlock, D.E., Lead compounds discovered from libraries, *Curr. Opin. Chem. Biol.*, 5, 273, 2001.

28. Lee, A. and Breitenbucher, J.G., The impact of combinatorial chemistry on drug discovery, *Curr. Opin. Drug Discovery Dev.*, 6, 494, 2003.

29. Rouhi, A.M., Rediscovering natural products, *Chem. Eng. News*, 81, 77, 2003.

30. (a) Ortholand, Y.-Y. and Ganesan, A., Natural products and combinatorial chemistry: back to the future, *Curr. Opin. Chem. Biol.*, 8, 271, 2004. (b) Ganesan, A., Integrating natural product synthesis and combinatorial chemistry, *Pure Appl. Chem.*, 73, 1033, 2001.

31. Rouhi, A.M., Moving beyond natural products, *Chem. Eng. News*, 81, 104, 2003.

32. Gordon, E.M. et al., Strategy and tactics in combinatorial organic synthesis: applications to drug discovery, *Chimia*, 51, 821, 1997.

33. Paululat, T. et al., Combinatorial chemistry: the impact of natural products, *Chimica Oggi*, 17, 52, 1999.

34. Watson, C., Polymer-supported synthesis of non-oligomeric natural products, *Angew. Chem., Int. Ed. Engl.*, 38, 1903, 1999.

35. (a) Boldi, A.M., Libraries from natural product-like scaffolds, *Curr. Opin. Chem. Biol.*, 8, 281, 2004. (b) Abreu, P.M. and Branco, P.S., Natural product-like combinatorial libraries, *J. Braz. Chem. Soc.*, 14, 675, 2003. (c) Hall, D.G., Manku, S., and Wang, F., Solution- and solid-phase strategies for the design, synthesis, and screening of libraries based on natural product templates: a comprehensive survey, *J. Comb. Chem.*, 3, 125, 2001. (d) Nielsen, J., Combinatorial synthesis of natural products, *Curr. Opin. Chem. Biol.*, 6, 297, 2002. (e) Wessjohann, L.A., Synthesis of natural product-based compound libraries, *Curr. Opin. Chem. Biol.*, 4, 303, 2000. (f) Ganesan, A., Integrating natural product synthesis and combinatorial chemistry, *Pure Appl. Chem.*, 73, 1033, 2001. (g) Wipf, P., Synthetic aspects of combinatorial chemistry, *Pharmaceutical News*, 9, 157, 2002. (h) Abel, U. et al., Modern methods to produce natural-product libraries, *Curr. Opin. Chem. Biol.*, 6, 453, 2002.

36. (a) Wilson, L.J., Recent advances in the solid-phase synthesis of natural products, in *Solid-Phase Organic Synthesis*, Burgess, K., Ed., John Wiley & Sons, New York, 2000, pp. 247–267. (b) Knepper, K., Gil, C., and Bräse, S., Natural product-like and other biologically active heterocyclic libraries using solid-phase techniques in the post-genomic era, *Comb. Chem. High Throughput Screen.*, 6, 673, 2003.

37. Ganesan, A., Natural products as a hunting ground for combinatorial chemistry, *Curr. Opin. Biotechnol.*, 15, 584, 2004.

38. (a) Rose, S. and Stevens, A., Computational design strategies for combinatorial libraries, *Curr. Opin. Chem. Biol.*, 7, 331, 2003. (b) Samiulla, D.S. et al., Rational selection of structurally diverse natural product scaffolds with favorable ADME properties for drug discovery, *Mol. Diversity*, 9, 131, 2005.

39. Feher, M. and Schmidt, J.M., Property distributions: differences between drugs, natural products, and molecules from combinatorial chemistry, *J. Chem. Inf. Comput. Sci.*, 43, 218, 2003.

40. Henkel, T. et al., Statistical investigation into the structural complementarity of natural products and synthetic compounds, *Angew. Chem., Int. Ed. Engl.*, 38, 643, 1999.

41. Lee, M.-L. and Schneider, G., Scaffold architecture and pharmacophoric properties of natural products and trade drugs: application in the design of natural product-based combinatorial libraries, *J. Comb. Chem.*, 3, 284, 2001.

42. Bajorath, J., Chemoinformatics methods for systematic comparison of molecules from natural and synthetic sources and design of hybrid libraries, *J. Comp. Aided Mol. Design*, 16, 431, 2002.

43. Liu, D.R. and Schultz, P.G., Generating new molecular function: a lesson from nature, *Angew. Chem., Int. Ed. Engl.*, 38, 36, 1999.

44. Hoess, R.H., Protein design and phage display, *Chem. Rev.*, 101, 3205, 2001.

45. (a) Powell, K.A. et al., Directed evolution and biocatalysis, *Angew. Chem., Int. Ed. Engl.*, 40, 3948, 2001. (b) Huisman, G.W. and Gray, D., Towards novel processes for the fine-chemical and pharmaceutical industries, *Curr. Opin. Biotech.*, 13, 352, 2002. (c) Huisman, G. and Sligar, S.G., *Curr. Opin. Biotech.*, 14, 357, 2003.

46. Khosla, C., Natural product biosynthesis: a new interface between enzymology and medicine, *J. Org. Chem.*, 65, 8127, 2000.

47. Coffen, D.L. and Xiao, X.-Y., A natural approach to combinatorial chemistry, *Chemistry Today*, 9–11, 2001.

48. (a) Xiao, X.-Y., Parandoosh, Z., and Nova, M.P., Design and synthesis of a taxoid library using radiofrequency encoded combinatorial chemistry, *J. Org. Chem.*, 62, 6029, 1997. (b) Xiao, X.-Y. et al., Solid-phase combinatorial synthesis using microkan reactors, Rf tagging, and directed sorting, *Biotechnol. Bioeng.*, 71, 44, 2000.

49. Frormann, S. and Jas, G., Natural products and combinatorial chemistry — the comeback of nature in drug discovery, *Bus. Briefing: Future Drug Discovery*, 84, 2002.

50. Johnson, C.R. and Zhang, B.R., unpublished results.

51. Boldi, A.M., Building discovery libraries from natural products, *ACS Prospectives Conference on Combinatorial Chemistry: New Methods, New Discoveries*, Leesburg, VA, September 21–24, 2003.

52. Tietze, L.F., Bell, H.P., and Chandrasekhar, S., Natural product hybrids as new leads for drug discovery, *Angew. Chem., Int. Ed. Engl.*, 42, 3996, 2003.

53. Chen, C. et al., Convergent diversity-oriented synthesis of small-molecule hybrids, *Angew. Chem., Int. Ed. Engl.*, 44, 2249, 2005.

54. Ulaczyk-Lesanko, A. and Hall, D.G., Wanted: new multicomponent reactions for generating libraries of polycyclic natural products, *Curr. Opin. Chem. Biol.*, 9, 266, 2005.

55. Gámez-Montaño, R. et al., Multicomponent domino process to oxa-bridged polyheterocycles and pyrrolopyridines, structural diversity derived from work-up procedure, *Tetrahedron*, 58, 6351, 2002.

56. Fayol, A. and Zhu, J., Synthesis of furoquinolines by a multicomponent domino process, *Angew. Chem., Int. Ed. Engl.*, 41, 3633, 2002.

57. Lavilla, R. et al., Dihydropyridine-based multicomponent reactions. Efficient entry into new tetrahydroquinoline systems through Lewis acid-catalyzed formal [4+2] cycloadditions, *Org. Lett.*, 5, 717, 2003.

58. (a) Tailor, J. and Hall, D.G., Tandem aza[4+2]/allylboration: a novel multicomponent reaction for the stereocontrolled synthesis of a a-hydroxyalkyl piperidine derivatives, *Org. Lett.*, 2, 3715, 2002. (b) Toure, B.B., Hoveyda, H.R., Tailor, J., Ulaczyk-Lesanko, A., and Hall, D.G., A three-component reaction for diversity-oriented synthesis of polysubstituted piperidines: solution and solid-phase optimization of the first tandem aza[4+2]/allylboration, *Chemistry*, 9, 466, 2003.

59. Lo, M.M.-C. et al., A library of spirooxindoles based on a stereoselective three-component coupling reaction, *J. Am. Chem. Soc.*, 126, 16077, 2004.

60. Messer, R., Fuhrer, C.A., and Häner, R, Natural product-like libraries based on non-aromatics, polycyclic motifs, *Curr. Opin. Curr. Biol.*, 9, 259, 2005.

61. Fujita, T. et al., Synthesis of the new mannosidase inhibitors, diversity-oriented 5-substituted swainsonine analogues, via a stereoselective Mannich reaction, *Org. Lett.*, 6, 827, 2004.

62. Ivachtchenko, A.V. et al., A parallel solution-phase synthesis of substituted 3,7-diazabicyclo[3.3.1]nonanes, *J. Comb. Chem.*, 6, 828, 2004.

63. Manku, S. and Curran, D.P., Fluorous mixture synthesis of 4-alkylidene cyclopentenones via a rhodium-catalyzed [2+2+1] cycloaddition of alkynyl allenes, *J. Comb. Chem.*, 7, 63, 2005.

64. Pirrung, M.C. and Liu, H., Modular, parallel synthesis of an illudinoid combinatorial library, *Org. Lett.*, 5, 1983, 2003.

65. Arya, P., Joseph, R., and Chou, D.T.H., Toward high-throughput synthesis of complex natural product-like compounds in the genomics and proteomics age, *Chem. Biol.*, 9, 145, 2002.

66. Breinbauer, R. et al., Natural product guided compound library development. *Curr. Med. Chem.*, 9, 2129, 2002.

67. Bunin, B.A. and Ellman, J.A., A general and expedient method for the solid-phase synthesis of 1,4-Benzodiazepine derivatives, *J. Am. Chem. Soc.*, 114, 10997, 1992.

68. Evans, B.E. et al., Methods for drug discovery: development of potent, selective, orally effective cholecystokinin antagonists, *J. Med. Chem.*, 31, 2235, 1988.

69. Sun, H.H., Barrow, C.J., and Cooper, R., Benzomalvin D, a new 1,4-benzodiazepine atropisomer, *J. Nat. Prod.*, 58, 1575, 1995.

70. Joshi, B.K. et al., Sclerotigenin: a new antiinsectan benzodiazepine from the sclerotia of *Penicillium sclerotigenum*, *J. Nat. Prod.*, 62, 650, 1999.

71. Sternbach, L.H., The benzodiazepine story, *J. Med. Chem.*, 22, 1, 1979.

72. Virgilio, A.A. and Ellman, J.A., Simultaneous solid-phase synthesis of β-turn mimetics incorporating side-chain functionality, *J. Am. Chem. Soc.*, 116, 11580, 1994.

73. Collins, P.W. and Djuric, S.W., Synthesis of therapeutically useful prostaglandin and prostacyclin analogs, *Chem. Rev.*, 93, 1533, 1993.

74. Chen, S. and Janda, K.D., Synthesis of prostaglandin E2 methyl ester on a soluble-polymer support for the construction of prostanoid libraries, *J. Am. Chem. Soc.*, 119, 8724, 1997.

75. Suzuki, M. et al., Three-component coupling synthesis of prostaglandins. A simplified, general procedure, *Tetrahedron*, 46, 4809, 1990.

76. Lee, K.J. et al., Soluble-polymer supported synthesis of a prostanoid library: identification of antiviral activity, *Org. Lett.*, 1, 1859, 1999.

77. Thompson, L.A. et al., Solid-phase synthesis of diverse E- and F-Series prostaglandins, *J. Org. Chem.*, 63, 2066, 1998.

78. Dragoli, D.R. et al., Parallel synthesis of prostaglandin E1 analogues, *J. Comb. Chem.*, 1, 534, 1999.

79. Johnson, C.R. and Braun, M.P., A two-step, three-component synthesis of PGE1: utilization of α-iodoenones in Pd(0)-catalyzed cross-couplings of organoboranes, *J. Am. Chem. Soc.*, 115, 11014, 1993.

80. (a) Leclercq, R. et al., *N. Engl. J. Med.*, 319, 157, 1988. (b) Tomasz, A., *N. Engl. J. Med.*, 330, 1247, 1994. (c) Hiramatsu, K., *Drug Resist. Updates*, 1, 135, 1998. (d) Nicolaou, K.C., Boddy, C.N.C., Bräse, S., and Winssinger, N., Chemistry, biology, and medicine of the glycopeptide antibiotics, *Angew. Chem. Int. Ed. Engl.*, 38, 2096, 1999.

81. Xu, R. et al., Combinatorial library approach for the identification of synthetic receptors targeting vancomycin-resistant bacteria, *J. Am. Chem. Soc.*, 121, 4898, 1999.

82. Nicolaou, K.C. et al., Solid- and solution-phase synthesis of vancomycin and vancomycin analogues with activity against vancomycin-resistant bacteria, *Chem. Eur. J.*, 7, 3798, 2001.

83. Jain, R.K., Trias, J., and Ellman, J.A., D-Ala-D-Lac binding is not required for the high activity of vancomycin dimers against vancomycin resistant *Enterococci*, *J. Am. Chem. Soc.*, 125, 8740, 2003 and references cited therein.

84. (a) Nicolaou, K.C. and Pfefferkorn, J.A., Solid-phase synthesis of natural products and natural product-like libraries, in *Handbook of Combinatorial Chemistry: Drugs, Catalysts, Materials*, Nicolaou, K.C., Hanko, R. and Hartwig, W., Eds., Wiley, Weinhaim, Germany, 2002, 613–642. (b) Nicolaou, K.C. and Pfefferkorn, J.A., Solid-phase synthesis of complex natural products and libraries thereof, *Biopolymers*, 60, 171, 2001.

85. (a) Nicolaou, K.C. et al., Total synthesis and chemical biology of the sarcodictyins, *Chem. Pharm. Bull.*, 47, 1199, 1999. (b) Nicolaou, K.C. et al., Solid and solution phase synthesis and biological evaluation of combinatorial sarcodictyin libraries, *J. Am. Chem. Soc.*, 120, 10814, 1998.

86. Nicolaou, K.C. et al., Synthesis of epothiolones A and B in solid and solution phase, *Nature*, 387, 268, 1997.

87. (a) Nicolaou, K.C. et al., Solid- and solution-phase synthesis of vancomycin and vancomycin analogues with activity against vancomycin-resistant bacteria, *Chem. Eur. J.*, 7, 3798, 2001. (b) Nicolaou, K.C. et al., Target-accelerated combinatorial synthesis and discovery of highly potent antibiotics effective against vancomycin-resistant bacteria, *Angew. Chem., Int. Ed. Engl.*, 39, 3823, 2000.

88. Nicolaou, K.C. et al., Solid phase synthesis of macrocycles by an intramolecular ketophosphonate reaction: synthesis of a (DL)-muscone library, *J. Am. Chem. Soc.*, 120, 5132, 1998.

89. (a) Nicolaou, K.C. et al., Combinatorial synthesis through disulfide exchange: discovery of potent psammaplin a type antibacterial agents active against methicillin-resistant *Staphylococcus aureus* (MRSA), *Chem. Eur. J.*, 7, 4280, 2001. (b) Nicolaou, K.C. et al., Optimization and mechanistic studies of psammaplin a type antibacterial agents active against methicillin-resistant *Staphylococcus aureus* (MRSA), *Chem. Eur. J.*, 7, 4296, 2001.

90. (a) Nicolaou, K.C. et al., Natural product-like combinatorial libraries based on privileged structures. 1. General principles and solid-phase synthesis of benzopyrans, *J. Am. Chem. Soc.*, 122, 9939, 2000. (b) Nicolaou, K.C. et al., Natural product-like combinatorial libraries based on privileged structures. 2. Construction of a 10,000-membered benzopyran library by directed split-and-pool chemistry using NanoKans and optical encoding, *J. Am. Chem. Soc.*, 122, 9954, 2000. (c) Nicolaou, K.C. et al., Natural product-like combinatorial libraries based on privileged structures. 2. The "libraries from libraries" principle for diversity enhancement of benzopyran libraries, *J. Am. Chem. Soc.*, 122, 9968, 2000.

91. Horton, D.A., Bourne, G.T., and Smythe, M.L., The combinatorial synthesis of bicyclic privileged structures or privileged substructures, *Chem. Rev., 103*, 893, 2003.

92. Nicolaou, K.C. et al., Combinatorial synthesis of novel and potent inhibitors of NADH: ubiquninone oxidoreductase, *Chem. Biol.*, 7, 979, 2000.

93. Fecik, R.A. et al., Use of combinatorial and multiple parallel synthesis methodologies for the development of anti-infective natural products, *Pure Appl. Chem.*, 71, 559, 1999.

94. Nicolaou, K.C. et al., Discovery of novel antibacterial agents active against methicillin-resistant *Staphylococcus aureus* from combinatorial benzopyran libraries, *Chembiochem*, 460, 2001.

95. Downes, M. et al., A chemical, genetic, and structural analysis of the nuclear bile acid receptor FXR, *Mol. Cell*, 11, 1079, 2003.

96. Nicolaou, K.C. et al., Discovery and optimization of non-steroidal FXR agonists from natural product-like libraries, *Org. Biomol. Chem.*, 1, 908, 2003.

97. Burke, M.D., Berger, E.M., and Schreiber, S.L., Generating diverse skeletons of small molecules combinatorially, *Science*, 302, 613, 2003.

98. Schreiber, S.L., Chemical genetics resulting from a passion for synthetic organic chemistry, *Bioorg. Med. Chem.*, 6, 1127, 1998.

99. Wong, J.C. et al., Modular synthesis and preliminary biological evaluation of stereochemically diverse 1,3-dioxanes, *Chem. Biol.,* 11, 1279, 2004.
100. Schmidt, D.R., Kwon, O., and Schreiber, S.L., Macrolactones in diversity-oriented synthesis: preparation of a pilot library and exploration of factors controlling macrocyclization, *J. Comb. Chem.,* 6, 286, 2004.
101. Krishnan, S. and Schreiber, S.L., Syntheses of stereochemically diverse nine-membered ring-containing biaryls, *Org. Lett.,* 6, 4021, 2004.
102. Spring, D.R. et al., Diversity-oriented synthesis of biaryl-containing medium rings using a one bead/one stock solution platform, *J. Am. Chem. Soc.,* 124, 1354, 2002.
103. Taylor, S.J., Taylor, A.M., and Schreiber, S.L., Synthetic strategy toward skeletal diversity via solid-supported, otherwise unstable reactive intermediates, *Angew. Chem., Int. Ed. Engl.,* 43, 1681, 2004.
104. Sello, J.K. et al., Stereochemical control of skeletal diversity, *Org. Lett.,* 5, 4125, 2003.
105. Kubota, H. et al., Pathway development and pilot library realization in diversity-oriented synthesis: exploring Ferrier and Pauson-Khand reactions on a glycal template, *Chem. Biol.,* 9, 265, 2002.
106. (a) Tan, D.S. et al., *J. Am. Chem. Soc.,* 121, 9073, 1999. (b) Tan, D.S. et al., *J. Am. Chem. Soc.,* 120, 8565, 1998.
107. Seitz, M. and Reiser, O., Synthetic approaches towards structurally diverse g-butyrolactone natural product-like compounds, *Curr. Opin. Chem. Biol.,* 9, 285, 2005.
108. Oguri, H. and Schreiber, S.L., Skeletal diversity via a folding pathway: synthesis of indole alkaloid-like skeletons, *Org. Lett.,* 7, 47, 2005.
109. Kumar, N. et al., Small-molecule diversity using a skeletal transformation strategy, *Org. Lett.,* 7, 2535, 2005.
110. (a) Reayi, A. and Arya, P., Natural product-like chemical space: search for chemical dissectors of macromolecular interactions, *Curr. Opin. Chem. Biol.,* 9, 240, 2005. (b) Arya, P. et al., Exploring new chemical space by stereocontrolled diversity-oriented synthesis, *Chem. Biol.,* 12, 163, 2005. (c) Arya, P. et al., Toward the library generation of natural product-like polycyclic derivatives by stereocontrolled diversity-oriented synthesis, *Pure Appl. Chem.,* 77, 163, 2005.
111. Couve-Bonnaire, S. et al., A solid-phase, library synthesis of natural product-like derivatives from an enantiomerically pure tetrahydroquinoline scaffold, *J. Comb. Chem.,* 6, 73, 2004.
112. Kane, T.R. et al., Solid-phase synthesis of 1,2,3,4-tetrahydroisoquinoline derivatives employing support-bound tyrosine esters in the Pictet–Spengler reaction, *J. Comb. Chem.,* 6, 564, 2004.
113. Arya, P. et al., Stereoselective diversity-oriented solution and solid-phase synthesis of tetrahydroquinoline-based polycyclic derivatives, *J. Comb. Chem.,* 6, 54, 2004.
114. Khadem, S. et al., A solution- and solid-phase approach to tetrahydroquinoline-derived polycyclics having a 10-membered ring, *J. Comb. Chem.,* 6, 724, 2004.
115. Arya, P. et al., Solution- and solid-phase synthesis of natural product-like tetrahydroquinoline-based polycyclics having a medium size ring, *J. Comb. Chem.,* 6, 735, 2004.
116. Arya, P. et al., A solid phase library synthesis of hydroxyindoline-derived tricyclic derivatives by Mitsunobu approach, *J. Comb. Chem.,* 6, 65, 2004.
117. Gan, Z. et al., Stereocontrolled solid-phase synthesis of a 90-member library of indoline-alkaloid-like polycyclics from an enantiorich aminoindoline scaffold, *Angew. Chem., Int. Ed. Engl.,* 44, 2, 2005.
118. (a) Koch, M.A. and Waldmann, H., Protein structure similarity clustering and natural product structure as guiding principles in drug discovery, *Drug Discovery Today,* 10, 471, 2005. (b) Dekker, F.J., Koch, M.A., and Waldmann, H., Protein structure similarity clustering (PSSC) and natural product structure as inspiration sources for drug development and chemical genomics, *Curr. Opin. Chem. Biol.,* 9, 232, 2005.
119. (a) Breinbauer, R., Vetter, I.R., and Waldmann, H., From protein domains to drug candidates-natural products as guiding principles in the design and synthesis of compound libraries, *Angew. Chem., Int. Ed. Engl.,* 41, 2879, 2002. (b) Breinbauer, R. et al., Natural product guided compound library development, *Curr. Med. Chem.,* 9, 2129, 2002.
120. (a) Brohm, D. et al., Solid-phase synthesis of dysidiolide-derived protein phosphatase inhibitors, *J. Am. Chem. Soc.,* 124, 13171, 2002. (b) Brohm, D. et al., Natural products are biologically validated starting points in structural space for compound library development: solid-phase synthesis of dysidiolide-derived phosphatase inhibitors, *Angew. Chem., Int. Ed. Engl.,* 41, 307, 2002.

121. Koch, M.A. et al., Compound library development guided by protein structure similarity clustering and natural product structure, *Proc. Natl. Acad. Sci. U.S.A.*, 101, 16721, 2004.

122. Boldi, A.M., Dener, J.M., and Hopkins, T.P., Solid-phase library synthesis of alkoxyprolines, *J. Comb. Chem.*, 3, 367, 2001.

123. Boldi, A.M., Hopkins, T.P., and Hu, C., unpublished results.

124. (a) Yan, B. et al., High-throughput purification of combinatorial libraries I: A high-throughput purification system using an accelerated retention window approach, *J. Comb. Chem.*, 6, 255, 2004. (b) Irving, M. et al., High-throughput purification of combinatorial libraries II: Automated separation of single diastereomers from a 4-amido-pyrrolidone library containing intentional diastereomer pairs, *J. Comb. Chem.*, 6, 478, 2004.

125. (a) Chemistry, Structures, & 3D Molecules @ 3Dchem.com. http//www.3dchem.com (accessed Feb 2006). (b) Hutchison, D.R. et al., Synthesis of *cis*-4a(*S*),8a(*R*)-perhydro-6(*2H*)-isoquinolines from quinine, *Org. Synth.*, 75, 223, 1997. (c)Kaufman, T.S. and Rúveda, E.A., The quest for quinine: Those who won the battles and those who won the war, *Angew. Chem. Int. Ed. Engl.*, 44, 854, 2005.

126. Marshall, D.L. and Liener, I.E., Modified support for solid-phase peptide synthesis which permits the synthesis of protected peptide fragments, *J. Org. Chem.*, 35, 867, 1970.

127. Padwa, A., Intramolecular 1,3-dipolar cycloaddition reactions, *Angew. Chem., Int. Ed. Engl.*, 15, 123, 1976.

128. Khan, N.M., Arumugam, V., and Balasubramanian S., Solid phase reductive alkylation of secondary amines, *Tetrahedron Lett.*, 37, 4819, 1996.

129. (a) Johnson, C.R. et al., Libraries of *N*-alkylaminoheterocycles from nucleophilic aromatic substitution with purification by solid-supported liquid extraction, *Tetrahedron*, 54, 4097, 1998. (b) Breitenbucher, J.G., Arienti, K.L., and McClure, K.J., Scope and limitations of solid-supported liquid-liquid extraction for the high-throughput purification of compound libraries, *J. Comb. Chem.*, 3, 528, 2001.

130. Darnell, J., Lodish, H., and Baltimore, D., *Molecular Cell Biology*, W.H. Freeman and Co., New York, 1990.

131. Woolard, F.X. and Brown, V., unpublished results.

132. Ono, N. et al., A convenient procedure for esterification of carboxylic acids, *Bull. Chem. Soc. Jpn.*, 51, 2401, 1978.

133. (a) Rao, C.G., A new rapid esterification procedure utilizing exceptionally mild reaction conditions, *Org. Prep. Procedures Int.*, 12, 225, 1980. (b) Czernecki, S. et al., Pyridinium dichromate oxidation: modifications enhancing its synthetic utility, *Tetrahedron Lett.*, 26, 1699, 1985.

134. Fang, L. et al., Kinetics study of amine cleavage reactions of various resin-bound thiophenol esters from Marshall linker, *J. Comb. Chem.*, 4, 362, 2002.

135. Irving, M.M. et al., Repeated use of solid supports in combinatorial synthesis: the case of Marshall resin recycling, *J. Comb. Chem.*, 3, 407, 2001.

136. Caron, P.R. et al., Chemogenomic approaches to drug discovery, *Curr. Opin. Chem. Biol.*, 5, 464, 2001.

137. Piggott, A.M. and Karuso, P., Quality, not quantity: the role of natural products and chemical proteomics in modern drug discovery, *Comb. Chem. High Throughput Screen.*, 7, 607, 2004.

138. Schreiber, S.L., Small molecules: the missing link in the central dogma, *Nat. Chem. Biol.*, 1, 64, 2005.

2 Natural Products and Combinatorial Chemistry — An Uneasy Past but a Glorious Future

A. Ganesan

CONTENTS

2.1 INTRODUCTION

Humans, being a part of nature, have traditionally relied on natural products to maintain their well-being. Although the degree of this dependence has steadily decreased over the ages, natural products remain the source of inspiration for a large proportion of our current therapeutic drugs. In the last two decades, the birth of combinatorial chemistry has significantly affected this connection, with both positive and negative repercussions. As combinatorial synthesis and other new technologies progress, the prospects for purely synthetic drugs appear even rosier. Under these circumstances, it is worthwhile to reassess the future role of natural products and the opportunities for its integration with combinatorial chemistry. This chapter is organized around three broad themes:

- A journey through time to recapitulate the past successes of natural products
- An examination of the distinguishing features of natural products and synthetic drugs
- Current and future trends of the mutually beneficial partnership between combinatorial chemistry and natural products

2.2 A BRIEF HISTORY OF NATURAL PRODUCTS AND DRUG DISCOVERY[1]

2.2.1 Prior to the 19th Century

During this long period, humankind engaged in a global combinatorial experiment to explore the flora and fauna in the environment for their healing properties. This was the age of traditional medicines, which were handed down from generation to generation either by local folklore or codification in pharmacopoeia. Some of these concoctions may have bordered on witchcraft and were of dubious value. However, the majority would not have withstood the test of time if they were without merit.

By the 19th century, organic chemistry had advanced to the stage of isolating the active principle from complex natural product extracts. Plant alkaloids such as morphine, quinine, and ephedrine were among the first to be obtained in crystalline form (Figure 2.1). Although their chemical constitution would continue to remain mysterious for some time, a crude extract was now replaceable by a pure compound. This marked a major milestone in medicinal chemistry, and we have made tremendous further strides since then.

It is worth remembering that these advances are enjoyed only by a privileged few. The WHO estimates that 80% of the world's population still relies on traditional medicines for their primary health care needs. These are invariably crude natural product extracts, with potentially large variations in composition from one batch to another, and that are not subjected to controlled clinical trials. It is also true that many of these traditional medicines, although consumed by millions of people who vouch for their efficacy, have not succumbed to Western-style inquisitions to determine their mechanism of action. Traditional medicines, contend their proponents, work by complex synergistic interactions between their constituents. Consequently, reductionist approaches that identify single active pharmaceutical ingredients are doomed to failure.

2.2.2 1900–1940

This phase saw continuous advances in chemistry and medicine. Organic chemistry was now capable of the synthesis of relatively complex targets by multistep sequences. The concept of systematically preparing a compound series and testing it for a biological effect became common, as exemplified in Ehrlich's quest for antisyphilitic agents. More and more pure natural products were identified and their structures elucidated. Synthetic mimics became a reality. For example, amphetamines were inspired by the natural product lead, ephedrine.

2.2.3 1940–1970

This was an extremely important era for biologically active natural products. Following the breathtaking success of penicillin as an antibacterial agent, thousands of soil microorganisms were investigated for their ability to inhibit the growth of other microorganisms. Indeed, most of our current antimicrobials come from this "Golden Age of Antibiotics," including many important compound classes such as β-lactams, aminoglycosides, macrolides, and tetracyclines. Also, the

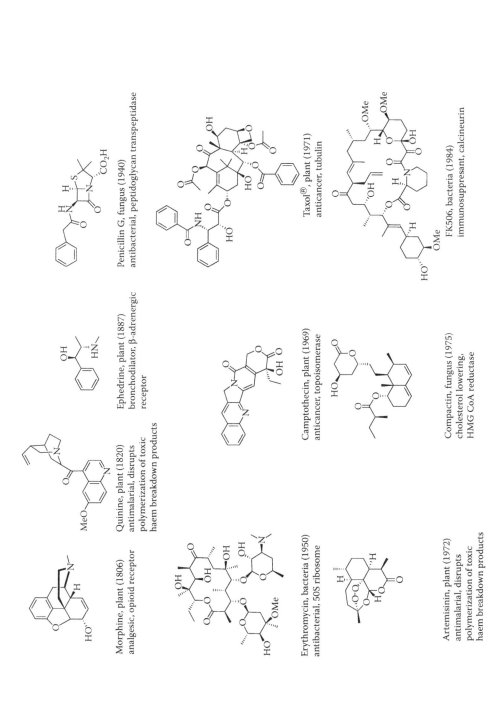

FIGURE 2.1 Ten important natural products in modern medicine. For each, the source, year of isolation, therapeutic area, and molecular target are indicated. For the antimalarial drugs, quinine and artemisinin, the precise mechanism of action is unclear.

U.S. government's "war on cancer" led to a massive screening program by the National Cancer Institute, which yielded important chemotherapeutic agents such as Taxol®, vinblastine, and camptothecin. Moreover, the cytotoxicity screens developed remain the most common primary assay for evaluating crude natural product extracts. At the same time, natural products were heavily exploited outside the antimicrobial and anticancer areas. Two classic examples are the medicinal chemistry endeavors devoted to the total and semisynthesis of thousands of compounds based on opium alkaloids as analgesics and steroids as hormones.

2.2.4 1970–1990

This marked the beginning of modern drug discovery as we know it. Thanks to advances in molecular biology, a plethora of new molecular targets became available and transformed drug discovery into much more of a rational science. Furthermore, due to recombinant DNA techniques, *in vitro* versions of these targets could be assayed on a large scale by high-throughput screening (HTS). Previously, screening predominantly depended on the whole organism or tissue slices.

Crude natural product extracts were a major source of samples for HTS campaigns. Among the natural products discovered by this process were blockbusters such as the statin HMG CoA reductase inhibitors and the immunosuppressant FK506. These compounds went straight into humans without further modification. In addition, mechanism-based screening identified numerous other natural products that are not themselves drugs but served as valuable leads for medicinal chemistry. Examples are the CCK antagonist asperlicin and the zaragozic acid squalene synthetase inhibitors. Besides random screening, focused investigations into particular natural product extracts of known potency were equally fruitful. For example, the venom of the pit viper yielded peptide inhibitors of the angiotensin-converting enzyme (ACE), which culminated in synthetic drugs of the captopril family. Investigations of the extract of *Artemisia annua* (wormwood), used in traditional Chinese medicine, led to the isolation of the antimalarial drug artemisinin.

Although this was an exciting time for natural products, purely synthetic molecules were an equally important source of drugs. Once an initial lead was found, whether from HTS, structure-based design, or simply by following up work from competitors or others, medicinal chemists believed they could optimize its properties by a fairly rational and iterative process. There was a growing sentiment that natural product screening could be eliminated if an alternative source of large numbers of compounds for HTS became available. The popularity of this philosophy varied from organization to organization. In big pharma, Merck, Sandoz (now Novartis), and Fujisawa (now Astellas) owed much of their pipeline to natural products and were positive about their potential. On the other hand, synthetic compounds were prevalent at Glaxo (later GlaxoWellcome, and now GlaxoSmithKline) and Pfizer. Smaller biotech companies typically never got involved with natural product screening.

2.2.5 1990–2000

By 1990, mechanism-based targets and HTS were the accepted paradigm for drug discovery. The HTS pace rapidly outstripped the size of internal compound archives, which further suffered from a lack of chemical diversity as they mainly originated from past medicinal chemistry programs. Meanwhile, combinatorial chemistry was proving capable of generating synthetic libraries ranging from hundreds to millions of compounds. Because these were compounds made by chemists, a route was in place for analog synthesis to follow up leads. In addition, the chosen scaffolds could have various filters built in order to maximize their lead-like or drug-like properties. Both these factors were considered attractive advantages of combinatorial chemistry over screening natural product extracts. Furthermore, as industry moved toward "blitz" HTS campaigns of short duration, assay support could no longer be guaranteed to identify an active compound and elucidate its structure from a crude extract. Although there were solutions for these perceived drawbacks, big pharma made the collective decision to downsize or phase out natural product screening.

2.2.6 THE EARLY 21ST CENTURY

The pharmaceutical and biotech industries had now espoused combinatorial chemistry for over a decade and poured billions of dollars into its implementation. Nevertheless, this neither severely reduced the timelines of drug discovery, the overall cost of the process, nor increased the rate of production of new chemical entities that met FDA approval. The reasons for this apparent failure are complex. First, combinatorial chemistry chiefly impacts the less expensive early stage of drug discovery, which is usually not the bottleneck. Combinatorial chemistry may do everything right in lead discovery and optimization, only to have a compound fail later for other reasons.

Second — and the combinatorial chemistry community is squarely to blame for this — we initially focused on the technology of how compounds are made and how they can be made rapidly, rather than on why they should be made. For logistic reasons, libraries beyond a certain size cannot be prepared as discrete entities, which provide the best opportunity for proper purification and characterization as well as the least potential for errors in HTS. Even for medium-sized libraries, the need for rapid and automated synthesis protocols placed severe restrictions on the types of organic reactions permissible. Time and again, we hear of libraries produced at high cost in those early days that are sitting on the shelves of companies or have been discarded, being of ill-defined composition and lacking in drug-like features.

The deficiencies of "first generation" combinatorial chemistry are obvious with hindsight and have been corrected. As a process, combinatorial chemistry is clearly highly valuable in speeding up analog synthesis during lead optimization. It is less clear whether it is an efficient engine for lead discovery. For this reason, there is an ongoing reappraisal of natural products that hopefully will be more rational than the recent abandonment.

2.3 THE VALUE OF NATURAL PRODUCTS IN DRUG DISCOVERY

In the last two decades, we have seen the birth of new disciplines that impact medicinal chemistry, including structure-based design, combinatorial chemistry, genomics, and proteomics. Each of these promised to revolutionize drug discovery, but to date they have not fulfilled their initial expectations. Nevertheless, they are rapidly maturing technologies that will play, working in concert, an increasing role in future synthetic drugs. As these approaches operate essentially independent of natural products, is it really necessary to continue prospecting the latter? Or to put it another way, what will we miss by omitting natural products from the drug discovery landscape?

2.3.1 SCAFFOLD DIVERSITY

Biosynthetic pathways make natural products, but laboratory chemists based on literature precedents make synthetic compounds. Although both strive for chemical diversity, each achieves this in fundamentally different ways. Biosynthesis uses a very limited set of building blocks, so diversity has to come from the branching out of early intermediates into different pathways (Figure 2.2). Synthetic chemists, however, have thousands of commercially available starting materials at their disposal. It is hence operationally easier to repeat the same synthesis with different reaction inputs to create many compounds by a combinatorial chemistry strategy. A recent computational study by Lee and Schneider[2] at Roche highlighted these differences. They found a total of 1,748 ring skeletons among 10,495 natural products in the BioscreenNP database, compared to only 807 ring skeletons from 5,757 trade drugs in the Derwent World Drug Index. There are many fused, angular, or bridged polycyclic ring systems unique to natural products. In another database comparison, bridgehead atoms occurred in 49% of natural products and in only 25% of trade drugs.[3]

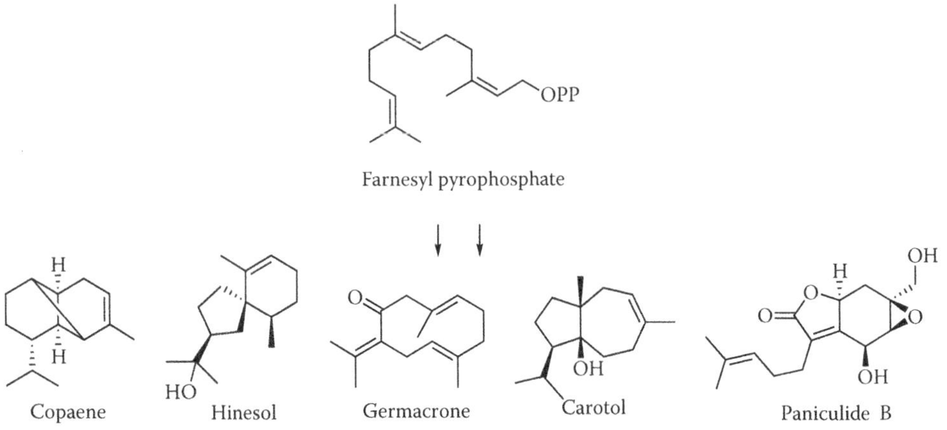

FIGURE 2.2 The biosynthesis of terpenoids provides an excellent example of achieving skeletal diversity. The single precursor farnesyl pyrophosphate gives rise to all sesquiterpenoids. Only five of the many known sesquiterpenoid ring skeletons are illustrated here.

Although both natural products and trade drugs may converge on the same objective of biological activity, they operate in almost parallel universes in chemical space. Lee and Schneider found that 83% of the natural product ring skeletons were absent among trade drugs, whereas 65% of the trade drug ring skeletons were absent in natural products. Nevertheless, both natural and synthetic compounds need to surmount the same barriers. In terms of Lipinski's "Rule of Five" parameters, the average natural product was not very different from the average trade drug, and both databases contained a similar proportion (~10%) of compounds with two or more rule violations. This smoothing out of averaged properties may not apply to narrower contexts. For example, a biologically active chemical space was recently defined from 531 structures based on 60 descriptors of drug-like properties.[4] However, when applied to natural product cyclooxygenase inhibitors, a significant proportion were outliers beyond the boundaries of the chemical space.

2.3.2 COMPLEMENTARITY IN ORGANIC SYNTHESIS

Many organic reactions employed by biosynthesis have an identical or close counterpart in organic synthesis. However, certain transformations are much more readily accomplished by nature, as illustrated[5] by the biosynthesis of Taxol® (Figure 2.3). First, there is the controlled cyclization of an acyclic precursor to the taxane ring skeleton. This and other cationic polyene cyclizations remain a daunting challenge to synthesis. Next, there is a series of oxygenations carried out by monoxygenases. At the present time, organic chemists can only dream of such site-selective functionalizations of C-H groups. Furthermore, biosynthetic reactions, carried out by chiral enzymes, are inherently stereodifferentiating, whereas asymmetric synthesis is a relatively recent development in organic synthesis. Finally, the functional groups are adjusted by biosynthesis to their correct oxidation state, followed by other transformations such as cyclization and esterification. Suffice it to say, in the total syntheses[6] of Taxol®, the same reactions rely on selective protection and deprotection of the varied oxygen-containing functional groups, and many steps are devoted to this process.

The preceding differences in building-block and reaction availability lead to structural differences between synthetic compounds and natural products. Feher and Schmidt compared databases of 10,968 drugs from the Chapman & Hall's *Dictionary of Drugs* against 670,536 combinatorial compounds (amalgamated from ChemBridge, ChemDiv, ComGenex, Maybridge, and SPECS collections) and 27,338 natural and seminatural compounds (from BioSPECS, ChemDiv, and Interbioscreen).[7] The last category was further refined to give 3,287 solely natural compounds. A

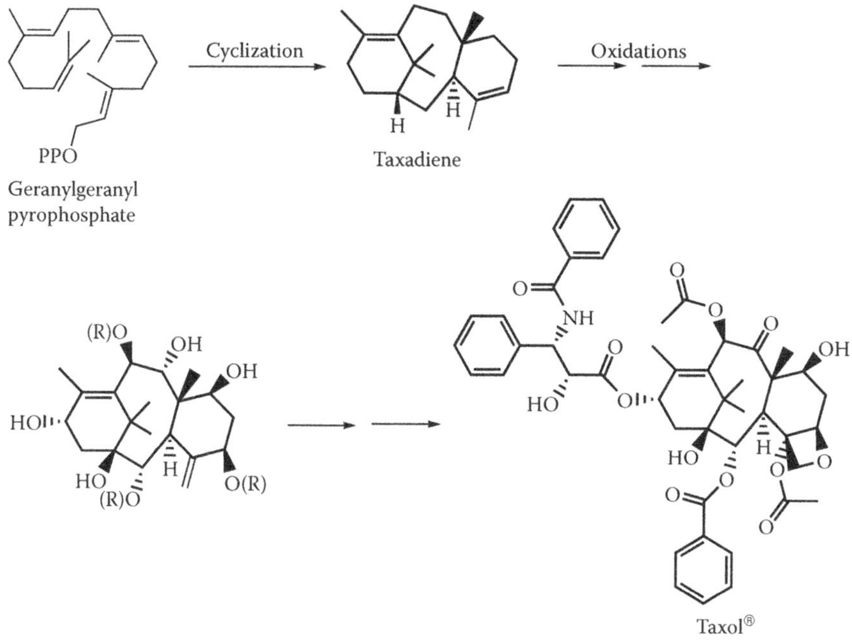

FIGURE 2.3 A simplified representation of the biosynthesis of Taxol®. The first committed step is the cyclization of geranylgeranyl pyrophosphate to the taxane skeleton. A series of site- and enantio-selective hydroxylations then takes place, followed by further transformations.

summary of the data revealed some interesting trends (Table 2.1). With regard to Lipinski-type parameters, natural products are higher in molecular weight and have more H-bond acceptors or donors than trade drugs. Probably, these features are compensated for by higher three-dimensional rigidity, as seen by the presence of more rings and fewer rotatable bonds. Natural products contain a much higher number of chiral centers and oxygen atoms, and a lower number of nitrogen, sulfur, or halogen atoms than synthetic compounds. There are also differences between the "natural and seminatural" column and "purely natural" products. As the latter are modified or serve as leads for synthetics, their properties converge closer to those of drugs.

2.3.3 CREATIVE AND UNEXPECTED SOLUTIONS

The preceding sections aside, the most novel feature of natural products is the wonderfully imaginative solutions they present for selective biological activity. Before their presence was observed in natural products, the chemistry of functional groups such as β-lactams, macrolides, and enediynes was virtually unknown, and they would never have been considered useful starting points for drug discovery. Natural products such as these have truly changed the way we view structure and reactivity. In addition, ingenious methods are employed in nature for cleverly masking reactive functional groups until necessary. The electrophilic character of the β-lactam antibiotics is only unleashed within the bacterial target enzyme's active site,[8] whereas the Bergman cyclization of enediyne antibiotics[9] requires intracellular activation.

Besides the use of novel functionality that is beyond rational design by synthetic chemists, nature is unsurpassed at the display of chemical information in three-dimensional space. For example, despite intensive efforts, there is no standard algorithm for translating a linear peptide into a small-molecule peptidomimetic. Meanwhile, numerous natural scaffolds perform this task efficiently, as in the opium alkaloids conformational mimicry of the enkephalins.[10] We will never know what revolutionary new scaffolds and pharmacophores will be lost if natural product screening is discontinued.

TABLE 2.1
Average Values (Rounded to Whole Integer unless Inappropriate) for Properties in Drugs, Natural, and Seminatural Compounds, Natural Products, and Combinatorial Compounds

Property	Drugs	Natural or Seminatural	Natural	Combinatorial
MW	340	381	414	393
Clog P	2	3	2	4
Lipinski-type acceptors	6	6	7	6
Lipinski-type donors	2	1	3	1
Rotatable bonds	6	5	4	6
Rings	3	4	4	3
Chiral centers	2	2	6	0
Heavy atoms	30	27	24	27
Carbon atoms	17	21	23	21
Nitrogen atoms	2	2	1	3
Oxygen atoms	4	4	6	3
Sulfur atoms	0.2	0.1	0.0	0.5
Halogen atoms	0.3	0.2	0.0	0.8

Source: Adapted from Feher, M. and Schmidt, J.M., Property distributions: differences between drugs, natural products, and molecules from combinatorial chemistry, *J. Chem. Inf. Comput. Sci.,* 43, 218, 2003.

Within the last 15 years alone, several unusual structures have progressed[1e] to clinical trials (Figure 2.4). Among myriad antitubulin agents, the structures of discodermolide and epothilone are unprecedented and certainly not predictable *a priori.* Similarly, the histone deacetylase inhibitor FK228 and the antibacterial GE2270A display unusual twists on a cyclic peptide or depsipeptide framework. The former functions as a relatively stable prodrug that undergoes intracellular disulfide bond cleavage to release a thiol that binds to the histone deacetylase's active site zinc. This is a very different approach from the nonselective inhibitors based on linear hydroxamic acids favored by medicinal chemists. In GE2270A, extensive heterocyclization of the peptide precursor has occurred, with undoubtedly important consequences for complementarity in shape for its binding to the elongation factor Tu and stability. A semisynthetic analog of GE2270A is in clinical trials, and an important part of the pipeline for the start-up Vicuron. Natural product-based drug discovery received a recent boost with the acquisition of Vicuron by Pfizer for $1.9 billion.

Natural products have been a particularly useful source of antimicrobial and anticancer drugs.[11] This is not coincidental, because such compounds work by differential inhibition or killing of the targeted cells in the presence of the host. In nature, it is likely that many secondary metabolites play a similar role in protecting the producing organism from ecological competition for resources or predation by other species. Nevertheless, natural products have also been a gold mine in other therapeutic areas. There is no advantage for a microorganism to produce a cholesterol-lowering agent, or a marine sponge to produce an antiviral molecule. Nevertheless, biological space is finite, and the macromolecules used by nature are invariant across species. Thus, a natural product may be active for one target in nature and, coincidentally, be active against another target screened by humans. As demonstrated by Waldmann, a search for similar protein folds and domains can be utilized to identify common ligands for disparate targets.[12]

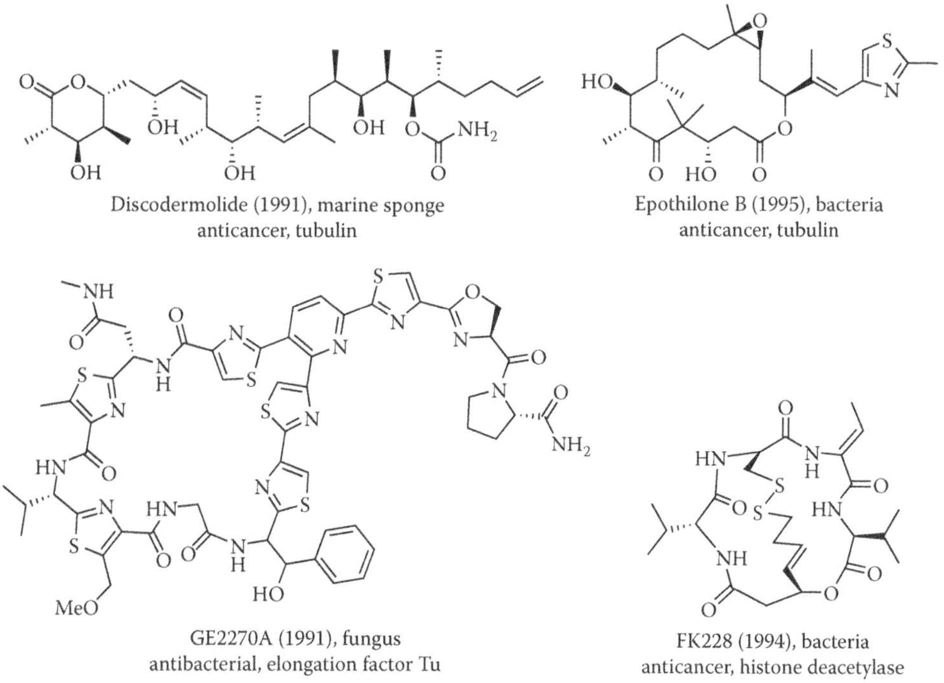

FIGURE 2.4 The structures of four recently discovered natural products with novel scaffolds that have become clinical candidates. The source, year of isolation, therapeutic area, and molecular target are indicated.

2.4 THE FUTURE OF NATURAL PRODUCTS RESEARCH AND ITS SYMBIOSIS WITH COMBINATORIAL CHEMISTRY

Natural products have played an unquestionably vital role in the discovery and development of our current medicines. Although there was a recent shift away from natural product screening in drug discovery for subjective reasons, the previous sections should have convinced the reader of its value. Although one cannot expect industry to simply reverse its decision and return to the pre-1990 *status quo*, indications are that natural products will have a secure place in future drug discovery. In these ventures, combinatorial chemistry[13] will function as an enabling rather than a rival technology in exploiting the potential of natural products.

2.4.1 NEW NATURAL PRODUCTS FOR THE NEW MILLENNIUM

The traditional pharma model of putting large collections of crude natural product extracts through HTS is no longer widely practiced. However, in reality, HTS has not been a huge source of natural products in the past. The reliance on specific mechanism-based assays means that highly active compounds will be missed if the appropriate target is not screened for. The growing shift from HTS to high-content screening (HCS) using cell-based phenotypic assays, for example, may be an improvement in this regard. Meanwhile, new natural products will continue to be isolated by academic groups. Some of these will be derived from traditional sources such as plants. Even today, it is estimated that only about 10% of all plants have been systematically examined for biological activity. At the same time, the net will be broadened to include more unusual sources, such as the marine environment, mammals and other animals, extremophiles, endophytic species, or ingredients present in our own food and drink.[14] Meanwhile, one must not overlook that careful examination of a single species can pay handsome dividends. Zeeck's group has shown[15] that a single bacterial

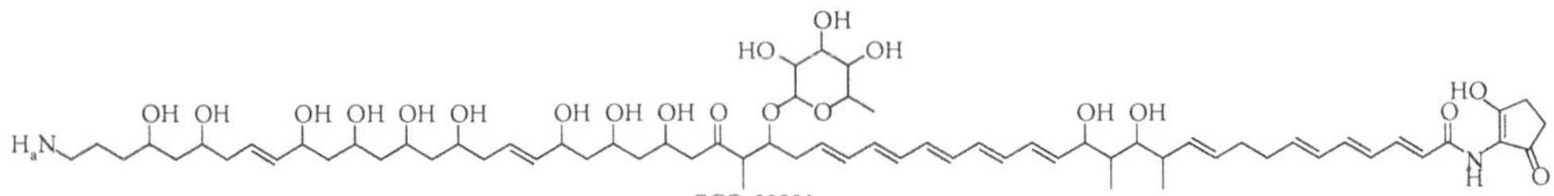

FIGURE 2.5 *In silico* natural product discovery. Genome scanning identified a type I polyketide synthetase in *Streptomyces azuinensis*, and the structure of the secondary metabolite produced was predicted. This was then verified by culture and isolation of the compound.

strain can produce a variety of secondary metabolites dependent upon culture conditions (see Chapter 5).

In the 21st century, molecular biology promises to radically alter the way we look for natural products or influence their production. The identification of biosynthetic gene clusters facilitates rational perturbation or expression in heterologous organisms as a prelude to the synthesis of hybrid metabolites that are not normally produced (see Chapter 4).[16] Similarly, genome sequencing will enable us to actively seek particular compounds, as illustrated by two recent examples. In the first, the gene cluster for the biosynthesis of sirodesmin, a fungal epipolythiodioxopiperazine of the gliotoxin family of alkaloids, was identified.[17] With this in hand, examination of other fungal species enabled the identification of additional putative clusters. We can expect this type of active prospecting for compounds of a particular class to increase in prominence and lead to the identification of new sources and new natural products. In the second example, genome scanning of *Streptomyces azuinensis,* a known bicyclomycin producer, revealed eleven other gene clusters. One of these coded for a type I polyketide synthetase. Owing to the knowledge accumulated with such modular enzyme complexes, it was actually possible to predict the complete structure of the secondary metabolite produced. This was confirmed by culturing the microorganism and isolating the compound ECO-02301, which possessed antifungal activity (Figure 2.5).[18]

Meanwhile, the complete genome of *Aspergillus fumigatus* was recently sequenced. For the first time, we can now enumerate all the gene clusters present in this fungus. Because the microbial production of secondary metabolites is extremely dependent on culture conditions, this may uncover new metabolites even from a well-known species such as *Aspergillus fumigatus,* which hitherto have never been observed in the laboratory. Another untapped source of microbial diversity is soil microorganisms, less than 1% of which have been successfully cultured. In the future, extraction of soil DNA and recombinant reconstitution of the metabolic pathways may offer the solution, effectively bypassing the need for a producing organism altogether.

2.4.2 COMBINATORIAL CHEMISTRY APPLIED TO NATURAL PRODUCTS: DISCOVERY LIBRARIES

The *de novo* synthesis of discovery libraries based around natural product scaffolds is a popular means of achieving biologically relevant chemical diversity. This very active area of research was recently reviewed,[19] and the following discussion will be restricted to the discussion of two privileged scaffolds, tetrahydro-β-carbolines and pyrazinoquinazolinediones, that are of personal interest. These scaffolds recur in a number of biologically active natural product alkaloids as well as synthetic drugs such as Lilly's PDE V inhibitor Cialis® (Figure 2.6).

In the laboratory, the tetrahydro-β-carboline ring system is readily accessed through the Pictet–Spengler reaction of tryptophan. Both solid- and solution-phase combinatorial syntheses were reported. The versatility of this scaffold can be seen in the leads for the different biological activities that were discovered from these endeavors by the groups of Ganesan,[20] Koomen,[21] and Kapoor (Figure 2.7).[22]

As for the pyrazinoquinazolinedione skeleton present in the fumiquinazoline family of antibiotics, precombinatorial chemistry approaches to this heterocycle involved lengthy sequences such

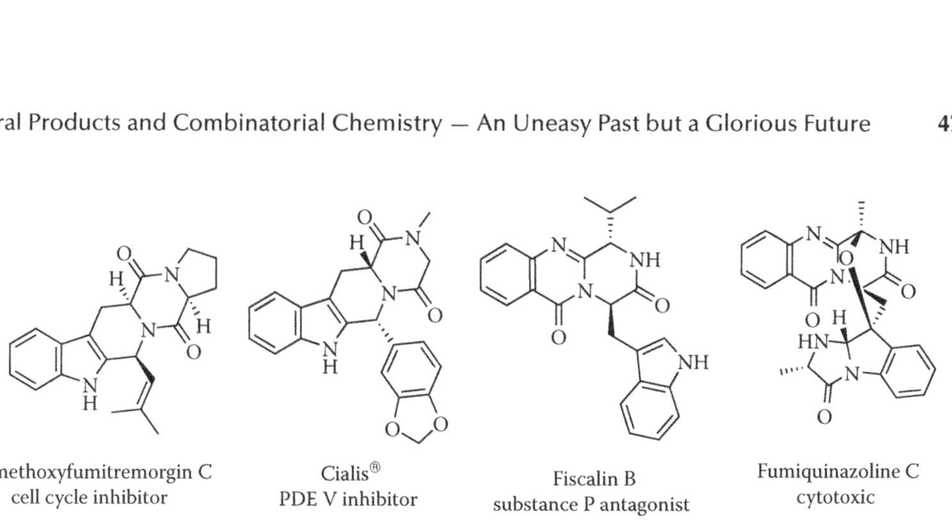

FIGURE 2.6 Examples of drugs and biologically active natural products containing either the tetrahydro-β-carboline or the pyrazinoquinazolinedione skeleton. Both of these can be considered as privileged scaffolds that are active against unrelated targets.

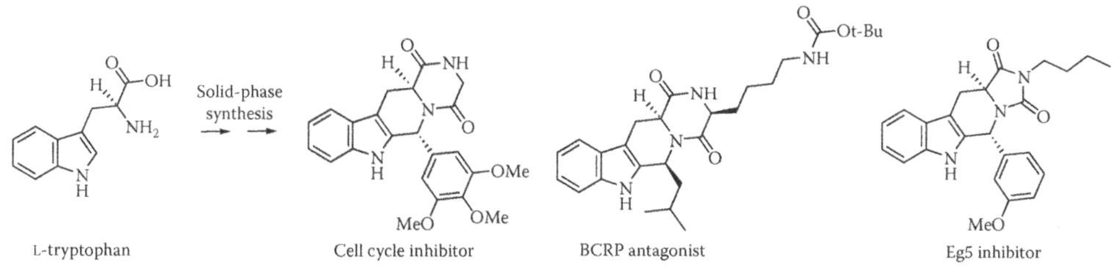

FIGURE 2.7 The Pictet–Spengler reaction on solid-phase provides access to libraries of compounds with the tetrahydro-β-carboline skeleton. This has been a fruitful area for lead discovery, as shown by the examples from three different groups.

as aza-Wittig reactions (Figure 2.8).[23] We devised a concise four-step route that was amenable to both solution- and solid-phase synthesis.[24] More recently, Arqule researchers reported a microwave-assisted three-component one-pot synthesis.[25] Although some epimerization of stereocenters was seen under the relatively harsh conditions, the one-step process used clearly facilitated library preparation.

2.4.3 COMBINATORIAL CHEMISTRY APPLIED TO NATURAL PRODUCTS: OPTIMIZATION LIBRARIES

Although natural products are the source for more than half of the current drugs, in many cases the original compound was modified. The natural product lead may not be optimal for the particular target of interest in humans, in terms of potency, selectivity, or pharmacokinetic properties such as ADME and toxicity. Such compounds are ripe for further investigation by analog array synthesis. Unlike the discovery libraries described in the preceding subsection, the goal here is to improve on the properties of the natural product against a single target.

The easiest optimization exercises, avoiding the need for total synthesis, are scaffold decoration or semisynthesis of a natural product available in bulk. A number of groups followed this strategy (examples are shown in Figure 2.9) with compounds such as yohimbine (Affymax),[26] scopolamine (Lambert, University of Melbourne),[27] the antibiotic GE2270A (Vicuron)[28] mentioned in Subsection 2.3.3, and the steroid estradiol (Poirier, Laval University).[29] In addition to using an intact natural product, degradation can provide unique building blocks for combinatorial libraries. For example, researchers recently reported the degradation of myxobacterial natural products isolated at the GBf

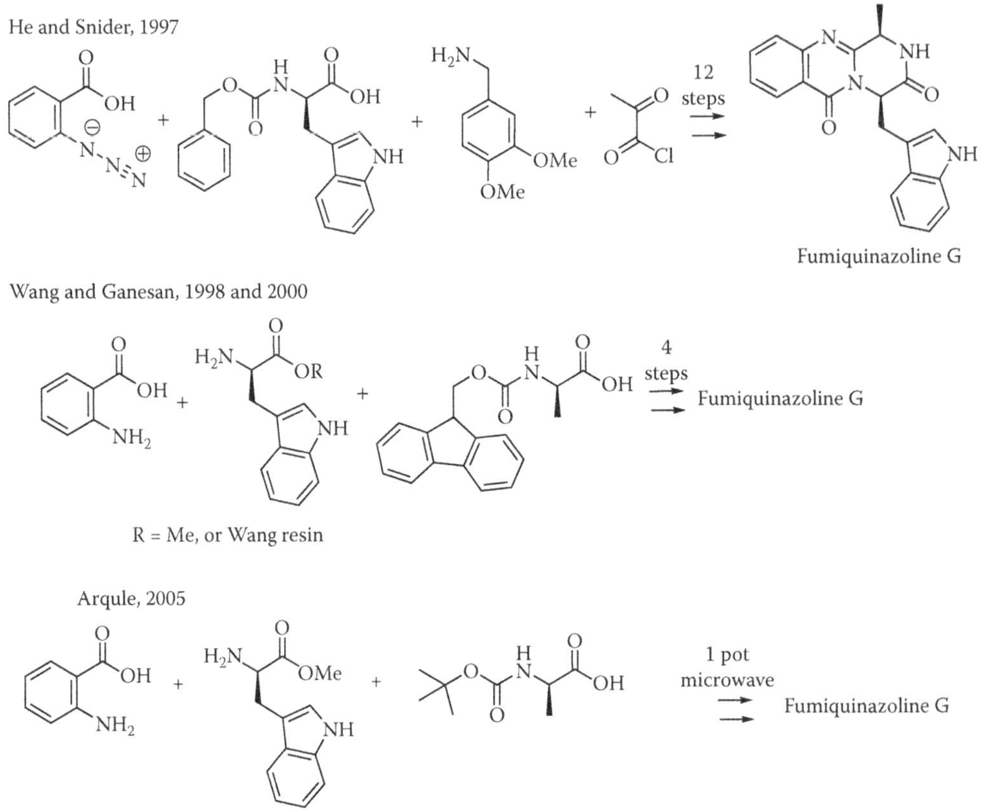

He and Snider, 1997

Fumiquinazoline G

Wang and Ganesan, 1998 and 2000

R = Me, or Wang resin

Arqule, 2005

FIGURE 2.8 The evolution of fumiquinazoline G syntheses over time. The initial, classical target-oriented synthesis by Snider is relatively long, compared to Ganesan's four-step combinatorial synthesis. Subsequently, Arqule reported a one-pot process that assembles the natural product in one step.

to give diol fragments for the synthesis of novel analogs. In one example,[30] a spiroketal from sorangicin was incorporated into a peptide hybrid.

If one is willing to invest time and resources on the total synthesis of a biologically active natural product, a method for analog preparation is available. Although the willingness of industry to tackle total syntheses of complex targets such as FK506 (Merck) and discodermolide (Novartis) should not be forgotten, such projects are particularly common in academia. In a recent example from my group, the antimycobacterial cyclic depsipeptide kahalalide A was prepared by solid-phase synthesis.[31] We employed the Kenner "safety-catch" sulfonamide linker, with cyclative cleavage[32] of the linear depsipeptide to the macrocycle as the final step (Figure 2.10). Although only a small number of analogs were tested for antimycobacterial activity, one was twofold more active than the natural product. Further improvements can be expected.

A second project involved the solution-phase total synthesis of spiruchostatin A,[33] a potent histone deacetylase inhibitor with structural similarities to FK228 (Figure 2.11). Although this was a challenging target that took one graduate student nearly three years to complete, we are now able to reap the harvest. The same methodology is currently being applied to the synthesis of analogs,[34] and a solid-phase route is under evaluation.

2.4.4 COMBINATORIAL CHEMISTRY WITH NATURAL PRODUCT-LIKE LIBRARIES

In Section 2.3, we discussed the unique properties of natural products and synthetic compounds. Now that we understand these design features, can we do away with natural products and construct synthetic libraries with similar attributes? This is the concept behind natural product-like libraries,

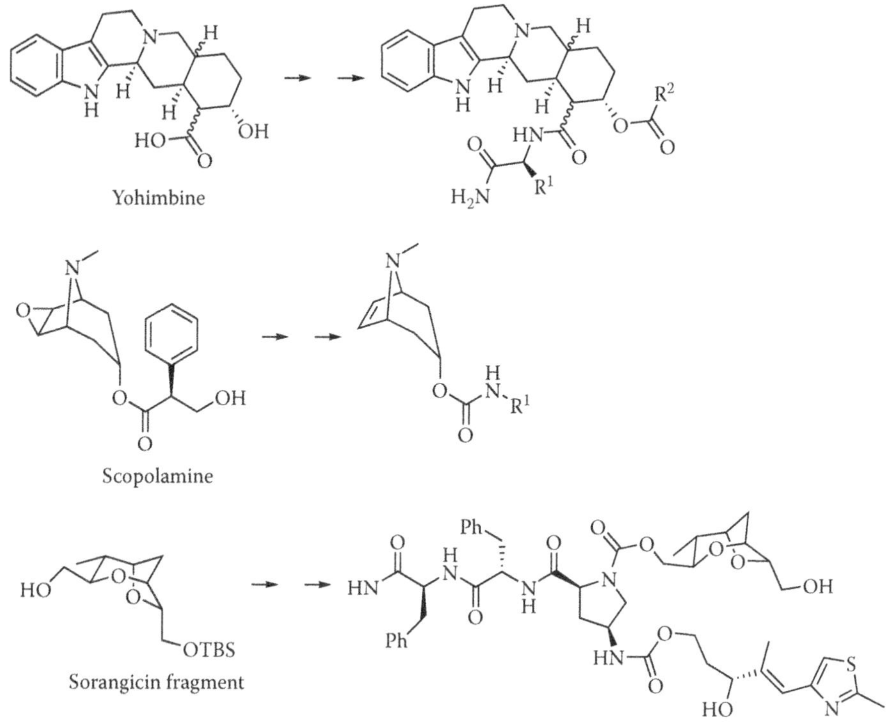

FIGURE 2.9 Three examples of combinatorial chemistry involving scaffold decoration of readily available natural products — yohimbine, scopolamine, and sorangicin (readily available to those who isolated it!).

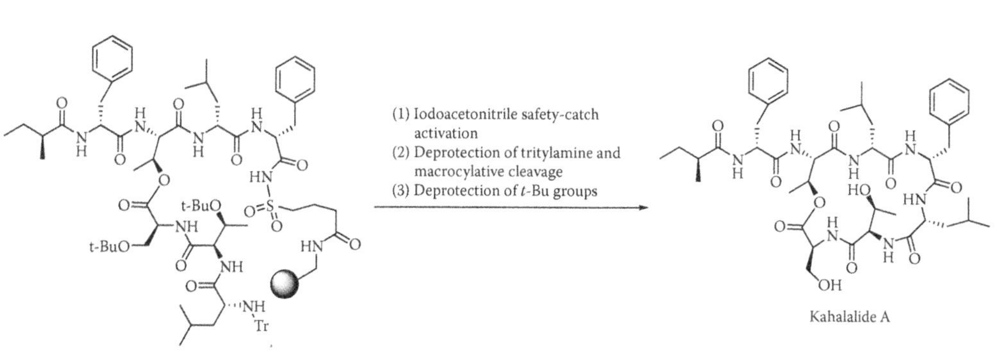

FIGURE 2.10 A solid-phase total synthesis of kahalalide A. The Kenner "safety-catch" sulfonamide linker was employed for backbone amide attachment. After assembly of the linear peptide, safety-catch activation resulted in macrocyclative cleavage from the resin.

which are currently very fashionable among academic practitioners of combinatorial chemistry.[35] These compounds often display high rigidity and are stereochemically rich, as in recent examples from Schreiber (Figure 2.12).[36] Only time will tell if such scaffolds will achieve the historical success of natural products against biological targets.

2.5 SUMMARY

Drug discovery is a highly complex and interdisciplinary affair with no foolproof guarantees of success. Furthermore, it is a continuously evolving and practical science that quickly embraces any

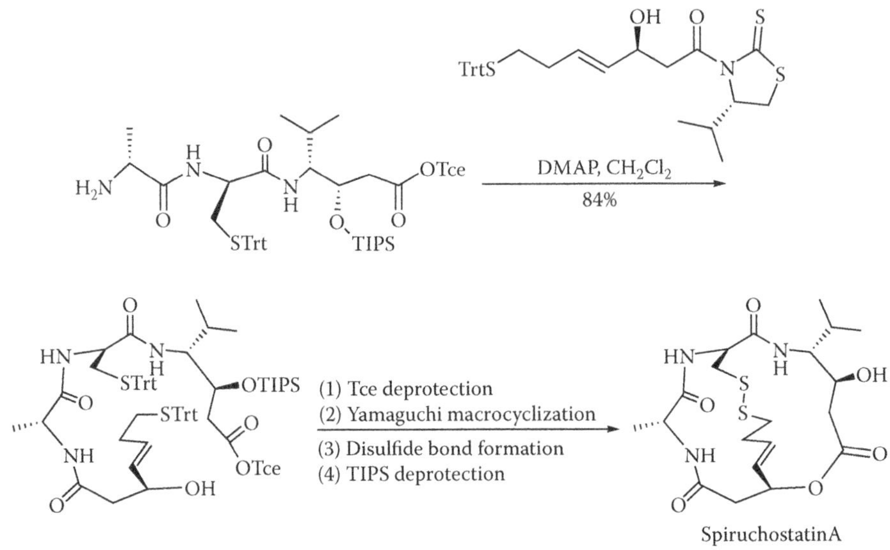

FIGURE 2.11 The strategy followed in the late stages of the total synthesis of spiruchostatin A, a histone deacetylase inhibitor. Cyclic depsipeptides such as these can be assembled in modular fashion, and readily lend themselves to analog preparation by variation of the building blocks.

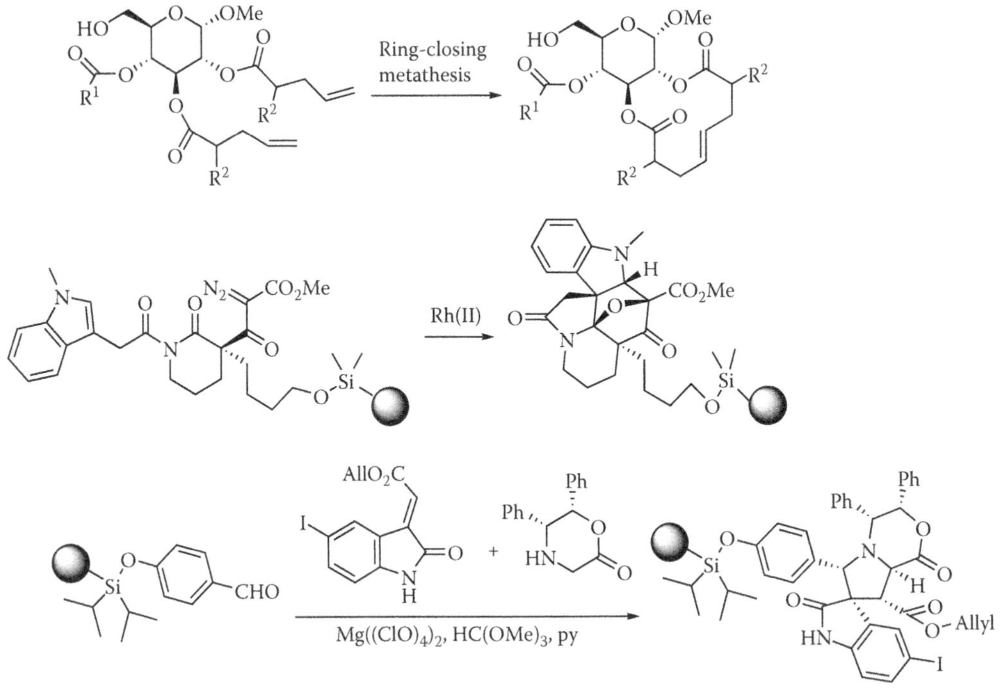

FIGURE 2.12 Three recent examples of natural product-like libraries from Schreiber's group. In the first, the library is prepared in mono- and bicyclic format to probe the importance of skeletal conformation. The next two examples illustrate natural product-like libraries made using 1,3-dipolar cycloaddition processes pioneered for total synthesis by Padwa and Williams, respectively.

tools that aid the long and arduous journey toward a drug. The past tells us that natural products research is an important source of medicines, whereas combinatorial chemistry is a relative new-comer. Acting together, these two disciplines have tremendous potential in evaluating the true worth of natural products. These may be marvelous compounds that are yet to be discovered, or simply chosen from the thousands of existing natural products whose biological properties have never been rationally investigated. We can safely say that such symbiotic projects integrating natural products and combinatorial synthesis have a bright future.

REFERENCES

1. For some recent excellent reviews on natural products and drug discovery, see (a) Shu, Y.Z., Recent natural products based drug development: a pharmaceutical industry perspective, *J. Nat. Prod.*, 61, 1053, 1998. (b) Newman, D.J., Cragg, G.M., and Snader, K.M., The influence of natural products upon drug discovery, *Nat. Prod. Rep.*, 17, 215, 2000. (c) Newman, D.J., Cragg, G.M., and Snader, K.M., Natural products as sources of new drugs over the period 1981–2002, *J. Nat. Prod.*, 66, 1022, 2003. (d) Butler, M.S., The role of natural product chemistry in drug discovery, *J. Nat. Prod.*, 67, 2141, 2004. (e) Butler, M.S., Natural products to drugs: natural product derived compounds in clinical trials, *Nat. Prod. Rep.*, 22, 162, 2005.
2. Lee, M.L. and Schneider, G., Scaffold architecture and pharmacophoric properties of natural products and trade drugs: application in the design of natural product-based combinatorial libraries, *J. Comb. Chem.*, 3, 284, 2001.
3. Henkel, T. et al., Statistical investigation into the structural complementarity of natural products and synthetic compounds, *Angew. Chem., Int. Ed. Engl.*, 38, 643, 1999.
4. Larsson, J. et al., Expanding the ChemGPS chemical space with natural products, *J. Nat. Prod.*, 68, 985, 2005.
5. Walker, K. and Croteau, R. *Proc. Natl. Acad. Sci. U.S.A.*, 97, 13591, 2000.
6. Nicolaou, K.C. and Sorensen, E.J., *Classics in Total Synthesis*, VCH, Weinheim, 1996.
7. Feher, M. and Schmidt, J.M., Property distributions: differences between drugs, natural products, and molecules from combinatorial chemistry, *J. Chem. Inf. Comput. Sci.*, 43, 218, 2003.
8. Scholar, E.M. and Pratt, W.B., Eds., *Antimicrobial Drugs*, 2nd ed., Oxford, New York, 2000.
9. Wolkenberg, S.E. and Boger, D.L., Mechanisms of in situ activation for DNA-targeting antitumor agents, *Chem. Rev.*, 102, 2477, 2002.
10. Waldhoer, M., Bartlett, S.E., and Whistler, J.L., Opioid receptors, *Annu. Rev. Biochem.*, 73, 953, 2004.
11. (a) Walsh, C., *Antibiotics: Actions, Origins, Resistance*, ASM, Washington, D.C., 2003. (b) Kingston, D.G.I. and Newman, D.J., The search for novel drug leads for predominantly antitumor therapies by utilizing mother nature's pharmacophoric libraries, *Curr. Opin. Drug Discovery Dev.*, 8, 207, 2005.
12. Dekker, F.J., Koch, M.A., and Waldmann, H., Protein structure similarity clustering (PSSC) and natural product structure as inspiration sources for drug development and chemical genomics, *Curr. Opin. Chem. Biol.*, 9, 232, 2005.
13. For earlier reviews on the interplay between combinatorial chemistry and natural products, see (a) Ortholand, J.-Y. and Ganesan, A., Natural products and combinatorial chemistry: back to the future, *Curr. Opin. Chem. Biol.*, 8, 271, 2004. (b) Ganesan, A., Natural products as a hunting ground for combinatorial chemistry, *Curr. Opin. Biotech.*, 15, 584, 2004.
14. (a) Cragg, G.M. and Newman, D.J., Biodiversity: a continuing source of novel drug leads, *Pure Appl. Chem.*, 77, 7, 2005. (b) Tulp, M. and Bohlin, L., Unconventional natural sources for future drug discovery, *Drug Discovery Today*, 9, 450, 2005.
15. Bode, H.B. et al., Big effects from small changes: possible ways to explore nature's chemical diversity, *ChemBioChem*, 3, 619, 2002.
16. Pelzer, S., Vente, A., and Bechthold, A., Novel natural compounds obtained by genome-based screening and genetic engineering, *Curr. Opin. Drug Discovery Dev.*, 8, 228, 2005.
17. Gardiner, D.M., Howlett, B.J. et al., The sirodesmin biosynthetic gene cluster of the plant pathogenic fungus *Leptosphaeria maculans*, *Mol. Microbiol.*, 53, 1307, 2004.
18. McAlpine, J.B. et al., Microbial genomics as a guide to drug discovery and structural elucidation: ECO-02301, a novel antifungal agent, as an example, *J. Nat. Prod.*, 68, 493, 2005.

19. (a) Abreu, P.M. and Branco, P.S., Natural product-like combinatorial libraries, *J. Braz. Chem. Soc.,* 14, 675, 2003. (b) Boldi, A.M., Libraries from natural product-like scaffolds, *Curr. Opin. Chem. Biol.,* 8, 281, 2004.

20. (a) Wang, H. and Ganesan, A., The *N*-acyliminium Pictet-Spengler condensation as a multicomponent combinatorial reaction on solid-phase and its application to the synthesis of demethoxyfumitremorgin C analogues, *Org. Lett.,* 1, 1647, 1999. (b) Wang, H. et al., Synthesis and evaluation of tryprostatin B and demethoxyfumitremorgin C analogues, *J. Med. Chem.,* 43, 1577, 2000. (c) Bonnet, D. and Ganesan, A., Solid-phase synthesis of tetrahydro-β-carboline-hydantoins via the *N*-acyliminium Pictet-Spengler reaction and cyclative cleavage, *J. Comb. Chem.,* 4, 546, 2002.

21. van Loevezijn, A. et al., Inhibition of BCRP-mediated drug efflux by fumitremorgin-type indolyl diketopiperazines, *Bioorg. Med. Chem. Lett.,* 11, 29, 2001.

22. Hotha, S., Renduchintala, K.V., Mayer, T.U., Kapoor, T.M. et al., HR22C16: a potent small-molecule probe for the dynamics of cell division, *Angew. Chem., Int. Ed. Engl.,* 42, 2379, 2003.

23. He, F. and Snider, B.B., Total synthesis of (+)-fumiquinazoline G and (+)-dehydrofumiquinazoline G, *Synlett,* 483, 1997.

24. (a) Wang, H. and Ganesan, A., Total synthesis of the quinazoline alkaloids (−)-fumiquinazoline G and (−)-fiscalin B, *J. Org. Chem.,* 63, 2432, 1998. (b) Wang, H. and Ganesan, A., Total synthesis of the fumiquinazoline alkaloids: solution-phase studies, *J. Org. Chem.,* 65, 1022, 2000. (c) Wang, H. and Ganesan, A., Total synthesis of the fumiquinazoline alkaloids: solid-phase studies, *J. Comb. Chem.,* 2, 186, 2000.

25. Liu, J.-F. et al., Three-component one-pot total syntheses of glyantrypine, fumiquinazoline F, and fiscalin B promoted by microwave irradiation, *J. Org. Chem.,* 70, 6339, 2005.

26. Atuegbu, A. et al., Combinatorial modification of natural products: preparation of unencoded and encoded libraries of *Rauwolfia* alkaloids, *Bioorg. Med. Chem.,* 4, 1097, 1996.

27. Aberle, N.S. et al., Parallel modification of tropane alkaloids, *Tetrahedron Lett.,* 42, 1975, 2001.

28. Clough, J. et al., Combinatorial modification of natural products: synthesis and in vitro analysis of derivatives of thiazole peptide antibiotic GE2270 A: A-ring modifications, *Bioorg. Med. Chem. Lett.,* 13, 3409, 2003.

29. Ciobanu, L.C. and Poirier, D., Solid-phase parallel synthesis of 17-substituted estradiol sulfamate and phenol libraries using the multidetachable sulfamate linker, *J. Comb. Chem.,* 5, 429, 2003.

30. Niggemann, J. et al., Natural product-derived building blocks for combinatorial synthesis. Part 1. Fragmentation of natural products from myxobacteria, *J. Chem. Soc., Perkin Trans. 1,* 2490, 2002.

31. Bourel, L. et al., Solid-phase total synthesis of kahalalide A and related analogues, *J. Med. Chem.,* 48, 1530, 2005.

32. For a review, see: Ganesan, A., Cyclative cleavage strategies for the solid-phase synthesis of heterocycles and natural products, in *Methods in Enzymology: Combinatorial Chemistry, Part B,* Morales, G.B. and Bunin, B.A., Eds., Academic Press, San Diego, CA, 2003, 369, 415.

33. Yurek-George, A. et al., Total synthesis of spiruchostatin A, a potent histone deacetylase inhibitor, *J. Am. Chem. Soc.,* 126, 1030, 2004.

34. Cecil, A., Wen, S., and Ganesan, A., unpublished results.

35. For reviews, see (a) Schreiber, S.L., Target-oriented and diversity-oriented organic synthesis in drug discovery, *Science,* 287, 1964, 2000. (b) Burke, M.D. and Schreiber, S.L., A planning strategy for diversity-oriented synthesis, *Angew. Chem., Int. Ed. Engl.,* 43, 46, 2004. (c) Reayi, A. and Arya, P., Natural product-like chemical space: search for chemical dissectors of macromolecular interactions, *Curr. Opin. Chem. Biol.,* 9, 240, 2005. (d) Shang, S. and Tan, D.S., Advancing chemistry and biology through diversity-oriented synthesis of natural product-like libraries, *Curr. Opin. Chem. Biol.,* 9, 248, 2005.

36. (a) Kim, Y. et al., Relationship of stereochemical and skeletal diversity of small molecules to cellular measurement space, *J. Am. Chem. Soc.,* 126, 14740, 2004. (b) Lo, M.M.-C. et al., A library of spirooxindoles based on a stereoselective three-component coupling reaction, *J. Am. Chem. Soc.,* 126, 16077, 2004. (c) Oguri, H. and Schreiber, S.L., Skeletal diversity via a folding pathway: synthesis of indole alkaloid-like skeletons, *Org. Lett.,* 7, 47, 2005.

3 Computational Analysis of Natural Molecules and Strategies for the Design of Natural Product-Based Compound Libraries

Florence L. Stahura and Jürgen Bajorath

CONTENTS

3.1 INTRODUCTION

Natural products continue to be an important resource for the discovery of therapeutically relevant molecules. In addition, naturally occurring molecules present a rich source of chemical diversity and information. This makes them an interesting target for chemoinformatic investigations, despite the fact that they are sometimes too large or chemically too complex for the analysis of structure–activity relationships, molecular similarity calculations, or other computational studies. Recently, efforts were made to statistically analyze collections of natural products, explore their chemical information content, and compare them to synthetic compounds.

Furthermore, a number of approaches were developed to generate compound libraries that incorporate natural product information. For example, natural molecules (specifically, selected natural structural motifs) were used as templates for synthetic diversification through combinatorial chemistry. In addition, molecules from natural and synthetic sources were combined in libraries. Many of these efforts were guided by synthetic and medicinal chemical knowledge and experience. The ensuing experimental strategies continue to play a major role in the generation of natural product-based libraries.

However, computational methods are also beginning to impact the design of such libraries, supported by insights gained from computational analyses of natural products and systematic comparison with synthetic compounds. For example, scaffolds obtained from natural molecules are subjected to combinatorial diversification and sampling of potential products. Furthermore, computational compound selection strategies can be applied to assemble molecules from natural and synthetic sources that are complementary to each other in chemical descriptor space (reflecting their chemical properties and molecular diversity). Moreover, molecular similarity methods are employed to identify synthetic mimics of natural molecules, which can then be used as chemically accessible starting points for library generation. This chapter focuses on computational approaches for the analysis of natural product collections and for comparison to synthetic molecules. Computer-aided strategies for the design of natural product-based libraries will also be described.

3.2 BACKGROUND

3.2.1 Natural Products in Drug Discovery

Despite the introduction of high-throughput technologies in pharmaceutical research, the importance of molecules isolated from natural products as a major drug source is undisputed.[1-3] Approximately 40% of the new drugs introduced between the mid-1980s and the 1990s were derived from natural products,[1] including 9 of the top 20 small-molecule drugs introduced in 1999.[2] In some therapeutic areas the predominance of natural product-based drugs is remarkable.[3] For example, 62% of current anticancer drugs come from natural products,[3] and their role in oncology is widely recognized.[4] Although 65% of antihypertensive drugs have originated from natural products,[3] this has been much less noticed. Figure 3.1 shows a few examples of biologically active and therapeutically relevant natural molecules.

Success stories of natural products as drug leads continue to be reported. For example, the recent discoveries of natural product-based tyrosine kinase inhibitors[5] or antagonists of the cancer-triggering Tcf–beta-catenin interactions illustrate this point.[6] The latter case study is particularly impressive because active compounds could be identified by screening a small natural product library of 7000 compounds. Screening of much larger synthetic libraries failed to produce any antagonists.[6]

3.2.2 The Popularity of Natural Products in Drug Discovery

Although the significance of such findings and the track record of natural molecules in drug discovery can hardly be disputed, it would be incorrect to assume that natural products have always been popular in the pharmaceutical industry. In fact, at least in recent years, quite the opposite has been true. In parallel with the advent of combinatorial chemistry and high-throughput screening about a decade ago, the focus on natural products as a viable drug source rapidly diminished in many pharmaceutical settings.[7] These developments were due in part to the belief that high-throughput technologies in combinatorial chemistry and biological screening would provide elegant and efficient means to increase the productivity of the pharmaceutical industry.[8] Consequently, the time-consuming and costly isolation of active compounds from natural sources[7,8] or the often-limited synthetic accessibility[4,9] would no longer require attention.

However, as is well known, research and development trends in the pharmaceutical industry are usually cyclical in nature and expectations are generally too high when new technologies are

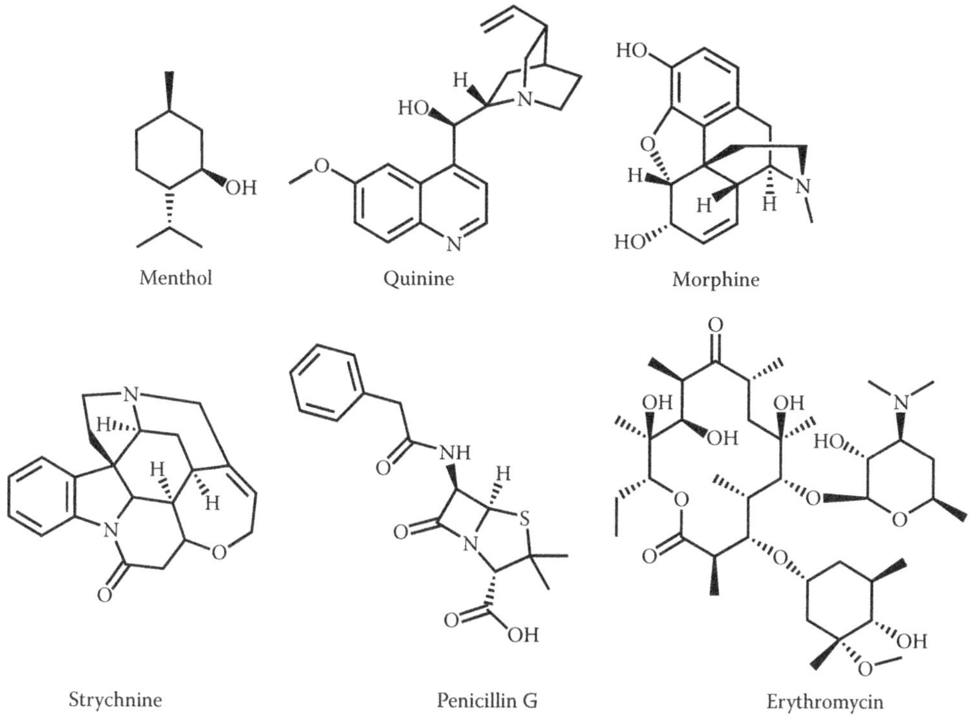

FIGURE 3.1 Examples of biologically active natural products.

introduced. Whereas high-throughput technologies and the resulting "numbers games" have thus far failed to dramatically increase the output of discovery programs (as was expected),[8] and their short-term impact is being questioned,[10] interest in natural products has yet again gradually increased in recent years.[11–13] Clearly, despite their fluctuating popularity and the relatively considerable efforts required to carry out natural product-based discovery programs, natural products are here to stay as fundamentally important sources for new drugs. This is not just because of their historic role and track record. For example, many natural product-derived leads exhibit very favorable pharmacokinetic profiles,[14,15] which may well compensate for the sometimes-limited oral availability.[16]

3.2.3 NATURAL MOLECULES AND COMPOUND LIBRARIES

One of the important reasons for the renaissance of natural products in pharmaceutical research is that natural products are no longer only regarded as leads or drug sources per se. In recent years, they are also increasingly being considered a source of largely unexplored chemical diversity and quality in compound library design and generation. Catalyzed by the availability of advanced analytical methodologies,[13] this represents an area in which synthetic chemistry, combinatorial chemistry, and natural product chemistry are beginning to merge[9,17] and in which significant progress is being made.[18,19] From a chemical perspective, the area of natural product-based libraries was recently extensively reviewed,[20–25] highlighting a number of approaches to natural product-oriented library design, ranging from total synthetic efforts[17–19] and semisynthetic[26] (and also biocatalytic[27]) derivatization or analoging to the use of preferred natural product scaffolds[28] for library synthesis. In particular, the selection of natural product scaffolds for combinatorial exploration and the generation of diverse libraries retaining natural product characteristics[18,19] have become focal points of current research and development in this area.[24,25]

3.2.4 *In Silico* Support

Given the substantial activities at the interface between natural products and combinatorial chemistry and the crucial role computational approaches play in the design of diverse combinatorial libraries,[29] one might anticipate that computational chemists and chemoinformaticians are already intensively investigating characteristic features of natural molecules and designing natural product-based libraries. This is not necessarily the case. Compared to other areas of chemoinformatics research and molecular design, the study of natural products using computational means is still a fairly underdeveloped field.[30] There are several potential reasons. Computational diversity analysis and design originally developed around synthetic reaction schemes and combinatorial approaches.[29,30] Collections of natural molecules and their structures (for example, the Chapman and Hall compendium[31]) were not part of the original knowledge base and analysis portfolio. In fact, computational diversity analysis and design became popular during the mid-1990s when interest in natural product-based discovery was rapidly diminishing in the pharmaceutical industry.

Moreover, the majority of available natural products screening libraries did not contain pure compounds with known structures but only unpurified or semipurified extracts.[32] This precluded computational analysis. In addition, the size and complexity of many natural molecules makes them difficult to study experimentally and by computational methods. For example, in some computational studies, similar to statistical analysis, unsuitable structures need to be removed from source databases prior to descriptor calculations and molecular comparisons. This is often an arduous, and thus rather unpopular, task.

Although the size and complexity of natural products can present some substantial problems for molecular diversity or similarity calculations, careful inspection of natural product collections has revealed that many natural molecules do not fundamentally differ from synthetic ones in molecular weight or in the presence of "exotic" chemical features.[30] Although, from a medicinal chemist's point of view, natural products generally are more difficult to synthesize and less drug-like,[16,33] systematic computational analyses of natural product structures are certainly feasible. In recent years, a number of chemoinformatics-type investigations focused on comparing the structures and the properties of natural products with synthetic compounds. Some of the rather interesting reported results will be discussed in the remainder of this chapter.

3.3 COMPUTATIONAL ANALYSIS OF NATURAL PRODUCTS

As discussed in this chapter, studies that target natural products using computational methods can be divided into different categories. Although the number is still fairly limited, at least some of the currently available reports introduced important concepts and laid the foundation for natural product-based library design.

3.3.1 Statistical Investigations

Statistical analyses of natural product data sets in comparison with synthetic compound collections represent the largest group of relevant studies (see Chapter 2, Section 2.3). One original report catalyzed further statistical analysis.[34] Upon the comparison of molecular core structures, functional groups, and physical or chemical properties in natural products, synthetic compounds, and drug-like molecules, the molecular and systematic differences between these compound sets were identified. Essentially, these authors calculated structural fragments and molecular properties, using descriptors, and compared their distributions in the different databases. The statistical data revealed some intuitive trends. The molecular weight of natural molecules was found to be, on average, higher than that of their synthetic counterparts. Moreover, drugs and natural products had higher oxygen content (i.e., alcohols, ethers, and esters), whereas synthetic and drug-like molecules were richer in nitrogen (i.e., amines, arenes, and amides), halogen, and sulfur atoms (see Chapter 1,

TABLE 3.1
Chemical Differences between
Natural and Synthetic Molecules
Revealed by Statistical Analyses

Property	Trends
Molecular weight	N > S
Halogen content	S > N
Oxygen content	N > S
Nitrogen content	S > N
Alcohols	N > S
Esters	N > S
Chemical saturation	N > S
Amines	S > N
Amides	S > N
Number of rings	N > S
Condensated rings	N > S
Aromatic character	S > N
Polarity	S > N
Chirality	N > S
H-bond donors	N > S
H-bond acceptors	N > S

Note: A summary of the results of statistical investigations (N = natural molecule, S = synthetic molecule).

Section 1.1). Synthetic and drug-like molecules also contained notably fewer hydrogen bond donor and acceptor functions than natural molecules. Regarding molecular fragments, approximately 40% of natural product scaffolds did not appear in synthetic molecules.[34] Table 3.1 summarizes some systematic property differences between natural and synthetic molecules.

Another study, subsequently published, compared the scaffold and R-group distributions in natural products and drug-like compounds and corroborated the earlier results.[35] Approximately 35% of complex ring systems found in drugs were also found in natural products, but only 17% of all unique ring structures in natural products were found in drugs. These investigators also carried out the first relatively systematic comparison of pharmacophore (topological) patterns in drugs and natural molecules.[35] Using a neural network technique, significant correspondence was demonstrated, consistent with the fact that natural products are a major source of trade drugs.

Another extensive statistical study compared the chemical properties of natural products, combinatorial libraries, and drugs.[36] Comparison of a number of chemical property distributions demonstrated that drugs cover an area, in global property space, overlapping with the interface between natural and combinatorial molecules. This is consistent with the fact that current drug molecules originate almost equally from natural and synthetic sources. Furthermore, a recent study analyzed in detail the statistical distribution of chemical properties and of structural fragments or motifs in a database of natural macrocycles.[37] Many natural macrocycles displayed similar intramolecular patterns of charge and polar groups, essentially resulting in distinct apolar and polar molecular sites. Studying such complex molecules in a systematic manner should be helpful in the design of new synthetic strategies.

Although the computational concepts underlying these types of statistically oriented analyses are not very challenging, they produced some important findings that correlated well with the chemical observations. For example, the observed differences in core structure distributions between natural products and synthetic molecules not only reflect the greater intrinsic chemical diversity of natural products but also help explain the difficulties in synthesizing many natural molecules, irrespective of their size or degree of complexity.

3.3.2 INFORMATION CONTENT ANALYSIS

Going beyond statistical analyses, information encoded in molecular descriptor value distributions in databases of natural or synthetic compounds was analyzed in quantitative terms by application of an entropy-based information-theoretic approach.[38,39] Descriptor value distributions, represented in histograms, can be reduced to their information content using Shannon entropy (SE) calculations.[38] Differences in information content between databases can be quantified using differential SE (DSE) analysis.[39,40] An extension of this approach, SE-DSE analysis,[41] makes it possible to classify molecular descriptors according to their relative information content in diverse databases and to identify those descriptors that are most responsive to systematic differences between compound databases.[39] Figure 3.2 illustrates the idea behind SE and DSE calculations. Using this approach, distributions of ~100 molecular property descriptors were compared in collections of ~200,000 synthetic compounds, ~116,000 natural products,[40,41] and ~12,000 drug-like molecules.[41]

In these calculations, a number of descriptors were identified that are very responsive to systematic differences between synthetic and natural molecules. Some of these descriptors

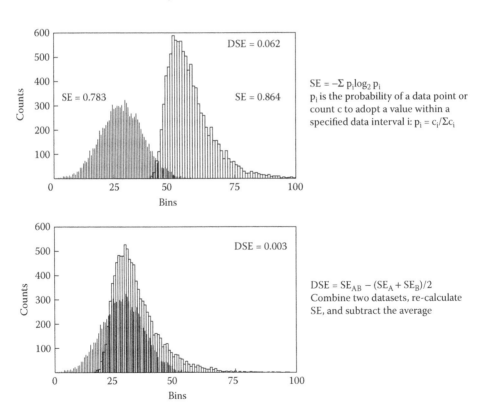

$$SE = -\Sigma\ p_i \log_2 p_i$$
p_i is the probability of a data point or count c to adopt a value within a specified data interval i: $p_i = c_i/\Sigma c_i$

$$DSE = SE_{AB} - (SE_A + SE_B)/2$$
Combine two datasets, re-calculate SE, and subtract the average

FIGURE 3.2 SE and DSE calculations. Histogram representations of values of a molecular descriptor with relatively high information content in two compound databases (A and B) and either distinct (top) or similar (bottom) value distributions. SE values are an entropic measure of information content. For the distributions, calculated scaled SE and DSE values are reported. DSE calculations add value range dependence as a parameter to information content analysis.

TABLE 3.2
Molecular Descriptors That Are Responsive to Systematic Differences between Natural and Synthetic and Natural and Drug-Like Molecules

Comparison	Molecular Descriptors
Natural vs. synthetic	Atomic polarizability of bonded atoms
	Connectivity indices
	Number of nonaromatic double bonds
	Number of hydrogen atoms
	Number of single bonds
	Number of hydrogen atoms
	Number of aromatic bonds
	Number of C-C double bonds
	Number of hydroxyl groups
	Number of halogen atoms
Natural vs. drug-like	Connectivity indices
	Number of single bonds
	Number of halogen atoms
	Partially charged surface area
	Number of basic atoms

Note: Descriptors reported according to the captured properties (specific names of implementations in modeling programs not indicated).

represented very complex formulations. For example, some composite descriptors map chemical properties on calculated molecular surfaces.[41] By contrast, others are rather intuitive. Examples of relatively simple descriptors that detect systematic differences between natural and other molecules are given in Table 3.2. Because significantly fewer drug-like molecules than synthetic molecules were available for comparison, systematic differences between natural and drug-like molecules were more difficult to detect. However, overlapping sets of descriptors were identified. Although information content analysis does not rely on direct structural comparisons and statistical criteria, the identified sensitive descriptors can easily be interpreted in chemical terms and reflect discrepancies between natural and synthetic compounds that were also revealed by statistical analysis (e.g., systematic differences in the degree of chemical saturation or heteroatom content). Although the computational analyses methods discussed are, in part, methodologically distinct, they have produced some similar findings, either by direct (statistical) or indirect means (information content analysis). The latter approach is significant for compound classification calculations because it can also identify mathematically complex descriptors that respond differently to properties of natural and synthetic molecules. These descriptors are not intuitively obvious or readily interpretable and are difficult to identify by structural comparisons.

3.3.3 COMPOUND CLASSIFICATION

A few computational studies attempted to systematically distinguish between compounds from natural and other sources and between active and inactive natural molecules. Applying a substructure-based clustering approach guided by genetic algorithm calculations, biological activity profiles for database compounds to distinguish active from inactive molecules were generated.[42] In these calculations, selected natural products were used in one of the test cases. Active natural products were predicted about seven times better than with methods based on random selections. Descriptor selection based on SE analysis provided a rational basis for QSPR-type calculations to predict database compounds from natural sources.[43] In these calculations, sets of descriptors with significant sensitivity to natural product properties, as identified by SE and variability analysis, were used to classify compounds as either natural or synthetic. With 80 to 90% prediction accuracy, randomly selected natural products were distinguished from synthetic database compounds, and natural molecules with known specific activity were differentiated from active synthetic compounds.[43]

In a recent classification study, scaffolds taken from natural products were subjected to a two-step analysis in order to identify a diverse set of natural product scaffolds with preferred ADME-properties.[44] Molecular diversity analysis and selection of natural product scaffolds were carried out using cell-based partitioning and the BCUT metrics.[45] To approximate ADME parameters, "rule-of-five"[46] calculations were carried out. A set of about 50 natural product scaffolds fulfilled the desired diversity and ADME criteria.[44]

3.3.4 VIRTUAL SCREENING

In computational screening of compound databases for active molecules, natural products are also considered. For example, a database only consisting of natural products was generated for structure-based virtual screening (docking),[47] a rather unusual source for docking calculations. Furthermore, a proprietary antagonist of the Bak–Bcl-x_L protein–protein interaction was identified in an experimental screen of a natural product collection.[48] Because this compound was difficult to synthesize, it was used as a starting point of a virtual screening effort using a combination of similarity searching and docking calculations. Through computational screening of synthetic compound libraries, the aim was to identify compounds with similar activity to the known antagonist and also those that were more synthetically accessible. Based on these calculations, five structurally distinct synthetic compounds were identified that displayed similar biological activity to the natural template molecules. The concept of "synthetic mimics" of selected natural products was validated.[48]

3.4 LIBRARY DESIGN

What is the impact of computational studies of natural molecules or natural product-based compounds on library design? Although still on a fairly limited scale, computational analyses aided in library design and made a verifiable impact. Essentially, two different types of approaches can be distinguished.

3.4.1 PRODUCT-BASED DESIGN

The product-based design strategy is the major focal point of computational studies using natural products. Regardless of the specifics, product-based library design generally attempts to use specifically selected scaffolds. In this case, natural molecules are the starting point and are subjected to combinatorial diversification. In the case of natural products, this is often done by a semisynthetic approach in which suitable functional groups in natural molecules are targeted as points for diversification. In this context, a key question is how to best select natural product scaffolds.

Computational investigations are beginning to provide guidance for such efforts. For example, studies designed to identify privileged natural product scaffolds that have a significant probability of acting against specific targets or target classes were reported.[28] Moreover, attempts to identify chemically diverse scaffolds with preferred ADME or other desired properties are carried out to provide a basis for the design of natural product-based screening libraries.[44] The observation that ring topology found in natural and synthetic molecules differs greatly whereas pharmacophores are much more similar, in principle, suggests that a rather promising library design strategy might be to combine natural scaffolds with synthetic pharmacophore-producing reagent sets.[35] Thus, computational studies provide several potential routes to improve natural product-based library design.

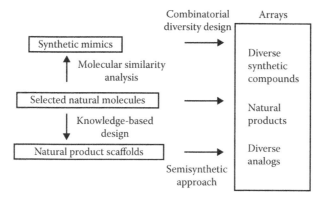

FIGURE 3.3 Summary of a computer-aided library design strategy that aims at the generation of hybrid arrays that are focused on selected natural products and contain diverse natural and synthetic compounds.

3.4.2 HYBRID LIBRARIES

A conceptually different approach derived from computational studies is the design of libraries that combine natural and synthetic molecules. To a large extent, the idea of finding synthetic mimics of natural products,[48,49] as mentioned earlier, and some product-based design elements[49] are central to such hybrid libraries. The computational design strategy is summarized in Figure 3.3. Arrays are focused on select natural products. The key computational component is a similarity-based expansion, using similarity-search tools or compound classification methods, to add similar synthetic compounds to select natural products.

In Figure 3.4, synthetic mimics of an active natural product, L-783,281, were identified by similarity searching and provided the basis for the design of a hybrid array.[49] Compound L-783,281, a quinone analog, is a fungal metabolite that was identified as a potent inhibitor of the insulin receptor tyrosine kinase.[50] Synthetic compounds reaching a chosen computational similarity threshold criterion can be further expanded by combinatorial diversification. Starting from the natural molecule template, scaffolds can be isolated and subjected to semisynthetic derivatization corresponding to product-based design. Ultimately, arrays are designed to combine select, active natural molecules and their analogs with a spectrum of synthetic, chemically accessible compounds that cover chemical space around them. The design of such hybrid arrays was introduced as an *in silico* concept to aid in the generation of novel natural product-based screening libraries.[49] As such, this strategy is clearly distinguished from the generation of libraries of hybrid molecules consisting of building blocks taken from different natural products.[51] The latter approach aims to exploit the high degree of chemical diversity encoded in natural products and to further extend, combinatorially, natural compound diversity.

3.5 CONCLUDING REMARKS

Occasionally articulated views that natural products are generally too large and complex for chemical and computational exploration represent an oversimplification. In fact, large numbers of naturally occurring molecules can be studied *in silico*, utilizing the same concepts, methods, and chemical descriptors developed for the analysis and design of synthetic compounds. The increasing availability of pure natural product collections is expected to generate further interest in computational analysis. Many current drugs that originated from natural products are now included in the computational study of drug-likeness.

Furthermore, although the generation of natural product-based compound libraries continues to be dominated by chemical considerations, computational studies of natural products are beginning

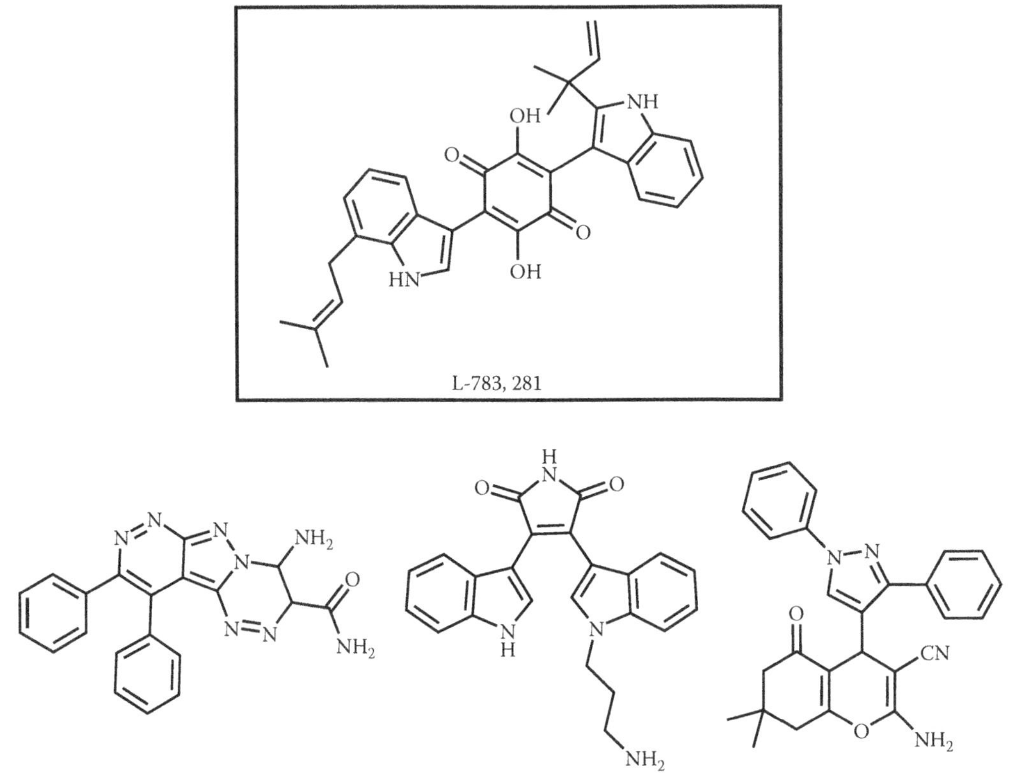

FIGURE 3.4 L-783,281 (top) and three synthetic mimics (bottom) selected on the basis of similarity-search calculations. The mimic in the center is a known inhibitor of protein kinase C.

to have an impact on this field. In recent years, a number of computational investigations have been reported that can provide guidance for library design efforts. Most of these computational studies focused on the comparative analysis of structural motifs and molecular properties in large collections of naturally occurring and synthetic compounds. Conclusions drawn from these studies are currently applied, for example, to answer the question of what natural product scaffolds to use for the generation of target-focused, drug-like, or lead-like libraries. Beyond scaffold investigations, compounds from natural and synthetic sources are complementary in terms of chemical property distributions, diversity, and accessibility. Thus, computational design strategies that combine natural and synthetic molecules provide another opportunity for the generation of natural product-based libraries.

REFERENCES

1. Cragg, G.M., Newman, D.J., and Snader, K.M., Natural products in drug discovery and development, *J. Nat. Prod.*, 60, 52, 1997.
2. Harvey, A., Strategies for discovering drugs from previously unexplored natural products, *Drug Discovery Today*, 5, 294, 2000.
3. Newman, D.J., Cragg, G.M., and Snader, K. M., Natural products as sources of new drugs over the period 1981–2002, *J. Nat. Prod.*, 66, 1022, 2003.
4. Cragg, G.M., Paclitaxel (Taxol®): a success story with valuable lessons for natural product drug discovery and development, *Med. Res. Rev.*, 18, 315, 1998.
5. Kissau, L. et al., Development of natural product-derived receptor tyrosine kinase inhibitors based on conservation of protein domain fold, *J. Med. Chem.*, 46, 2917, 2003.

6. Lepourcelet, M. et al., Small-molecule antagonists of the oncogenic Tcf/beta-catenin protein complex, *Cancer Cell*, 5, 91, 2004.
7. Strohl, W.R., The role of natural products in a modern drug discovery program, *Drug Discovery Today*, 5, 39, 2000.
8. Drews, J., Drug discovery: a historical perspective, *Science*, 287, 1960, 2000.
9. Nicolaou, K.C. et al., The art and science of total synthesis at the dawn of the twenty-first century, *Angew. Chem., Int. Ed. Engl.*, 39, 44, 2000.
10. Lahana, R., How many leads from HTS?, *Drug Discovery Today*, 4, 447, 1999.
11. Lawrence, R.N., Rediscovering natural product biodiversity, *Drug Discovery Today*, 4, 449, 1999.
12. Rouhi, A.M., Rediscovering natural products, *Chemical and Engineering News*, 81, 77, 2003.
13. Koehn, F.E. and Carter, G.T., The evolving role of natural products in drug discovery, *Nat. Rev. Drug Discovery*, 4, 206, 2005.
14. Lipinski, C.A., Drug-like properties and the causes of poor solubility and permeability, *J. Pharmacol. Toxicol. Methods*, 44, 235, 2000.
15. Tupl, M. and Bohlin, L., Unconventional natural sources for future drug discovery, *Drug Discovery Today*, 9, 450, 1999.
16. Veber, D.F. et al., Molecular properties that influence the oral availability of drug candidates, *J. Med. Chem.*, 45, 2615, 2002.
17. Burke, M.D. and Schreiber, S.L., Discovery of small molecules: a planning strategy for diversity-oriented synthesis, *Angew. Chem., Int. Ed. Engl.*, 43, 46, 2004.
18. Tan, D.S. et al., Stereoselective synthesis of over two million compounds having structural features both reminiscent of natural products and compatible with miniaturized cell-based assays, *J. Am. Chem. Soc.*, 120, 8565, 1998.
19. Nicolaou, K.C. et al., Natural product-like combinatorial libraries based on privileged structures. 1. General principles and solid-phase synthesis of benzopyrans, *J. Am. Chem. Soc.*, 122, 9939, 2000.
20. Wessjohann, L.A., Synthesis of natural product-based compound libraries, *Curr. Opin. Chem. Biol.*, 4, 303, 2000.
21. Hall, D.G., Manku, S., and Wang, F., Solution- and solid-phase strategies for the design, synthesis, and screening of libraries based on natural product templates: a comprehensive survey, *J. Comb. Chem.*, 3, 125, 2001.
22. Breinbauer, R. et al., Natural product guided compound library development, *Curr. Med. Chem.*, 9, 2129, 2002.
23. Ortholand, J.Y. and Ganesan, A., Natural products and combinatorial chemistry: back to the future, *Curr. Opin. Chem. Biol.*, 8, 271, 2004.
24. Boldi, A.M., Libraries from natural product scaffolds, *Curr. Opin. Chem. Biol.*, 8, 281, 2004.
25. Tan, D.S., Current progress in natural product-like libraries for discovery screening, *Comb. Chem. High Throughput Screen.*, 7, 631, 2004.
26. Höfle, G. and Sefkow, M., Substitutions at the thiazole moiety of epothilone, *Heterocycles*, 48, 2485, 1998.
27. Turner, N. and Schneider, M., Biocatalysis — molecular, structural, and synthetic advances, *Curr. Opin. Chem. Biol.*, 4, 65, 2000.
28. Breinbauer, R., Vetter, I.R., and Waldmann, H., From protein domains to drug candidates — natural products as guiding principles in the design and synthesis of compound libraries, *Angew. Chem., Int. Ed. Engl.*, 41, 2879, 2002.
29. Martin, Y.C. et al., Diverse viewpoints on computational aspects of molecular diversity, *J. Comb. Chem.*, 3, 231, 2001.
30. Bajorath, J., Chemoinformatics methods for systematic comparison of molecules from natural and synthetic sources and design of hybrid libraries. *Mol. Diversity*, 5, 305, 2002.
31. Chapman and Hall, *Dictionary of Natural Products*, CRC Press LLC, NW Corporate Boulevard, Boca Raton, FL.
32. Bindseil, K.U., et al., Pure compound libraries: a new perspective for natural product-based drug discovery, *Drug Discovery Today*, 6, 840, 2001.
33. Muegge, I., Selection criteria for drug-like compounds, *Med. Res. Rev.*, 23, 302, 2003.
34. Henkel, T. et al., Statistical investigation into the structural complementarity of natural products and synthetic compounds, *Angew. Chem., Intl. Ed. Engl.*, 38, 643, 1999.

35. Lee, M.L. and Schneider, G., Scaffold architecture and pharmacophoric properties of natural products and trade drugs: application in the design of natural product-based combinatorial libraries, *J. Comb. Chem.*, 3, 284, 2001.

36. Feher M. and Schmidt J.M., Property distributions: differences between drugs, natural products, and molecules from combinatorial chemistry, *J. Chem. Inf. Comput. Sci.*, 43, 218, 2003.

37. Wessjohann, L.A. et al., What can a chemist learn from nature's macrocycles? — a brief, conceptual view, *Mol. Diversity*, 9, 171, 2005.

38. Godden, J.W. and Bajorath, J., Shannon entropy — a novel concept in molecular descriptor and diversity analysis, *J. Mol. Graph. Model.*, 18, 73, 2000.

39. Godden, J.W. and Bajorath, J., An information-theoretic approach to descriptor selection for database profiling and QSAR modeling, *QSAR Comb. Sci.*, 22, 487, 2003.

40. Godden, J.W. and Bajorath, J., Differential Shannon entropy as a sensitive measure of differences in database variability of molecular descriptors, *J. Chem. Inf. Comput. Sci.*, 41, 1060, 2001.

41. Godden, J.W. and Bajorath, J., Chemical descriptors with distinct levels of information content and varying sensitivity to differences between selected compound databases identified by SE-DSE analysis, *J. Chem. Inf. Comput. Sci.*, 42, 87, 2001.

42. Gillet, V.J., Willett, P., and Bradshaw, J., Identification of biological activity profiles using substructural analysis and genetic algorithms, *J. Chem. Inf. Comput. Sci.*, 38, 165, 1998.

43. Stahura, F.L. et al., Distinguishing between natural products and synthetic molecules by descriptor Shannon entropy analysis and binary QSAR calculations, *J. Chem. Inf. Comput. Sci.*, 40, 1245, 2000.

44. Samiulla, D.S. et al., Rational selection of structurally diverse natural product scaffolds with favorable ADME properties for drug discovery, *Mol. Diversity*, 9, 131, 2005.

45. Pearlman, R.S. and Smith, K.M. Novel software tools for chemical diversity, *Perspect. Drug Discovery Design*, 9, 339, 1998.

46. Lipinski, C.A. et al., Experimental and computational approaches to estimate solubility and permeability in drug discovery and development settings, *Adv. Drug Delivery Rev.*, 23, 3, 1997.

47. Shen, J. et al., Virtual screening on natural products for discovering active compounds and target information, *Curr. Med. Chem.*, 10, 2327, 2003.

48. Stahura, F.L. et al., Methods for compound selection focused on hits and application in drug discovery, *J. Mol. Graph. Model.*, 20, 439, 2002.

49. Stahura, F.L. et al., Design of array-type compound libraries that combine information from natural products and synthetic molecules, *J. Mol. Model.*, 6, 550, 2000.

50. Zhang, B. et al., Discovery of a small molecule insulin mimetic with antidiabetic activity in mice, *Science*, 284, 974, 1999.

51. Tietze, L.F., Bell, H.P., and Chandrasekhar, S., Natural product hybrids as new leads for drug discovery, *Angew. Chem., Int. Ed. Engl.*, 42, 3996, 1999.

4 Accessing Expanded Molecular Diversity through Engineered Biosynthesis of Natural Products

Steven D. Dong and David C. Myles

CONTENTS

4.1 INTRODUCTION

Natural products have always played a major role in the discovery of new therapeutic agents for the treatment of disease.[1] These compounds show enormous structural variety and complexity (Figure 4.1).[2] Correspondingly, the biological activities of these agents span a very broad range, serving functions as divergent as antibiotics to anticancer treatments. Because of this combination of structural diversity and biological activity, it is not surprising that natural products have historically played an important role in the pharmaceutical industry. Between the years 1982 and 2002,

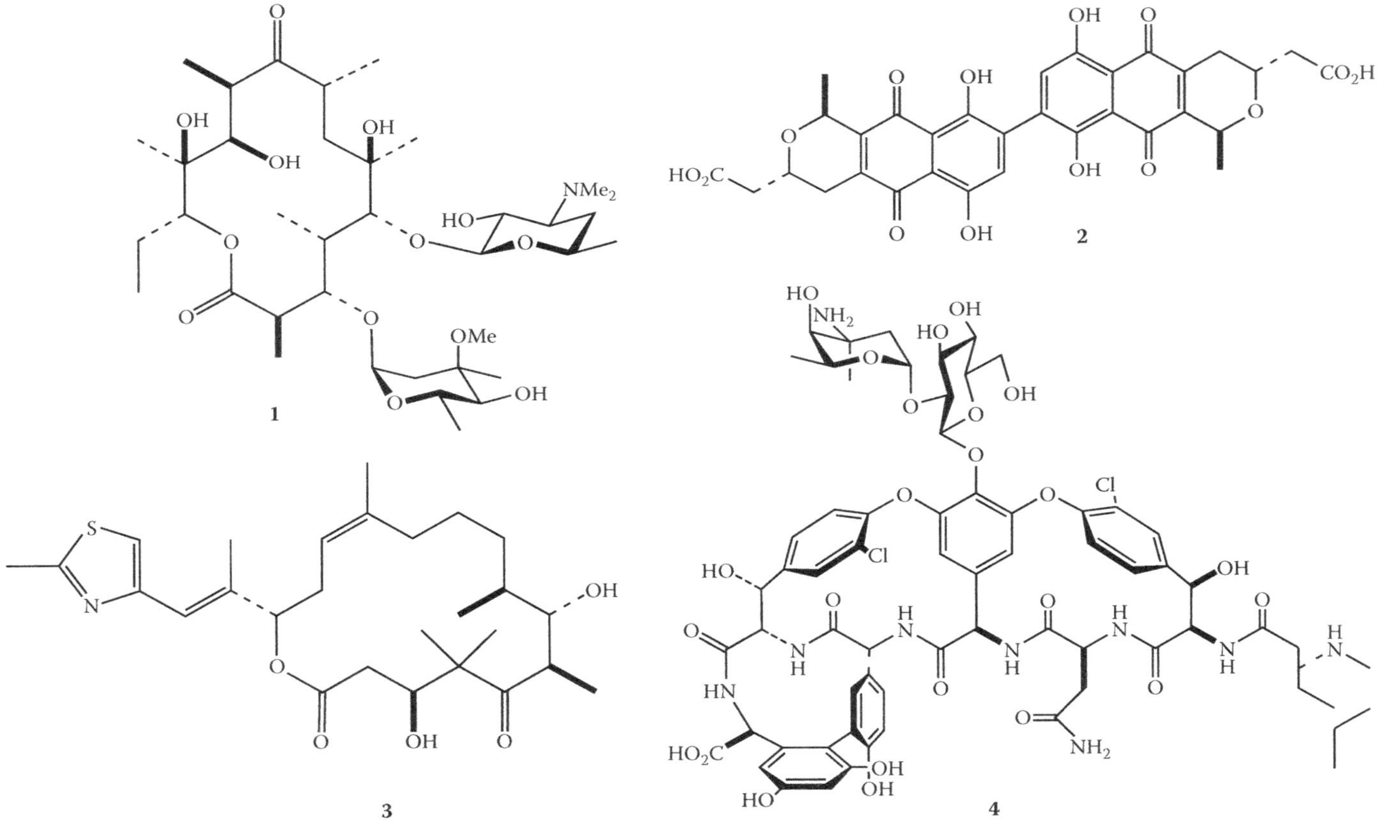

FIGURE 4.1 Representative examples of the diverse structure and functionality achievable from PKS- and NRPS-mediated biosynthesis: erythromycin A (**1**), actinorhodin (**2**), epothilone D (**3**), and vancomycin (**4**).

natural products accounted for 33% of the new chemical entities identified.[3] The impact of natural products is found to be even greater when compounds inspired by natural products are included in such an analysis.

The value of natural products is clear. Unfortunately, natural products' discovery and development has historically relied on screening methods that are often capricious, and rational approaches for accessing the structural diversity and complexity of these compounds have been elusive. Thus, the pharmaceutical industry turned to the power of modern synthetic methods and combinatorial chemistry to generate huge libraries of new compounds for the identification of new lead structures and new chemical entities (NCEs). Although this approach is now beginning to bear fruit, it has not yet lived up to its early promise of supplanting completely the other sources of structural diversity for new lead compounds.

Meanwhile, over the last 20 years, an interdisciplinary approach employing biochemistry, bacterial genetics, and synthetic chemistry has begun to reveal a pathway for the rational access of natural products' structural diversity. It has been discovered that many natural products are produced by biological "machinery" that is both highly modular and processive. Thus, in the same way that a chemist might "mix and match" chemical functionality to generate libraries, it is becoming possible to take a "combinatorial biosynthetic" approach with the proteins responsible for natural product metabolism.[4,5]

In this review, we will outline the basic organizational logic underpinning three of the most important and best-understood biosynthetic processes responsible for bioactive natural products. We will then consider how an understanding of the biochemistry and genetics involved in these processes has led to initial success and also revealed future possibilities in combinatorial biosynthesis. Specifically, we first discuss the engineering of the biosynthetic machinery that produces polyketides (PKs, **1** and **2**, Figure 4.1) and the nonribosomal peptides (NRPs, **4**, Figure 4.1). These two classes of natural products are biosynthesized by modular arrays of enzymes that function cooperatively to fashion the desired structures. Employing a combinatorial approach, nature often combines both PK and NRP functionality together in the same natural product (**3**, Figure 4.1). Frequently, many PK and NRP compounds undergo further modification or "tailoring" by other enzymes following the initial bond-forming reactions. All of the enzymatic functions in this cascade are potential targets for engineering. The third section of the review discusses the engineering of two important tailoring pathways: P450-mediated hydroxylation and glycosylation.

The combinatorial potential of the PK and NRP classes of compounds is extraordinary. To date, over 10,000 PKs have been discovered from natural sources.[6] However, recent theoretical calculations employing stoichiometric balances and combinatorial theory suggest that there are well over a billion possible structures.[6] Understanding how these compounds are biosynthesized and how to manipulate the genetics of these systems may allow researchers to begin to tap into this rich source of structural diversity.

4.2 PK-DERIVED AGENTS

The first class of natural products we will consider is biosynthetically derived PKs. Although the value of this class of compounds as medicinal agents was established with the discovery of erythromycin (**1**, Figure 4.1) in 1952, an understanding of the biology behind PK biosynthesis only began to emerge in the late 1980s with the work of Malpartida and Hopwood on the biosynthesis of actinorhodin (**2**, Figure 4.1).[7] In the early 1990s, the genes responsible for the biosynthesis of erythromycin A were isolated and sequenced independently by Tuan et al.[8] and Cortes et al.[9]

Since these initial groundbreaking discoveries, an increasingly deeper understanding of the logic and possibilities of PK biosynthesis has emerged. A large number of excellent reviews cover many aspects of this work.[4,10–12] In this section, we will focus on opportunities for creating PK structural diversity through biosynthesis and chemobiosynthesis.

4.2.1 POLYKETIDE SYNTHASES (PKSS) AND PK BIOSYNTHESIS

All PKs are united by a basic structural pattern originating from two-carbon units containing alternating oxygenated (carbonyl, hydroxyl) and unoxygenated carbons. This fundamental repetition found in all PK scaffolds is a consequence of how they are assembled biosynthetically by proteins known as polyketide *synthases* (PKSs). Researchers have begun to take advantage of the emerging understanding of how the PKSs are organized and operate in order to engineer increased chemical diversity into PK structures.

The biosynthesis of erythromycin is illustrative of the logic of how PKSs operate. As with many PKs, the genes coding the erythromycin PKS (*eryAI*, *eryAII*, and *eryAIII*) are found clustered (Figure 4.2a) in the genome of the producing organism — in this case, the soil bacterium *Saccharopolyspora erythraea*. Each of these three genes is about 10 kbp in size and, correspondingly, they encode for three equally enormous multifunctional proteins or subunits (Figure 4.2b) with weights of approximately 350 kDa. In this particular example, the PK product of these three subunits is the erythromycin precursor 6-deoxyerythronolide (6-dEB). Consequently, these subunits are named collectively DEBS1, 2, and 3 (Figure 4.2b). The complete DEBS PKS is a dimer with a molecular weight greater than 2 MDa, a size roughly commensurate with a complete ribosome.

Each DEBS protein is responsible for a portion of the chain extension in the biosynthesis of 6-dEB. As first proposed by Donadio et al.,[13] the activities of each DEBS protein can be further broken down into modules (Figure 4.2c). Each module is responsible for the addition of one building block and any subsequent processing. The unique activities of these modules lie in the combination of catalytic domains that constitute them (Figure 4.2d, Table 4.1). The three essential domains required for a round of chain extension are a ketosynthase (KS), acyltransferase (AT), and acyl carrier protein (ACP).

The sequence of steps involved in a round of chain extension by the KS, AT, and ACP domains is illustrated for module 4 of DEBS2 in Figure 4.3. The growing PK enters the cycle tethered to the ACP of the previous module via a thioester bond to a 20-Å-long phosphopantetheinyl prosthetic group, itself covalently attached to a serine of the ACP. The long phosphopantetheinyl group

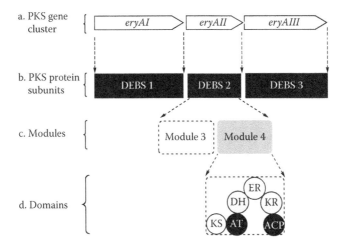

FIGURE 4.2 Hierarchy of the archetypical 6-deoxyerythronolide B PKS gene cluster. The three *ery* genes (a) encode a total of three subunits (b) that can be reduced further to independent modules. (c) In a type I PKS, each nonloading or nonterminating module is responsible for a single two-carbon extension to the growing PK through the action of three essential domains (d), the KS, AT, and ACP. The KR, DH, and ER domains are optional members of the PKS that control the oxidation state at the incoming β–keto carbon.

TABLE 4.1
Comparison of PKS and NRPS Domains

PKS		NRPS	
Domain Type	Function	Domain Type	Function
AT	Acyl transferase	A	Adenylation
ACP	Acyl carrier protein	PCP	Peptide carrier protein
KS	Ketosynthase	C	Condensation
KR	Ketoreductase	E	Epimerase
ER	Enoylreductase	O	Oxidase
DH	Dehydratase	R	Reductase
TE	Thioesterase	TE	Thioesterase

provides a flexible tether that allows the PK to access the different domains of the module. The PK is first transferred to the active-site cysteine of the downstream KS (Figure 4.3a) via transthioesterification. Meanwhile, the AT of module 4 selects and loads a specific diacid-acyl CoA (in this case, methylmalonyl CoA) via ester formation to an active-site serine (Figure 4.3a). Once loaded,

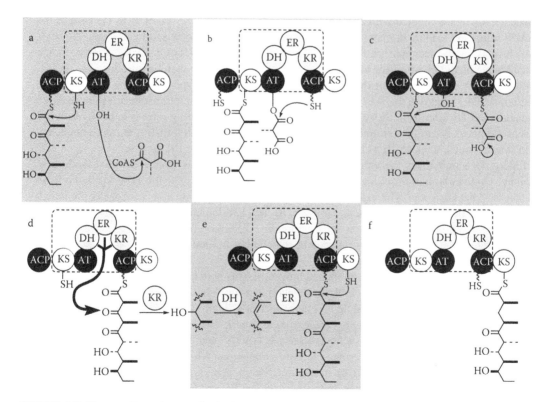

FIGURE 4.3 The coordinated steps of a single round of chain extension as illustrated by module 4 (indicated by the dashed-line box) of DEBS2: (a) the completed chain on the ACP of module 3 of DEBS2 is transferred to the KS of module 4; meanwhile, the module 4 AT selects methylmalonylCoA as the next two-carbon extender unit; (b) the extender unit is transferred to the module 4 ACP; (c) the KS catalyzes the Claisen condensation of the extender unit onto the growing chain; (d) still positioned on the ACP, the β-carbonyl is reduced to the methylene via the reductive module 4 KR, DH, and ER domains; (f) finally, the extended PK is transferred to the KS of the next module in preparation for the next chain elongation step.

the AT-methylmalonyl thioester is transferred to the phosphopantetheinyl group of the module 4 ACP (Figure 4.3b). Module 4 is now primed for the critical chain extension event. The KS catalyzes the decarboxylation-driven Claisen condensation of the ACP-loaded methylmalonic acid on the KS-polyketide thioester (Figure 4.3c) resulting in a two-carbon elongation of the PK chain. Importantly, the PK remains tethered to module 4 via the phosphopantetheinyl group of the ACP.

An opportunity for structural diversity arises from the makeup of the PKS module at this point. Up to three optional but distinct domains may be found in a module, and such is the case with module 4 of DEBS2. Each of these reductive domains is responsible for tailoring the oxidation state of the β-keto position of the tethered PK (Figure 4.3d and Figure 4.3e). The NADPH-dependent ketoreductase (KR) acts first to reduce the β-keto group stereoselectively to the alcohol. If a dehydratase (DH) domain is present, the elimination of the alcohol is promoted to provide the unsaturated thioester. Finally, if an enoylreductase (ER) is also present, conjugate double-bond reduction occurs, fully saturating the α and β positions. If branching is present at the α-position, the ER effects this reduction stereoselectively. Once the oxidation state of the β-position is settled, the extended PK is ready for transfer to the KS of the next downstream module to begin the process again (Figure 4.3f).

In addition to the three essential and three optional domains found in PKS modules, there are two other distinct domains present in PKSs. The first is the loading domain, which often comprises an AT and an ACP (Figure 4.4). Other loading domain configurations that have been identified include a CoA ligase coupled with an ACP as well as the modular KS°-AT-ACP. In the case of the AT-ACP setup, the loading domain AT, as is true of all PKS ATs, selects and loads a specific acyl-CoA. Just as in the example of the DEBS2 module 4, the AT-thioester is transferred to the phosphopantethinyl group of the loading ACP domain, where it can then be introduced to the KS of module 1. At the other end of the PKS is the 25- to 35-kDa polypeptide segment known as a thioesterase (TE) (Figure 4.4). The TE utilizes an active-site serine in a transesterification reaction to generate the TE-ester from the ACP-bound PK thioester. The resulting ester is either hydrolyzed to yield the linear PK as the terminal free acid or, as in the case with 6-dEB, the ester is attacked by an internal nucleophile such as the 13-hydroxy group to yield the PK lactone.

Figure 4.4 summarizes the entire process of 6-dEB biosynthesis from the initial loading of propionyl-CoA starter unit to six rounds of chain elongation and, finally, to the TE-catalyzed release and macrocyclization event. The oxidation states of the various β-carbon atoms are easily predicted, based solely on the presence and absence of KR, DH, and ER domains in each module (Table 4.1). For example, the β-keto group remains intact in module 3 because there are no reductive domains present, whereas the β-keto group is reduced to the alcohol in modules 1, 2, 5, and 6 because of the presence of a solitary KR domain.

What is most striking about PKS-mediated biosynthesis is both its processivity and modularity. The process has been likened to assembly line manufacturing.[11] In the case of 6-dEB, the process is noniterative, inasmuch as each module is responsible for a single step in the PK chain elongation. Historically, such noniterative PKSs are categorized as type I. Two other classes of iterative PKSs have also been identified and have been cataloged as "type II" and "type III."[12,14] These PKSs use single modules to catalyze multiple steps of chain elongation. Whereas the type II relies on the now-familiar ACP domain to play a role in this iterative process, type III PKSs dispense with the ACP altogether, utilizing acyl-CoA substrates instead. The PK products most associated with type II and III PKSs are phenolic aromatics compounds such as actinorhodin (**2**, Figure 4.1). However, as more examples of PKSs that display overlapping or "transitional" behavior between the different historical types have emerged, it has been argued that such a classification system may no longer be appropriate.[15]

Nonetheless, because of their iterative nature, type II and III PKSs provide less flexible platforms for combinatorial biosynthesis. Thus, to date, the majority of PKS genetic engineering efforts have focused on type I PKS systems. The remainder of this section will focus on developments in the engineering of noniterative type I PKSs.

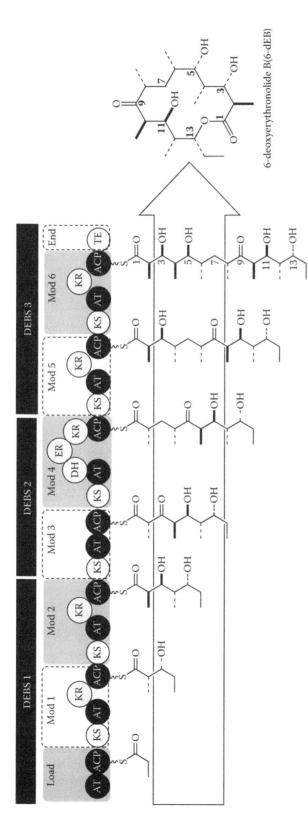

FIGURE 4.4 The entire sequence of 6-dEB PKS-mediated biosynthesis. As noted in the text and Figure 4.2, each module is responsible for a two-carbon chain extension to the growing PK. Additionally, the loading AT domain of DEBS1 is responsible for loading a specific starter unit — in this case, propionate. The ending domain in DEBS3 utilizes a thioesterase to promote the macrocyclization of the C13-hydroxyl with the C1-ester to yield the 6-dEB lactone. The modularity and repetition involved in noniterative type I PKS biosynthesis is readily apparent.

4.2.2 Opportunities for Engineering PK Structural Diversity

Once the organization of PKS-mediated biosynthesis was recognized, researchers sought ways to take advantage of the PKS modularity in order to engineer PKSs capable of making new PKs. To date, the most successful strategies have included (1) domain substitution and modification, (2) module substitution, and (3) precursor-directed biosynthesis. Examples of each of these approaches will be presented as well as a brief discussion of the current limitations and future outlook.

4.2.2.1 Domain Substitution and Modification

One of the most useful current approaches to PKS engineering is the substitution of one domain for another. This approach has been most widely demonstrated for AT swaps (Figure 4.5a).[10] The native AT domain, which selectively loads starter or extender units, may be replaced with another AT domain that has different starter or extender unit specificity. For example, the AT in module 6 of DEBS3 (Figure 4.5a) specifically loads methylmalonylCoA. Kato and coworkers replaced this native AT with the methoxymalonylCoA-specific AT8 from the FK520 PKS gene cluster resulting in the production of the novel 2-desmethyl-2-methoxy-6-DEB.[16] ATs specific for malonylCoA have been substituted for each of the ATs in modules 1–6 yielding six new *nor*-methyl 6dEB analogs.[17–20] Similar AT swaps substituting malonate for methylmalonate have been demonstrated with spiramycin,[21] FK520,[22] rapamycin,[10] and geldanamycin.[10]

Changing domain activity is not limited to AT swaps. "Gain-of-function" and "loss-of-function" mutagenesis have proved to be powerful means of increasing structural diversity. For example, McDaniel et al. added additional reductive functionality to module 2 of DEBS1 by swapping out the native KR for the KR, DH, and ER domains from module 1 of the rapamycin PKS (RAPS, Figure 4.5b).[23] Alternatively, one can use deletion mutagenesis to change the biosynthetic pathway. Substituting the KR of module 5 of DEBS with an inactive peptide resulted in the loss of reductive activity and a macrolide that contains a ketone in place of a hydroxyl (Figure 4.5c).[13,23]

Where changes in separate domains are orthogonal, they can be potentially combined to further broaden the structural space. In this case, both the gain-of-function mutagenesis at module 2 of DEBS1 could be combined with the deletion mutagenesis of module 5 of DEBS3, resulting in the macrolide illustrated in Figure 4.5d. Taking just such an approach of mixing and matching domain changes, McDaniel et al. constructed a library of > 50 macrolides consisting of 6-dEB analogs with 1, 2, and 3 altered carbon centers.[23] It has been estimated that if one had 2 possible ATs and 5 different β-carbon modifiers to choose from, substituting them into the DEBS PKS could lead to 10^7 possible PK structures. A similar approach to the much larger RAPS PKS has the theoretical possibility of yielding 10^{14} structures.[23]

4.2.2.2 Module Swapping and Hybrid PKs

An alternative to changing individual domains of a module to obtain new PK structures is the swapping of entire modules. Gokhale and coworkers constructed a PKS fusion between subunit DEBS1 and the TE domain of DEBS3, which produces a triketide lactone (Figure 4.6a and Figure 4.6b).[24] In place of module 2, they introduced either module 3 of DEBS2 (Figure 4.6a) or module 5 from the rifamycin PKS cluster (Figure 4.6b). The results were lactones differing in their β-carbon oxidation states. Ultimately, module 5 from rifamycin was introduced into the complete DEBS PKS cluster (Figure 4.6c), resulting in an active mutant capable of producing 6-dEB with a nonnative module.

In terms of molecular diversity, the ability to prepare truncated PKSs from larger PKS gene clusters as shown in Figure 4.6a and Figure 4.6b provides an interesting opportunity to generate small molecules with dense patterns of stereochemistry and oxygenation. The triketide lactones in Figure 4.6a and Figure 4.6b are excellent examples. Besides direct investigation as possible drug

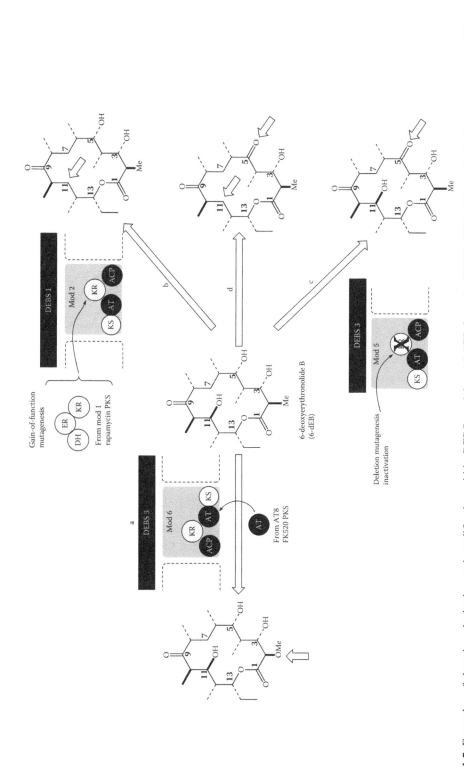

FIGURE 4.5 Examples of domain substitution and modification within DEBS modules. (a) AT8 from the FK520 PKS with specificity for methoxymalonylCoA is substituted for the methylmalonylCoA-specific AT of module 6 of DEBS3. The resulting macrolide incorporates a methoxyl group at C2. (b) A gain-of-function mutation in module 2 of DEBS1 in which three reductive domains from module 1 of RAPS are substituted for the native KR found in module 2 of DEBS1. The resulting analog is fully reduced to the methylene at C11. (c) A loss-of-function mutation in which the KR of module 5 of DEBS3 is inactivated results in the C6-keto analog. (d) Combining the gain-of-function mutation in (b) and loss-of-function mutation in (c) leads to an analog containing two simultaneous changes.

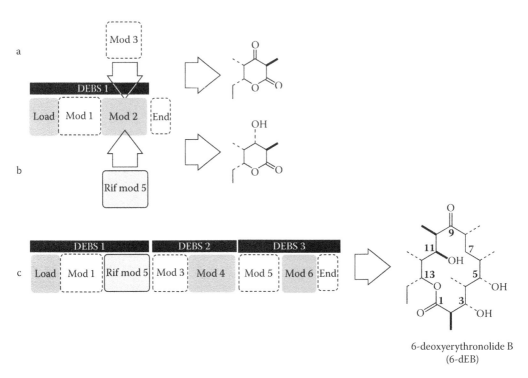

FIGURE 4.6 Three examples of module swapping within the DEBS subunits. (a, b) A construct linking DEBS1 with the DEBS3 TE was prepared. Either (a) module 3 of DEBS2 or (b) module 5 from the rifamycin PKS gene cluster was substituted for module 2 of DEBS1, resulting in triketide lactones with differing levels of oxidation at C3. (c) Module 5 from the rifamycin PKS was substituted for module 2 of DEBS1 within the context of the entire DEBS PKS, yielding 6-dEB.

candidates, these small PKs can, more importantly, be used as building blocks or advanced intermediates in the preparation of natural product libraries.[25]

Within families of related PKSs, this approach can be extended even further to the generation of hybrid PKs through the construction of hybrid sets of modules or subunits. One example is illustrated in Figure 4.7a to Figure 4.7c. The 16-membered macrolide aglycones, chalcolactone (Figure 4.7a) and tylactone (Figure 4.7b), utilize similar genes with a nearly identical organization. Taking advantage of this commonality, Reeves et al. constructed a hybrid macrolide through complementation by introducing the first two modules from the chalcomycin PKS cluster into a special tylosin-producing organism engineered for a loss of KS activity (KS1° null mutation).[26] The result is a hybrid macrolide whose "left" side is derived from chalcolactone and "right" side from tylactone.

4.2.2.3 Precursor Feeding

Despite the variety of ways to take advantage of the modular organization of PKSs through genetic manipulation, the structural diversity depends solely on the number and intrinsic activities of the seven fundamental domains: AT, ACP, KS, KR, DH, ER, and TE (Table 4.1). Ideally, one could expand the palette of chemical functionality even further to include moieties not normally found in nature. To date, the best solution to this problem is the utilization of unnatural precursor feeding. This process is illustrated in Figure 4.8. Using loss-of-function mutagenesis, the active site Cys[729] of the module 1 KS of DEBS1 was mutated to an Ala, resulting in loss of KS activity (KS1° null mutation) and the inability to load and initiate PK biosynthesis from the normal starter units.[27] The mutated DEBS can then incorporate exogenously introduced "diketides" (a four-carbon PK

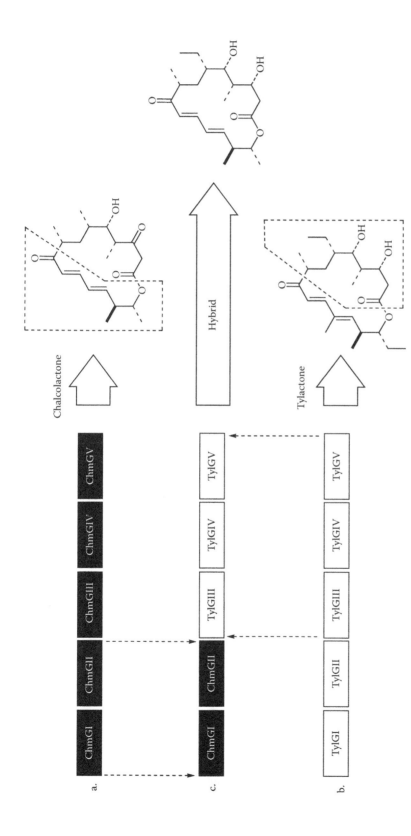

FIGURE 4.7 The (a) chalcomyin PKS and (b) tylosin PKS genes encode similar 16-membered macrolides. (c) Complementation of *chmGI-II* into tylosin-producing *S. fradiae* engineered with a KS1° *tyl* PKS produces the hybrid 16-membered macrolide.

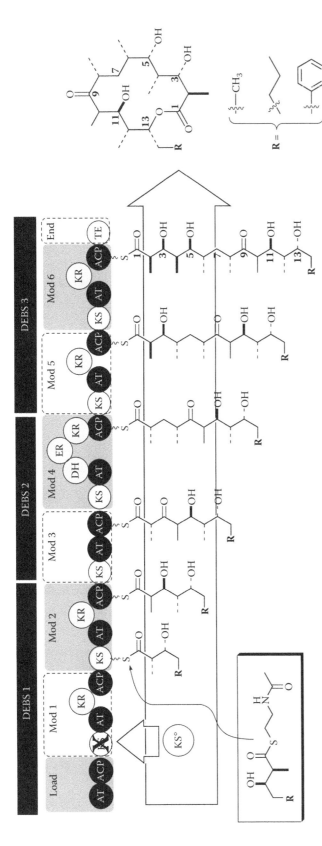

FIGURE 4.8 Precursor-directed biosynthesis. A KS1° mutation is introduced in module 1 of DEBS1. Exogenously fed diketide SNACs with varying side chains are incorporated into the macrolide.

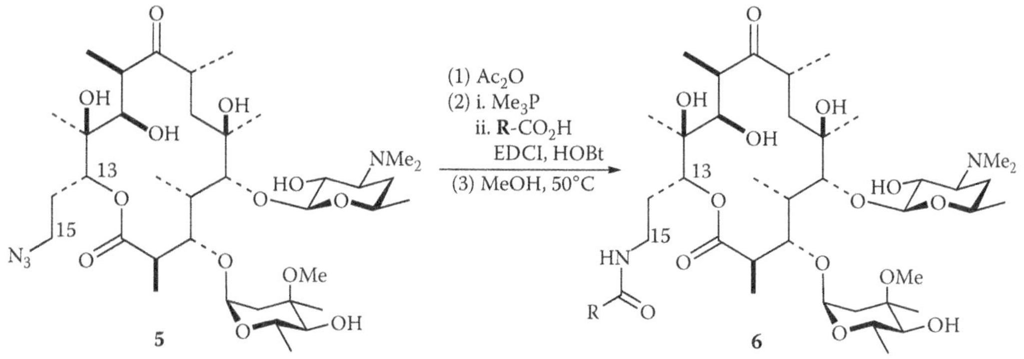

FIGURE 4.9 15-Azido-erythromycn A (**5**) is prepared via precursor feeding and bioconversion. Staudinger reduction of the azido group followed by *in situ* acylation of the resulting ylide provides access to a library of 15-amidoerythromycin macrolides (**6**).

equivalent to one round of chain elongation) that are activated as the thioester of *N*-acetylcysteamine (SNAC) onto the next downstream KS. The SNAC is a biomimetic analog of the normal pantethienyl tether.

In the context of the KS1° null DEBS PKS, Jacobsen et al. fed a variety of structurally dissimilar diketide SNACs (Figure 4.8).[27] A key question was whether the downstream modules could successfully process the growing PK containing unnatural and bulky side chains. In all three cases, the diketide SNAC was incorporated into the resulting macrolide. This approach has now been utilized in the production of several novel 14- and 16-membered lactones that were supplied either with diketide or triketide SNACs.[24,28,29]

Furthermore, the tolerance of the downstream modules to different side chains has proved to be fairly robust. Thus, SNACs containing halogenated side chains as well as other synthetically useful handles have subsequently been employed.[10] Once such nonnaturally occurring handles are incorporated into the macrolide, they permit an even more expansive set of possible structural diversity if semisynthesis is employed after the biosynthesis is complete. For example, Shaw and coworkers used a chemobiosynthetic approach to prepare an array of 15-amido-substituted erythromycin A compounds (Figure 4.9).[30] Precursor feeding and bioconversion were employed to generate 15-azido-erythromycin A (**5**).[31,32] Subsequently, trimethylphosphine-mediated reduction of the azide, followed by *in situ* coupling of the resulting ylide to a library of carboxylic acids yielded 19 different 15-amidoerythromycins **6**. In screening this library, it was discovered that benzylamide members of this new class of macrolides had antibacterial activity that was equivalent to or better than that of the parent erythromycin A against macrolide-resistant bacterial strains.

4.2.3 CURRENT LIMITATIONS AND FUTURE OPPORTUNITIES

Great progress has been made in the field of PKS engineering in the last 10 years. However, there are a number of obstacles yet to be overcome. One of the most limiting aspects of this technology is the often low titers of the engineered PKs produced from fermentation. A number of approaches currently address this issue, including the identification of industrial bacterial strains that are already highly optimized to produce PKs in high titers.[10] Additionally, heterologous expression of PKS genes in nonnative strains such as *E. coli*, which are much more amenable to genetic manipulation, may permit further metabolic engineering to increase titers.[10,33,34]

Currently, a major focus in the PKS engineering field is to achieve a better understanding of the role protein–protein interactions play in the efficiency and activity of PKSs.[11] It has become evident that the growing PK, while tethered to the ACP, must be properly oriented with respect to the other domains if efficient chain extension and modification is to occur.[35] Relative structural perturbations often arise when domains or modules are changed. In particular, these disturbances can stem from nonoptimized intra- and intermodular linkers that normally mediate the interactions between domains and modules. Fortunately, it appears that module–module interaction can be engineered independently of module catalytic function.[36–39]

Menzella et al. have recently reported a novel approach to address both the heterologous expression of PKS genes as well as the identification of promiscuous intermodular linkers.[40] Using bioinformatic analysis, all known two-carbon PK units and their associated modules were cataloged from a wide variety of PKs. The genes encoding this set of modules were then redesigned and codon-optimized for expression in *E. coli*. Included in this *de novo* design process was the installation of a "common set" of restriction sites. The result was a universal design of PK genes comprising "synthetic building blocks," flanked by unique restriction sites that permitted facile mixing and matching of these synthetic modules.

To demonstrate the power of this system, 14 optimized modules from 8 different PKS clusters were prepared, and these modules were combined in 154 bimodular combinations. Remarkably, 72 out of the 154 combinations produced the corresponding triketide lactone (see Figure 4.6a). This screen revealed a universal donor module, a highly promiscuous TE module and, importantly, a generic intermodular linker that was always present in all the productive combinations.

This *de novo* design presents a new means of preparing combinatorial PKS libraries. It takes advantage of the potentially high production levels of *E. coli* and the panoply of genetic tools available to manipulate *E. coli*. Future work will seek to prepare larger and more complex PKs. Using this system as well as other approaches,[41] a greater understanding of the rules that govern linkers and PKS protein–protein interactions may lead to engineered PKSs with improved efficiencies and output.

4.3   ENGINEERED NRP BIOSYNTHESIS

The second class of natural products we shall discuss are NRPs. Similar to the PK natural products, NRPs have found widespread clinical use. Representative members of this class of natural products include: the antibiotics vancomycin (**4**, Figure 4.1), penicillin, bacitracin, and actinomycin D; the antitumor drug bleomycin; the immunosuppressant cyclosporine A; and the siderophore enterobactin.[42]

One of the many unique features found in NRPs is that they are not limited to the 20 "standard" common amino acids. Rather, NRPs often contain unusual structural elements including D-amino acids, peptide-derived heterocycles, and *N*-methyl groups.[43] Access to these moieties is the consequence of the way in which NRPs are biosynthesized by nonribosomal peptide synthetases (NRPSs). Perhaps not surprisingly, the molecular logic underpinning NRP synthesis is very similar to that in the PKS systems.[4]

Several excellent reviews detail much of the work on the biosynthesis of NRPs.[4,43–45] The next section will illustrate the biosynthetic similarities between NRPSs and PKSs. We will then highlight the opportunities to introduce diversity in NRPs through genetic engineering as well as chemoenzymatic approaches.

4.3.1   THE NRPS AND NRP BIOSYNTHESIS

The logic and hierarchy of NRP biosynthesis is highly complementary to that of the PKS system. Similar to a peptide or protein prepared by "conventional" biosynthesis mediated by the ribosome, the fundamental moiety uniting all NRPs is the peptide bond. However, instead of being coded directly from DNA, NRPs are coded by the NRPSs that biosynthesize them.

Similar to PKSs (refer to Figure 4.2), NRPS genes are usually found in clusters. In turn, these NRPSs code for large, multifunctional subunit proteins. These subunits contain modules, each one of which is responsible for one round of elongation. Within each module are functional domains that are responsible for facilitating each elongation step. Table 4.1 illustrates the logical similarity between the domains found in PKSs and NRPSs.

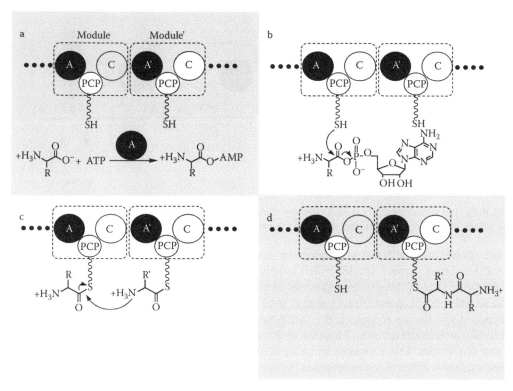

FIGURE 4.10 The coordinated steps for the synthesis of a single peptide bond in an NRPS. (a) The adenylation domain catalyzes the reaction between ATP and an amino acid. (b) The thiol of the PCP is acylated. (c) The condensation domain promotes the formation of the peptide bond. (d) The upstream module is now ready for acylation.

Because the nature of the bond formation process is different in PKs and NRPs (Claisen condensations vs. peptide bond formation), NRPSs necessarily utilize different domains. Specifically, the three essential domains of the NRPS are an adenylation (A), peptide carrier protein (PCP), and condensation (C) domain. NRP biosynthesis begins with activation of the amino acid through adenylation of the carboxylic acid group by the A domain, a process facilitated by the presence of Mg^{2+}. In the adenylation reaction, ATP reacts with the selected amino acid to form the activated phosphoester aminoacyl adenylate and pyrophosphate (Figure 4.10a). The A domain is also a key point of selectivity during NRP biosynthesis as it is responsible for the selection of the particular amino acid to be activated.

In contrast to the high selectivity frequently observed in the analogous AT domain in a PKS module, A domains of the NRP do not always show high selectivity for a particular amino acid. In some cases, NRPs can be found as mixtures of closely related congeners. A notable example of this phenomenon is the immunosuppressive agent cyclosporine, for which approximately 24 distinct natural analogs have been isolated.[46] Such natural "libraries" of compounds are directly analogous to peptide libraries that might be synthesized using combinatorial chemical synthetic strategies. The promiscuity found in the cyclosporine A domain illustrates a naturally occurring example of structural diversity derived from NRPSs that can be exploited for the creation of NCEs.

Activation of the amino acid in the A domain sets the stage for transfer of the acyl group to the PCP. In anticipation of this event (and in analogy to the PKS) the *apo* domain must be converted to its *holo* form by transfer of a phosphopantathienyl moiety to a conserved serine on the enzyme. This homologation is promoted by a phosphopantathienyl transferase in the presence of coenzyme A.[47] The terminal thiol group of the cysteine serves as the nucleophile, attacking the aminoacyl

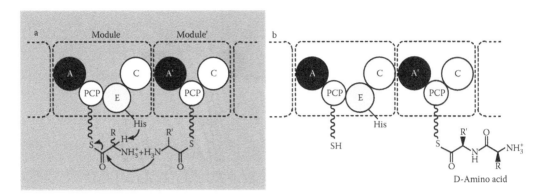

FIGURE 4.11 The epimerase domain facilitates the rapid equilibration of the α-carbon. The condensation domain selects the D isomer to form the oligopeptide containing the D-amino acid residue.

adenylate, displacing AMP, and generating the thioester (Figure 4.10b). Adjacent modules are activated in a similar manner. Guided by the condensation domain, the downstream amino moiety displaces the thiolate of the upstream thioester to generate the peptide bond (Figure 4.10c).

The flexible phosphopantathienyl linker of the PCP allows the tethered growing peptide chain to gain access to tailoring domains if they are present in the module. A hallmark of NRPs is the presence of D-amino acid residues. Thus, the most common tailoring activity found in NRPS is the epimerase (E) domain. These domains affect rapid epimerization of the α-carbon of the amino acid to afford a mixture of D and L stereoisomers (Figure 4.11). Interestingly, this rapidly equilibrating mixture is subsequently resolved kinetically by the downstream condensation domain during the formation of the peptide bond.[48] Alternatively, the adenylation domain may select for a D-amino acid that has been epimerized by an external racemase.

A number of other modifications of the peptide chain are observed in NRPs. Commonly found modifications include N-methylation and heterocyclization. Less common alterations include the addition of fatty acid residues, carbohydrates, and carboxylic acids to the NRP scaffold.[43] Methylation of the α-amino group is accomplished by a methyl transferase nested in a modified A domain, with S-adenosylmethionine serving as the source of the electrophilic methyl group. In contrast, the condensation domain itself is modified by insertion of a cyclization (Cy) domain to promote the heterocyclization of cysteine or serine to form thiazoline or oxazoline heterocycles (Figure 4.12).[49] Once formed, the heterocycles may also be oxidized (Ox) to the corresponding thiazole (as in the epothilones)[50] or oxazole. Alternatively, the rings can be reduced (R) to form the thiazolidine (as in yersiniabactin[51]) or oxazolidine ring (Figure 4.12).

Upon completion of the biosynthesis, the oligopeptide must be released by the enzyme complex. This transformation is achieved by a TE domain located after the PCP domain in the final module of the NRPS. After initial transfer of the terminal carboxyl function from the thiol of the PCP to a serine residue in the TE, the ester is either hydrolyzed to yield an acyclic product or engaged by a nucleophile on the peptide chain to afford a cyclic product. In the case of cyclic products, the exact nature of the cycle (i.e., ring size) is governed by features in the TE.

4.3.2 OPPORTUNITIES FOR ENGINEERING NRP STRUCTURAL DIVERSITY

The modular nature of NRPS, similar to the PKS, invites the opportunity for domain swapping as well as the alteration or elimination of the tailoring enzymes. Thus, engineering the NRPS provides the possibility of generating altered biosynthetic apparatus with the capacity of creating new polypeptides. All aspects of the biosynthetic process, including initial acylation, chain elongation and, ultimately, chain termination are targets for engineering of the biosynthetic process. Examples of these types of engineering include: (1) domain substitution and (2) module deletion.

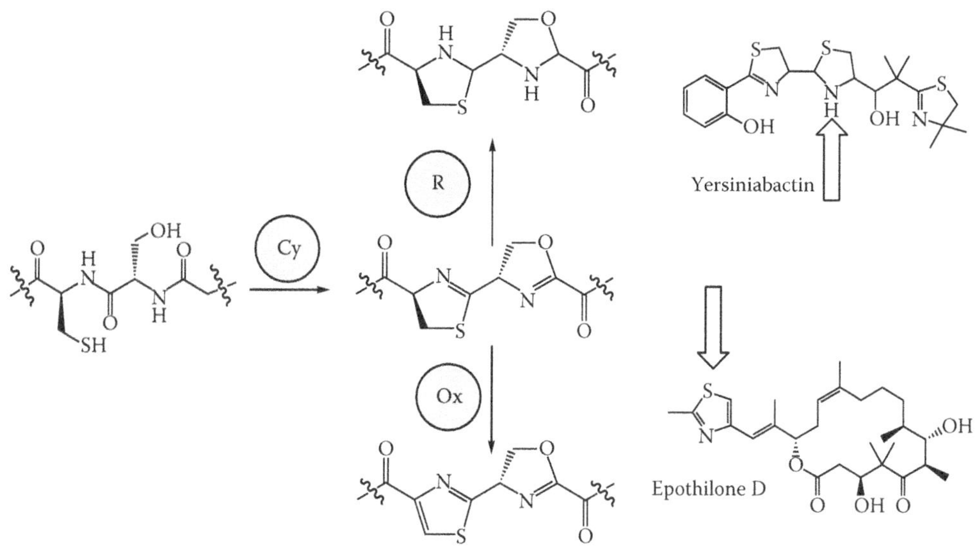

FIGURE 4.12 A modified condensation domain facilitates the cyclization of serine or cysteine residues to the corresponding oxazoline or thiazoline. These heterocycles may remain untouched for the remainder of the biosynthesis or they may undergo enzymatic oxidation to the thiazole as in epothilone D or undergo enzymatic reduction to the thiazolidine as in yersiniabactin.

4.3.2.1 Domain Substitution

Surfactin is a potent antifungal and antibacterial NRP agent (Figure 4.13),[52] and the reengineering of the surfactin NRPS is an early example of generating NRP structural diversity through domain swapping. The surfactin macrocyclic structure is synthesized by a seven-module NRPS terminated by a TE domain that releases the final product. In analogy to the swapping of AT domains to alter the branching pattern of a PK produced by an engineered PKS, the A-PCP didomains of an NRPS may be exchanged to alter the amino acid sequence of the product polypeptide. The final module of this NRPS, containing the C-A-PCP-TE domains, introduces a leucine as the final amino acid residue and then closes the macrocycle via engagement of the hydroxyl residue of the 3-hydroxy-butyrate loaded in module 1.[52]

To alter the amino acid at this position, A-PCP domains selecting for other amino acids from unrelated biosynthetic pathways were inserted into the surfactin NRPS.[53] The result of this effort was the creation of analogs of surfactin containing ornithine, valine, or phenylalanine in place of leucine (Figure 4.13). These modified surfactin analogs all retained biological activity. However, the reengineered compounds were biosynthesized by the producing organism in minute quantities. One explanation for the poor productivity may be the low tolerance of the subsequent modules for the modified NRP substrate. Interestingly, the TE was found to be tolerant of the unnatural side chains introduced by the domain substitution. Rather, the poor productivity of the engineered strains was due to the high specificity of the C domain in module 7 for the leucine side chain of the native polypeptide.

4.3.2.2 Combined PKS and NRPS

In another example of genetically engineered structural diversity, the NRPS of yersiniabactin, an iron-chelating siderophore found in *Y. pestis*, has been used to study TE portability, PCP domain swapping, and issues relating to module–module fusion (Figure 4.14).[51] The biosynthesis of yersiniabactin is achieved via a combined NRPS and PKS consisting of 5 single-function proteins (28 to 57 kDa) and two larger (229 and 349 kDa) multifunctional proteins. YbtS and YbtE likely carry out the synthesis and activation of the salicylate building block. After transfer of the aryl

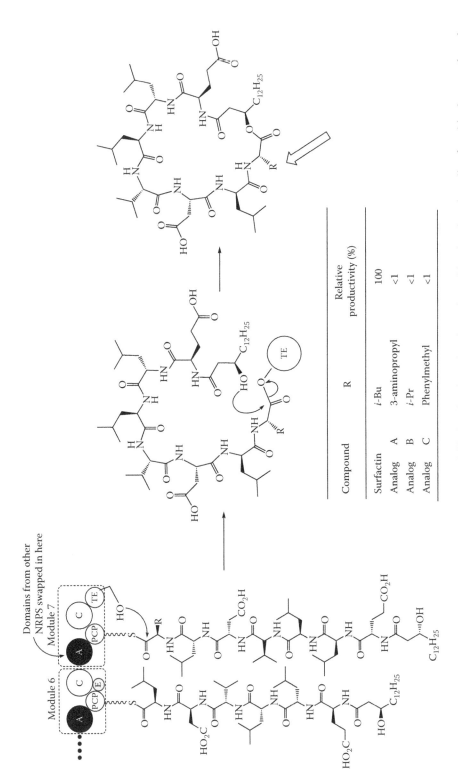

Compound		R	Relative productivity (%)
Surfactin		*i*-Bu	100
Analog	A	3-aminopropyl	<1
Analog	B	*i*-Pr	<1
Analog	C	Phenylmethyl	<1

FIGURE 4.13 Analogs of surfactin are synthesized by swapping A-PCP domains that select for amino acids other than the valine found in the natural product.

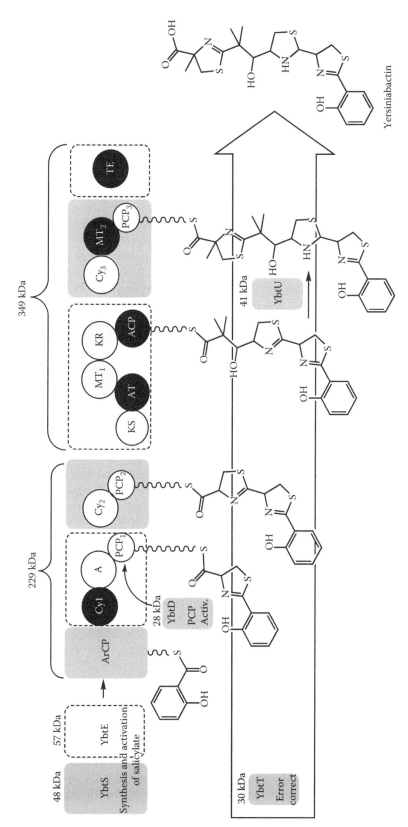

FIGURE 4.14 The combined NRPS and PKS of yersiniabactin.

residue to the 229-kDa fragment, two elongation steps are achieved, incorporating cysteines that are subsequently cyclized to the corresponding thiazolines. The first 5 domains of the 349-kDa polypeptide comprise the PKS portion of the biosynthetic machinery. In addition to the typical PKS domains — KS, AT, KR, and ACP — this module contains a methyl transferase required for the introduction of the *gem*-dimethyl moiety found in yersiniabactin. The final NRPS module also contains a methyl transferase — in this case, responsible for the methylation of the thiazoline ring. During the later stage of the biosynthesis, the reductase YbtU reduces the central thiazoline to the thiazolidine found in the natural product. The TE domain releases the yersiniabactin from the NRPS. The remaining two modules serve important supporting roles in the biosynthesis. YbtD is a PPT transferase responsible for activation of the carrier domains, and YbtT is a TE domain that is thought to facilitate error correction in the biosynthetic process.

To test the potential of domain swapping in the yersiniabactin NRPS/PKS, a series of fusion proteins containing portions of the biosynthetic apparatus were cloned (Figure 4.15). These fusion proteins were incubated *in vitro* in the presence of salicylate, cysteine, CoASH, and YbtE, and the resulting products of the enzymatic process were studied. For example, a fusion protein comprising Cy1, A1, and PCP3 followed by the TE domain produced the expected carboxylic acid (Figure 4.15a). The product corresponded to one peptide extension with cysteine, followed by cyclization to the thiazoline and release by the TE. This experiment demonstrates not only that PCP3 can accept a nonnative substrate but also that the TE will operate on a truncated NRP.

Perhaps not surprisingly, engineered NRPSs are not always so readily compliant. For instance, when the fusion protein comprised (1) ArCP, Cy1, A, and PCP1 (all adjacent modules in the normal yersiniabactin NRPS); (2) Cy2 paired with PCP3; and (3) the TE, it failed to produce the expected carboxylic acid (Figure 4.15b). A number of other examples of domain swapping in the yersinia-bactin NRPS/PKS were also investigated. For combinations that produced products, the efficiency of the biosynthesis was three- to eightfold lower than the wild-type producer.

4.3.2.3 Chemoenzymatic Approach to Diversity

A key structural trait of many biologically active NRPs and PKs is the fact that they are macrocycles. A consequence of such structural constraint is that the molecule is often "locked" in a conformation that has higher affinity for its target owing to the resulting reduction in entropy. Although many useful synthetic macrocyclization methods have been developed, they still often suffer from poor yields as well as poor scalability due to their high-dilution conditions.[43] Thus, it was recognized that the TE found in NRPS systems could play a valuable role as a synthetic mediator of macro-cyclization.

To this end, researchers prepared TEs, excised from the rest of the NRPS, as isolated enzymes and screened their ability to cyclize substrates outside their native synthetase context.[54] For example, the TE domain from the tyrocidine NRPS (Figure 4.16) has been isolated and shown to retain activity.[55] As has been previously discussed, conventional substrates for TE are activated as the phosphopantetheinyl thio ester of a PCP (Figure 4.16a). Analogous to diketide-feeding experiments with PKSs, Trauger and coworkers demonstrated that the phosphopantetheinyl-PCP could be replaced with a simple SNAC thioester and still be utilized by the TycC TE (Figure 4.16b).[55] Thus, the TycC TE will tolerate changes in the activated linker.

Following up on this result, Kohli and coworkers derivatized PEGA resin with a biomimetic phosphopantetheinyl linker.[56] Extending from this linker, they synthesized the acyclic tyrocidine A peptide precursor via standard solid-phase peptide synthesis methods (Figure 4.16c). The resin-bound peptide could be cleaved and macrocyclized by the TycC TE to generate tyrocidine A. Using these two approaches, a library of peptides, varying in length and side chain were prepared, and the library was screened against the TycC TE. The screening process revealed that the TycC TE was highly promiscuous, tolerating changes in positions 2 to 8 (Figure 4.16). The TE was most sensitive to changes at the *C*- and *N*-terminal residues, particularly D-Phe[1]. More recently, the

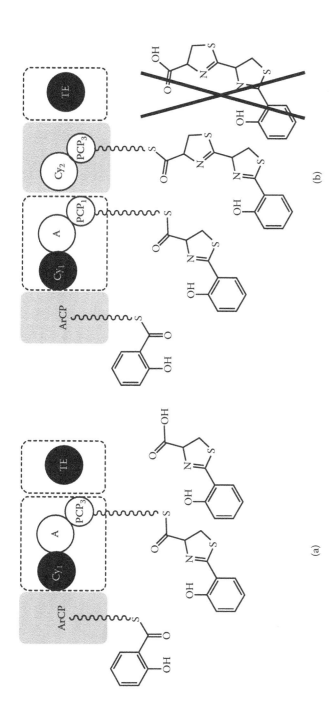

FIGURE 4.15 (a) The fusion protein consisting of ArCP, Cy1, A fused to PCP3, and the TE produces the expected NRPS product. (b) The fusion protein consisting of the first five domains capped by PCP3 and TE fails to produce the expected product.

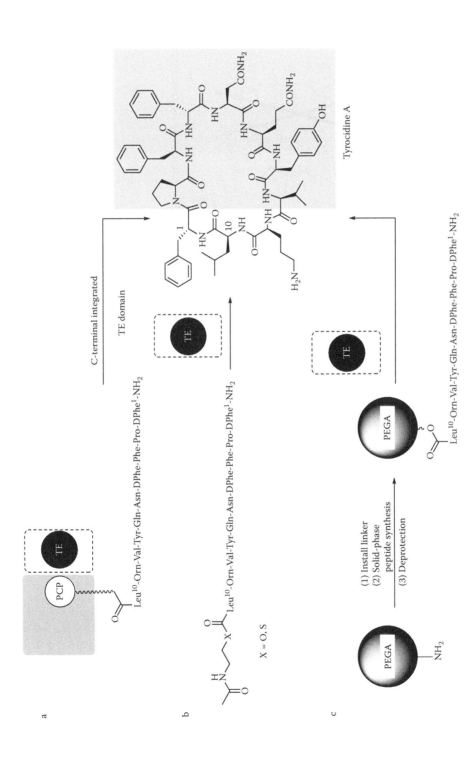

FIGURE 4.16 Isolated NRPS TEs can catalyze the macrocyclization of NRPs such as tyrocidine A that are shaded gray. (a) Conventional NRP macrocyclization following NRPS-mediated elongation. (b) Chemical synthesis of the 10-mer acyclic peptide followed by activation as the thioester SNAC for macrocyclization by the isolated TycC TE. (c) Installation of the linker on PEGA resin, followed by solid-phase peptide synthesis and deprotection to yield the immobilized acyclic NRP. The TycC TE accepts the resin-bound peptide as a substrate for macrocyclization.

Walsh group has reported preparing hybrid acyclic peptide–PK molecules by solid-phase synthesis.[57] These hybrid molecules have also been successfully macrocyclized both off- and on-resin.

This combination of solid-phase peptide synthesis followed by TE-mediated macrocyclization mimics the function of NRPSs, in which a peptide chain is first elongated and then cleaved and often cyclized. The applicability of this approach for the discovery of NCEs is apparent: established synthetic combinatorial methods could be employed in concert with the enzymatic machinery of the NRPS. In the case of the tyrocidine A library, two analogs were discovered that displayed both improved antibiotic activity and a better *in vitro* therapeutic index (human erythrocytes vs. Gram positive/negative bacteria) relative to the parent tyrocidine A.

4.4 GLYCOSYLATION AND TAILORING

A wide variety of structural diversity is available from PKS- and NRPS-mediated biosynthesis. However, in numerous cases, additional functionality is required for biological activity. Thus, genes encoding tailoring enzymes to add additional functionality are often found within or near the gene clusters of PKSs and NRPSs. Two of the most common tailoring enzymes encountered in both classes of natural products are (1) cytochrome P450 heme protein monooxygenase and (2) glycosyltransferases. The roles of these two enzymes in the maturation of 6-dEB into erythromycin are illustrated in Figure 4.17.

4.4.1 Cytochrome P450 Enzymes

Both PKSs and NRPSs employ cytochrome P450 enzymes to introduce oxygenation at positions in the molecules not normally accessible to the PKS and NRPS enzymes. For example, during the post-PKS maturation of 6-dEB into erythromycin, the $NADPH/O_2$-dependent P450 enzyme EryF introduces a 6-β-OH group regio- and stereospecifically (Figure 4.17). Later, EryK introduces a second β-hydroxyl at C12 to yield erythromycin C.

The family of P450 oxygenases offer a powerful means of adding functionality for the diversification of natural products' structures.[5] A key component of their use in a combinatorial biosynthetic setting is that they need to be somewhat promiscuous with respect to the substrate. For example, EryK has a very stringent selectivity for 14-membered macrolides bearing a C3-sugar.[58] However, PicK, a P450 monooxygenase from the related 14-membered macrolide picromycin PKS, does not require the presence of the C3-sugar.[59] Identifying more promiscuous P450s, coupled with P450 enzyme engineering and evolution, may lead to effective new ways of introducing oxygenation into PKS and NRPS natural products.

4.4.2 Glycosylation

The role and importance of glycosylation in the biological activity of many natural products has become increasingly apparent over the last decade. For example, erythromycin A is a PK antibiotic that binds to the ribosome and inhibits bacterial peptide synthesis. Erythromycin contains two sugars: L-cladinose at C3 and D-desosamine at C5 (Figure 4.17). Both sugars are required for antibacterial activity; if either carbohydrate is missing, a significant reduction in activity is observed. Recent co-crystal structures of macrolide antibiotics with the ribosome provide a structural basis for this dependence on the presence of the carbohydrates for activity.[60,61] The D-desosamine makes specific and tight contacts with the ribosome-binding pocket containing A2058.

With the realization that sugars are often more than mere decoration or solubilization groups, researchers have begun to identify ways of incorporating the wide variety of sugars and sugar structures into varying natural and unnatural product backbones. To address this challenge of expanding glycosylation structural space, two major hurdles must be addressed: (1) identify glycosyltransferases (Gtfs) that are promiscuous enough to tolerate differences in both the sugar donor

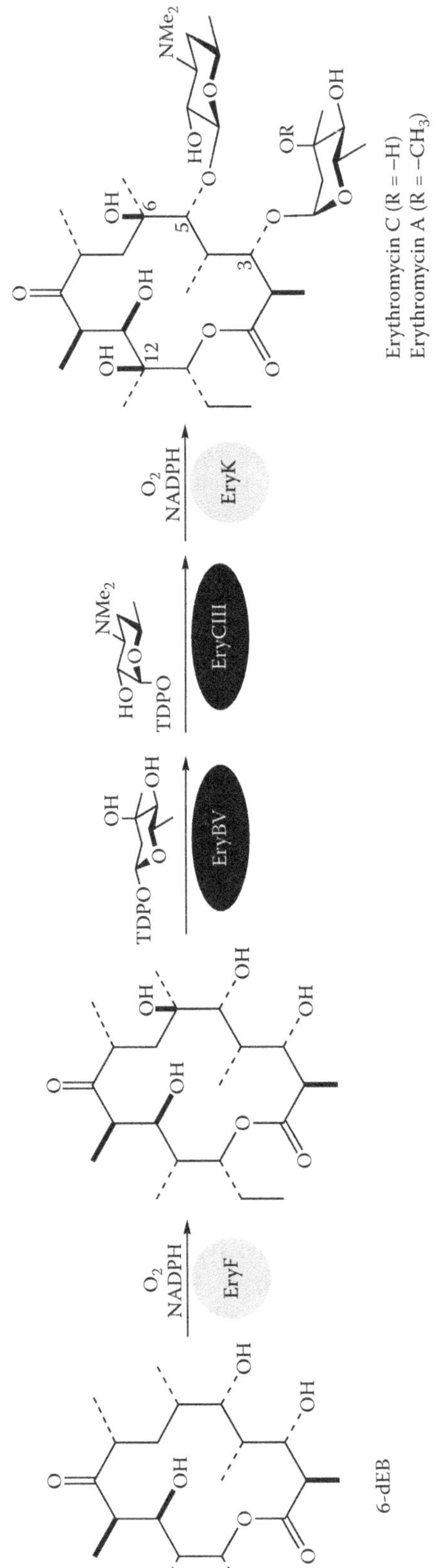

FIGURE 4.17 Tailoring enzymes involved in the maturation of 6-dEB into erythromycin.

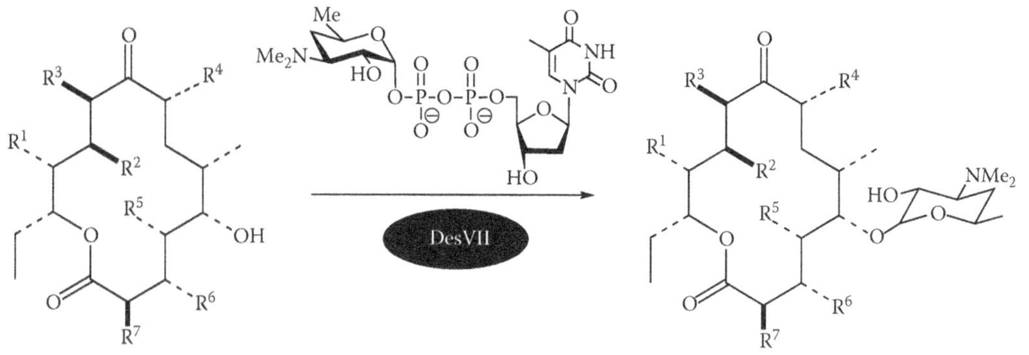

$R^1 = -H, -CH_3; R^2 = -OH, -H; R^3 = -H, -CH_3; R^4 = -H, -CH_3$
$R^5 = -H, -CH_3; R^6 = -OH, =O; R^7 = -H, -CH_3$

FIGURE 4.18 Promiscuous glycosyltransferase DesVII is capable of transferring D-desosamine to a library of 6-dEB analogs generated through PKS engineering.

(either a TDP- or UDP-activated sugar) and the acceptor (the natural product) and (2) develop a means of making diverse arrays of the sugar donors available, many of which are often unstable and difficult to synthesize chemically.

4.4.2.1 Gtf Promiscuity in PKS Systems

As was previously described, PKS engineering of the DEBS gene cluster, utilizing multiple genetic modifications, resulted in a library of > 50 6-dEB analogs.[23] However, the expression and production method employed did not provide for their subsequent glycosylation. To address this issue, Tang and McDaniel[62] isolated the seven des genes (desI-VII) responsible for the biosynthesis of the TDP-activated form of D-desosamine. They also isolated the Gtf desVIII that catalyzes the transfer of TDP-desosamine to 6-dEB. All eight genes were expressed in *S. lividans*, and individual members of the library of 6-dEB analogs were then fed to this strain. More than 20 of the available 14-membered macrolide 6-dEB analogs were glycosylated with D-desosamine (Figure 4.18), and many displayed antibiotic activity. The DesVII Gtf displays the promiscuity necessary to recognize multiple sugar acceptors.

4.4.2.2 Gtf Promiscuity in NRPS Systems

Similar to the sugars on erythromycin, the carbohydrates found on the glycopeptide vancomycin (Figure 4.19) play a role in its antibacterial activity, particularly with regard to overcoming resistance mechanisms.[63,64] Vancomycin is commonly referred to as the "antibiotic of last resort" because of its powerful antibacterial activity against Gram-positive pathogens (see Chapter 1 [Section 1.3] and Chapter 11, Subsection 11.5.1). Its primary mechanism of action is the inhibition of bacterial cell wall biosynthesis. Vancomycin binds to peptidyl cell wall precursors and prevents their incorporation into the growing cell wall leading to catastrophic wall weakening and, ultimately, cell death. Bacteria develop resistance to vancomycin by reprogramming the structure of the cell wall precursors such that they have lowered affinity for vancomycin. However, lipoglycopeptides, such as teicoplanin (Figure 4.19), as well as semisynthetic analogs of vancomycin in which hydrophobic groups are appended to the disaccharide, regain activity against these resistant strains. Although the exact mechanism behind this change in behavior is still under investigation, it is clear that modifications of the vancomycin sugars can lead to important improvements in activity against vancomycin-resistant bacteria.

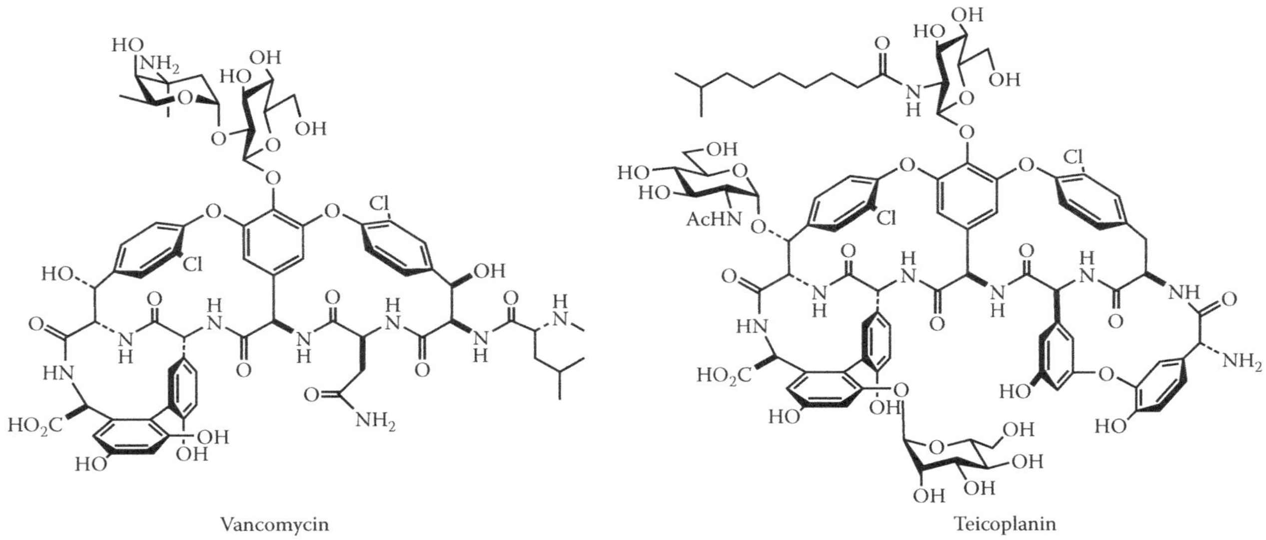

FIGURE 4.19 The glycopeptide antibiotic vancomycin and lipoglycopeptide antibiotic teicoplanin.

Losey and coworkers cloned and expressed the two Gtfs, GtfE, and GtfD, responsible for adding glucose and vancosamine sequentially to the vancomycin aglycone (Figure 4.20).[65] They showed that GtfE, similar to DesVIII, was tolerant of changes in the acceptor substrate.[66] GtfE transferred UDP-glucose to both the native vancomycin aglycone as well as the agylcone of teicoplanin. Additionally, GtfE was promiscuous enough to accept a variety of glucose-derived sugars (Figure 4.20), including deoxy and amino sugars. The presence of the amino group in particular provides a tantalizing handle for further semisynthetic modifications and diversification.

The second Gtf, GtfD, also displays tolerance for both acceptor and donor (Figure 4.21).[66] GtfD was found to glycosylate two unnaturally monoglycoslated vancomycin aglycones in which either a deoxy or amino glucose replaces the native glucose. GtfD utilized the unnatural sugar epimer UDP-4-*epi*-vancosamine. Thus, both GtfE and GtfD display the necessary qualities of donor and acceptor promiscuousness required to develop combinatorial glycosylation systems.

4.4.2.3 Expanding Diversity in Sugar Donors

A hindrance to exploring the diversity of sugar incorporation is the relative paucity of available UDP- or TDP-sugars. These activated sugars can be difficult to prepare and are often unstable.[67] A biosynthetic approach to this problem is the development of nucleotidytransferase enzymes capable of generating a variety of activated nucleotidyldiphospho sugars. To this end, Barton et al. have reported the structure-based reengineering of glucosephosphate thymidyltransferase E_p.[68] The reengineered E_p transferase can convert a variety of sugar phosphates to their activated TDP form.

Another option for increasing sugar structural diversity is the incorporation of chemical handles within the sugar. Fu and coworkers have taken advantage of the tolerance of GtfE to incorporate 6-azido-glucose on the vancomycin aglycone (Figure 4.22).[69] With the orthogonal azidosugar present, the researchers were able to prepare a library of 24 vancomycin analogs by performing copper-mediated Hüisgen 1,3-dipolar cycloadditions between the azide and functionalized alkynes. A number of the derivatives had antibacterial activity comparable to vancomycin itself.

4.5 CONCLUSIONS

Nature has many elegant systems for the generation of structural diversity within natural products. PKs and NRPs are particularly rich sources of structural diversity and have historically been important in the realm of clinically important agents. Research over the last 25 years has revealed how these natural products are biosynthesized. The biosynthesis involves a highly processive and modular process that readily lends itself to genetic engineering. Manipulation of the PK and NRP biosynthetic machinery through genetics has led to new compounds through mixing and matching of appropriate subunits, modules, and domains. More remains to be understood about how these systems interact with each other, in order to overcome problems such as low compound production levels. Nonetheless, when coupled with changes to tailoring enzymes such as P450s and Gtfs, as well as the utilization of unnatural synthetically derived building blocks, the potential for rapid access to NCEs and expansion of structural diversity is clear.

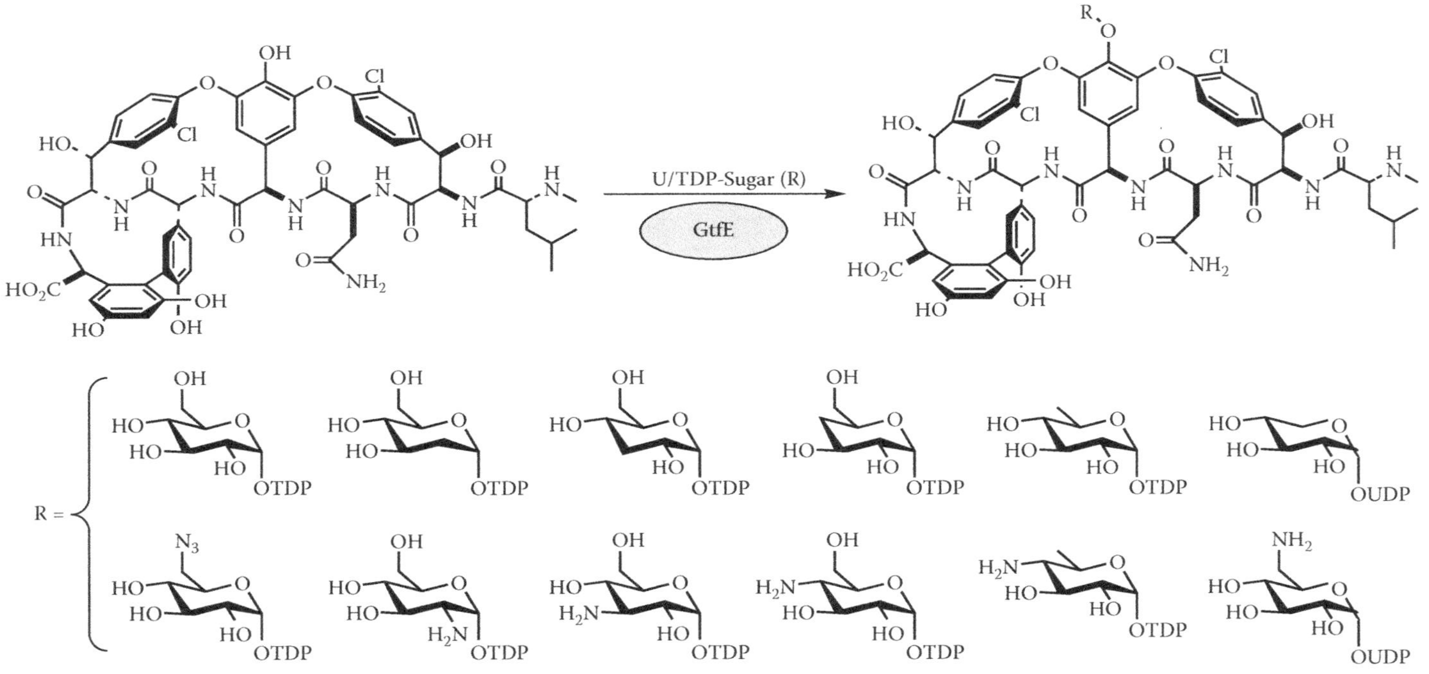

FIGURE 4.20 GtfE from the vancomycin biosynthetic gene cluster is promiscuous enough to utilize a variety of glycosyl donors. GtfE is also tolerant of structural diversity in the aglycone.

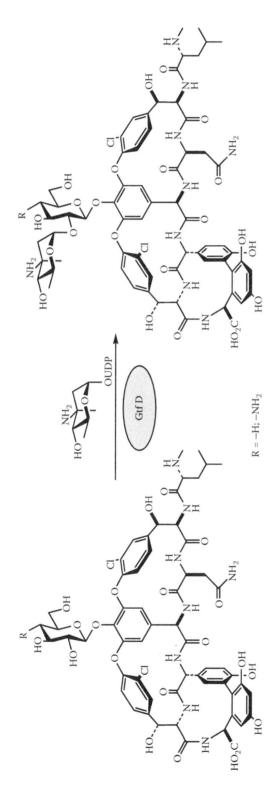

FIGURE 4.21 The second glycosyltransferase in the vancomycin biosynthesis gene cluster GtfD can utilize both an unnatural donor (UDP-*epi*-vancosamine) and unnatural acceptor (4-deoxy and 4-amino glucose).

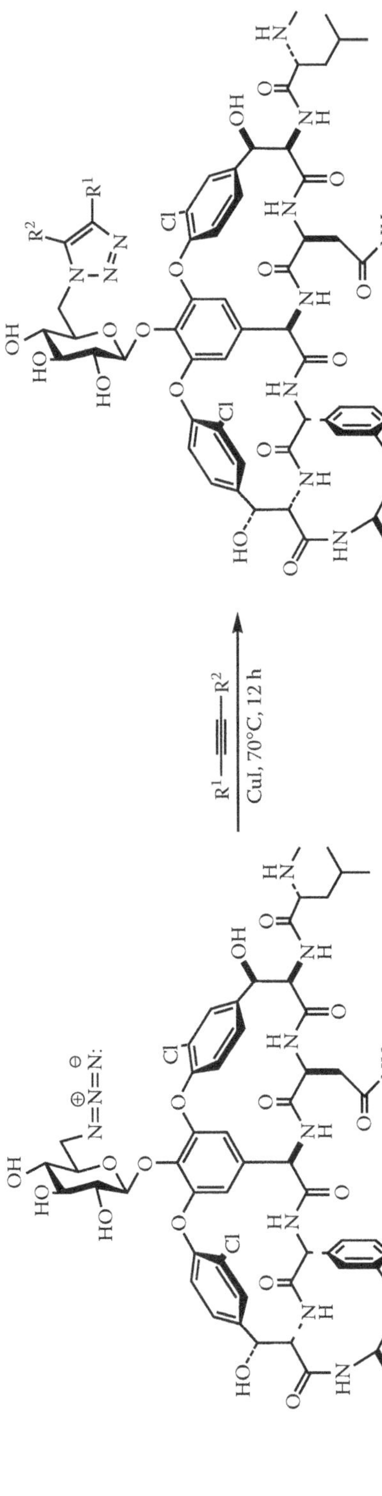

FIGURE 4.22 Incorporation of unnatural 6-azido-gluose onto the vancomycin aglycone provides a synthetic handle for 1,3-dipolar cycloadditions.

REFERENCES

1. Newman, D.J., Cragg, G.M., and Snader, K.M., The influence of natural products upon drug discovery, *Nat. Prod. Rep.*, 17, 215, 2000.
2. Clardy, J. and Walsh, C., Lessons from natural molecules, *Nature*, 432, 829, 2004.
3. Newman, D.J., Cragg, G.M., and Snader, K.M., Natural products as sources of new drugs over the period 1981–2002, *J. Nat. Prod.*, 66, 1022, 2003.
4. Cane, D.E., Walsh, C.T., and Khosla, C., Harnessing the biosynthetic code: combinations, permutations, and mutations, *Science*, 282, 63, 1998.
5. Walsh, C.T., Combinatorial biosynthesis of antibiotics: challenges and opportunities, *Chembiochem*, 3, 125, 2002.
6. Gonzalez-Lergier, J., Broadbelt, L.J., and Hatzimanikatis, V., Theoretical considerations and computational analysis of the complexity in polyketide synthesis pathways, *J. Am. Chem. Soc.*, 127, 9930, 2005.
7. Malpartida, F. and Hopwood, D.A., Molecular cloning of the whole biosynthetic pathway of a Streptomyces antibiotic and its expression in a heterologous host, *Nature*, 309, 462, 1984.
8. Tuan, J.S. et al., Cloning of genes involved in erythromycin biosynthesis from *Saccharopolyspora erythraea* using a novel actinomycete-*Escherichia coli* cosmid, *Gene*, 90, 21, 1990.
9. Cortes, J. et al., An unusually large multifunctional polypeptide in the erythromycin-producing polyketide synthase of *Saccharopolyspora erythraea*, *Nature*, 348, 176, 1990.
10. McDaniel, R., Welch, M., and Hutchinson, C.R., Genetic approaches to polyketide antibiotics. 1, *Chem. Rev.*, 105, 543, 2005.
11. Weissman, K.J., Polyketide biosynthesis: understanding and exploiting modularity, *Philos. Trans. Ser. A Math Phys. Eng. Sci.*, 362, 2671, 2004.
12. Staunton, J. and Weissman, K.J., Polyketide biosynthesis: a millennium review, *Nat. Prod. Rep.*, 18, 380, 2001.
13. Donadio, S. et al., Modular organization of genes required for complex polyketide biosynthesis, *Science*, 252, 675, 1991.
14. Shen, B., Polyketide biosynthesis beyond the type I, II, and III polyketide synthase paradigms, *Curr. Opin. Chem. Biol.*, 7, 285, 2003.
15. Muller, R., Don't classify polyketide synthases, *Chem. Biol.*, 11, 4, 2004.
16. Kato, Y. et al., Functional expression of genes involved in the biosynthesis of the novel polyketide chain extension unit, methoxymalonyl-acyl carrier protein, and engineered biosynthesis of 2-desmethyl-2-methoxy-6-deoxyerythronolide B, *J. Am. Chem. Soc.*, 124, 5268, 2002.
17. Katz, L. and McDaniel, R., Novel macrolides through genetic engineering, *Med. Res. Rev.*, 19, 543, 1999.
18. Ruan, X. et al., Acyltransferase domain substitutions in erythromycin polyketide synthase yield novel erythromycin derivatives, *J. Bacteriol.*, 179, 6416, 1997.
19. Liu, L. et al., Biosynthesis of 2-nor-6-deoxyerythronolide B by rationally designed domain substitution, *J. Am. Chem. Soc.*, 119, 10553, 1997.
20. Petkovic, H. et al., A novel erythromycin, 6-desmethyl erythromycin D, made by substituting an acyltransferase domain of the erythromycin polyketide synthase, *J. Antibiot. (Tokyo)*, 56, 543, 2003.
21. Kuhstoss, S. et al., Production of a novel polyketide through the construction of a hybrid polyketide synthase, *Gene*, 183, 231, 1996.
22. Revill, W. P. et al., Genetically engineered analogs of ascomycin for nerve regeneration, *J. Pharmacol. Exp. Ther.*, 302, 1278, 2002.
23. McDaniel, R. et al., Multiple genetic modifications of the erythromycin polyketide synthase to produce a library of novel "unnatural" natural products, *Proc. Natl. Acad. Sci. U.S.A.*, 96, 1846, 1999.
24. Gokhale, R.S. et al., Dissecting and exploiting intermodular communication in polyketide synthases, *Science*, 284, 482, 1999.
25. Shaw, S.J. et al., A fragment assembly approach to discodermolide: synthesis of the C-1 to C-8 portion from a triketide lactone, manuscript in preparation, 2005.
26. Reeves, C.D. et al., Production of hybrid 16-membered macrolides by expressing combinations of polyketide synthase genes in engineered Streptomyces fradiae hosts, *Chem. Biol.*, 11, 1465, 2004.

27. Jacobsen, J.R. et al., Precursor-directed biosynthesis of erythromycin analogs by an engineered polyketide synthase, *Science,* 277, 367, 1997.
28. Jacobsen, J.R., Cane, D.E., and Khosla, C., Spontaneous priming of a downstream module in 6-deoxyerythronolide B synthase leads to polyketide biosynthesis, *Biochemistry,* 37, 4928, 1998.
29. Jacobsen, J.R. et al., Precursor-directed biosynthesis of 12-ethyl erythromycin, *Bioorg. Med. Chem.,* 6, 1171, 1998.
30. Shaw, S.J. et al., 15-amido erythromycins: synthesis and in vitro activity of a new class of macrolide antibiotics, *J. Antibiot. (Tokyo),* 58, 167, 2005.
31. Carreras, C. et al., *Saccharopolyspora erythraea*-catalyzed bioconversion of 6-deoxyerythronolide B analogs for production of novel erythromycins, *J. Biotechnol.,* 92, 217, 2002.
32. Ashley, G. et al., Preparation of new macrolide antibiotics by chembiosynthesis, manuscript in preparation, 2005.
33. Pfeifer, B.A. et al., Biosynthesis of complex polyketides in a metabolically engineered strain of *E. coli, Science,* 291, 1790, 2001.
34. Murli, S. et al., Metabolic engineering of *Escherichia coli* for improved 6-deoxyerythronolide B production, *J. Ind. Microbiol. Biotechnol.,* 30, 500, 2003.
35. Hans, M. et al., Mechanistic analysis of acyl transferase domain exchange in polyketide synthase modules, *J. Am. Chem. Soc.,* 125, 5366, 2003.
36. Wu, N. et al., Assessing the balance between protein-protein interactions and enzyme-substrate interactions in the channeling of intermediates between polyketide synthase modules, *J. Am. Chem. Soc.,* 123, 6465, 2001.
37. Tsuji, S.Y., Cane, D.E., and Khosla, C., Selective protein-protein interactions direct channeling of intermediates between polyketide synthase modules, *Biochemistry,* 40, 2326, 2001.
38. Tsuji, S.Y., Wu, N., and Khosla, C., Intermodular communication in polyketide synthases: comparing the role of protein-protein interactions to those in other multidomain proteins, *Biochemistry,* 40, 2317, 2001.
39. Wu, N., Cane, D.E., and Khosla, C., Quantitative analysis of the relative contributions of donor acyl carrier proteins, acceptor ketosynthases, and linker regions to intermodular transfer of intermediates in hybrid polyketide synthases, *Biochemistry,* 41, 5056, 2002.
40. Menzella, H. G. et al., Combinatorial polyketide biosynthesis by *de novo* design and rearrangement of modular polyketide synthase genes, *Nat. Biotechnol.,* 23, 1171, 2005.
41. Kumar, P. et al., Intermodular communication in modular polyketide synthases: structural and mutational analysis of linker mediated protein-protein recognition, *J. Am. Chem. Soc.,* 125, 4097, 2003.
42. Hahn, M. and Stachelhaus, T., Selective interaction between nonribosomal peptide synthetases is facilitated by short communication-mediating domains, *Proc. Natl. Acad. Sci. U.S.A.,* 101, 15585, 2004.
43. Sieber, S.A. and Marahiel, M.A., Molecular mechanisms underlying nonribosomal peptide synthesis: approaches to new antibiotics, *Chem. Rev.,* 105, 715, 2005.
44. Walsh, C.T., Polyketide and nonribosomal peptide antibiotics: modularity and versatility, *Science,* 303, 1805, 2004.
45. Schwarzer, D., Finking, R., and Marahiel, M.A., Nonribosomal peptides: from genes to products, *Nat. Prod. Rep.,* 20, 275, 2003.
46. Weber, G. et al., The peptide synthetase catalyzing cyclosporine production in *Tolypocladium niveum* is encoded by a giant 45.8-kilobase open reading frame, *Curr. Genet.,* 26, 120, 1994.
47. Stein, T. et al., Detection of 4-phosphopantetheine at the thioester binding site for L-valine of gramicidinS synthetase 2, *FEBS Lett.,* 340, 39, 1994.
48. Belshaw, P.J., Walsh, C.T., and Stachelhaus, T., Aminoacyl-CoAs as probes of condensation domain selectivity in nonribosomal peptide synthesis, *Science,* 284, 486, 1999.
49. Schneider, T.L., Shen, B., and Walsh, C.T., Oxidase domains in epothilone and bleomycin biosynthesis: thiazoline to thiazole oxidation during chain elongation, *Biochemistry,* 42, 9722, 2003.
50. Altmann, K.H., The merger of natural product synthesis and medicinal chemistry: on the chemistry and chemical biology of epothilones, *Org. Biomol. Chem.,* 2, 2137. [Epub July 14, 2004.]
51. Suo, Z., Thioesterase portability and Peptidyl carrier protein swapping in Yersiniabactin synthetase from *Yersinia pestis, Biochemistry,* 44, 4926, 2005.

52. Kluge, B. et al., Studies on the biosynthesis of surfactin, a lipopeptide antibiotic from *Bacillus subtilis* ATCC 21332, *FEBS Lett.,* 231, 107, 1988.
53. Stachelhaus, T., Schneider, A., and Marahiel, M.A., Rational design of peptide antibiotics by targeted replacement of bacterial and fungal domains, *Science,* 269, 69, 1995.
54. Kohli, R.M. et al., Generality of peptide cyclization catalyzed by isolated thioesterase domains of nonribosomal peptide synthetases, *Biochemistry,* 40, 7099, 2001.
55. Trauger, J.W. et al., Peptide cyclization catalyzed by the thioesterase domain of tyrocidine synthetase, *Nature,* 407, 215, 2000.
56. Kohli, R.M., Walsh, C.T., and Burkart, M.D., Biomimetic synthesis and optimization of cyclic peptide antibiotics, *Nature,* 418, 658, 2002.
57. Kohli, R.M. et al., Chemoenzymatic route to macrocyclic hybrid peptide/polyketide-like molecules, *J. Am. Chem. Soc.,* 125, 7160, 2003.
58. Lambalot, R.H. et al., Overproduction and characterization of the erythromycin C-12 hydroxylase, EryK, *Biochemistry,* 34, 1858, 1995.
59. Betlach, M.C. et al., Characterization of the macrolide P-450 hydroxylase from *Streptomyces venezuelae* which converts narbomycin to picromycin, *Biochemistry,* 37, 14937, 1998.
60. Schlunzen, F. et al., Structural basis for the interaction of antibiotics with the peptidyl transferase centre in eubacteria, *Nature,* 413, 814, 2001.
61. Hansen, J.L. et al., The structures of four macrolide antibiotics bound to the large ribosomal subunit, *Mol. Cell,* 10, 117, 2002.
62. Tang, L. and McDaniel, R., Construction of desosamine containing polyketide libraries using a glycosyltransferase with broad substrate specificity, *Chem. Biol.,* 8, 547, 2001.
63. Dong, S.D. et al., The structural basis for induction of VanB resistance, *J. Am. Chem. Soc.,* 124, 9064, 2002.
64. Kahne, D. et al., Glycopeptide and lipoglycopeptide antibiotics, *Chem. Rev.,* 105, 425, 2005.
65. Losey, H.C. et al., Tandem action of glycosyltransferases in the maturation of vancomycin and teicoplanin aglycones: novel glycopeptides, *Biochemistry,* 40, 4745, 2001.
66. Losey, H.C. et al., Incorporation of glucose analogs by GtfE and GtfD from the vancomycin biosynthetic pathway to generate variant glycopeptides, *Chem. Biol.,* 9, 1305, 2002.
67. Oberthur, M., Leimkuhler, C., and Kahne, D., A practical method for the stereoselective generation of beta-2-deoxy glycosyl phosphates, *Org. Lett.,* 6, 2873, 2004.
68. Barton, W.A. et al., Expanding pyrimidine diphosphosugar libraries via structure-based nucleotidylyltransferase engineering, *Proc. Natl. Acad. Sci. U.S.A.,* 99, 13397, 2002.
69. Fu, X. et al., Diversifying vancomycin via chemoenzymatic strategies, *Org. Lett.,* 7, 1513, 2005.

5 Natural Product-Based, Chemically and Functionally Diverse Libraries

Peter Eckard, Ulrich Abel, Hans-Falk Rasser, Werner Simon, Bernd Sontag, and Friedrich G. Hansske

CONTENTS

5.1 INTRODUCTION

The substantial decline in the number of new chemical entities (NCEs) discovered per year by the pharmaceutical industry is completely in contrast to the promises of top executives that timelines for drug discovery will be shortened dramatically with the introduction of high-throughput technologies, namely target-oriented high-throughput screening, in combination with the huge numbers of combinatorially derived synthetic chemicals. The "one target–one compound–one disease" hypothesis proved false for almost all complex, multifunctional diseases. Many new targets identified from genomics called for an increase of the chemical diversity of the in-house compound libraries. These were the result of decades of classical medicinal chemistry supporting developmental projects. Every synthetic chemist knew that the huge output in sheer numbers would undoubtedly come at the expense of quality. The extremely low success rate of randomly synthesized screening libraries was the result of a lack of knowledge about the relationship between structure (chemistry) and function (physiology). In addition, the limited experience of the chemists, the lack of feasibility of the synthetic methods applied, and the very narrow selection of scaffolds aggravated the situation.

Inspiration from highly functional and chemically diverse natural compounds from various biosynthetic pathways and somewhat prescreened by protein motifs, many having similarities in human systems, for their specific biological interactions (defense, communication, and modulation) is needed to boost the success of lead discovery. The economic generation of small and smart collections of chemically and functionally diverse compound libraries with distinct biological function, oral bioavailability, and limited toxicity is needed for new pharmaceuticals. Because most if not all natural products have passed a functional filter, selected natural scaffolds are ideal starting points for focused libraries addressing target families as well as for generic libraries that supplement general screening. Applying the methodologies of parallel synthesis and semisynthesis with polyfunctional natural scaffolds will have a dramatic impact on the generation of new leads and developmental candidates.

We will cover the various methods of accessing attractive scaffolds: by screening against single targets or target families, by literature mining, by *in silico* methods (see Chapter 3), and by chemical screening. After describing the production of substantial amounts of starting materials, general scaffold-based libraries will be described. Examples of various natural product-based libraries will demonstrate the excellent opportunities for drug discovery.

What are the advantages of natural products as starting materials (scaffolds) for focused libraries? First of all, they are privileged structures with a high content of information (three-dimensional structure), and they are not randomly synthesized. They are not only biosynthetically derived and encoded in conservative gene clusters but also interact with a broad variety of proteins, which enables pharmaceutical and therapeutic potential. The unique chemical diversity can be leveraged with combinational libraries. In addition, this strategy can be supported by specific biotransformations as well as scaffold screening in taxonomically closely related species.

A major strategic focus is the supply issue. Sourcing limitations for scaffolds of choice can be tackled by classical as well as genetic strain optimization methods and comparatively low-cost scale-up fermentations. Once a derivative of a natural scaffold is identified by its promising screening results, it might be strategically helpful to explore and develop a more economical source, i.e., total synthesis. Successful natural product-based library design requires a highly efficient multidisciplinary team of microbiologists, screening experts, natural compound chemists, spectroscopists, and chemists with experience in semisynthesis. Diversity-oriented synthesis starting from optimized and privileged natural products will generate smart collections of highly diverse libraries, which will ultimately have the potential to shorten substantially the drug discovery timeline.

5.2 SELECTION OF SCAFFOLDS

A scaffold is the basic structural element used as the starting point for the generation of chemical libraries. Such a scaffold should provide opportunities for chemical modification. With the ultimate goal of identifying lead compounds for preclinical development, attachment of new functionality and pharmacophoric groups onto natural products will give rise to compound libraries that complement a company's compound collection.

There are two basic scaffold types, the *functional scaffold* and the *structural scaffold*. The functional scaffold is a molecule with specific biological activity directed toward a molecular target (e.g., kinase inhibition or receptor antagonism) or a specific indication (e.g., antiviral or proapoptotic). A functional scaffold is a lead compound used for the generation of focused libraries to optimize specific lead properties (e.g., potency, selectivity, or bioavailability) while maintaining the basic activity. Such a library can also be employed as a target-class-specific focused library for hit and lead discovery. Because active molecules are the ultimate goal of every lead or drug discovery project, libraries based on functional scaffolds with proven activity are the most useful. On the other hand, functional scaffold-based libraries with their target class-specific activity bias may have limited utility in the discovery of leads for a broad range of targets or indications.

The structural scaffold is a molecule with certain structural features (e.g., specific ring systems, chiral centers, heteroatoms, three-dimensional structures, or functional groups). Structural scaffold libraries may be more complementary to an existing generic library for hit and lead search over a wide range of target classes. These libraries can increase available chemical space, fill structural gaps, or add larger multifunctional molecules. Structural scaffold-based libraries have no activity bias and may therefore be more universally applicable than functional scaffold-based libraries.

Synthetic compounds as scaffolds for combinatorial libraries have not fulfilled initial expectations with regard to facilitating the discovery of new drugs. The differences of properties of drugs, natural products, and combinatorial compounds, which have been discussed by Feher and Schmidt,[1] may be the reason. Important parameters for the success of drugs are potency and selectivity, which are determined to a large extent by stereochemistry and rigidity. Chiral centers contribute to selectivity through stereospecific binding sites on proteins. Rigidity increases binding affinity because of the lower entropic losses upon binding to the target protein. Interestingly, natural compounds have a comparatively high number of chiral centers and exhibit greater rigidity than synthetic compounds. The biosynthetic origin and many functions of natural products require interactions with proteins. Natural products have evolved with a functional filter, which makes them ideal candidates for interactions with target proteins in drug discovery.

The structural and functional features of natural compounds offer some unique opportunities for using them as scaffolds for library design compared to synthetic compounds. The basic requirement for the selection of suitable natural compound scaffolds is the knowledge of an experienced medicinal chemist. Some general guidelines to help prioritize structures are available. A preferred structure has the following properties:

- It is amenable to chemical modification and introduction of functional groups.
- It allows avoidance of protecting groups during synthesis.
- It has a molecular weight below 500.
- It has no highly reactive groups.

- It is chemically stable as a solid and in solution (e.g., stability is unaffected by oxidation, temperature, weak acids and bases, structural rearrangements).
- It is available by fermentation (with a yield greater than 10 mg/l) or from plants (with a yield greater than 100 mg/kg from fresh plant material).
- It is isolated by a simple procedure, preferably without chromatographic steps.
- It has a solubility in standard organic solvents that is greater than 10 mg/ml.

To qualify as a structural scaffold, a compound must have the following attributes:

- Display reasonable decoration with moderately reactive groups (e.g., OH, COOH, NH_2, NHR, NR_2, SH, C=O, C=C, $CONH_2$, epoxide, electrophilic aromatic positions, Michael systems)
- Have no (or only a few) equally reactive groups, in order to avoid the generation of different isomers during synthesis
- Have a rigid scaffold (e.g., polycyclic structures)
- Display a suitable molecular topology and distribution of chiral centers and heteroatoms
- Show a diversity that is complementary to the existing compound library

In addition to the first set of criteria, a functional scaffold also must display suitable biological activity and selectivity.

A variety of sources and selection methods for suitable natural compound scaffolds is available and is described in the text that follows.

5.2.1 SELECTION BASED ON FUNCTIONAL PROPERTIES

The main sources for the identification of natural compound functional scaffolds are the public literature, database mining, screening or bioassays, and virtual docking of natural compounds into three-dimensional structures of target proteins (see Chapter 3).

5.2.1.1 Public Literature or Database Mining

Information about the functional properties of natural compounds are available from several sources (e.g., literature databases such as PubMed and natural compound databases such as the *Dictionary of Natural Products* from Chapman and Hall/CRC). For example, a search for kinase inhibitors in the *Dictionary of Natural Products* identified approximately 100 natural compounds as potential functional scaffolds. Wan et al. reported the use of hymenialdisine (**1**, Figure 5.1), originally isolated from marine sponges, as a scaffold for a kinase inhibitory library. Hymenialdisine (**1**) is a nanomolar inhibitor of several kinases (e.g., CDKs and MEK1). Among the 56 hymenialdisine (**1**) analogs, compounds with enhanced kinase selectivity or improved cellular potency were identified.[2] Nakijiquinone C (**2**, Figure 5.1), isolated from a marine sponge, was shown to inhibit EGFR, ERBB2 (HER2/neu), and PKC.[3] Kissau et al. constructed a library based on nakijiquinone C (**2**) in their search for novel inhibitors of related kinases.[4] Of the 74 compounds synthesized in a library, 7 compounds had micromolar IC_{50} values in a panel of 6 kinases. Nakijiquinone C (**2**) was inactive against Tie-2, but one of the reported analogs inhibited Tie-2 with an IC_{50} value of 18 μM. Saframycin A (**3**, Figure 5.1), isolated from *Streptomyces lavendulae*,[5] was shown to have antitumor properties.[6] The recently demonstrated antiproliferative activity of (**3**) involves interaction with duplex DNA and GAPDH.[7] A library of more than 70 analogs was synthesized and tested for activity against melanoma and lung cancer cell lines. In comparison to saframycin A (**3**), half of the new compounds showed increased antiproliferative activity — some of them with a 20-fold improvement.[8]

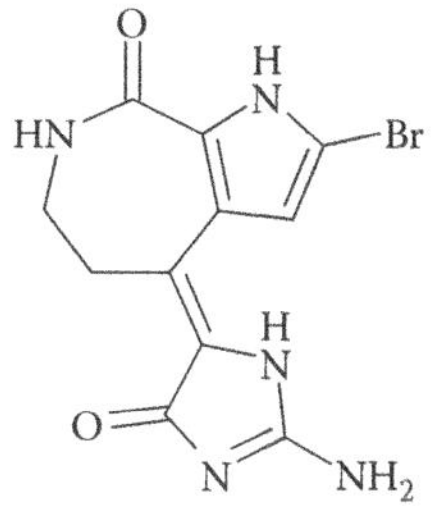

Hymenialdisine (**1**)

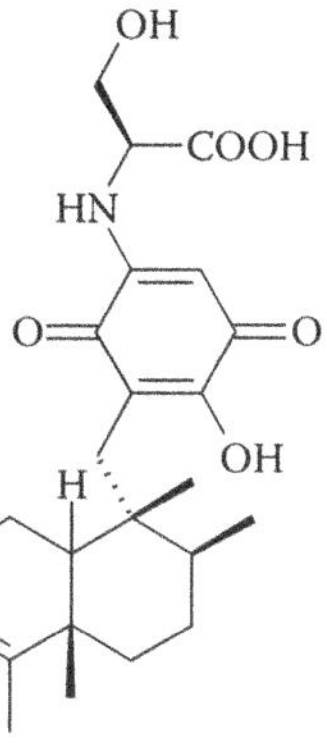

Nakijiquinone C (**2**)

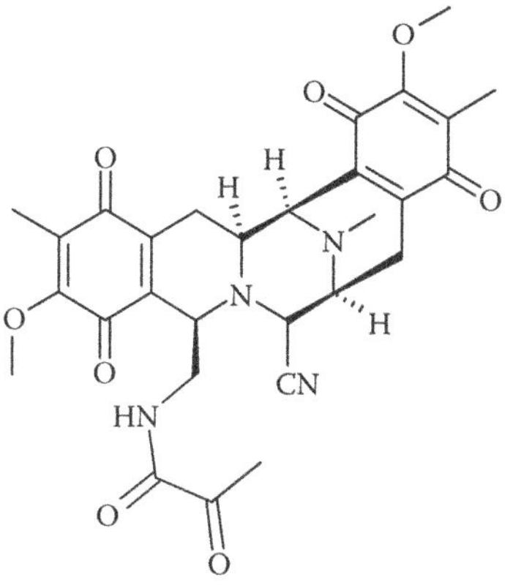

Saframycin A (**3**)

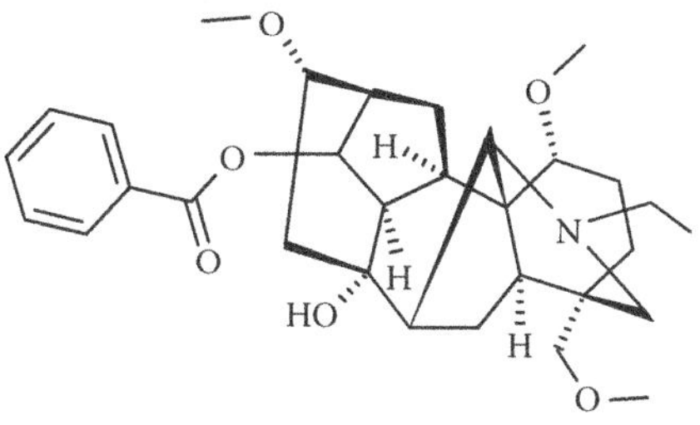

14-Benzoyl-talatisamine (**4**)

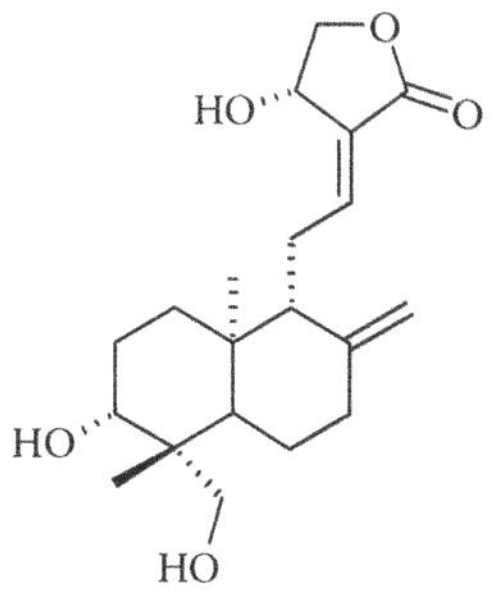

Andrographolide (**5**)

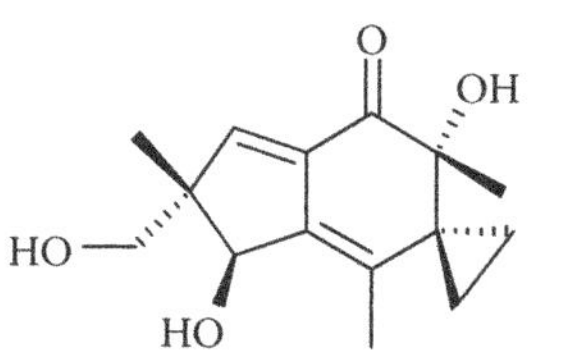

Illudin S (**6**)

FIGURE 5.1 Selected examples of natural product scaffolds.

5.2.1.2 Screening or Bioassays

Screening of proprietary libraries provides a unique opportunity for the identification of novel scaffolds. Such libraries contain either isolated natural products or complex samples based on fermentation broths. These libraries can be screened in biological assays to identify active molecules. Such compounds can then be evaluated as possible scaffolds for libraries. Today a wide variety of screening methods is available to identify bioactive molecules in low, medium, or high throughput. Such assays can be performed with isolated molecular targets or with whole cells. However, the compatibility of natural product-based libraries with screening assays may be an issue. Libraries of isolated natural compounds are compatible with current screening assays and formats and may be used for the identification of functional scaffolds in a similar fashion as synthetic chemical libraries.

Crude extracts of microbial broths or from plant material are frequently used to identify new biologically active molecules. This procedure was quite successful in the past in identifying antibiotic, antifungal, or antitumor compounds, with assays based on microbial growth or cellular proliferation. Examples are the identification of the antitumor compound fredericamycin A by screening of fermentation broths for cytotoxic activity with tumor cell lines[9] and the discovery of the potent immunosuppressive compound FK-506 in a screen for the inhibitory effects on IL-2 production.[10] Crude extracts from fermentation broths are complex mixtures and, therefore, not ideally suited for today's lead discovery efforts. Moreover, the use of a low total-compound concentration to avoid artifacts results in the relatively low concentration of individual compounds and leads to low assay sensitivity. Synergistic effects of mixture components that show activity which is lost upon fractionation, and the chance of masking of agonistic activity by antagonistic compounds in the same sample, is higher in more complex mixtures. The long timelines required for the process, from hit identification to the isolation of the natural compounds, also complicate lead discovery.

Samples with reduced complexity (e.g., chromatographic fractions from fermentation broths) are better suited for screening assays but require more time for sample preparation and screening. Prefractionated samples (subfractions), described by Abel et al., represent an ideal compromise between assay throughput and mixture complexity.[11] After filtration of microbial fermentation broths, the filtrate is fractionated by HPLC into nine subfractions. The mycelium is extracted and represents the tenth subfraction. Such subfractions, containing on average 15 to 25 compounds, are dissolved in DMSO to a predefined concentration, stored in microplates, and are ready for screening. Subfractions were successfully and efficiently used for the identification of bioactive molecules and scaffolds in several projects. In addition to the reduced complexity of subfraction libraries, an efficient and reproducible process is required for the isolation of an active natural compound after the initial subfraction screening hit.

5.2.1.3 Virtual Screening and Docking

Virtual screening methods using a computational approach to identify molecules that fit into the active site of a target protein are also useful. Although there remain several challenges in virtual screening (e.g., sampling the various conformations of flexible molecules or calculating absolute binding energies in an aqueous environment), some successful examples were recently published.[12] Based on a model of the thyroid hormone receptor ligand-binding domain, virtual screening of more than 250,000 compounds of the Available Chemicals Directory database provided 75 hits for testing in a cell-based assay. The best compound acted as a receptor antagonist with an IC_{50} value of 1.5 μM.[13]

Despite such reported successes, wet screening is preferred and is much more reliable than virtual screening, provided assays are available and compounds are physically accessible. In case of natural compounds, where more than 150,000 structures have been published (though most of them are not easily obtainable for screening), virtual screening is an important approach to identifying functional scaffolds. All, or a large subset, of the published structures can be screened virtually, and a few scaffolds with high docking scores can then be used in wet screening. Upon activity confirmation, libraries can be designed around the functional scaffolds. For example, 14-benzoyltalatisamine (**4**, Figure 5.1) was identified in a virtual screening approach from the China Natural Products Database as a potassium-channel blocker, inhibiting Iκ with an IC_{50} value of 3.8 μM in a whole-cell voltage-clamp assay.[14]

5.2.2 Selection Based on Structural Properties

The main sources for the identification of natural compound structural scaffolds are the public literature, database mining, and chemical screening.

5.2.2.1 Public Literature or Database Mining

The same sources of information described in Subsection 5.2.1.1 can be used with a different set of search criteria. The most promising approach is the visual inspection of structures (or a subset of structures resulting from a search with some simple exclusion criteria such as molecular weight or the presence of toxic groups) by an experienced medicinal chemist and a microbiologist. Two examples of promising and versatile structural scaffolds are the diterpenoid andrographolide (**5**), a constituent of *Andrographis paniculata*,[15] and illudin S (**6**), isolated from *Clitocybe illudens*,[16] shown in Figure 5.1 (see Chapter 9, Section 9.3).

5.2.2.2 Chemical Screening

Chemical screening is a methodology to characterize, identify, or find new natural compounds directly from biological extracts, using thin-layer chromatography (TLC) and various staining reagents.[17,18] Separation of an extract by TLC and staining results in a characteristic pattern, the so-called metabolic fingerprint. This can be used to compare the spectrum of secondary metabolites of different organisms and to identify interesting or unique natural compounds. TLC behavior and staining depend on compound properties such as the presence of certain functional groups, ring systems, chromophores, and compound polarity. The quantity of a natural compound in the extract, another scaffold criterion, can be roughly estimated from the stained TLC spot. Careful examination of a metabolic fingerprint may therefore lead to the discovery of unique structural scaffolds. Important prerequisites for a successful application of this method are extensive experience in assessing the TLC pattern, suitable image analysis software, and an appropriate database that links TLC properties to natural compounds or structural features.

During the last few years, many natural products were detected by their chemical reactivity and color formation (e.g., the antiproliferative jenamidines,[19] the anthelminthic macrolide clonostachydiol,[20] or the nematocidal elaiophylin[21]). Chemical screening by itself is not able to deliver functional scaffolds with a previously defined biological activity. A modified technique, the biomolecular chemical screening, combines chemical screening with an assessment of the affinity to a certain target.[22] Here, TLCs are run with and without a binding target (e.g., DNA), and binding hits are identified by differential analysis of the two corresponding TLC patterns.[23,24]

5.3 PRODUCTION OF SCAFFOLDS

Once a scaffold is selected, the desired compound is produced in gram quantities in order to support feasibility studies and library generation. About 1 to 5 g of scaffold is usually sufficient for the initial evaluation of synthetic possibilities and the generation of a small library. In general, there are four main sources of scaffold supply:

- Production by fermentation of microorganisms
- Production by semisynthesis from an easily available natural precursor
- Isolation from natural sources such as plants or marine invertebrates
- Production of the natural compound scaffold by total synthesis

For each scaffold a thorough evaluation is mandatory to find an efficient and economic method for its production. Total syntheses are often described for natural products, but they frequently consist of many synthetic steps, require excessive use of protecting groups, employ expensive or toxic reagents, or exhibit low overall yields. Total synthesis, therefore, is usually not the method of choice for the production of complex natural compounds. However, Smith et al. achieved a gram-scale synthesis of (+)-discodermolide and demonstrated the feasibility of using total synthesis to generate useful quantities of natural products for drug discovery.[25] More recently, in a series of publications, researchers at Novartis reported the synthesis of 60 g of (+)-discodermolide.[26]

In spite of this noteworthy achievement in synthesis, many natural products can be obtained by isolation methods in sufficient quantities for library synthesis. For example, 0.02 to 0.2 percentage by weight of ginkgolides are obtained from *Ginkgo biloba* leaves (see Subsection 5.4.1) and 0.025 percentage by weight of betulinic acid is extracted from *Betula alba* (see Subsection 5.3.4.1). If a natural compound originates from plants or animals, access may be restricted: The natural producer may not be readily available, it may be a rare or protected organism, or the production rate may be too low. In such cases, total synthesis or semisynthesis may overcome the shortage. A well-known example is paclitaxel (Taxol®), originally isolated from the stem bark of the pacific yew, *Taxus brevifolia*. Today, paclitaxel is prepared by semisynthesis, starting from 10-deacetylbaccatin III, which is readily available from the leaves of *Taxus baccata* (common or European yew).

Alternatively, if the producer of the natural compound is a microorganism, fermentation is usually the method of choice. Today, many important pharmaceuticals are isolated from fermentation broths, including the immunosuppressives cyclosporin A, tacrolimus (FK-506), and sirolimus (rapamycin); the antibiotics erythromycin and vancomycin; and human insulin. By genetic engineering, microorganisms can be optimized for the production of one single compound, and this can result in an increased product titer and a facilitated isolation procedure. Subsection 5.3.1 to Subsection 5.3.4 will focus on the production of natural product scaffolds by fermentation of microorganisms.

5.3.1 Sources and Selection of Producer Microorganisms

The availability of strains and an appropriate technical infrastructure are the prerequisites for the production of natural scaffolds in amounts necessary for library synthesis. For our in-house scaffold selection, we screen subfractions and compounds that originate from our own highly diverse microbial strain collection, containing more than 45,000 actinomycetes and 8,000 fungi. If scaffolds are identified from literature or database mining, it may be difficult to acquire the producing strain.

Generally, microorganisms can be obtained from public strain collections such as the American Type Culture Collection (ATCC; U.S.), Deutsche Sammlung von Mikroorganismen und Zellkulturen GmbH (DSMZ, German Collection of Microorganisms and Cell Cultures; Braunschweig, Germany), or Centraalbureau voor Schimmelcultures (CBS, Fungal Biodiversity Center; Utrecht; Netherlands). If a literature-described microorganism is not available for cultivation, alternative strains may be found (e.g., several related strains of the same genus). Even if the original producer strain is available, acquisition of some additional related strains from the same genus or even samples of the same species from different origins is desirable. Frequently, such alternative organisms produce different amounts of the scaffold or may produce different analogs in addition to the desired compound. It is absolutely essential to work with a genetically stable production strain. Different reliable preservation methods should be used for each strain (e.g., freeze-drying or storage in liquid nitrogen) to minimize the risk of genetic instability.

5.3.2 STRAIN CULTIVATION

The production of secondary metabolites by microorganisms is strongly influenced by growth conditions and nutritional factors such as carbon source, nitrogen source, minerals, trace elements, and vitamins. Compound precursors added to the medium may influence product formation. The same organism cultivated in two different media may produce different analogs of a compound as the major product. Without prior knowledge about the preferred growth conditions for a given microorganism or the production of a special compound, fermentation broths often lack relevant levels of natural product scaffolds. Strain cultivation typically starts with numerous fermentation experiments to determine the medium composition and growth parameters for high scaffold yields. Scaffold titer should be quantified twice daily, preferably after HPLC separation with MS-, UV-, and/or ELSD detection.

There are a few generally accepted strategies to select suitable growth conditions. Tormo et al. reported a computer-aided HPLC-based method for the selection of production media for actinomycete strains.[27] Another strategy is to grow the strains in a multitude of different media or to cultivate them with different fermentation conditions (e.g., of aeration, stirring, and pH). The starting point for media optimization can be the completely random selection of media or the selection of specific media with the most diverse compositions possible.

We found that starting with a set of 4 very different culture media and fermentation in 1-l Erlenmeyer shaking flasks with 250-ml culture volume usually gives some initial data about the production rate and suggests some routes for further optimization of growth conditions. The next steps would be additional media variations and/or scaling up with 15-l laboratory or 100-l pilot-scale fermenters. Frequently, however, the production rates in fermenters differ markedly from those in shaking flasks or other microcultures. Here, additional optimization work is necessary. Sometimes a classical or molecular biology mutational approach may be required to increase production rates to the levels suitable for library synthesis.[28] Time invested in the optimization of culture conditions always pays off during the next downstream step, scaffold isolation.

5.3.3 SCAFFOLD ISOLATION

The detailed isolation process is dictated to a large extent by the scaffold's chemical properties. Nevertheless, we provide a general outline in the text that follows and a few examples in Subsection 5.3.4. After termination of the fermentation process, the whole broth, preferably with a product titer greater than 10 mg/l, is freeze-thawed in order to facilitate the separation of the mycelial fraction from the culture supernatant. After the thawing and an optional centrifugation step, the culture broth is filtered through depth filter sheets. At this point, the amount of target compound should be quantified in the filtrate and the mycelium, and a part with less than 10% of the total quantity is discarded.

The desired compounds are then extracted from the filtrate by solid-phase extraction with polymer resins such as Amberlite® XAD-16 HP (Rohm and Haas) or Diaion® HP20 (Mitsubishi Chemical) and/or from the mycelium with suitable solvents such as methanol, acetone, methylethylketone, butanol, or ethyl acetate. A first purification step of the resulting extracts is performed preferably using size-exclusion chromatography on Sephadex® LH-20 (Pharmacia) with methanol, acetone, or ethyl acetate as eluents, depending on the polarity of the compounds. Alternatively, a liquid–liquid extraction step may be considered with high-speed counter-current chromatography (HSCCC) or fast centrifugal partition chromatography (FCPC). A final preparative-scale HPLC on reversed-phase C_{18}-material usually delivers the desired compound in a sufficiently pure form. Medium pressure liquid chromatography (MPLC) on silica gel can be considered before HPLC if there are still too many impurities present. The purity of the resulting natural product scaffolds should be > 90% (HPLC) before starting the different synthetic steps for library generation.

5.3.4 Examples of the Production of Natural Product Scaffolds

5.3.4.1 Betulinic Acid

The bark of the white birch, *Betula alba*, contains only minor amounts (0.025 percentage by weight) of betulinic acid, a pentacyclic lupane-type triterpene (**17**, Figure 5.8). Because vast quantities of white birch bark are available, betulinic acid (**17**) is available in large quantities. However, the methods used to isolate betulinic acid (**17**) from the bark are laborious and expensive. Recently, a process to manufacture betulinic acid (**17**) economically from the precursor betulin was developed. The alcohol betulin is a cheap and abundant starting material for the preparation of betulinic acid (**17**). Betulin constitutes 25 percentage by weight of white birch bark and is readily isolated. Betulin is converted in good yields to betulinic acid (**17**) by two different oxidation methods followed by a stereoselective reduction of the ketone group.[29] The properties of betulinic acid (**17**) and its libraries are described in Subsection 5.6.1.

5.3.4.2 Fredericamycin A

To supply enough fredericamycin A (**18**, Figure 5.9) for semisynthetic experiments, the producing strain *Streptomyces griseus* was mutated in-house by different methods to increase the production rate from 10 mg/l to 100 mg/l. An additional increase was achieved by media variation and resulted in a final titer of approximately 700 mg/l. For the isolation of fredericamycin A (**18**) from the fermentation broth, we changed the standard procedure described earlier because fredericamycin A has low solubility in almost all common solvents. We developed a new and efficient purification process that consisted of extraction, dissolution, and repeated precipitation without any chromatographic steps.[30] In this way, we isolated 65 g of fredericamycin A (**18**) from a 100-l fermentation. The unique structure of fredericamycin A (**18**), its antitumor properties, and focused libraries are described in detail in Subsection 5.6.2.

5.3.4.3 Borrelidin

Borrelidin (**7**, Figure 5.2) is a structurally unique macrocyclic natural compound isolated from *Streptomyces rochei*[31] and related *Streptomyces* species. Borrelidin (**7**) inhibits the cyclin-dependent kinase Cdc28/Cln2[32] and threonyl tRNA synthetase.[33] Borrelidin (**7**) exhibits antiviral,[34] antibacterial,[35] and antimalarial[36] properties, but most intriguing is the potent inhibition of angiogenesis *in vitro* in the aorta tube formation assay (with an IC_{50} value of 0.8 n*M*)[37] and *in vivo* in the mouse dorsal air sac model.[38]

We generated a semisynthetic library with borrelidin (**7**) as the starting point in our studies. By optimization of media and growth conditions, we increased the production rate of our strain from 10 mg/l to over 200 mg/l. We isolated 10 g of borrelidin (**7**) from a 80-l fermentation, using the standard workup procedure (extraction, flash chromatography on silica gel, and preparative HPLC) for our synthetic program. In addition, we isolated three new borrelidin derivatives in substantial amounts for initial QSAR studies.

5.3.4.4 Irumamycin

Irumamycin (**8**, Figure 5.2) is a macrolide antibiotic, isolated by Omura et al. in 1982,[39] which is related to the structurally similar venturicidines[40] and aabomycins.[41] We isolated irumamycin (**8**) during an activity-guided antifungal screen and decided to generate a small focused library. After extensive media variation, we found an optimized medium in which the production rate was ten times higher than in standard media (approximately 400 mg/l). For an initial pilot study, we isolated 1 g of irumamycin (**8**), following the standard procedure described earlier, along with 1 known and 4 new derivatives in quantities of 100 to 200 mg each.

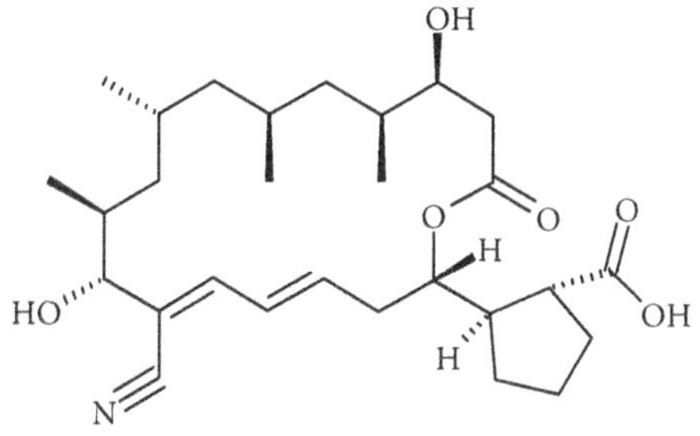

Borrelidin (**7**)

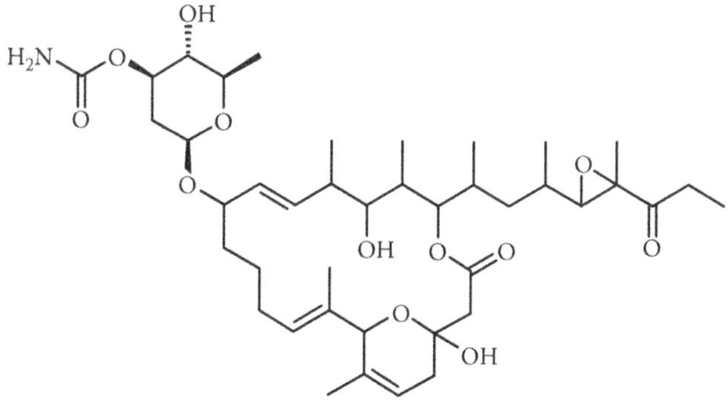

Irumamycin (**8**)

FIGURE 5.2 Structures of borrelidin (**7**) and irumamycin (**8**).

5.4 ENDOGENOUS OR BIOSYNTHETIC NATURAL PRODUCT LIBRARIES

The biosynthetic route to secondary metabolites is based on the action of specific enzymes. Depending on the availability of nutrients, the growth conditions, and the regulation of the expression of biosynthetic genes, at different points in time a cell contains different biosynthetic precursors, the final natural product, derivatives thereof, and degradation products. Beginning compound isolation interrupts the biosynthetic process and reveals a picture of the natural product pattern at a single point in time. Therefore, a fermentation broth often contains a small focused library of related natural products. Such endogenous or biosynthetic natural product libraries frequently allow for a rapid initial study of structure–activity relationships (SAR) before the synthesis of larger compound libraries. A few examples of endogenous natural compound libraries are presented in the following text.

5.4.1 GINKGOLIDE LIBRARY

Ginkgo biloba represents the oldest living tree species on earth. Preparations of the nuts or leaves of this tree are an important part of traditional Chinese medicine and, presently, *Ginkgo* leave extracts are among the best-selling phytomedicinal drugs worldwide. They are used for the enhancement of memory and blood circulation and for delaying the symptoms of Alzheimer's disease.[42]

Structurally the most intriguing as well as the pharmacologically most active are the terpene trilactones, ginkgolides A, B, C, J, and M (**9a** to **9e**, Figure 5.3), and bilobalide A (**10**, Figure 5.3), all of which together make up a small endogenous library of natural products.

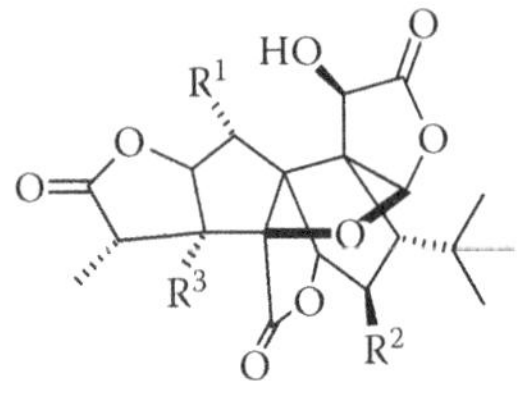

	R^1	R^2	R^3
Ginkgolide A (**9a**)	H	H	OH
Ginkgolide B (**9b**)	OH	H	OH
Ginkgolide C (**9c**)	OH	OH	OH
Ginkgolide J (**9d**)	H	OH	OH
Ginkgolide M (**9e**)	OH	OH	H

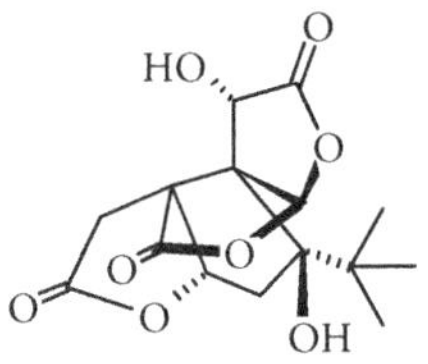

FIGURE 5.3 Endogenous library from *Ginkgo biloba*: ginkgolides (**9a** to **9e**) and bilobalide (**10**).

The ginkgolides were isolated for the first time in 1932 by Furukawa from the stem bark of *G. biloba*[43] and were structurally elucidated in 1967 by Maruyama et al.[44] Today, most pharmacologically active preparations (such as EGb 761) are produced from the dried leaves of *G. biloba*, which contain 0.02 to 0.2 percentage of weight of ginkgolides and 0.5 to 2 percentage of weight of bilobalid A (**10**). The terpene trilactones are antagonists of the platelet-activating factor receptor[45] and ligands of glycine receptors[46] and of peripheral benzodiazepine receptors.[47] Owing to their unique structural and pharmacological properties, a great deal of synthetic development was done to achieve the total synthesis of the ginkgolides. A semisynthetic library based on the ginkgolide scaffold was prepared and pharmacologically evaluated by Braquet at al. (**11a** to **11j**, Table 5.1).[48]

5.4.2 COLLYBOLIDE LIBRARY

An example of an endogenous natural product library of fungal origin are the collybolides, isocollybolides, and their derivatives, isolated from fruiting bodies of *Collybia maculata* and *C. peronata* (Basidiomycetes).[49,50] During our ongoing search for new active natural products, we detected strong activity of the collybolides as L-type calcium-channel inhibitors or modulators, making them potentially useful for the treatment of diseases such as hypertension, migraine, stroke, asthma, or Alzheimer's disease.[51] To date, eight different structures have been published[52] (**12a** to **12h**, Figure 5.4) and four new compounds have been isolated in our own laboratories (**12i** to **12l**, Figure 5.4) from fruiting bodies; fermentation broths were used for initial SAR studies.

5.4.3 LIBRARY GENERATION BY BIOTRANSFORMATION

Some structures are difficult to obtain by chemical synthesis. An alternative method is based on the ability of microorganisms to degrade and modify exogenous compounds, and this can be used

TABLE 5.1
Semisynthetic Library Based on the Ginkgolide
Scaffold and SAR Studies as Platelet-Activating
Factor Receptor Antagonists

Compound	R^1	R^2	R^3	IC$_{50}$ [μM]
Ginkgolide A (**9a**)	H	H	H	0.74
Ginkgolide B (**9b**)	OH	H	H	0.25
Ginkgolide C (**9c**)	OH	H	OH	7.1
11a	H	CH$_3$	H	13
11b	OH	CH$_3$	H	0.29
11c	OH	CH$_3$	OH	3.0
11d	H	CH$_2$CH$_3$	H	62
11e	OH	CH$_2$CH$_3$	H	7.2
11f	OH	CH$_2$CH$_3$	OH	9.3
11g	OCH$_3$	H	H	0.66
11h	OCH$_3$	H	OH	4.2
11i	OCH$_2$CH$_3$	H	H	1.1
11j	OCH$_2$CH$_3$	H	OH	8.5

Adapted from Stromgaard, K. and Nakanishi, K., *Angew. Chem.*,
116, 1682, 2004. With permission.

for biotransformations. For almost every type of chemical reaction, enzymes can be found that act as catalysts to complement synthetic modification methods. Under mild biotransformation reaction conditions, microbial systems have been used successfully to introduce chiral centers, resolve racemates, differentially convert identical functional groups with similar reactivities, functionalize a nonactivated carbon regioselectively, and to convert labile molecules.[53] Some disadvantages include the high expenditure for the development of biotransformation processes, long reaction times, and comparatively low substrate and product concentrations.

Whereas intensive research on stereoselective hydroxylation and other biotransformations in the fields of small synthetic molecules[54] and steroids[55] has been performed, there are only a few reports on the transformation of other natural products such as macrocyclic molecules.[56–62] A very interesting example, described by Kuhnt et al.,[63] demonstrated that cyclosporin A (**13**, Figure 5.5) is hydroxylated when added to cultures of *Nonomuraea dietziae*. Interestingly, the particular hydroxylation position depended on the cultivation conditions. Using another strain of *Nonomuraea*, we added 1 g/l cyclosporin A (**13**) to the culture medium and found, in addition to the 4-(γ-hydroxy)-derivative (**13a**, Figure 5.5), a 9-(γ-hydroxy)-compound (**13b**, Figure 5.5) as the major biotransformation product. Both compounds are already known as degradation products found in patients treated with cyclosporin A (**13**), which is metabolized by the human hepatic cytochrome P450 III monooxygenase A system.[64] The hydroxylated compounds had drastically different properties compared to cyclosporin A (**13**). The immunosuppressive activity of 4-(γ-hydroxy)-cyclosporin A

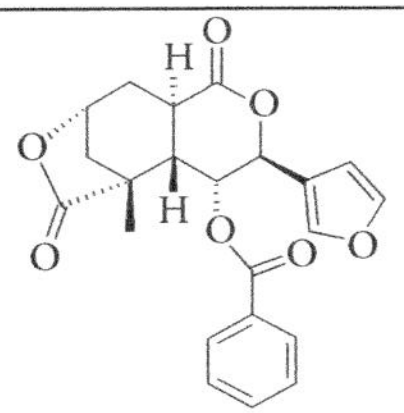	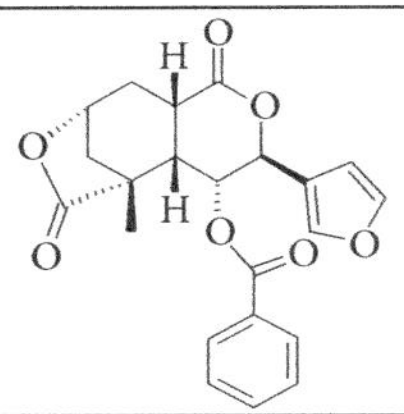	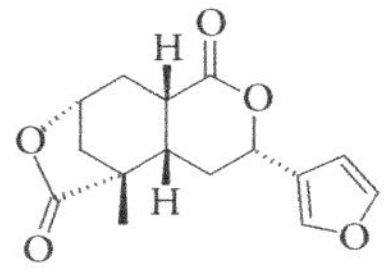	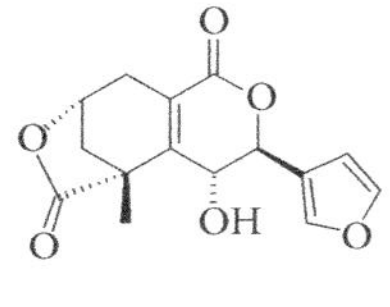
Collybolide = 6-benzoylcollybolidol **(12a)** (*C. maculata*)	Isocollybolide = 9-*epi*-collybolide **(12b)** (*C. maculata*)	6-desoxycollybolidol (*C. peronata*) **(12c)**	5, 9-didehydrocollybolidol (*C. maculata*) **(12d)**
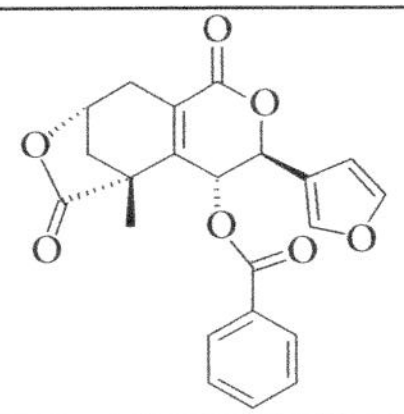	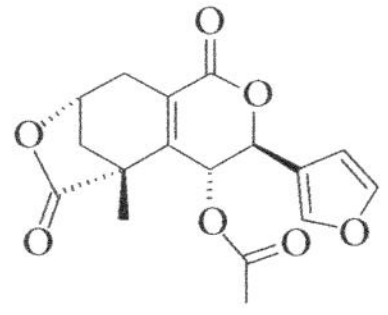	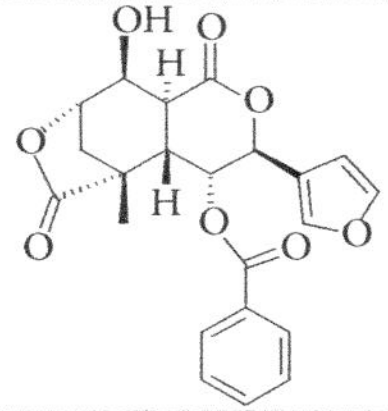	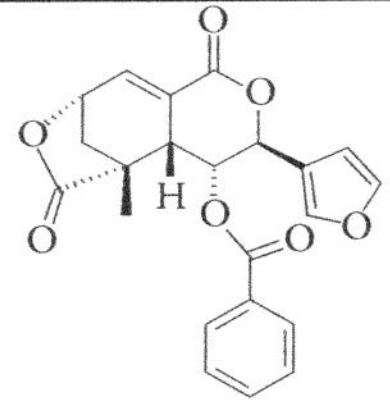
5, 9-didehydrocollybolide **(12e)** (*C. maculata*)	6-acetyl-5, 9-didehydrocollybolidol **(12f)** (*C. maculata*)	1-hydroxycollybolide **(12g)** (*C. maculata*)	1, 9-didehydrocollybolide **(12h)** (*C. maculata*)
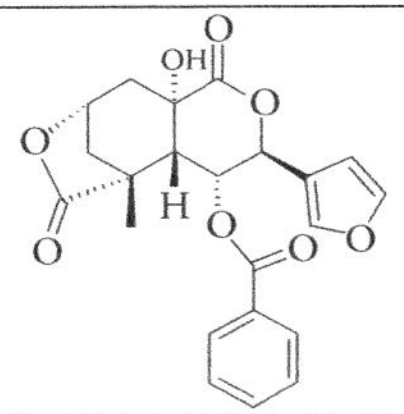	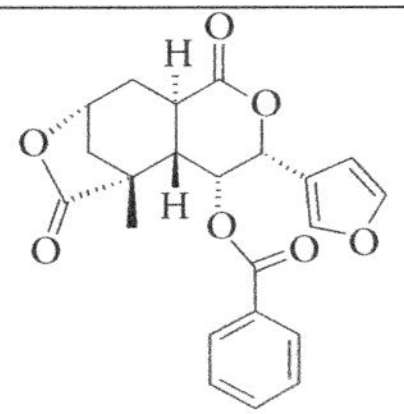	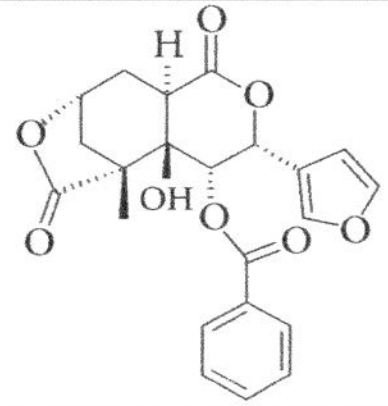	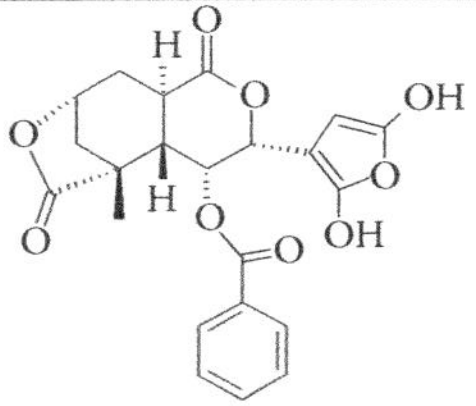
9-hydroxycollybolide **(12i)** (*C. maculata*)	7-*epi*-collybolide **(12j)** (*C. maculata*)	5-hydroxycollybolide **(12k)** (*C. maculata*)	12, 13-dihydroxycollybolide **(12l)** (*C. maculata*)

FIGURE 5.4 Endogenous library of collybolides from *Collybia sp.*

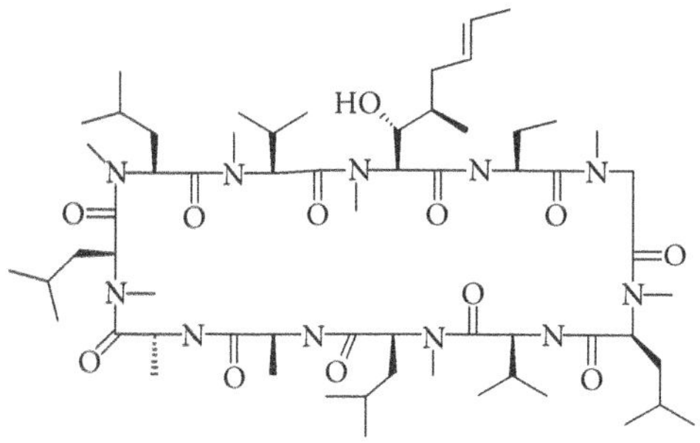

Cyclosporin A (**13**)

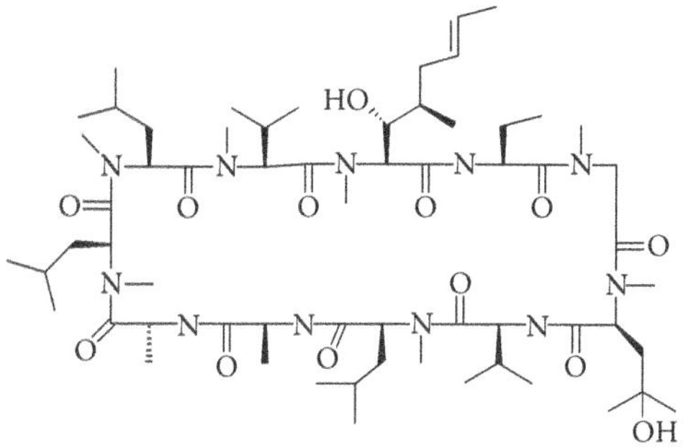

4-(γ-hydroxy)-cyclosporin A (**13a**)

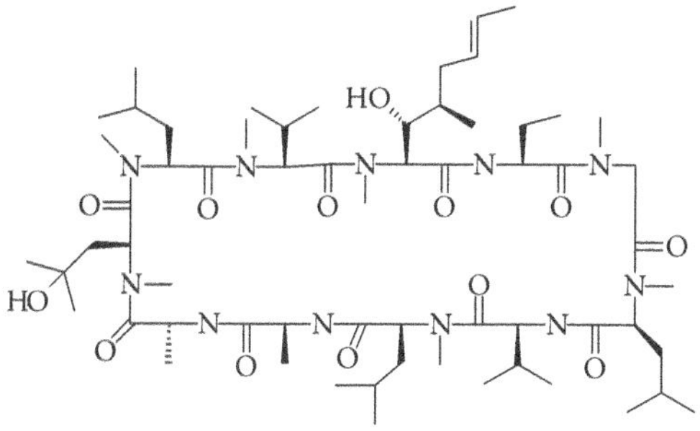

9-(γ-hydroxy)-cyclosporin A (**13b**)

FIGURE 5.5 Cyclosporin A (**13**) and two hydroxylated analogs (**13a** and **13b**), produced by biotransformation in *Nonomurea* sp.

(**13a**), as measured in the mouse mixed lymphocyte assay, was reduced 100-fold and the binding affinity to cyclophilin was increased by a factor of 2.

5.5 SYNTHETIC STRATEGIES FOR NATURAL PRODUCT-BASED LIBRARIES

The isolation of natural products from organisms generally yields compound libraries covering large areas in diversity space.[65,66] In some cases, sets of structurally related natural compounds are isolated (see Section 5.4) that are suitable for an initial analysis of SAR. Mostly, however, further structural variations and focused libraries are required for in-depth determination of SAR and lead optimization. The new paradigm for creating such small-molecule libraries by combinatorial and parallel synthesis has been successfully applied to natural product synthesis and enabled chemists to build arrays of derivatives based on a common natural product template.[67–69]

Compound libraries can generally be categorized by their origin (starting materials), by their synthetic strategy (synthesis or semisynthesis), or by their ultimate use (focused or generic libraries) (Figure 5.6). In the case of semisynthetic libraries (path A), the starting material is produced by biological sources such as plants or microorganisms and either used directly as a synthesis scaffold

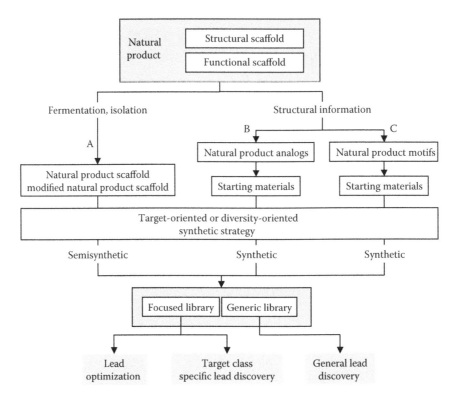

FIGURE 5.6 General classification of synthetic strategies and natural product-based libraries.

that is then decorated by substituents or it is synthetically modified in order to facilitate chemoselective transformations. The library concept can be target oriented if a certain SAR is integrated into the design or diversity oriented in the absence of any SAR information. Synthetic libraries based on natural products may be designed either as natural product analogs produced by a total synthesis approach (path B) or as a library integrating only certain motifs of a natural product into the final compounds (path C). Examples of path B are Nicolaou's epothilone[70,71] and Waldmann's dysidiolide[72] libraries. Path C is exemplified by Schreiber's approach. He generated diverse synthetic libraries based on natural-product-like precursors using complexity-generating reactions and diversity-oriented synthesis (DOS).[73–76] Another example of path C is Nicolaou's synthesis of large generic libraries based upon the benzopyran privileged structure, a core motif that shows especially high occurrence in known bioactive natural compounds.[77–79]

There are three main strategies and concepts for the transformations of natural compound-based scaffolds for library generation: scaffold decoration, core extension, and pharmacophore introduction (Figure 5.7). Scaffold decoration creates diversity by modifying positions around the scaffold (6-hydroxyhupercin A and analog, **14** and **14a**, Figure 5.7). A very large number of analogs (decoration library) is generated. Core extension of natural compound scaffolds (hydrogrammic acid and analog, **15** and **15a**, Figure 5.7) is a substantial synthetic process that leads to natural-compound-like templates suitable for chemical modifications. Introduction of pharmacophore groups or molecule parts (secologanin aglycon and analog, **16** and **16a**, Figure 5.7) aims at increasing the drug-likeness of the natural product scaffold.

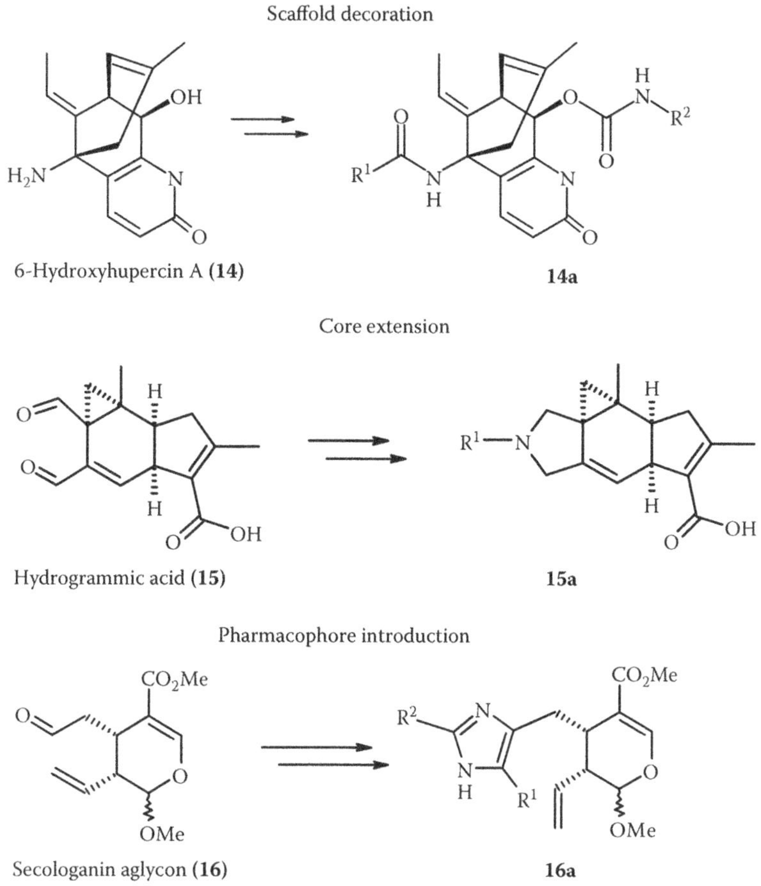

Scaffold decoration

6-Hydroxyhupercin A (14)   14a

Core extension

Hydrogrammic acid (15)   15a

Pharmacophore introduction

Secologanin aglycon (16)   16a

FIGURE 5.7 The synthetic strategies of scaffold decoration, core extension, and pharmacophore introduction.

5.6 EXAMPLES OF NATURAL PRODUCT-BASED LIBRARIES

5.6.1 BETULINIC ACID-BASED FUNCTIONAL LIBRARY

A functional library based on betulinic acid (17, Figure 5.8), a scaffold isolated from plants, is an example of using a synthetic strategy that involves semisynthetic modifications and DOS by scaffold decoration. Kashiwada et al. described the library design and evaluation of betulinic acid derivatives as anti-HIV (human immunodeficiency virus) compounds.[80]

5.6.1.1 Properties of Betulinic Acid

Betulinic acid (17) has been shown to exhibit a variety of biological activities, including anti-inflammatory[81] and antimalarial[82] activity. Recent work, however, focused on the anti-HIV and antitumor activity of betulinic acid (17). Betulinic acid (17) inhibits HIV replication in H9 lymphocyte cells,[83] blocks HIV-1 entry into cells,[84] and inhibits DNA polymerase β.[85] Synthetically modified derivatives of betulinic acid (17) were investigated as inhibitors of HIV-1 (see following text). In addition, betulinic acid (17) showed selective and pronounced cytotoxicity against cultured human melanoma cells (MEL-2).[86] Betulinic acid (17) and some closely related derivatives induce apoptosis (programmed cell death) in cancer cells.[87] Owing to the observed specificity for melanoma cells, betulinic acid (17) seems to be a promising anticancer agent.

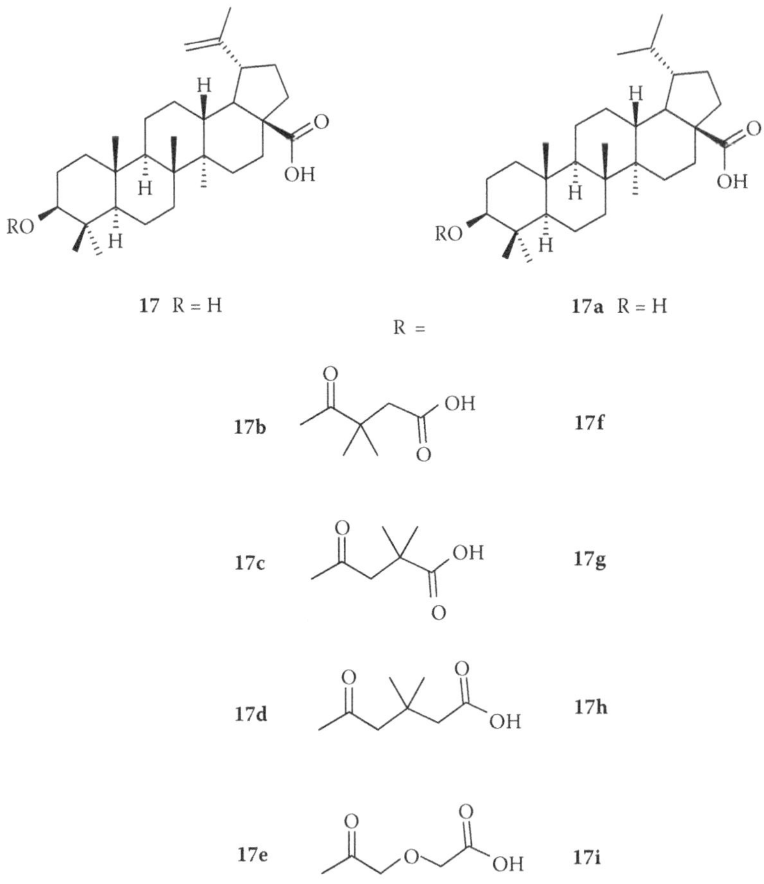

17 R = H   **17a** R = H

R =

17b   17f

17c   17g

17d   17h

17e   17i

FIGURE 5.8 Library based on the scaffold of betulinic acid (**17**).

5.6.1.2 Scaffold Production

Betulinic acid (**17**) is made available in sufficient quantities for library synthesis using a semisynthetic process starting from betulin, an abundant component of the bark of the white birch as described in Subsection 5.3.4.1.

5.6.1.3 Chemistry, Library Design, and Synthesis

The C3 hydroxy, C17 carboxylic acid, and C20 exomethylene groups in betulinic acid are easily modified (Figure 5.8). The modification focused on the introduction of an acyl group at the C3 position of betulinic acid (**17**) and dihydrobetulinic acid (**17a**), a modified scaffold prepared from betulinic acid. Betulinic acid (**17**) and dihydrobetulinic acid (**17a**) were treated with 3,3-dimethylglutaric anhydride or diglycolic anhydride in pyridine and in the presence of dimethylaminopyridine (DMAP) to furnish the corresponding 3-*O*-acyl derivatives **17d**, **17e**, **17h**, and **17i**. Similar treatment of betulinic acid (**17**) and dihydrobetulinic acid (**17a**) with dimethylsuccinic anhydride resulted in a mixture of 3-*O*-(2′,2-dimethylsuccinyl)- and 3-*O*-(3′,3′-dimethylsuccinyl) betulinic acid (**17b** and **17c**) and the corresponding dihydrobetulinic acids **17f** and **17g**.[88]

TABLE 5.2
Anti-HIV Activity of Betulinic Acid and
Dihydrobetulinic Acid Derivatives

Compound	IC_{50} (μM)[a]	EC_{50} (μM)[b]	Therapeutic Index[c]
17	13.0	1.4	9.3
17a	12.6	0.9	14
17b	15.9	2.7	5.9
17c	7.0	<0.00035	>20,000
17d	4.5	0.0023	1,974
17e	11.7	0.01	1,172
17f	7.7	0.56	13.8
17g	4.9	<0.00035	>14,000
17h	5.8	0.0057	1,017
17i	13.1	0.0056	2,344
AZT[d]	1875	0.15	12,500

[a] Concentration that is toxic to 50% of mock-infected H9 cells.
[b] Concentration that inhibits viral replication by 50%.
[c] Therapeutic index = IC_{50} value divided by EC_{50} value.
[d] AZT = azidothymidine.

Reprinted with permission from Kashiwada, Y. et al., *J. Med. Chem.*, 39, 1017, 1996. Copyright 2005 American Chemical Society.

5.6.1.4 Library Evaluation

To improve the comparatively weak inhibition of viral replication and the low therapeutic index (TI) of betulinic acid (**17**) and dihydrobetulinic acid (**17a**), the small set of hemiesters of low-molecular-weight dicarboxylic acids described earlier were prepared and evaluated for activity.[80] The anti-HIV assay indicated that 3-*O*-(3′,3-dimethylsuccinyl) betulinic acid (**17c**) and 3-*O*-(3′,3-dimethylsuccinyl) dihydrobetulinic acid (**17g**) were both extremely potent anti-HIV agents in acutely infected H9 lymphocytes with EC_{50} values of less than 0.35 nM. These compounds exhibited a remarkable TI of greater than 20,000 and 14,000, respectively. Some of the other library compounds (e.g., **17d**, **17h**, and **17i**) also exhibited potent anti-HIV activity in combination with a high TI (Table 5.2). In summary, the functional anti-HIV library based on the betulinic acid (**17**) scaffold provided improved lead compounds and valuable SAR information. The anti-HIV-1 activity of betulinic acid and dihydrobetulinic acid hemiesters increased in the following order: oxalylic < malonyl < succinyl <glutaryl hemiesters. With the exception of **17i**, dimethyl groups at the 3′ position of the succinyl and glutaryl part are required for maximal activity.

5.6.2 Fredericamycin A-Based Functional Library

Here we describe an in-house functional library based on fredericamycin A (**18**, Figure 5.9), a scaffold isolated from the fermentation broth of an actinomycete. The focused semisynthetic library used the DOS and scaffold decoration approaches.

5.6.2.1 Fredericamycin A Properties

Fredericamycin A (**18**) is a structurally unique[89] and potent antitumor antibiotic isolated from the fermentation broth of a *Streptomyces griseus* strain. It is active *in vitro* against fungi, Gram-positive bacteria, and tumor cell lines and shows *in-vivo* activity against P388 leukemia, CD8F mammary,

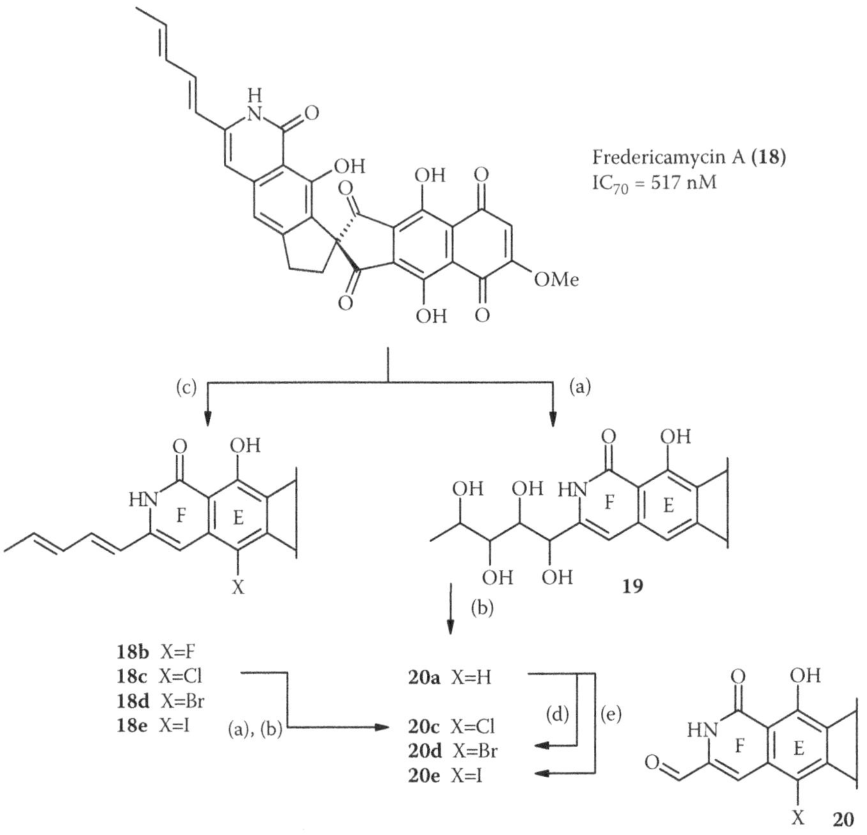

FIGURE 5.9 Fredericamycin E and F ring modifications.
(a) NMO, OsO$_4$, DCM, MeOH, H$_2$O, 87% from **18**, 75% from **18c**; (b) NaIO$_4$, DMF, H$_2$O, 100% yield for **20a**, 90% yield for **20c**; (c) for X = F: Selectfluor®, DMF, 14% yield for **18b**; for X = Cl: NCS, DMF, 30% yield for **18c**; for X = Br: NBS, DMF, 32% yield for **18d**; for X = I: NIS, DMF, 17% yield for **18e**; (d) Br$_2$, DMF, 83% yield; (e) ICl, DMF, 80%.

and DU-145 prostate tumor xenografts in mice.[90,91] Fredericamycin A (**18**) is a low micromolar inhibitor of both topoisomerase I and II and inhibits peptidyl-prolyl *cis-trans* isomerase PIN1 with a K$_i$ value of 0.82 μM.[92,91] Both the antitumor activity and the unique spiro structure of fredericamycin A (**18**) attracted our interest in its use as a scaffold for chemically diverse libraries.

5.6.2.2 Scaffold Production

Multigram quantities of pure fredericamycin A (**18**) were obtained using an optimized isolation and purification process from the fermentation broth of *Streptomyces griseus* strain mutants without any chromatographic steps (see Subsection 5.3.4.2).

5.6.2.3 Chemistry, Library Design, and Synthesis

The unprecedented architecture and biological activity of fredericamycin A (**18**) gave rise to numerous efforts toward the total synthesis of racemic[93–103] as well as of enantiopure (*S*)-frederi-camycin A (**18**).[104,105] Nevertheless, these synthesis pathways have rarely been used for the generation of fredericamycin A analogs or for SAR studies. We developed a semisynthetic strategy to prepare fredericamycin A derivatives with improved potency and tumor selectivity.[106–108] Because we had no initial SAR in hand, the design of the derivatives was mainly driven by chemical

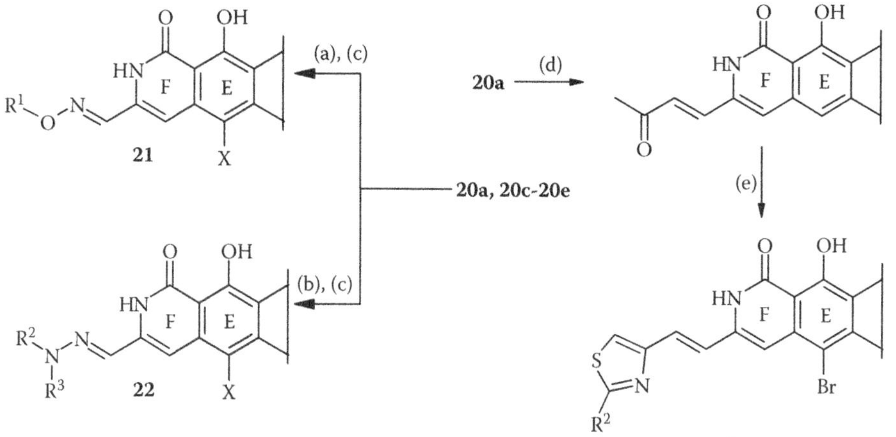

FIGURE 5.10 Side-chain functionalization of aldehydes **20**.
(a) $RONH_2$, TFA, DMF; (b) $R_2R_3NNH_2$, TFA, DMF; (c) polystyrene-WangCHO, polystyrene-PhSO$_2$NNH$_2$, 90–95% yield; (d) MeC(O)CH$_2$P(O)OEt$_2$, 1,1,3,3-tetramethylguanidine, 69% yield; (e) 1. Br$_2$, DMF, 87% yield and 2. R^{2}C(S)NH$_2$, 62% yield for **24**, 93% yield for **25**, 39% yield for **26**.

feasibility and with the overall goal of introducing polar and low-molecular-weight groups in order to increase polarity and solubility. Our initial E and F ring modifications began with degradation of the F-ring diene side chain in order to lower the molecular weight and gain a reactive functional group (Figure 5.9). Osmium tetroxide catalyzed *bis*-dihydroxylation of the pentadiene side chain of **18**, and this resulted in tetrahydroxylated fredericamycin A (**19**). The relative and absolute stereochemistry of the tetrol side chain could not clearly be assigned by ^{1}H NMR but both ^{1}H-NMR and LC-MS analysis indicated the existence of a single stereoisomer. This intermediate was subjected to diol cleavage, using sodium periodate to give aldehyde **20a**. Conversion of **18**, using either electrophilic fluorine donor Selectfluor® or halosuccinimides as the halogen source, resulted in E ring halogenated products **18b** to **18e**. A direct bromination or iodination of fredericamycin aldehyde **20a** used bromine or iodinechloride in DMF, respectively, to give aldehydes **20d** and **20e**. The E-ring chlorinated aldehyde **20c** was prepared by *bis*-dihydroxylation and subsequent diol cleavage of chlorofredericamycin **18c**.

Treatment of fredericamycin aldehydes **20** using a slight excess of *O*-alkylhydroxylamines or hydrazines under acidic conditions generally led to the formation of oximes **21** and hydrazones **22**, respectively (Figure 5.10). Clean conversions were observed in most cases, and unreacted reagent or aldehyde starting material were easily removed by sequestration using Wang-aldehyde polystyrene and sulfonylhydrazide polystyrene scavenger resin, an ideal prerequisite for parallel synthesis. Wittig–Horner olefination of aldehyde **20a** with diethyl 2-oxopropylphosphonate and excess tetramethylguanidine as a base gave F ring butenone **23**. When **23** was treated with bromine, bromination of the phenolic E ring as well as of the butenone side chain occurred. The intermediate α-bromoketone was further converted into thiazole-substituted bromofredericamycines **24** to **26** upon reaction with substituted or unsubstituted thioamides (Figure 5.10).

The A ring was modified by exchanging the methoxy group with alcohols **30** and **33** or amines **27** to **29**, **31**, and **32** (Figure 5.11). Conversion with amines proceeded smoothly at room temperature in DMF, whereas reaction with alcohols required the use of excess reagent, elevated temperatures, and prolonged reaction times. Testing of some initially synthesized examples of derivatives **21** and **22** for antiproliferative properties revealed IC$_{70}$ values in the submicromolar range. This finding triggered further SAR exploration by producing and screening a 115-member library

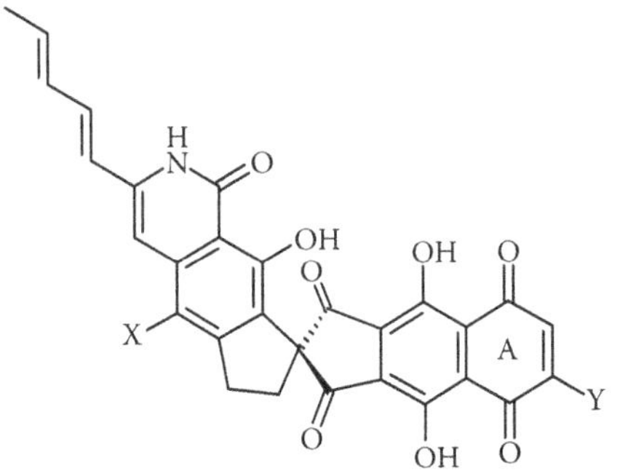

X=H
18 Y=OMe
(a) → **27** Y=morpholine
28 Y=NHi-Pr
29 Y=NMe_2
(b) → **30** Y=$O(CH_2)_2NMe_2$

X=Br
18d Y=OMe
(a) → **31** Y=NHc-Pr
32 Y=NHn-Pr
(b) → **33** Y=OEt

FIGURE 5.11 Nucleophilic replacement of A-ring methoxy group. (a) Amine, DMF r.t.; each 99% yield; (b) KOAc, ROH as solvent, 80°C, 36% yield for **30**, 71% yield for **33**.

(fredericamycin A Library A) consisting of fredericamycin A-oximes and fredericamycin A-hydrazones with optional halogenation on the phenolic E ring. As F ring substituents, a set of 23 acyl hydrazines, 7 hydrazines, and 9 *O*-alkylhydroxylamines was selected for the synthesis of Library A, based on commercial availability and molecular weight. These building blocks were reacted combinatorially with three fredericamycin-A-derived aldehydes, **20a**, **20c**, and **20d**, respectively. Library synthesis was performed on a 14-μmol per compound scale, using inexpensive standard 96-deep well plates and multichannel pipetting. Unreacted building blocks and reagents were scavenged using solid-phase reagents, which were then removed by filtration using a 96-well filter plate and a vacuum manifold. A complete LC-MS analysis of library members revealed a quantitative synthesis success rate with a wide range of yields (25 to 100%) and high product purities (85 to 100%). Without further purification, the reaction solutions were evaporated, weighed, redissolved in DMSO, and tested for their antiproliferative activity.

As a result of the observed Library A SAR, a 29-member follow-up library (fredericamycin A Library B) was designed using fredericamycin A aldehyde building blocks **20a**, **20c**, **20d**, **20e**, and *O*-alkyl hydroxylamines and acyl hydrazines. The diversity focused around substituents found in Library A exhibiting the highest activity. Not all hydroxylamines and hydrazines were combined with each of the four aldehydes for Library B, because some of the combinations were already included in Library A. Furthermore, a third and final set of 107 fredericamycin derivatives (fredericamycin A Library C) related to the analogs described in Libraries A and B was synthesized. The synthesis of Library C was based mainly on the procedures described for Libraries A and B.

5.6.2.4 Library Evaluation

The 3 libraries of 251 structurally diverse new fredericamycin derivatives were evaluated in antitumor assays. Antitumor properties of the libraries were tested in cell viability assays, using a modified propidium iodide assay with human tumor cell lines (Oncotest GmbH, Freiburg, Germany).[109] Primary screening was performed at 1 µg/ml against a 10-cell-line panel (lung: LCL H460, LXFL 529L, LXFA 629L; breast: MACL MCF7, MAXF 401NL; melanoma: MEXF 462NL,

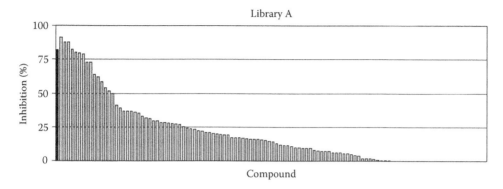

FIGURE 5.12 Potency (inhibition of cell viability) of the 115-member Library A of fredericamycin derivatives (gray bars) in comparison to fredericamycin (**18**, black bar).

MEXF 514L; renal: RXF 944L, RXF 486L; uterus: UXF 1138L) and delivered inhibition (%) of cell viability compared to controls in the absence of the test compound. Secondary screening was performed as a dose–response experiment from 0.1 to 1000 ng/ml against the 10-cell-line panel and 2 prostate cell lines, (PRCL PC3M, PRCL 22RV1) and provided IC_{70} values against individual cell lines, a mean IC_{70} across all cell lines, and information on tumor selectivity. Tumor selectivity was calculated as the number of cell lines that show an IC_{70} lower (i.e., activity higher) than one third of the mean IC_{70} as a percentage, allowing an estimate of general cell toxicity vs. tumor-cell-specific toxicity.

Primary screening of the 115-member initial Library A (E and F ring modifications) provided the potencies summarized in Figure 5.12. Seven of the derivatives exhibited a potency similar to fredericamycin A (**18**). SAR analysis demonstrated that six of the seven most active derivatives all carried an E ring bromo substitution as well as F ring oxime substituents. As seen in Table 5.3, secondary screening revealed benzyloxime **21a** as the most active derivative, having a mean IC_{70} value of 50 n*M* compared to 517 n*M* for fredericamycin A (**18**). However, the selectivity of **21a** was only slightly improved compared to fredericamycin A (**18**).

As seen in Figure 5.13, primary screening of the 29-member follow-up Library B (E and F ring modifications **21** and **22**, Figure 5.10) revealed that 21 of 29 compounds showed good activity compared to fredericamycin A (**18**). In the secondary assay, 8 of the most active compounds (**21b** to **21i**) showed IC_{70} values between 39 n*M* and 133 n*M* compared to 517 n*M* for fredericamycin A (**18**). Two compounds had improved tumor selectivity of 42% and 29% compared to 19% for fredericamycin A (**18**). Compound **21b** showed the highest potency and the best tumor selectivity of 39 n*M* and 42%, respectively (Table 5.3 and Figure 5.14).

A fredericamycin library consisting of 251 compounds (Library A, Library B, and the additional compounds of Library C) was synthesized. Of the most interesting 82 compounds after secondary screening, 50 compounds showed increased potency, 21 compounds showed increased tumor selectivity, and 17 compounds showed both increased potency and tumor selectivity (Figure 5.15). The most potent compound was bromofredericamycin **18d** (mean IC_{70} 8 n*M*, selectivity 26%), whereas the most selective compound was the oxime of fredericamycin **21b** (tumor selectivity 42%, mean IC_{70} 39 n*M*). These compounds revealed similar or better properties *in vitro* than some of the reference compounds (e.g., camptothecin 9 n*M* and 38%, adriamycin 39 n*M* and 38%; Table 5.3 and Figure 5.15).

In summary, starting from fermentation-derived fredericamycin A (**18**), a semisynthetic compound library with a total of 251 derivatives was produced and tested for antiproliferative activity against human tumor cell lines *in vitro*. Compounds with low nanomolar potency and increased tumor selectivity were identified. These results exemplify the power of semisynthetic derivatization as an optimization method of natural product drug leads.

TABLE 5.3
Potency and Tumor Selectivity of Fredericamycin A Derivatives

Compound	R^1	R^2	R^3	Inhibition at 1 μg/ml (%)[a]	Mean IC$_{70}$ (μM)[b]	Tumor Selectivity (%)[c]
Paclitaxel	—	—	—	n.d.	0.0007	65
Camptothecin	—	—	—	n.d.	0.0086	38
Adriamycin	—	—	—	n.d.	0.0387	38
18	CH$_3$CH=CH-CH=CH-	H	OCH$_3$	82	0.5171	19
18b	CH$_3$CH=CH-CH=CH-	F	OCH$_3$	91	0.0197	16
18c	CH$_3$CH=CH-CH=CH-	Cl	OCH$_3$	91	0.0139	16
18d	CH$_3$CH=CH-CH=CH-	Br	OCH$_3$	92	0.0081	26
18e	CH$_3$CH=CH-CH=CH-	I	OCH$_3$	93	0.0105	37
19	CH$_3$CHOHCHOHCHOH-	H	OCH$_3$	85	0.1860	16
20a	CHO	H	OCH$_3$	31	n.d.	n.d.
20c	CHO	Cl	OCH$_3$	25	n.d.	n.d.
20d	CHO	Br	OCH$_3$	42	n.d.	n.d.
20e	CHO	I	OCH$_3$	39	n.d.	n.d.
21a	PhCH$_2$ON=CH-	Br	OCH$_3$	91	0.0496	23
21b	(CH$_3$)$_2$CHON=CH-	H	OCH$_3$	87	0.0394	42
21c	PhCH$_2$ON=CH-	I	OCH$_3$	92	0.0710	16
21d	(CH$_3$)$_2$CHON=CH-	Br	OCH$_3$	91	0.0737	29
21e	(morpholine–ethyl–O–N=CH– structure)	I	OCH$_3$	90	0.1023	4
21f	CH$_3$(CH$_2$)$_5$ON=CH-	H	OCH$_3$	92	0.1199	17
21g	3-F-C$_6$H$_3$-CH$_2$ON=CH-	H	OCH$_3$	88	0.1217	13
21h	CH$_3$(CH$_2$)$_5$ON=CH-	I	OCH$_3$	92	0.1253	8
21i	4-F-C$_6$H$_3$-CH$_2$ON=CH-	I	OCH$_3$	91	0.1333	4
22a	(4-methylpiperazin-1-yl–N=CH– structure)	Br	OCH$_3$	73	0.6583	7
23	CH$_3$COCH=CH-	H	OCH$_3$	59	n.d.	n.d.
24	(2-methylthiazol-4-yl–CH=CH–CH$_3$ structure)	Br	OCH$_3$	85	0.1244	6
25	(2-aminothiazol-4-yl–CH=CH–CH$_3$ structure)	Br	OCH$_3$	77	0.6120	11
26	(2-acetamidothiazol-4-yl–CH=CH–CH$_3$ structure)	Br	OCH$_3$	83	0.5122	14

TABLE 5.3 (CONTINUED)
Potency and Tumor Selectivity of Fredericamycin A Derivatives

Compound	R^1	R^2	R^3	Inhibition at 1 μg/ml (%)[a]	Mean IC_{70} (μM)[b]	Tumor Selectivity (%)[c]
27	CH₃CH=CH-CH=CH-	H	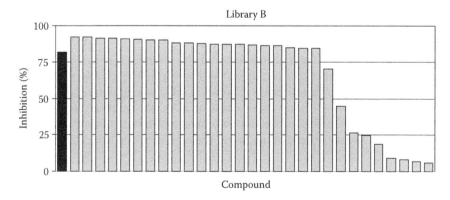	98	0.0774	21
28	CH₃CH=CH-CH=CH-	H	-NHCH(CH₃)₂	91	0.6336	29
29	CH₃CH=CH-CH=CH-	H	-N(CH₃)₂	96	0.3059	29
30	CH₃CH=CH-CH=CH-	H	-OCH₂CH₂N(CH₃)₂	94	1.0191	4
31	CH₃CH=CH-CH=CH-	Br	-NH-cyclopropyl	97	0.0964	26
32	CH₃CH=CH-CH=CH-	Br	-NHCH₂CH₂CH₂CH₃	94	0.3017	26
33	CH₃CH=CH-CH=CH-	Br	OCH₂CH₃	97	0.1597	15

Note: n.d.: Not determined.

[a] Potency: inhibition of cell viability (100-test/control* 100).

[b] Potency: mean IC_{70} of the cell panel.

[c] Tumor selectivity: percentage of cell lines where individual cell IC_{70} < 1/3 of the mean IC_{70} of the cell panel.

FIGURE 5.13 Potency (inhibition of cell viability) of the 29-member follow-up Library B of fredericamycin derivatives (gray bars) in comparison to fredericamycin A (**18**, black bar).

5.7 CONCLUDING REMARKS AND OUTLOOK

The rich structural diversity and complexity of natural products have inspired chemists and pharmacologists since the beginning of the pharmaceutical industry. Even today, many drugs are natural products or natural product derivatives. Natural scaffolds, meticulously assembled in the molecular

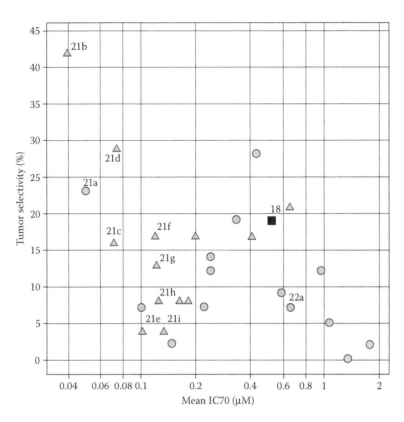

FIGURE 5.14 Tumor selectivity (%) and potency (mean IC_{70}) of selected fredericamycin derivatives of libraries A (grey circles) and B (grey triangles) in comparison to fredericamycin A (**18**, black square); labels are oriented top right of the corresponding symbol.

logic of biosynthetic pathways, will play an important role as the central design element in synthetic and semisynthetic ligands, which ultimately become developmental candidates and therapeutic agents to fulfill the unmet medical needs in disease management. The message from the dramatic decline in the number of new chemical entities in recent years should trigger a revival in the interest in unique natural product leads as starting scaffolds for the generation of focused and generic libraries for early-stage drug discovery. The semisynthetic approach offers exciting new possibilities to exploit the unique natural chemical diversity of low-molecular-weight libraries in the quest for new drugs. The DOS approach is an ancient method used by all organisms with pronounced secondary metabolism. But in contrast to the random synthetic approach, natural compounds are the result of a biological interaction and optimization between compounds and targets: chemical biology at its best.

REFERENCES

1. Feher, M. and Schmidt, J.M., Property distributions: differences between drugs, natural products, and molecules from combinatorial chemistry, *J. Chem. Inf. Comput. Sci.*, 43, 218, 2003.
2. Wan, Y. et al., Synthesis and target identification of hymenialdisine analogs, *Chem. Biol.*, 11, 247, 2004.
3. Kobayashi, J., Madono, T., and Shigemori, H., Nakijiquinones C and D, new sesquiterpenoid quinones with a hydroxy amino acid residue from a marine sponge inhibiting c-erbB-2 kinase, *Tetrahedron*, 51, 10867, 1995.
4. Kissau, L. et al., Development of natural product-derived receptor tyrosine kinase inhibitors based on conservation of protein domain fold, *J. Med. Chem.*, 46, 2917, 2003.

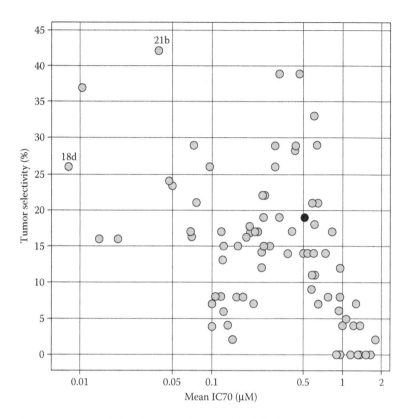

FIGURE 5.15 Tumor selectivity (%) and potency (mean IC_{70}) of the best 82 compounds of the 251-member fredericamycin libraries A, B, and C (gray circles) in comparison to fredericamycin A (**18**, black circle); labels are oriented top right of the corresponding symbol.

5. Arai, T., Takahashi, K., and Kubo, A., New antibiotics saframycins A, B, C, D, and E, *J. Antibiot. (Tokyo)*, 30, 1015, 1977.

6. Arai, T. and Kubo, A., *The Alkaloids*, Vol. 21, Brossi, A., Ed., Academic Press, New York, 1983, chap. 3.

7. Xing, C. et al., Identification of GAPDH as a protein target of the saframycin antiproliferative agents, *Proc. Natl. Acad. Sci. U.S.A.*, 101, 5862, 2004.

8. Myers, A.G. and Plowright, A.T., Synthesis and evaluation of bishydroquinone derivatives of (-)-saframycin A: identification of a versatile molecular template imparting potent antiproliferative activity, *J. Am. Chem. Soc.*, 123, 5114, 2001.

9. Pandey, R.C. et al., Fredericamycin A, a new antitumor antibiotic. I. Production, isolation and physicochemical properties, *J. Antibiot. (Tokyo)*, 34, 1389, 1981.

10. Kino, T. et al., FK-506, a novel immunosuppressant isolated from a Streptomyces. I. Fermentation, isolation, and physicochemical and biological characteristics, *J. Antibiot. (Tokyo)*, 40, 1249, 1987.

11. Abel, U. et al., Modern methods to produce natural product libraries, *Curr. Opin. Chem. Biol.*, 6, 453, 2002.

12. Shoichet, B.K., Virtual screening of chemical libraries, *Nature*, 432, 862, 2004.

13. Schapira, M. et al., Discovery of diverse thyroid hormone receptor antagonists by high-throughput docking, *Proc. Natl. Acad. Sci. U.S.A.*, 100, 7354, 2003.

14. Liu, H. et al., Structure-based discovery of potassium channel blockers from natural products: virtual screening and electrophysiological assay testing, *Chem. Biol.*, 10, 1103, 2003.

15. Puri, A. et al., Immunostimulant agents from Andrographis paniculata, *J. Nat. Prod.*, 56, 995, 1993.

16. Anchel, M., Hervey, A., and Robbins, W.J., Antibiotic substances from Basidiomycetes. VII. Clitocybe illudens, *Proc. Natl. Acad. Sci. U.S.A.*, 36, 300, 1950.

17. Zähner, H., Drautz, H., and Weber, W., Novel approaches to metabolite screening, in *Bioactive Microbial Products: Search and Discovery*, Bu'Lock, J.D. et al., Eds., Academic Press, London, 1982, p. 51.

18. Grabley, S., Thiericke, R., and Zeeck, A., The chemical screening approach, in *Drug Discovery from Nature*. Grabley, S. and Thiericke, R., Eds., Springer-Verlag, Heidelberg, 1999, p. 124.

19. Hu, J.F. et al., Jenamidines A to C: Unusual alkaloids from *Streptomyces sp.* with specific antiproliferative properties obtained by chemical screening, *J. Antibiot. (Tokyo)*, 56, 747, 2003.

20. Grabley, S. et al., Secondary metabolites by chemical screening. 21. Clonostachydiol, a novel anthelmintic macrodiolide from the fungus Clonostachys cylindrospora (strain FH-A 6607), *J. Antibiot. (Tokyo)*, 46, 343, 1993.

21. Hammann, P., Kretzschmar, G., and Seibert, G., Secondary metabolites by chemical screening. 7. I. Elaiophylin derivatives and their biological activities, *J. Antibiot. (Tokyo)*, 43, 1431, 1990.

22. Maier, A. et al., Biomolecular-chemical screening: a novel screening approach for the discovery of biologically active secondary metabolites. I. Screening strategy and validation, *J. Antibiot. (Tokyo)*, 52, 945, 1999.

23. Maier, A. et al., Biomolecular-chemical screening: a novel screening approach for the discovery of biologically active secondary metabolites. II. Application studies with pure metabolites, *J. Antibiot. (Tokyo)*, 52, 952, 1999.

24. Maul, C. et al., Biomolecular-chemical screening: a novel screening approach for the discovery of biologically active secondary metabolites. III. New DNA-binding metabolites, *J. Antibiot. (Tokyo)*, 52, 1124, 1999.

25. Smith, A.B., III et al., Evolution of a gram-scale synthesis of (+)-discodermolide, *J. Am. Chem. Soc.*, 122, 8654, 2000.

26. (a) Mickel, S.J. et al., Large-scale synthesis of the anti-cancer marine natural product (+)-discodermolide. Part 1: synthetic strategy and preparation of a common precursor, *Org. Proc. Res. Dev.*, 8, 92, 2004. (b) Mickel, S.J. et al., Large-scale synthesis of the anti-cancer marine natural product (+)-discodermolide. Part 2: Synthesis of fragments C1-6 and C9-14, *Org. Proc. Res. Dev.*, 8, 101, 2004. (c) Mickel, S.J. et al., Large-scale synthesis of the anti-cancer marine natural product (+)-discodermolide. Part 3: Synthesis of fragment C15-21, *Org. Proc. Res. Dev.*, 8, 107, 2004. (d) Mickel, S.J. et al., Large-scale synthesis of the anti-cancer marine natural product (+)-discodermolide. Part 4: Preparation of fragment C7-24, *Org. Proc. Res. Dev.*, 8, 113, 2004. (e) Mickel, S.J. et al., Large-scale synthesis of the anti-cancer marine natural product (+)-discodermolide. Part 5: Linkage of fragments C1-6 and C7-24 and finale, *Org. Proc. Res. Dev.*, 8, 122, 2004.

27. Tormo, J.R. et al., A method for the selection of production media for Actinomycete strains based on their metabolite HPLC profiles, *J. Ind. Microbiol. Biotechnol.*, 30, 582, 2003.

28. Weist, S. and Süssmuth, R.D., Mutational biosynthesis — a tool for the generation of structural diversity in the biosynthesis of antibiotics, *Appl. Microbiol. Biotechnol.*, epub, 2005.

29. Pezzuto, J.M. and Kim, D.S.H.L., Int. Pat. Appl., WO 9843936A2, 1998.

30. Sontag, B., Int. Pat. Appl., WO2004024696A1.

31. Berger, J., Jampolsky, L.M., and Goldberg, M.W., Borrelidin, a new antibiotic with anti-Borrelia activity and penicillin enhancement properties, *Arch. Biochem.*, 22, 476, 1949.

32. Tsuchiya, E. et al., Borrelidin inhibits a cyclin-dependent kinase (CDK), Cdc28/Cln2, of Saccharomyces cerevisiae, *J. Antibiot. (Tokyo)*, 54, 84, 2001.

33. Gantt, J.S., Bennett, C.A., and Arfin, S.M., Increased levels of threonyl-tRNA synthetase in a borrelidin-resistant Chinese hamster ovary cell line, *Proc. Natl. Acad. Sci. U.S.A.*, 78, 5367, 1981.

34. Lumb, M. et al., Isolation of vivomycin and borrelidin, two antibiotics with anti-viral activity, from a species of Streptomyces (C2989), *Nature*, 206, 263, 1965.

35. Buck, M., Farr, A.C., and Schnitzer, R.J., The anti-Borrelia effect of borrelidin, *Trans. N. Y. Acad. Sci.*, Ser. 11, 207, 1949.

36. Otoguro, K. et al., In vitro and in vivo antimalarial activities of a non-glycosidic 18-membered macrolide antibiotic, borrelidin, against drug-resistant strains of Plasmodia, *J. Antibiot. (Tokyo)*, 56, 727, 2003.

37. Wakabayashi, T. et al., Borrelidin is an angiogenesis inhibitor; disruption of angiogenic capillary vessels in a rat aorta matrix culture model, *J. Antibiot. (Tokyo)*, 50, 671, 1997.

38. Funahashi, Y. et al., Establishment of a quantitative mouse dorsal air sac model and its application to evaluate a new angiogenesis inhibitor, *Oncol. Res.*, 11, 319, 1999.
39. Omura, S. et al., Irumamycin, a new antibiotic active against phytopathogenic fungi, *J. Antibiot. (Tokyo)*, 35, 256, 1982.
40. Brufani, M. et al., Concerning venturicidin B, botrycidin and the sugar components of venturicidins A and B., *Helv. Chim. Acta*, 51, 1293, 1968.
41. Akita, H. et al., Identity of aabomycin A with venturicidins, *Agric. Biol. Chem.*, 54, 2465, 1990.
42. Stromgaard, K. and Nakanishi, K., Chemistry and biology of terpene trilactones from *Ginkgo biloba*, *Angew. Chem., Int. Ed. Engl.*, 43, 1640, 2004.
43. Furukawa, S., *Sci. Pap. Inst. Phys. Chem. Res. (JPN)*, 19, 27, 1932.
44. Maruyama, M. et al., The ginkgolides. III. The structure of the ginkgolides, *Tetrahedron Lett.*, 8, 309, 1967.
45. Braquet, P., The Ginkgolides: Potent platelet activating factor antagonists isolated from *Ginkgo biloba* L: chemistry, pharmacology, and clinical applications, *Drugs of the Future*, 12, 642, 1987.
46. Betz, H. et al., Structure and functions of inhibitory and excitatory glycine receptors, *Ann. N. Y. Acad. Sci.*, 868, 667, 1999.
47. Gavish, M. et al., Enigma of the peripheral benzodiazepine receptor, *Pharmacol. Rev.*, 51, 629, 1999.
48. Braquet, P.G. et al., BN 52021 and related compounds: a new series of highly specific PAF-acether receptor antagonists isolated from *Ginkgo biloba* L., *Blood Vessels*, 16, 558, 1985.
49. Pascard-Billy, C., Structure cristalline et moléculaire de l'isocollybolide, *Acta Crystl.* B28, 331, 1972.
50. Bui, A.M. et al., Isolement et analyse structurale du collybolide, nouveau sesquiterpene extrait de *Collybia maculata* alb. et sch. ex fries (basidiomycetes), *Tetrahedron*, 30, 1327, 1974.
51. Gerlitz, M., German patent application filed.
52. Castronovo, F. et al., Fungal metabolites, Part 45: the sesquiterpenes of Collybia maculata and Collybia peronata, *Tetrahedron,* 57, 2791, 2001.
53. Leuenberger, H.G.W., Interrelations of chemistry and biotechnology — I. Biotransformation — a useful tool in organic chemistry, *Pure Appl. Chem.*, 62, 753, 1990.
54. Breuer, M. et al., Industrial methods for the production of optically active intermediates, *Angew. Chem., Int. Ed. Engl.*, 43, 788, 2004.
55. Mahato, S.B. and Garai, S., Advances in microbial steroid biotransformation, *Steroids*, 62, 332, 1997.
56. Chen, T.S. et al., Microbial transformation of immunosuppressive compounds. I. Desmethylation of FK506 and immunomycin (FR 900520) by *Actinoplanes sp.* ATCC 53771, *J. Antibiot. (Tokyo)*, 45, 118, 1992.
57. Chen, T.S. et al., Microbial transformation of immunosuppressive compounds. II. Specific desmethylation of 13-methoxy group of FK 506 and FR 900520 by an unidentified Actinomycete ATCC 53828 [corrected], *J. Antibiot. (Tokyo)*, 45, 577, 1992.
58. Petuch, B.R. et al., Microbial transformation of immunosuppressive compounds. III. Glucosylation of immunomycin (FR 900520) and FK 506 by Bacillus subtilis ATCC 55060, *J. Ind. Microbiol.*, 13, 131, 1994.
59. Chen, T.S. et al., Microbial transformation of immunosuppressive compounds. IV. Hydroxylation and hemiketal formation of ascomycin (immunomycin) by *Streptomyces sp.* MA 6970 (ATCC No. 55281), *J. Antibiot. (Tokyo)*, 47, 1557, 1994.
60. Schulman, M. et al., Microbial conversion of avermectins by *Saccharopolyspora erythrea*: hydroxylation at C28, *J. Antibiot. (Tokyo)*, 46, 1016, 1993.
61. Nishida, H. et al., Generation of novel rapamycin structures by microbial manipulations, *J. Antibiot. (Tokyo)*, 48, 657, 1995.
62. Pacey, M.S. et al., Biotransformation of selamectin with *Streptomyces lydicus* SX-1298 using a novel static agar fermentation system with Reemay mesh, *J. Antibiot. (Tokyo)*, 54, 448, 2001.
63. Kuhnt, M. et al., Microbial biotransformation products of cyclosporin A, *J. Antibiot. (Tokyo)*, 49, 781, 1996.
64. Lensmeyer, G.L., Wiebe, D.A., and Carlson, I.H., Identification and analysis of nine metabolites of cyclosporine in whole blood by liquid chromatography. 1: Purification of analytical standards and optimization of the assay, *Clin. Chem.*, 33, 1841, 1987.

65. Lee, M.L. and Schneider, G., Scaffold architecture and pharmacophoric properties of natural products and trade drugs: application in the design of natural product-based combinatorial libraries, *J. Comb. Chem.*, 3, 284, 2001.

66. Henkel, T. et al., Statistical investigation into the structural complementarity of natural products and synthetic compounds, *Angew. Chem., Int. Ed. Engl.*, 38, 643, 1999.

67. Hall, D.G., Manku, S., and Wang, F., Solution- and solid-phase strategies for the design, synthesis, and screening of libraries based on natural product templates: a comprehensive survey, *J. Comb. Chem.*, 3, 125, 2001.

68. Wessjohann, L.A., Synthesis of natural product-based compound libraries, *Curr. Opin. Chem. Biol.*, 4, 303, 2000.

69. Dolle, R.E., Comprehensive survey of combinatorial library synthesis: 2003, *J. Comb. Chem.*, 6, 623, 2004.

70. Nicolaou, K.C. et al., Synthesis of epothilones A and B in solid and solution phase, *Nature*, 387, 268, 1997.

71. Nicolaou, K.C. et al., Designed epothilones: combinatorial synthesis, tubulin assembly properties, and cytotoxic action against taxol-resistant tumor cells, *Angew. Chem., Int. Ed. Engl.*, 36, 2097, 1997.

72. Brohm, D. et al., Natural products are biologically validated starting points in structural space for compound library development: solid-phase synthesis of dysidiolide-derived phosphatase inhibitors, *Angew. Chem., Int. Ed. Engl.*, 41, 307, 2002.

73. Burke, M.D. and Schreiber, S.L., A planning strategy for diversity-oriented synthesis, *Angew. Chem., Int. Ed. Engl.*, 43, 46, 2004.

74. Lee, D., Sello, J.K., and Schreiber, S.L., Pairwise use of complexity-generating reactions in diversity-oriented organic synthesis, *Org. Lett.*, 2, 709, 2000.

75. Chen, C. et al., Convergent diversity-oriented synthesis of small-molecule hybrids, *Angew. Chem., Int. Ed. Engl.*, 117, 2289, 2005.

76. Schreiber, S.L., Target-oriented and diversity-oriented organic synthesis in drug discovery, *Science*, 287, 1964, 2000.

77. Nicolaou, K.C. et al., Natural product-like combinatorial libraries based on privileged structures. 1. General principles and solid-phase synthesis of benzopyrans, *J. Am. Chem. Soc.*, 122, 9939, 2000.

78. Nicolaou, K.C. et al., Natural product-like combinatorial libraries based on privileged structures. 2. Construction of a 10,000-membered benzopyran library by directed split-and-pool chemistry using NanoKans and optical encoding, *J. Am. Chem. Soc.*, 122, 9954, 2000.

79. Nicolaou, K.C. et al., Natural product-like combinatorial libraries based on privileged structures. 3. The "libraries from libraries" principle for diversity enhancement of benzopyran libraries, *J. Am. Chem. Soc.*, 122, 9968, 2000.

80. Kashiwada, Y. et al., Betulinic acid and dihydrobetulinic acid derivatives as potent anti-HIV agents, *J. Med. Chem.*, 39, 1016, 1996.

81. Yogeeswari, P. and Sriram, D., Betulinic acid and its derivatives: a review on their biological properties, *Curr. Med. Chem.*, 12, 657, 2005.

82. Bringmann, G. et al., Betulinic acid: isolation from *Triphyophyllum peltatum* and *Ancistrocladus heyneanus*, antimalarial activity, and crystal structure of the benzyl ester, *Planta Med.*, 63, 255, 1997.

83. Fujioka, T. et al., Anti-AIDS agents, 11. Betulinic acid and platanic acid as anti-HIV principles from *Syzigium claviflorum*, and the anti-HIV activity of structurally related triterpenoids, *J. Nat. Prod.*, 57, 243, 1994.

84. Mayaux, J.F. et al., Triterpene derivatives that block entry of human immunodeficiency virus type 1 into cells, *Proc. Natl. Acad. Sci. U S A*, 91, 3564, 1994.

85. Ma, J., Starck, S.R., and Hecht, S.M., DNA polymerase beta inhibitors from *Tetracera boiviniana*, *J. Nat. Prod.*, 62, 1660, 1999.

86. Pisha, E. et al., Discovery of betulinic acid as a selective inhibitor of human melanoma that functions by induction of apoptosis, *Nat. Med.*, 1, 1046, 1995.

87. Fulda, S. et al., Betulinic acid triggers CD95 (APO-1/Fas)- and p53-independent apoptosis via activation of caspases in neuroectodermal tumors, *Cancer Res.*, 57, 4956, 1997.

88. Lee, K.-H. et al., Int. Pat. Appl., WO 9639033A1, 1996.

89. Misra, R. and Pandey, R.C., Fredericamycin A, an antitumor antibiotic of a novel skeletal type, *J. Am. Chem. Soc.*, 104, 4478, 1982.

90. Warnick-Pickle, D.J. et al., Fredericamycin A, a new antitumor antibiotic. II. Biological properties, *J. Antibiot. (Tokyo)*, 34, 1402, 1981.

91. Lu, K.P. and Fischer, G., Int. Pat. Appl., WO 2004002429A2.

92. Latham, M.D. et al., Inhibition of topoisomerases by fredericamycin A, *Cancer. Chemother. Pharmacol.*, 24, 167, 1989.

93. Kelly, T.R. et al., Synthesis of (±)-fredericamycin A, *J. Am. Chem. Soc.*, 108, 7100, 1986.

94. Kelly, T.R. et al., Synthesis of (±)-fredericamycin A, *J. Am. Chem. Soc.*, 110, 6471, 1988.

95. Clive, D.L.J. et al., Total synthesis of crystalline (±)-fredericamycin A. Use of radical spirocyclization, *J. Chem. Soc., Chem. Commun.*, 1489, 1992.

96. Clive, D.L.J. et al., Total synthesis of crystalline (±)-fredericamycin A, Use of radical spirocyclization, *J. Am. Chem. Soc.*, 116, 11275, 1994.

97. Rama Rao, A.V. et al., Synthesis of (±) fredericamycin A, *Tetrahedron Lett.*, 34, 2665, 1993.

98. Rama Rao, A.V. et al., Synthesis of (±)-fredericamycin A, *Heterocycles*, 37, 1893, 1994.

99. Saint-Jalmes, L. et al., *Bull. Soc. Chim. Fr.*, 130, 447, 1993.

100. Wendt, J.A., Gauvreau, P.J., and Bach, R.D., Synthesis of (±)-fredericamycin A, *J. Am. Chem. Soc.*, 116, 9921, 1994.

101. Boger, D.L. et al., Total synthesis of natural and ent-fredericamycin A, *J. Am. Chem. Soc.*, 117, 11839, 1995.

102. Boger, D.L., Azadiene Diels-Alder reactions: scope and applications. Total synthesis of natural and ent-fredericamycin A, *J. Heterocyc. Chem.*, 33, 1510, 1996.

103. Kita, Y. et al., Total synthesis of the antitumor antibiotic (±)-fredericamycin A by a linear approach, *Chem. Eur. J.*, 6, 3897, 2000.

104. Kita, Y. et al., Asymmetric total synthesis of fredericamycin A, *Angew. Chem., Int. Ed. Engl.*, 38, 683, 1999.

105. Kita, Y. et al., Enantioselective total synthesis of a potent antitumor antibiotic, fredericamycin A, *J. Am. Chem. Soc.*, 123, 3214, 2001.

106. Simon, W. and Abel, U., Int. Pat. Appl., WO2004004713A1.

107. Simon, W. and Abel, U., Int. Pat. Appl., WO03087060A1.

108. Abel, U. and Simon, W., Int. Pat. Appl., WO03080582A2.

109. Dengler, W.A. et al., Development of a propidium iodide fluorescence assay for proliferation and cytotoxicity assays, *Anticancer Drugs*, 6, 522, 1995.

6 The Use of Polymer-Supported Reagents and Scavengers in the Synthesis of Natural Products

Steven V. Ley, Ian R. Baxendale, and Rebecca M. Myers

CONTENTS

6.1 INTRODUCTION

The laboratory preparation of complex natural products is a daunting task. Yet despite its many challenges, architectures of increasing complexity are succumbing to the skills of the synthetic chemist. Synthetic chemistry can provide natural products in quantities far exceeding nature's supply as well as labeled materials, structural analogs, novel scaffolds, and mimetics. These all enhance our knowledge of the natural world. Although truly impressive in terms of delivering target molecules, the power of modern-day synthesis has a long way to go. This is especially true when it comes to preparing compounds while minimizing the generation of waste by-products and spent reagents. Indeed, we are nowhere near the levels of atom efficiency necessary for a sustainable future. Complex synthesis remains a challenging occupation requiring an exceptional level of experimental skill, extensive knowledge of both mechanistic and molecular reactivity, and a bold, inventive, and creative spirit. It is the combination of these qualities that transforms the synthesis process from one of simple logistics to an art form.

The world is ever hungry for new compounds, particularly those displaying important biological or material properties. Consequently, the speed with which we synthesis chemists are able to generate new compounds has become a significant factor in the discovery process. However, the desire for large numbers of compounds often restricts the choice of synthetic methods; this is unacceptable if real structural diversity is the goal. Considerable improvements in reagents and new reaction types are urgently needed to effect new and strategically important chemical transformations. We must develop new generations of catalysts, recyclable reagents, and purification procedures. We must also employ less solvent, use cleaner chemistries, utilize lower toxicity processes and, in general, through increased levels of automation, maximize efficiency at all levels. Modern-day natural product synthesis should be aware of these issues and strive to advance the science of molecular assembly. After all, it is the strategies, concepts, and new reagents that have a more enduring impact than any individual molecule synthesis. Having said this, new healing drugs based on the remarkable biological properties of certain natural products will undoubtedly benefit mankind for many years to come.

In this chapter, we only concentrate on one aspect of natural product synthesis development, namely, the use of polymer-supported reagents,[1] scavengers, and various catch-and-release techniques to prepare pure natural products. The aim of this approach is to achieve complex molecule synthesis in a multistep mode without using conventional workup procedures such as extensive use of silica gel chromatography, distillation, or crystallization.[2] This alternative method also encourages the greater use of robotics owing to the compatibility of the solid reagents with standard automated and repetitive procedures (e.g., filtration and washing cycles). In addition, reaction optimization can be more easily facilitated using intelligent feedback loops. These factors mean that processes can be adapted quite readily to flow reactor devices. When applied to natural product synthesis, these methods not only furnish specific target molecules more rapidly but can also, after appropriate functional group manipulation, generate many related structural scaffolds.[3] By splitting intermediate batches, large collections of new compounds can be obtained on a scale useful for biological studies, thereby enabling extensive investigation of structure–activity profiles.[4]

Immobilizing a reagent onto a polymer support allows for spent reagents to be easily removed by filtration, regenerated, and subsequently recycled. This is undoubtedly attractive where expensive ligands or catalysts are used in the synthesis process. Another desirable feature becomes apparent in situations in which highly toxic or obnoxious materials have to be used. Immobilization of otherwise unpleasant materials makes them largely benign and thus easier and safer to handle. Because the process of supporting reagents in this way makes them site-isolated, it then becomes possible to conduct more than one transformation in a single reaction vessel with combinations of otherwise mutually incompatible reagents. Separation problems such as complex mixtures or co-running materials can be resolved by the judicious use of scavenging and catch-and-release protocols. In this latter case, polymer-supported species can be designed to react specifically with the desired synthesis product (the capturing process). Then, by filtration and washing, the final pure material can be displaced from the support (the releasing process).

A significant advantage compared to conventional on-bead synthesis is that reactions can be monitored in real time by standard solution-phase techniques at all stages during the synthesis process. By the application of design of experiments (DoE) software, or by employing other parallel reaction techniques, one can also optimize the reaction schemes rapidly.[5] It is foreseeable that the use of informatic feedback mechanisms will allow self-optimization of chemical reactions. Clearly, these techniques are not limited to long linear routes, but may also be applied to more efficient convergent routes.

Although the preceding features are strong endorsements for the use of polymer-supported reagents, scavengers, and catch-and-release methods in multistep synthesis of natural products, there are issues that need to be addressed. The first of these relates to cost. Although some supported reagents are cheap and have been used on the ton scale (especially where efficient recycling has been possible), others are currently expensive. When it comes to the preparation of certain natural products, the present cost of the reagents and scavengers is less important than the natural product itself. The limit of nature's supply of a given natural product and its related extraction costs can, in certain circumstances, be prohibitively expensive both in financial and environmental terms.

The other concern relating to the use of supported systems is that they are often, but not always, slower to react than their solution counterparts. This can be problematic, but can often be overcome by a number of methods. An obvious way to enhance reactivity is by improving the surface topology of the supported material. The use of focused microwaves, ultrasound, or novel solvents such as ionic liquids,[6] or indeed combinations of these can also be useful in increasing reactivity and reaction rates.[7]

6.2 THE POLYMER-SUPPORTED REAGENT APPROACH TO NATURAL PRODUCT SYNTHESIS

6.2.1 ALKALOIDS

6.2.1.1 Oxomaritidine and Epimaritidine

Supported reagents have been used in synthesis since 1946,[8] but their application in natural product synthesis has been very limited. Indeed, where they had been used, it was only to effect a specific transformation where conventional systems had previously failed. Our group, however, has sought to develop the application of immobilized systems to complex synthetic challenges in a multistep mode.[9] The first serious application of supported reagents for natural product synthesis was published in 1999.[10] Here, concise routes to two amaryllidaceae alkaloids, oxomaritidine (**1**) and epimaritidine (**2**), in just five and six steps, respectively, were reported (Scheme 6.1).

Supported reagents, featured in all of the steps, were used in a sequential fashion and led to pure products following a simple filtration to remove spent reagents. The route began with the quantitative oxidation of 3,4-dimethoxybenzyl alcohol (**3**) to the corresponding aldehyde **4** using polymer-supported perruthenate (PSP), a catalytic oxidant previously developed by our group.[11] Next, the reductive amination of aldehyde **4** was achieved, first by coupling with the phenolic amine **5** and then by reduction using a polymer-supported borohydride reagent[12] to furnish the corresponding secondary amine **6** under previously optimized conditions.[13] Amine **6** was protected as the trifluoroacetate (**7**) by treatment with trifluoroacetic anhydride and immobilized dimethylaminopyridine (PS-DMAP) as the catalyst and base. The resulting product **7** underwent smooth oxidative coupling to the spirodienone **8** using polymer-supported hypervalent iodine diacetate, a reagent developed specifically for the task.[14] In order to guarantee good conversion in this process, it was essential to use trifluoroethanol as the solvent. Finally, a wet carbonate ion-exchange resin simultaneously deprotected and initiated cyclization (via conjugate addition) to give the first of the natural products, oxomaritidine (**1**), in essentially quantitative yield after filtration and solvent evaporation. Subsequent stereoselective reduction of oxomaritidine (**1**), using an immobilized copper boride (or nickel boride) equivalent, delivered the second natural product epimaritidine (**2**) in an excellent 50% overall yield over the six-step sequence. The synthesis was scalable to deliver gram quantities of the products.

In order to extend the versatility of this route, a portion of oxomaritidine was progressed through a hydrogenation step to remove the double bond yielding **9**. Subsequent reductive amination provided a small series of amine analogs based on structure **10**. Obviously, other interesting intermediates such as spirodienone **8** could also be channeled into other combinatorial chemistry programs providing for further structural modification.

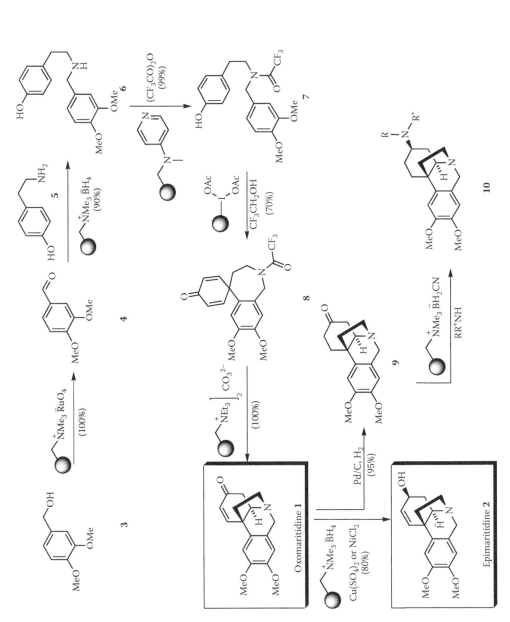

SCHEME 6.1 The first orchestrated and successive application of supported reagents in the synthesis of the amaryllidaceae alkaloids oxomaritidine and epimaritidine.

6.2.1.2 Norarmepavine

Using supported reagent and scavenger methods, we more recently reported the first stereoselective synthesis of the tetrahydroisoquinoline alkaloid (–)-norarmepavine (**11**) (Scheme 6.2).[15] The first step of this short synthesis employed the solvent-free focused microwave coupling of amine **12** with protected phenolic ester **13** to give the amide **14**. Amide **14** was taken on into the next step without further purification and was converted to the Bischler–Napieralski product **15** using a dehydrating reagent prepared from triflic anhydride and PS-DMAP. This was followed by scavenging with polymer-supported N-(2-aminoethyl)aminomethyl polystyrene to remove any excess anhydride. Using a polymer-supported phosphazene base (PS-BEMP), the intermediate triflate was converted to the free base **15**. Asymmetric reduction of **15** was achieved using a proline-derived reductant **16**. The resulting product was rapidly purified using a catch-and-release technique with polymer-supported sulfonic acid to effect the capture and a methanolic ammonia release step to give pure amine **17**. After a standard benzyl deprotection by hydrogenolysis using palladium hydroxide on carbon, the alkaloid (–)-norarmepavine (**11**) was obtained. This synthesis again nicely illustrates the power of these methods for natural product synthesis. Normal procedures would have required extensive chromatography to obtain pure materials at intermediate stages, but here only scavenging and solvent evaporation were necessary.

6.2.1.3 Epibatidine

The potent analgesic alkaloid epibatidine (**18**),[16] a compound isolated from the Ecuadorian poison frog *Epipedobates tricolour*, was a more challenging compound to synthesize using immobilization techniques (Scheme 6.3). This synthesis involved the orchestrated employment of ten polymer-supported reagents and scavengers to give epibatidine (**18**) in greater than 90% purity without any chromatographic steps. The synthesis began with the transformation of the commercially available acid chloride **19** to aldehyde **20**. This was achieved in a two-step process by reduction of the acid chloride to the intermediate alcohol with polymer-supported borohydride, then partial reoxidation by the PSP reagent to deliver aldehyde **20** in excellent yield. Alternative immobilized oxidants such as supported permanganate[17] and diacetoxyiodobenzene[14,18] were equally efficient. Oxidants such as Magtrieve[19] (magnetized CrO_2 and MnO_2), although adequate, were less suitable as they required considerably longer reaction times. Aldehyde **20** then underwent straightforward conversion to the nitrostyrene **21**. A Henry reaction promoted by the basic Amberlite resin (IRA 420 OH form) and followed by elimination of an intermediate trifluoroacetate using polymer-bound diethylamine as a base gave **21**. By NMR analysis at each stage, products were shown to be better than 95% pure. In an important modification to the preceding sequence of reactions, it was found that by the incorporation of supported reagents contained in sealed porous polymer pouches,[20] the conversion of chloride **19** to nitrostyrene **21** was possible in a one-pot operation. Thus, when an individual reaction was deemed to be complete (by thin-layer chromatography [TLC] and liquid chromatography-mass spectroscopy [LC-MS]) the pouch was simply removed, washed with solvent, and the next reagent pouch added to the flask. This process obviated the need for filtration between individual steps.

In the next phase of the synthesis, a regioselective Diels–Alder reaction of nitrostyrene **21** with 2-*tert*-butyldimethylsilyloxybutadiene in a sealed tube at 120°C and workup with a volatile acid (TFA) to hydrolyze the intermediate enol ether gave ketone **22** exclusively as the *trans*-substituted product. Stereoselective reduction of the carbonyl and corresponding mesylate formation gave **23**, again using an immobilized suite of reagents. Selective reduction of the nitro **23** to amine **24** in the presence of other sensitive functionalities (e.g., the 2-chloropyridine unit) was next achieved using polymer-supported nickel boride in excellent yield. Treatment of amine **24** with PS-BEMP initiated an intramolecular displacement of the mesylate which, following a scavenging step with a polystyrene aminomethyl resin to remove excess mesylate, yielded the natural product precursor

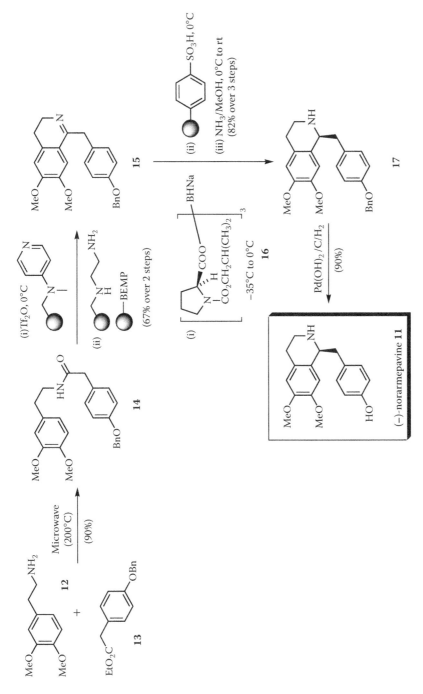

SCHEME 6.2 Enantioselective synthesis of the isoquinoline alkaloid (−)-norarmepavine using polymer-supported reagents.

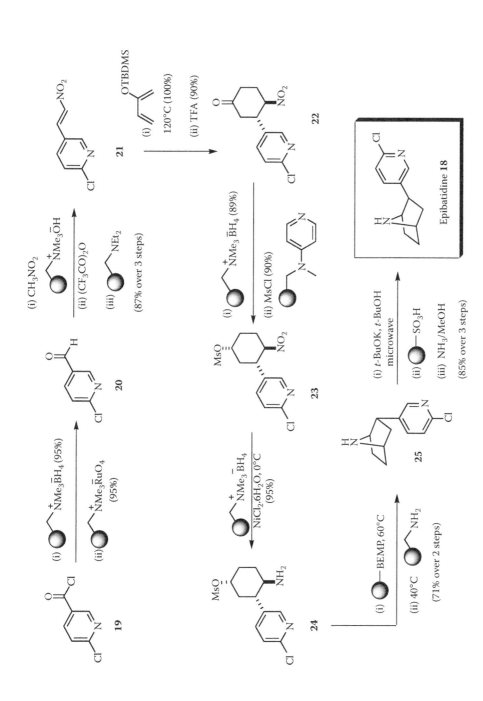

SCHEME 6.3 Synthesis of the potent analgesic compound (±)-epibatidine using an orchestrated multistep sequence of polymer-supported reagents.

25. Epimerization to epibatidine (**18**) was readily achieved by focused microwave heating in the presence of potassium *tert*-butoxide. The natural product was ultimately isolated by a catch-and-release technique as discussed earlier.

6.2.1.4 Nornicotine and Nicotine

Related to epibatidine are the alkaloids nornicotine (**25**) and nicotine (**26**). The pharmacological effects of these compounds in binding nicotinic acetylcholine receptors (nAChR) and their resulting improvements of cognitive function and neuroprotection are well documented.[21] Accordingly, they have become attractive targets for synthesis and the construction of derivative libraries. One such preparation of both nornicotine (**25**) and nicotine (**26**) exploited the use of immobilized reagents in a multistep sequence of reactions (Scheme 6.4).[22]

An interesting aspect of this work was the use of polymer-supported phosphine[23] and carbon tetrabromide to effect the conversion of alcohols to bromides. Furthermore, the use of an ion-exchange azide resin[24] minimized potential hazards in azide isolation, because they can be obtained *in situ* and directly hydrogenated to an innocuous amine product. The two catch-and-release steps utilized in the route were particularly convenient in the isolation of pure products. The conversion of nornicotine (**25**) to nicotine (**26**) used polymer-supported cyanoborohydride to bring about the reductive amination with formaldehyde. This work was described in a full paper in which detailed experimental conditions were provided to help those less familiar with the use of immobilized reagents. This project was developed further by generating a number of analogs. By reacting nornicotine (**25**) under a variety of conditions, ureas, sulfonamides, amides, and alkylated derivatives were made, again employing reagent immobilization concepts (Scheme 6.5).

6.2.1.5 2-Aryl-Substituted Pyrroles

Modification of the initial route also allowed the preparation of 2-aryl-substituted pyrroles. The aryl pyrrole motif occurs in various natural products such as in the broad-spectrum antibiotic pyrrolonitrin and in the insecticide dioxapyrrolomycin.[25] Therefore, new routes to these structural types are always important. For example, immobilized reagents and scavengers may conveniently be used to prepare a library of 3-aryl-substituted pyrroles as shown in Scheme 6.6.

6.2.1.6 Plicamine

Much of what has been discussed so far constitutes the preparation of relatively simple structures. However, these syntheses actually helped prepare the ground for more challenging targets such as the synthesis of the alkaloid (+)-plicamine (**27**) (Scheme 6.7).[26] Extensive use of parallel optimization methods and focused microwave techniques, to achieve fast reaction times, were used to complete the total synthesis in just 6 weeks without rehearsal of any reactions using conventional solution-phase methods or separation techniques. In this respect this synthesis stands out. Rather than discussing the details of all the steps involved in this route, as many follow the principles and concepts already established in this review, only the new aspects are highlighted. The reader is encouraged to read the full details of the work,[27] which describes how the route (in multigram quantities) can be modified to afford analogs or generate related structural scaffolds for further chemical decoration. Also of note is that because this synthesis relied on a single asymmetric center in the starting material to control all the others, by use of the opposite enantiomer, the unnatural (−)-plicamine enantiomer can also be prepared and examined for biological activity.

The first observation in the plicamine synthesis of note is that the polymer-supported hypervalent iodine reagent, mentioned earlier, again performed well to convert phenol **28** to spirodienone **29**. Nafion-H (fluorosulfonic acid resin) catalyzed the final cyclization of **29** to form the tricyclic core of the natural product in virtually quantitative yield. After stereo- and regioselective reduction of **30** using supported borohydride, the highly hindered intermediate alcohol was methylated by

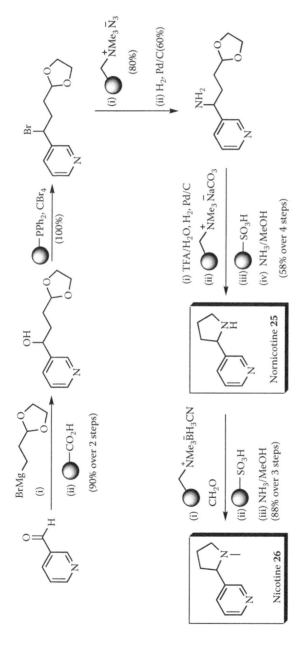

SCHEME 6.4 Multistep synthesis of nornicotine and nicotine.

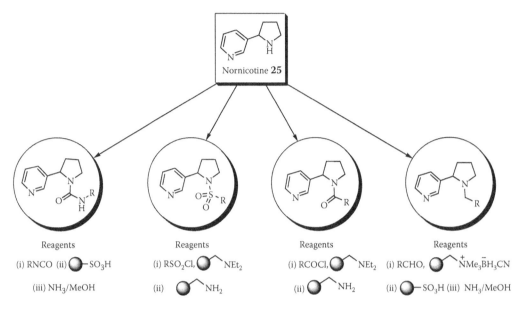

SCHEME 6.5 Combinatorial decoration of the nornicotine core.

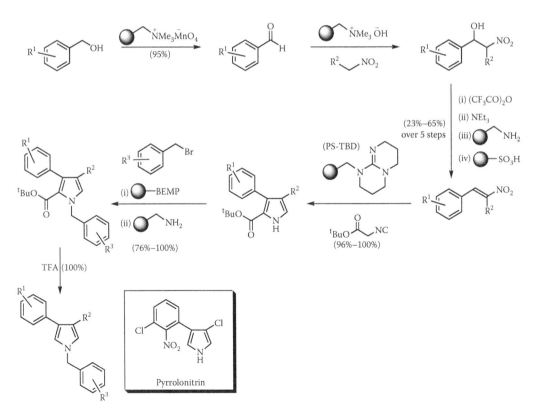

SCHEME 6.6 Preparation of a library of 3-aryl-substituted pyrroles.

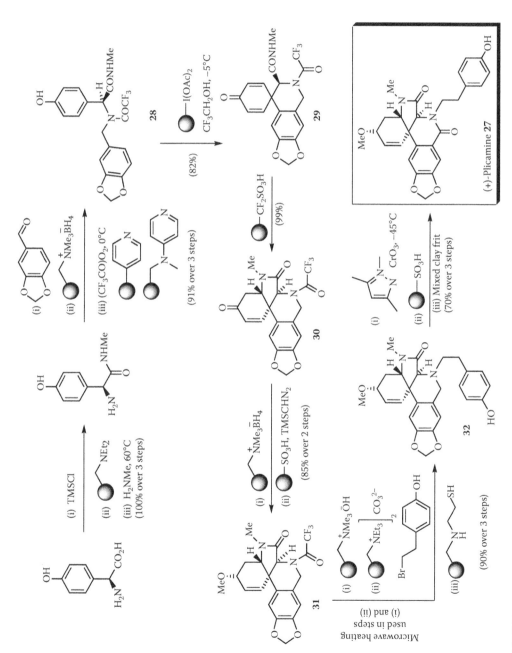

SCHEME 6.7 Synthesis of the alkaloid (+)-plicamine using polymer-supported reagents.

treatment with trimethylsilyl diazomethane and supported sulfonic acid resin to give **31**. This process is very mild and is recommended in difficult situations. The steps to **32** from **31** were relatively straightforward. However, the final oxidation of amine **32** to (+)-plicamine required significant development. This was eventually achieved using CrO_3 and 3,5-dimethylpyrazole[28] followed by scavenging with Amberlyst 15 resin. The chromium salts were efficiently removed by filtration through a mixed bed containing Varian Chem Elut CE 1005 and Montmorillonite K10 clay to give (+)-plicamine (**27**).

The route to plicamine has also been diverted at an earlier stage to afford two further natural products, plicane (**33**) and obliquine (**34**) (Scheme 6.8).[29] The ability to divert material in this way is attractive and was facilitated by using immobilized reagent methods. These more advanced syntheses clearly illustrate the opportunities created by embracing these methods for multistep transformations.

6.2.1.7 Trichostatin A

The hydroxamic acid natural product trichostatin A (**35**) is a potent inhibitor of histone deacetylase (HDAc) (Scheme 6.9).[30] An imbalance in histone acetylation is associated with various cancers, and there is also increasing evidence that acetylation and the corresponding deacetylation of proteins are important regulatory modifications in cellular events.[31] HDAc inhibitors cause a reversal of transcriptional repression and, consequently, upregulation of tumor suppressors. It is this behavior that has made them attractive drug candidates. In terms of synthesis, the hydroxamic functional group can cause problems owing to its propensity to strongly bind metal atoms that may have been used during its preparation. Therefore, heavy metal contamination of final products can be a serious concern as levels over ten parts per million are not acceptable in drug substances. In order to construct chemical libraries based on hydroxamic acids, careful attention to any metal contamination is required. The usually required extensive chromatography is clearly very time consuming during any compound library synthesis.

To overcome some of these problems, a five-step process was designed to allow the preparation of a focused array of compounds with three levels of structural diversity. Opportunity for diversification arose from the initial inputs and the later alkylation step to provide compounds related to trichostatin A (Scheme 6.9).[32] The first two steps of the synthesis followed well-established chemistry, namely sulfonylation of an excess of *p*-iodoaniline in the presence of PS-DMAP. The excess aniline being scavenged using a sulfonic acid in the normal way. The acidity of the resulting sulfonamide that meant it was readily alkylated using alkyliodides and the PS-BEMP as the base. The next reaction involved Heck coupling of the aryliodide with acrylic acid facilitated by palladium(0). It is this step in which solution-phase palladium-mediated cross coupling reactions are problematic and where product contamination normally occurs. However, after screening various catalysts, the polyurea microencapsulated palladium catalysts (Pd-EnCat®) that we had developed for other coupling reactions (e.g., Suzuki and Stille reactions)[33] performed exceptionally well to give the coupled products **36** with very low levels of palladium contamination after direct filtration to remove the spent encapsulated catalyst. The Heck reaction was optimized using a ReactArray SK233 automated reaction workstation to expedite the operation. Using a 2-(1*H*-benzotriazol-1-yl)-1,1,3,3-tetramethyl uranium hexafluoroantimonate derivative on polystyrene resin (PS-HBTU), the acid (**36**) was captured and simultaneously activated. After washing the resin to remove contaminants, a solution of *O*-THP-protected hydroxylamine was added in order to react with the captured active ester without the addition of any other reagents. This approach of "catch then react" to release a different product is an extremely useful tactic and one we have used in other syntheses such as the preparation of the erectile dysfunction drug sildenafil (Viagra®).[34] The last step of the HDAc inhibitor synthesis used Amberlyst 15, a sulfonic acid resin, to promote *O*-deprotection of the tetrahydropyran ether and to afford the sulfonamide hydroxamic acid derivatives **38**. Recent improvements to the autopreparation of these potential HDAc inhibitors have been reported.[35]

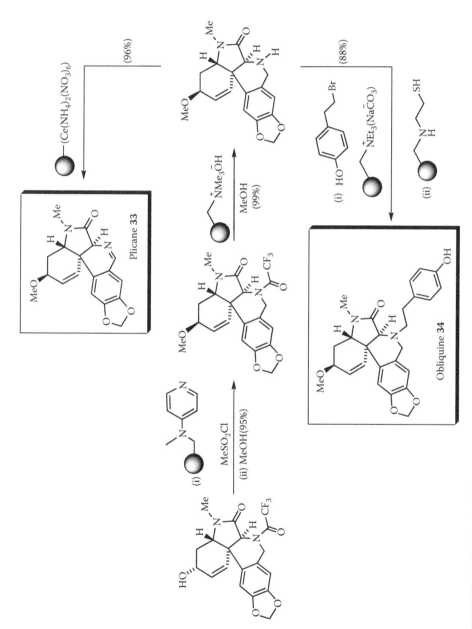

SCHEME 6.8 Synthesis of plicane and obliquine.

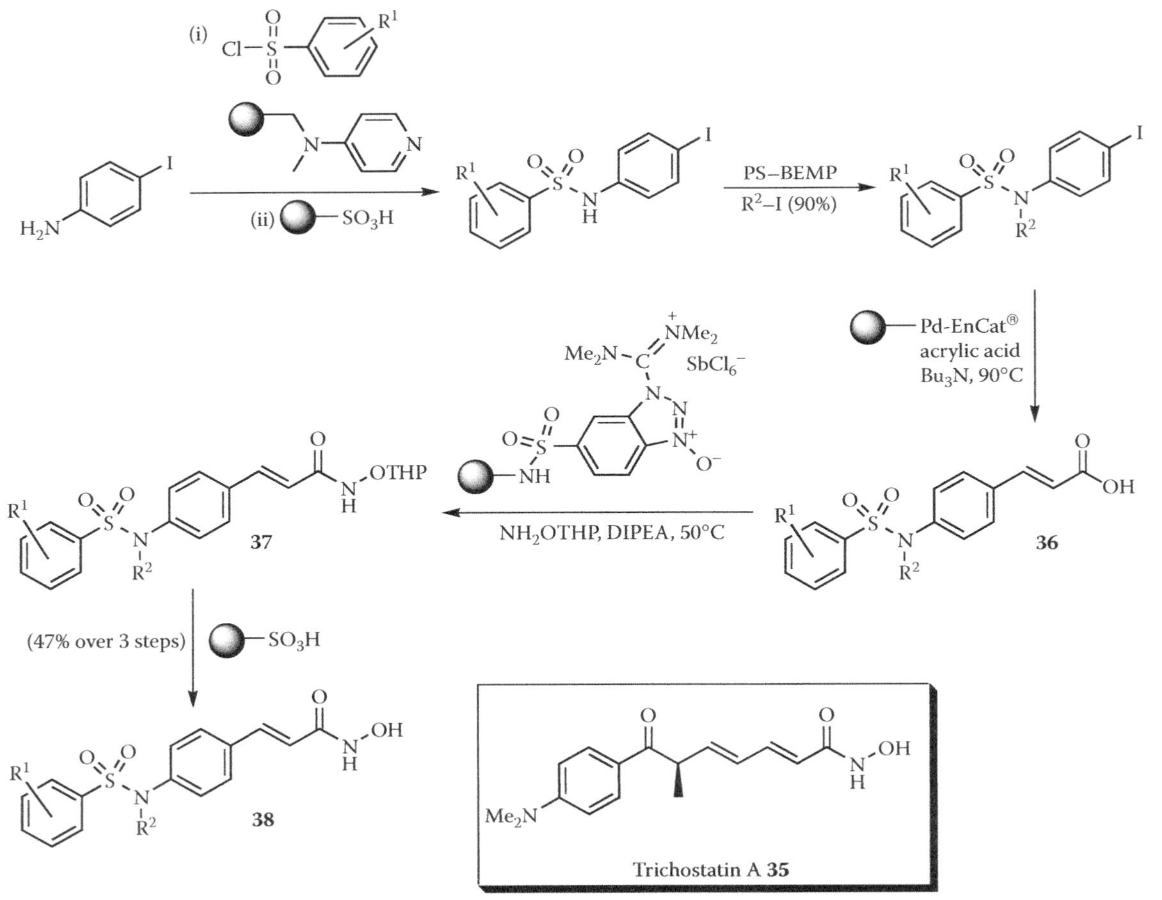

SCHEME 6.9 Synthesis of histone deacetylase inhibitors based on the structure of trichostatin A.

6.2.2 γ-Aminobutyric Acid Analogs

As an important natural neurotransmitter,[36] γ-aminobutyric acid (GABA) further illustrates how these immobilized chemical tools are useful in synthesis. GABA analogs are necessary for probing structure–activity relationships and, again, immobilized methods have shown themselves to be highly compatible with the isolation of these otherwise highly polar molecules. A route to conformationally restricted GABA analogs was developed (Scheme 6.10). It began with the cyclopropane *meso*-diester **39,** which was terminally differentiated in an asymmetric fashion using a commercially available Eupergit-supported pig liver esterase (PLE). The general availability of immobilized enzymes makes them attractive alternatives in multistep synthesis, especially when designing routes based on supported species. In the future, as additional enzymes with specific properties become available from directed evolution programs, we can expect their even greater incorporation into synthesis. The subsequent synthetic steps of the cyclopropyl GABA analog **40** made use of conventional but unpleasant reagents such as borane dimethyl sulfide and thionyl bromide. Nevertheless, in these reactions, product cleanup was easily accomplished using polymer-supported scavengers and quenching agents. One also sees further application of the resin-bound azide reagent during this synthesis. By use of this resin, hazardous azido products were not isolated but simply reduced and Boc-protected *in situ*. At the end of the synthesis an acid "catch-and-release" afforded the hydrolyzed hydrochloride salt of the cyclopropanyl GABA analog **40**. It was also possible to prepare the corresponding 4-, 5- and 6-membered ring analogs.[37]

6.2.3 Lignins

6.2.3.1 Carpanone

The deceptively complex natural product carpanone (**41**) was made from commercially available sesamol in a relatively simple set of reactions using immobilized reagents (Scheme 6.11). Several interesting aspects came to light as a result of this synthesis. After allylation of sesamol using allyl bromide and PS-BEMP, the product underwent an extremely clean Claisen rearrangement by using a combination of toluene and an ionic liquid to absorb the energy from an external microwave source (after 3×15 min reaction periods at 220°C). These binary conditions, following from earlier work, were simple to operate under, because after heating and then cooling, the mixture was separated using a simple liquid handler to remove the product-containing toluene layer. The ionic liquid can also be recovered and reused in further experiments. The toluene fraction containing **42** was then used directly in the double-bond isomerization reaction to yield conjugated compound **43**. This was achieved stereoselectively (11:1 *trans*:*cis*) by use of a new immobilized iridium catalyst **44**[38] developed specifically for this project. However, it has also been shown to be general for other double-bond isomerizations at room temperature.[39] Lastly, following original work by Chapman using other oxidants, the phenolic styrene was converted to carpanone (**41**) by application of a modified Jacobsen Salen cobalt complex under catalytic conditions with molecular oxygen.[40] After scavenging with a carbonate resin to remove unreacted phenols and a trisamine resin to remove an unwanted aldehyde by-product, the crystalline natural product was obtained in excellent yield and purity. Mechanistically, carpanone was finally formed by oxidative dimerization through carbon coupling of the phenol (**43**) followed by a highly stereoselective intramolecular Diels–Alder reaction.

6.2.3.2 Polysyphorin

A further interesting application of the polymer-supported iridium catalysts **44** in natural product synthesis was found during the preparation of several neolignin derivatives such as polysyphorin, rhaphidecursinol B, surinamensin, and oxosurinamensin (Figure 6.1).

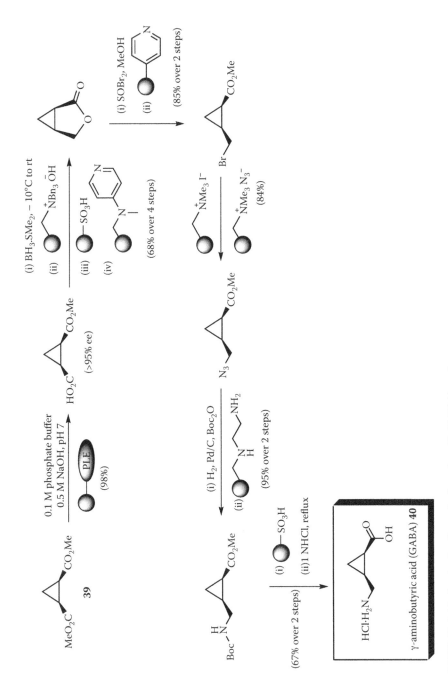

SCHEME 6.10 Synthesis of γ-aminobutyric acid (GABA) analogs.

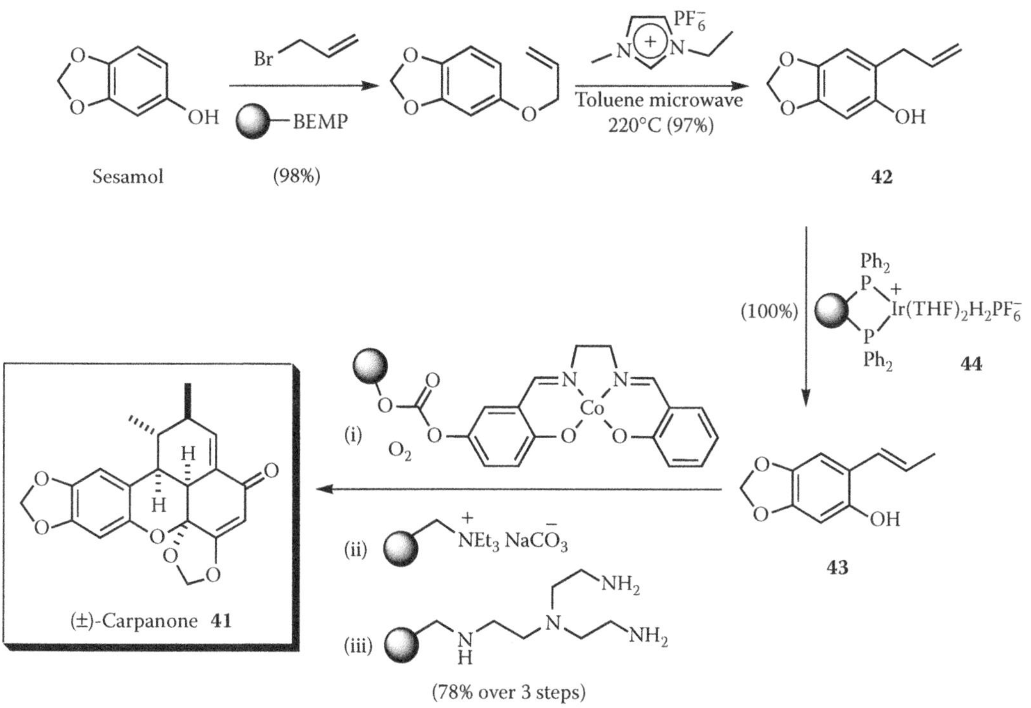

SCHEME 6.11 Synthesis of carpanone.

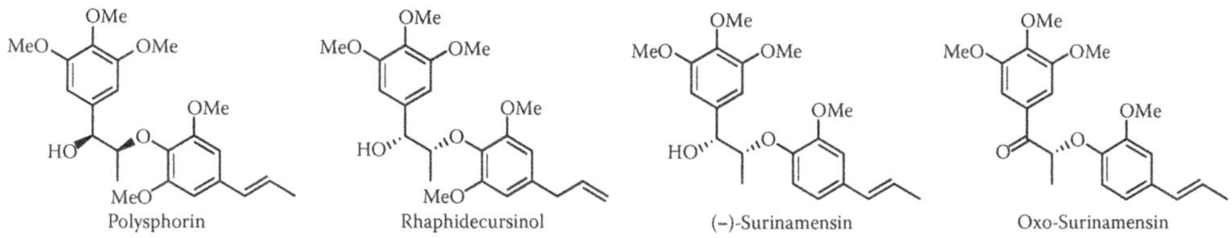

FIGURE 6.1 Other natural neolignins and analogs prepared using the protocols developed for the synthesis of polysyphorin.

The route to polysyphorin (**45**) began with the double-bond isomerization of vinyl intermediate **46** to the *trans*-phenolic alkene **47** using the supported iridium catalyst **44** (Scheme 6.12). Following a standard methylation with MeI and PS-BEMP as the base, the resulting alkene was subjected to an asymmetric dihydroxylation under the Sharpless conditions to give diol **48**. Although this reaction proceeds well, extensive workup is usually necessary, and products often require chromatography. In order to overcome these problems, we found that diol **48** could in fact be cleanly captured onto a boronic acid resin[41] and then released in pure form by hydrolysis in acetone. The next interesting step made use of dicyanodichloroquinone (DDQ) as a selective oxidant to convert the benzylic hydroxyl group of diol **48** to the corresponding ketone **49**. This reaction is notorious in that it generates many highly colored by-products. However, by using immobilized ascorbic acid[42] in the workup as well as other scavengers, a completely clear and colorless solution of the pure product **49** was obtained. The rest of the synthesis, involving tosylation, coupling, and selective reduction, proceeded in a straightforward fashion (Scheme 6.12). Given that these neolignins display significant antifungal and antimalarial activity, the preparative methods described earlier were also extended to generate a number of new analogs that are currently undergoing biological evaluation.

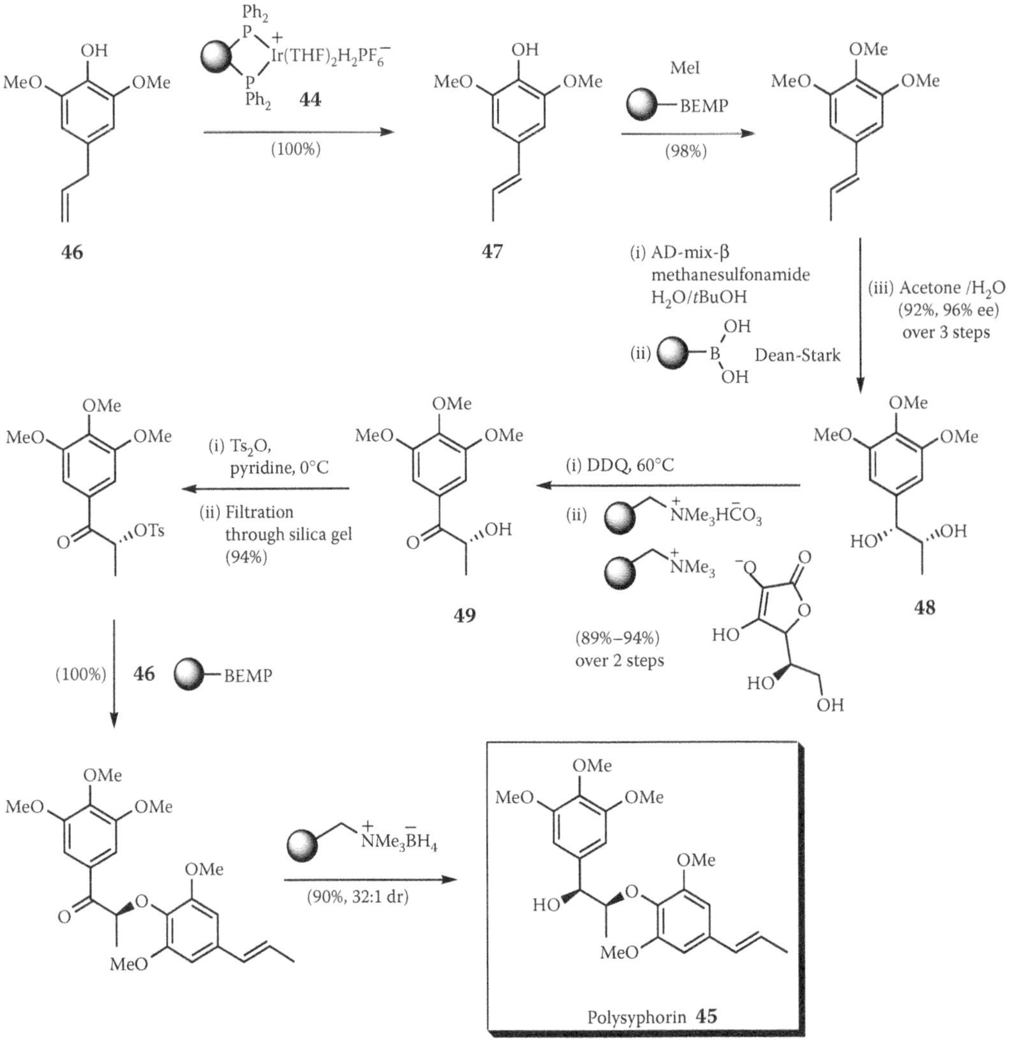

SCHEME 6.12 Synthesis of polysyphorin utilizing the polymer-supported iridium catalyst also used in a key step in the synthesis of carpanone.

6.2.4 Benzofurans

The benzofuran core is another structural motif widely prevalent in various natural product classes. As a common structure in many pharmaceutical compounds, methods for accessing this class of compound are desirable. Supported reagents again demonstrated their effectiveness in constructing benzofuran derivatives (Scheme 6.13, Sequence (a)). Benzofurans with wide functional diversity were prepared by the reaction of phenolic precursors with α-bromoketones (Scheme 6.13).[43] Monobromination of commercially available acetophenones using polymer-supported pyridinium perbromide provided α-bromo starting materials. These are versatile building blocks for a large range of heterocyclic compounds. For example, these products may be coupled with phenols where the R groups can range between H, Me, Br, OMe, CF$_3$, and NO$_2$. The most effective base used in these reactions was supported 1,5,7-triazabicyclo[4.4.0]dec-5-ene (PS-TBD), which enabled the formation of the keto-ether precursors **50**. Cyclization of these to the corresponding substituted benzofurans **51** employed catalytic sulfonic acid resin at 57°C. If *ortho*-cyanophenol is used as the coupling partner, then the alternative aminoacyl benzofurans are generated upon treatment with

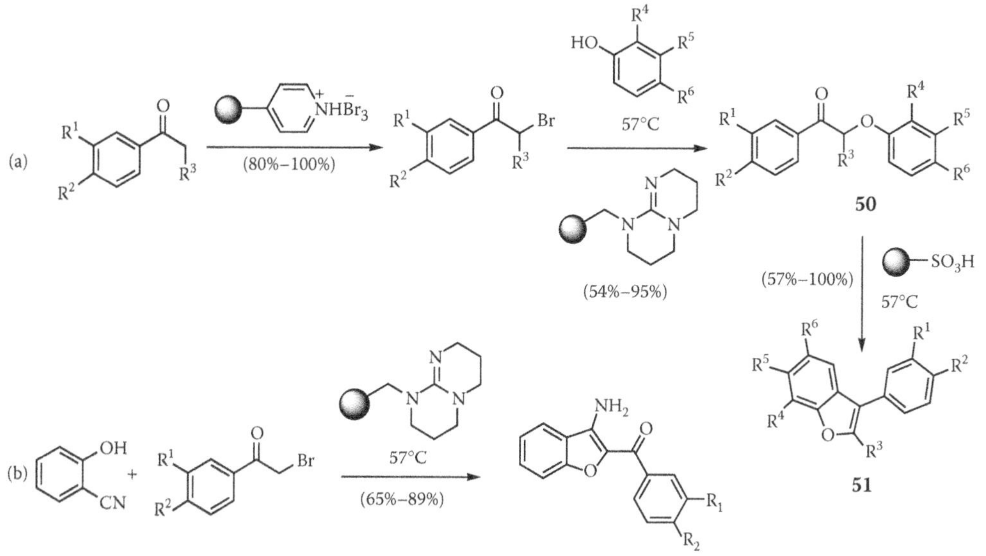

SCHEME 6.13 Synthesis of an array of substituted benzofurans.

PS-TBD (Scheme 6.13, Sequence (b)). If catachols are used, then 1,4-dioxanes are the preferred products.[44]

6.2.5 BENZOPYRANS

The Nicolaou group examined other natural product libraries based on privileged structures such as the benzopyrans (see Chapter 1 and Chapter 11). These ring structure motifs are found in a range of biologically active natural products and, like the benzofurans, they feature prominently in many pharmaceutical agents and possess impressive biological functions (Figure 6.2).[45] In the case of the benzopyran libraries, substrates were immobilized onto solid supports using reactions such as cycloadditions (Scheme 6.14). The supported entities were then elaborated "on-bead" with a succession of steps, and ultimately released by elimination or reduction reactions. Very large libraries (10,000 members) were generated by the application of IRORI NanoKan technology using optical encoding to assist in the sorting of the library members. The process described led to "libraries from libraries" as a method for enhancing structural diversity.

The synthetic approach to these molecules made use of an immobilized selenyl bromide reagent that reacted with *ortho*-allyl phenols **52** to give the corresponding supported benzopyran scaffolds **53** (Scheme 6.14). These were further elaborated through a variety of reactions whereby the various R group were modified. Following oxidation and *syn*-elimination from the support, a large number of benzopyrans based on the structure of **54** were released in such high purity that they could immediately be screened for biological activity. Although this scheme is greatly simplified, it is important to emphasize the variation actually achieved during the chemical elaboration stage. In addition, the released products can be modified further still by supported-reagent mediated processes such as epoxidation and ring-opening to give whole new sets of library members. The real beauty of this work lies in its adaptability; the chemistry is not restricted to intramolecular ring-forming reactions through oxygen atoms. Nitrogen substituents also perform similar processes when treated with *ortho*-allyl anilines **55** and the selenium-based resin (Scheme 6.15). This led to rapid access to immobilized indolines **56**. The resulting scaffolds can undergo numerous transformations. By cleaving the products in a traceless manner, a library of natural product-like molecules (**57** and **58**) can be obtained.[46]

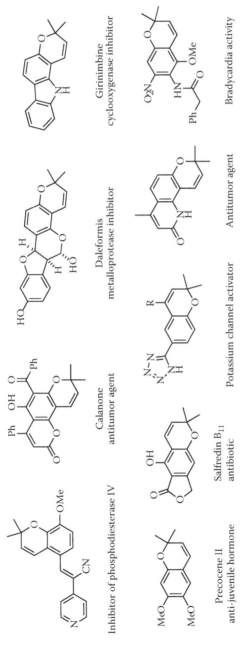

FIGURE 6.2 Pharmaceutical agents containing benzopyrans.

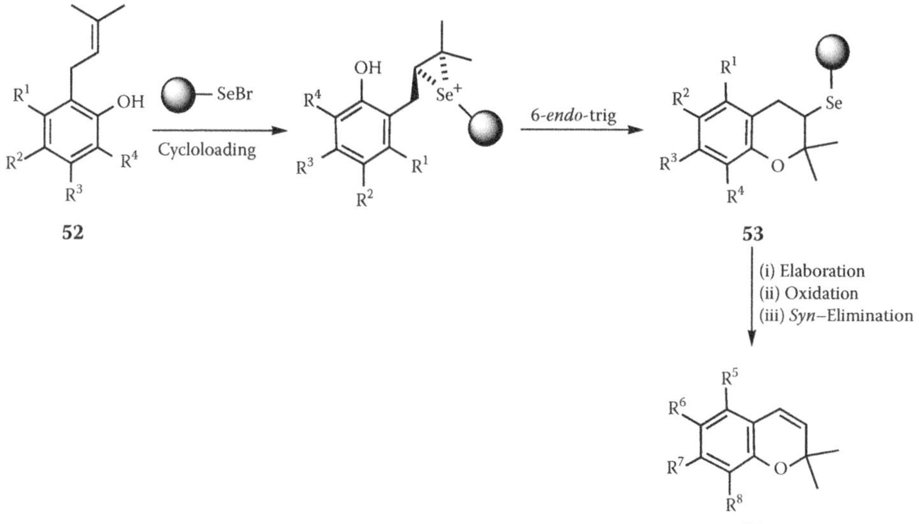

SCHEME 6.14 Synthesis of a library of benzopyrans employing the use of a polymer-supported selenium resin.

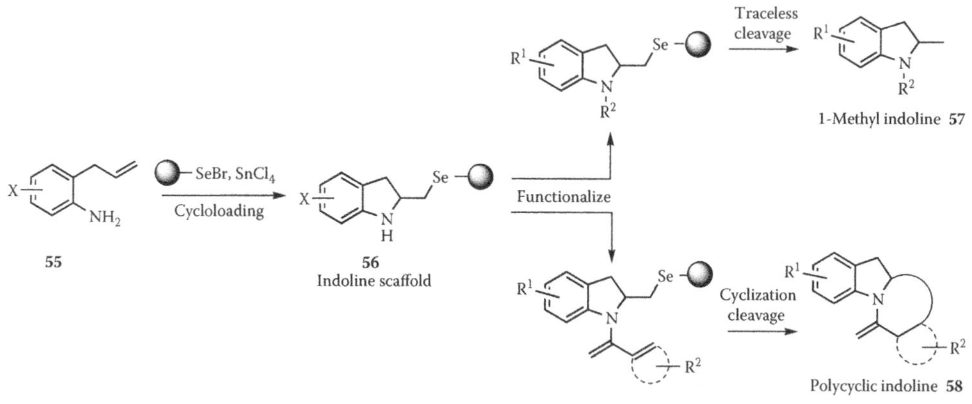

SCHEME 6.15 Synthesis of a library of bicyclic and polycyclic indolines.

6.2.6 OLIGOSACCHARIDES

Complex oligosaccharides, both in natural and unnatural arrangements, have long been a source of intellectual stimulation for organic chemists because of their challenging structures. The larger oligosaccharides can be linear or multibranched, and precise anomeric control of the glycosyl bond construction is always challenging. Kirschning developed a general strategy for solution-phase glycosylation in the formation of glycoconjugates and oligosaccharides using polymer-assisted reactions.[47] Carbohydrates containing anomeric acetates were activated using a supported silyl triflate to provide the glycosyl donor; these were then coupled with a range of sugars in an iterative process, giving remarkably high selectivity and good yields, a process that can be readily extended to trisaccharide synthesis (Scheme 6.16).[48]

Protected glycal **59** was transformed to iodogylcosyl acetate **60** using an immobilized *bis*-acetoxy iodate resin. This was then progressed through a series of coupling steps involving integration with some classical solution-phase techniques to afford the trisaccharide **61**. Kirschning

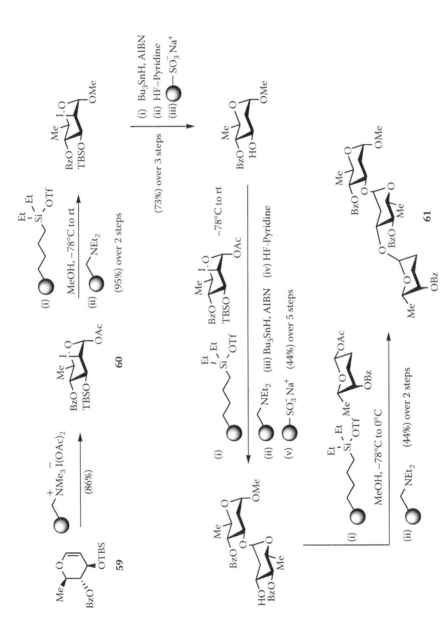

SCHEME 6.16 Synthesis of trisaccharides using polymer-supported silyl triflate as an activator.

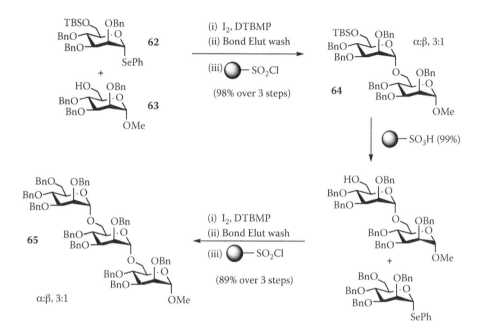

SCHEME 6.17 Trisaccharide synthesis using polymer-supported tosyl chloride.

also demonstrated that glycosyl phenylsulfides[49] may be activated using the hypervalent iodine resin by coupling a number of hindered alcohols using these methods.[50] Selenophenyl glycosyl donors are useful for oligosaccharide synthesis as they can be activated in a number of ways depending on the appropriate coupling environment. However, nearly all of the methods generated obnoxious by-products, which must be removed by chromatography or other techniques requiring significant practical expertise.

The conditions developed by Field et al.[51] using iodine activation are useful when integrated with new workup methods involving polymer-supported scavenging. Using a sulfonyl chloride resin to remove unwanted hydroxyl-containing by-products or excess starting glycosyl acceptor, trisaccharides were prepared without any chromatography (Scheme 6.17).[52] Simply treating TBDMS-protected selenyl donor **62** with iodine, a glycosyl donor **63**, and the hindered organic base di-*tert*-butyl-4-methylpyridine (DTBMP), the desired product was formed in a 3:1 α:β ratio. Passing the mixture through a Bond Elut cartridge and then treating it with polymer-supported tosyl chloride scavenged the hydroxylated materials and furnished disaccharide **64** in 98% yield very cleanly and without chromatography. The silicon protecting group was then removed using polymer-supported sulfonic acid to give pure disaccharide. The disaccharide was sequentially coupled with the next glycosyl selenyl donor using the same procedure to provide the trisaccharide **65**, also in excellent yield. Global deprotection using palladium on carbon in the usual way removed the benzyl protection. Obviously, these methods can be adapted to produce a range of oligosaccharides in good α:β ratios, in excellent yields and high purity.

6.2.7 EPOTHILONES

The epothilones have generated wide interest in the scientific community owing to their ability to inhibit tumor cell proliferation by inducing mitotic arrest through microtubule stabilization.[53] Owing to this activity, they have also become principal targets for many synthesis groups. Indeed, they provide an excellent platform to explore the full armory of supported reagents and scavengers for complex molecule synthesis. The epothilones lend themselves well to convergent synthetic approaches as they can be easily disconnected to more manageable fragments for analog

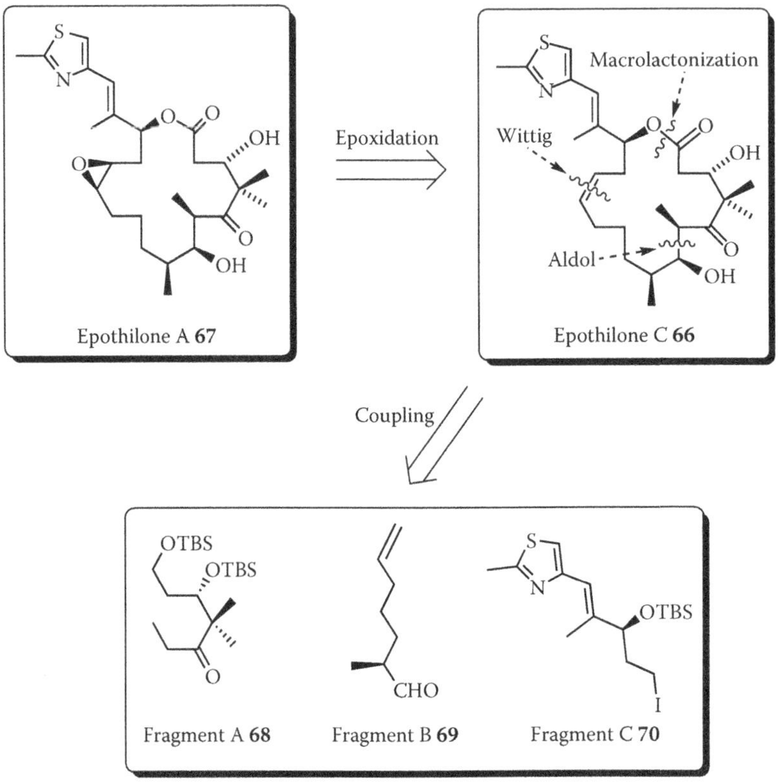

SCHEME 6.18 A retrosynthetic analysis of epothilone A and C.

development, a crucial aspect of pharmaceutical drug programs. For the synthesis of epothilone C (**66**) and epothilone A (**67**) by epoxidation, a convergent synthesis plan was devised that would require the coupling of three major fragments **68, 69,** and **70** (Scheme 6.18). An important criterion for this route was conducting the work in an efficient and clean manner using only immobilized reagents, scavengers, and catch-and-release techniques. The overall aim was to avoid chromatography, crystallization, distillation, and water washes, which are common in conventional approaches to these molecules. Consequently, using only immobilized reagents, scavengers, and catch-and-release techniques can achieve this goal.[54]

In the full paper on this work, a number of alternative routes to the various fragments were reported.[55] Three routes to fragment A (**68**) were investigated, but the one shown in Scheme 6.19 was the shortest, the most efficient, and also proved to be the most easily scaled. The route was conceptually similar to syntheses developed previously by Mulzer and Taylor.[56] The key feature was the formation of the C2-C3 bond with concomitant introduction of the desired C3 stereocenter by application of an asymmetric Mukaiyama aldol reaction.[57] In this synthesis, the aldol reaction progressed in excellent yield to give the alcohol **71** with an enantiomeric excess of 92%. This was achieved using a complex of borane with N-tosylphenylalanine as the chiral ligand. Workup of this reaction required the addition of a minimum amount of water and a boron-selective scavenger, Amberlite IRA-743, to quench the reaction and remove the contaminating boric acid. Filtration and solvent removal produced a suspension of amino acid and aldol product **71**. However, the insolubility of the N-protected amino acid in nonpolar solvents allowed dissolution of the desired aldol product. Subsequent filtration enabled the amino acid to be recovered and recycled, while concentration of the filtrate gave the purified alcohol **71**. After protection as its *tert*-butyldimethylsilylenol ether **72** and reaction with (trimethylsilylmethyl)lithium followed by a scavenger quench using a carboxylic acid resin, direct filtration and solvent removal yielded the ketone **73**.

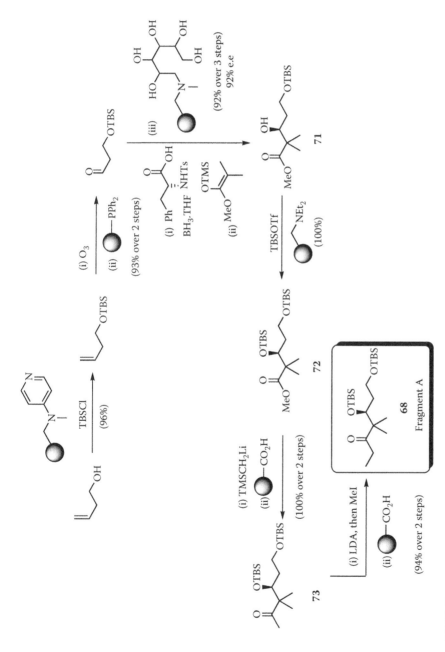

SCHEME 6.19 Synthesis of fragment A of epothilone A.

α-Methylation of ketone **73** via its lithium enolate and workup again with a polymer-supported carboxylic acid gave fragment A (**68**) in just six steps from commercially available starting material (Scheme 6.19).

The preparation of the second key coupling partner fragment B (**69**) was achieved in just five steps from the commercially available bromide (-)-(*R*)-3-bromo-2-methyl-1-propanol (**74**) (Scheme 6.20). Protection of the alcohol as its THP-ether using a polymeric sulfonic acid followed by Finkelstein halide exchange gave iodide **75**. Homologation of alkyl iodide **75** with a cuprate derived from 3-butenylmagnesium bromide produced the corresponding alkene **76**. Addition of a carboxylic acid resin and the trisamine resin quenched the reaction and scavenged dissolved copper salts. Finally, after deprotection of the THP acetal, the resulting alcohol was oxidized to the fragment B aldehyde **69**. Several oxidants were investigated for this process, but the most expedient proved to be pyridinium chlorochromate PCC on basic alumina.

The final fragment C (**70**) was constructed in a convergent fashion from (*S*)-α-hydroxyl-γ-lactone (**77**) and the chloromethyl triazole hydrochloride (**78**) (Scheme 6.21). Lactone **77** was therefore elaborated to ketone **79**. Ketone **79** was coupled with a phosphonate derived from thiazole **78** via a highly stereoselective Horner–Wadsworth–Emmons reaction. A polymer-supported aldehyde was used to scavenge any excess phosphonate from this reaction to yield the *bis*-TBS-protected adduct **80**. TBS-protected diol **80** was taken through to fragment C by previously described steps. Ultimately, iodide **70** was captured onto a polymer-supported triphenylphosphine to produce Wittig salt **81** as the coupling precursor (Scheme 6.21).

With all the fragments now in hand, the final fusion of the components began (Scheme 6.22). Previously, the stereoselective aldol coupling of fragments A and B to form the C6-C7 bond was shown to be highly sensitive to proximal and remote functionality in both fragments. In this work, the coupling of ketone **68** with the aldehyde **69** although bearing close similarity to previous studies, was a novel combination. Using LDA as a base, this aldol coupling proceeded in quantitative yield with better than 13:1 stereoselectivity for the desired product. An acetic acid quench was followed by treatment with a diamine functionalized polymer. This resin served two purposes, to remove the excess acid and to sequester a small amount of unreacted aldehyde.

The elaborated adduct **82** was then silyl-protected and the double bond cleaved by ozonolysis. Workup of the intermediate ozonide was achieved by application of immobilized triphenylphosphine. This is an excellent procedure as the triphenylphosphine oxide produced in normal solution workups and in Wittig chemistry causes practical difficulties that often require multiple chromatographic separations to obtain pure material. Here, simple filtration was sufficient. The resin-bound phosphonium salt **81** derived from fragment C (Scheme 6.21) was treated with an excess of sodium hexamethyldisilazide (NaHMDS) followed by washing with anhydrous THF to give an isolable salt-free ylide. This was coupled stereoselectively to give the *cis*-olefin, which required application of a dilute methanolic solution of camphorsulfonic acid to effect selective removal of the primary TBS-protecting group to give the free alcohol **84**. This process required a separate scavenging step using an immobilized carbonate resin to remove the acid with the volatile MeOTBS by-product being removed under reduced pressure.

Catalytic *tetra*-*N*-propylammonium perruthenate (TPAP) oxidation of alcohol **84** followed by filtration through a pad of silica gel to remove morpholine and ruthenium by-products gave an intermediate aldehyde, which was immediately oxidized to the corresponding acid using a previously developed modified Pinnick procedure.[58] Finally, selective desilylation prior to macrocyclization was finally achieved using tetrabutyl ammonium fluoride (TBAF) solution to afford intermediate **85**. Immobilized versions of fluoride or other acidic resins led to complex mixtures. As a consequence of using the TBAF procedure, an aqueous extraction was necessary, and this gave pure deprotected alcohol **85**. This was the first and only water wash used in the whole synthesis up to this point. The final steps to epothilone A (**67**) simply required application of the Yamaguchi macrolactonization procedure using PS-DMAP and a catch-and-release purification to produce epothilone C. Upon epoxidation with DMDO, epothilone A was obtained (Scheme 6.22).

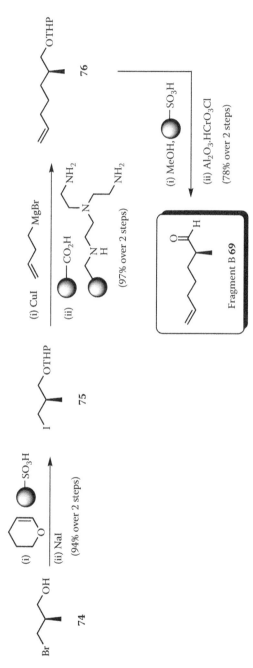

SCHEME 6.20 Synthesis of fragment B of epothilone A.

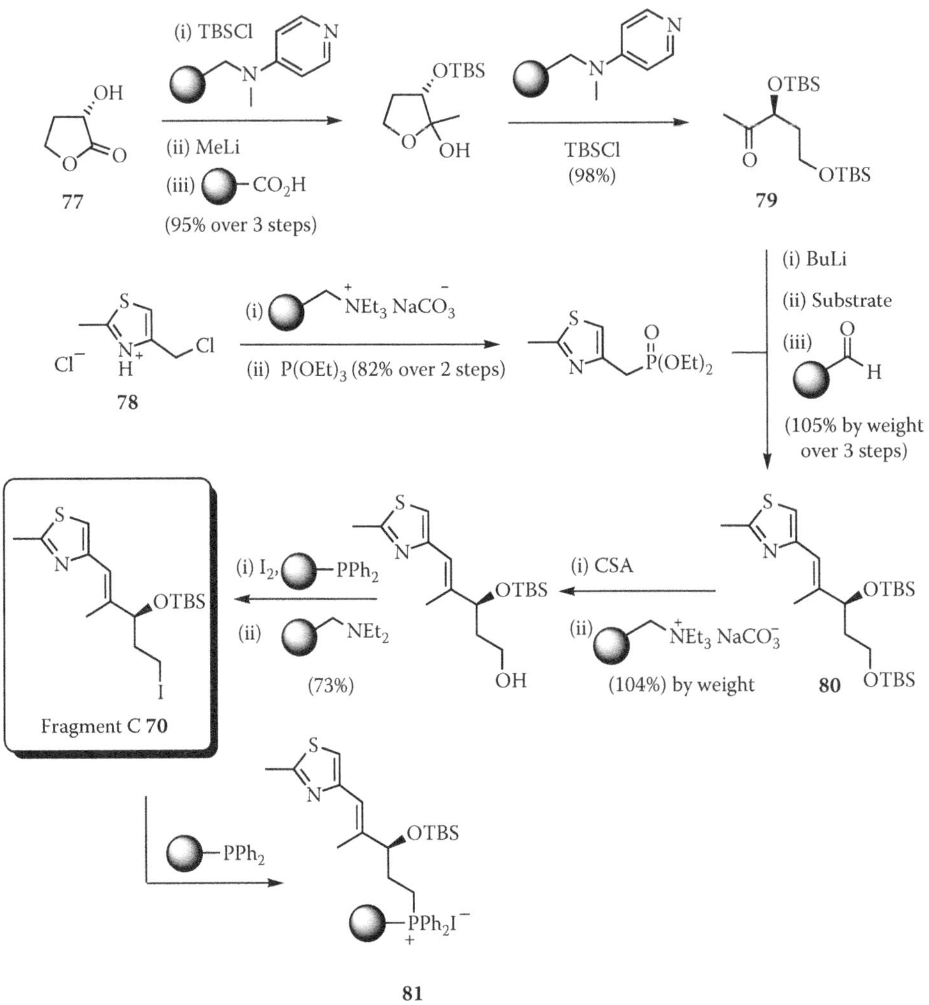

SCHEME 6.21 Synthesis of fragment C of epothilone A.

Without a doubt, this synthesis constitutes a triumph for the utility of supported reagents and scavenging techniques for multistep complex molecule assembly. The high stereoselectivity and overall yield in this synthesis of epothilone A compares well with the best of all of the previous and conventional routes. By the discussed route, the target molecule was delivered in 29 steps, with the longest linear sequence being only 17 steps from readily available materials.

6.3 CONCLUSION

This review focused principally on the multistep application of immobilized reagents, scavengers, and catch-and-release techniques to the synthesis of natural products and their related structural analogs. However, the methods described have a much broader impact and are particularly well suited for use in parallel format for chemical library generation or for the clean generation of drug substances. The techniques are also readily adapted to flow reactors and are compatible with many existing automation methods. In terms of their general use in organic synthesis programs, we hope that they would be integrated along with conventional methods. However, we are mindful that no single technology will solve all the diverse synthesis problems that are required to construct complex molecular entities.

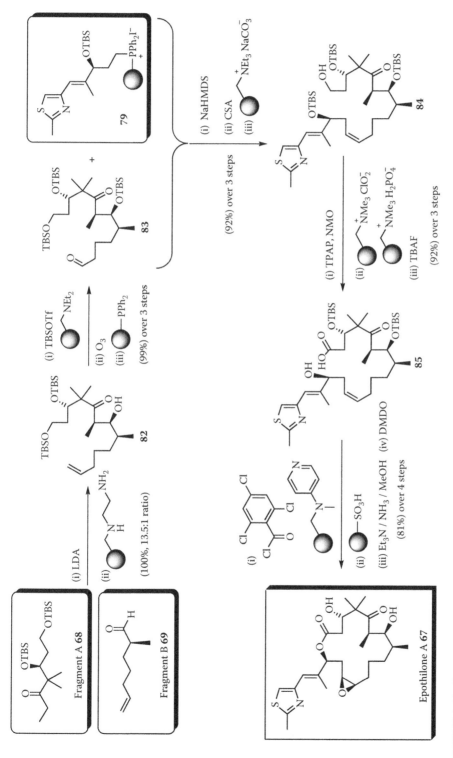

SCHEME 6.22 Coupling the three epothilone fragments in the synthesis of epothilone A using polymer-supported reagents.

REFERENCES

1. Ley, S.V., Baxendale, I.R., Bream, R.N., Jackson, P.S., Leach, A.G., Longbottom, D.A., Nesi, M., Scott, J.S., Storer, R.I., and Taylor, S.J., Multi-step organic synthesis using solid-supported reagents and scavengers: a new paradigm in chemical library generation, *J. Chem. Soc., Perkin Trans.* 1, 3815, 2000.

2. (a) Ley, S.V. and Baxendale, I.R., New tools and concepts for modern organic synthesis, *Nat. Rev. Drug Discovery*, 1, 573, 2002. (b) Baxendale, I.R., Storer, R.I., and Ley, S.V., Supported reagents and scavengers in multi-step organic synthesis in *Polymeric Materials in Organic Synthesis and Catalysis*, Buchmeiser, M.R., Ed., VCH, Berlin, 2003, p. 53.

3. (a) Wessjohann, L.A., Synthesis of natural product-based compound libraries, *Curr. Opin. Chem. Biol.*, 4, 303, 2000. (b) Thompson, L.A., Recent applications of polymer-supported reagents and scavengers in combinatorial, parallel, or multistep synthesis, *Curr. Opin. Chem. Biol.*, 4, 324, 2000. (c) Ganesan, A., Integrating natural product synthesis and combinatorial chemistry, *Pure Appl. Chem.*, 73, 1033, 2001. (d) Nielsen, J., Combinatorial synthesis of natural products, *Curr. Opin. Chem. Biol.*, 6, 297, 2002. (e) Nicolaou, K.C., *Chem. Eur. J.*, 7, 3798, 2001. (f) Arya, P. and Baek, M.-G., Natural product-like chiral derivatives by solid-phase synthesis, *Curr. Opin. Chem. Biol.*, 5, 292, 2001. (g) Abel, U. et al., Modern methods to produce natural-product libraries, *Curr. Opin. Chem. Biol.*, 6, 453, 2002. (h) Hodge, P., Organic synthesis using polymer-supported reagents, catalysts and scavengers in simple laboratory flow systems, *Curr. Opin. Chem. Biol.*, 7, 362, 2003. (i) Ortholand J.–Y. and Ganesan, A., Natural products and combinatorial chemistry: Back to the future, *Curr. Opin. Chem. Biol.*, 8, 271, 2004. (j) Boldi, A.M., Libraries from natural product-like scaffolds, *Curr. Opin. Chem. Biol.*, 8, 281, 2004.

4. (a) Habermann, J., Ley, S.V., and Scott, J.S., Clean six-step synthesis of a piperidino-thiomorpholine library using polymer-supported reagents, *J. Chem. Soc., Perkin Trans.* 1, 3127, 1998. (b) Ley, S.V. and Massi, A., Polymer supported reagents in synthesis: preparation of bicyclo[2.2.2]octane derivatives via tandem Michael addition reactions and subsequent combinatorial decoration, *J. Comb. Chem.*, 2, 104, 2000. (c) Ley, S.V., Bream, R.N., and Procopiou, P.A., Synthesis of the β_2-agonist (*R*)-salmeterol using a sequence of supported reagents and scavenging agents, *Org. Lett.*, 4, 3793, 2002.

5. (a) Jamieson, C., Congreave, M.S., Emiabata-Smith, D.F., and S.V. Ley, A rapid approach for the optimisation of polymer supported reagents in synthesis, *Synlett*, 1603, 2000. (b) Jamieson, C., Congreave, M.S., Emiabata-Smith, D.F., Ley, S.V. and Scicinski, J.J., Application of ReactArray robotics and design of experiments techniques in optimisation of supported reagent chemistry, *Org. Proc. Res. Dev.*, 6, 823, 2002.

6. Ley, S.V., Leach, A.G., and Storer, R.I., A polymer-supported thionating reagent, *J. Chem. Soc., Perkin Trans.* 1, 358, 2001.

7. Westman, J., *Personal Chemistry I*, Uppsala AB, WO 00/72956 A1.

8. Sussman, S., Catalysis by acid-regenerated cation exchangers, *Ind. Eng. Chem.*, 38, 1228, 1946.

9. (a) Ley, S.V., Baxendale, I.R., Brusotti, G., Caldarelli, M., Massi, A., and Nesi, M., Solid-supported reagents for multi-step organic synthesis: preparation and application, *Il Farmaco*, 57, 321, 2002. (b) Ley, S.V. and Baxendale, I.R., Organic synthesis in a changing world, *Chem. Rec.*, 2, 377, 2002.

10. Ley, S.V., Schucht, O., Thomas, A.W., and Murray, P.J., Synthesis of the alkaloids (±)-oxomaritidine and (±)-epimaritidine using an orchestrated multi-step sequence of polymer supported reagents, *J. Chem. Soc., Perkin Trans.* 1, 1251, 1999.

11. (a) Hinzen, B. and Ley, S.V., Polymer supported perruthenate (PSP): a new oxidant for clean organic synthesis, *J. Chem. Soc., Perkin Trans. 1*, 1907, 1997. (b) Hinzen, B., Lenz, R., and Ley, S.V., Polymer supported perruthenate (PSP): clean oxidation of primary alcohols to carbonyl compounds using oxygen as cooxidant, *Synthesis*, 977, 1998.

12. Hutchins, R.O., Natale, N.R., and Taffer, I.M., Cyanoborohydride supported on an anion exchange resin as a selective reducing agent, *J. Chem. Soc., Chem. Commun.*, 1088, 1978.

13. Ley, S.V., Bolli, M., Hinzen, B., Gervois, A.-G., and Hall, B.J., Use of polymer supported reagents for clean *multi-step* organic synthesis: preparation of amines and amine derivatives from alcohols for use in compound library generation, *J. Chem. Soc., Perkin Trans.*, 1, 2239, 1998.

14. Ley, S.V., Thomas, A.W., and Finch, H., Polymer-supported hypervalent iodine reagents in "clean" organic synthesis with potential application in combinatorial chemistry, *J. Chem. Soc., Perkin Trans.* 1, 669, 1999.

15. Baxendale, I.R., Davidson, T.D., Ley, S.V., and Perni, R.H., Enantioselective synthesis of tetrahydrobenzylisoquinoline alkaloid (–)-norarmepavine using polymer-supported reagents, *Heterocycles*, 60, 2707, 2003.

16. Habermann, J., Ley, S.V., and Scott, J.S., Synthesis of the potent analgesic compound (±)-epibatidine using an orchestrated multi-step sequence of polymer supported reagents, *J. Chem. Soc., Perkin Trans.* 1, 1253, 1999.

17. Caldarelli, M., Habermann, J., and Ley, S.V., Clean five-step synthesis of an array of 1,2,3,4-tetrasubstituted pyrroles using polymer-supported reagents, *J. Chem. Soc., Perkin Trans. 1*, 107, 1999.

18. Togo, H., Nogami, G., and Yokoyama, M., Synthetic application of poly[styrene(iodoso diacetate)], *Synlett*, 534, 1998.

19. Lee, R.A. and Donald, D.S., Magtrieve™ an efficient, magnetically retrievable and recyclable oxidant, *Tetrahedron Lett.*, 38, 3857, 1997.

20. Reagents were encapsulated into pouches by Cambridge Combinatorial, Cambridge, U.K.

21. (a) McDonald, I.A., Cosford, N., and Vernier, J.-M., *Annual Reports in Medicinal Chemistry*, Bristol, J.A., Ed., Academic Press, San Diego, CA., 30, 41, 1995. (b) Tomizawa, M. et al., Pharmacological characteristics of insect nicotinic acetylcholine-receptor with its ion-channel and the comparison of the effect of nicotinoids and neonicotinoids, *J. Pest. Sci.*, 20, 57, 1995. (c) Holladay, M.W., Dart, M.J., and Lynch, J.K., Neuronal nicotinic acetylcholine receptors as targets for drug discovery, *J. Med. Chem.*, 40, 4169, 1997.

22. Baxendale, I.R., Brusotti, G., Matsuoka, M., and Ley, S.V., Synthesis of nornicotine, nicotine and other functionalised derivatives using solid-supported reagents and scavengers, *J. Chem. Soc., Perkin Trans.* 1, 143, 2002.

23. Triphenylphosphine polymer bound < 3 mmol/g, available from Fluka Cat. No. 93093.

24. Azide exchange resin: azide on Amberlite IRA-400 available from Aldrich Cat. No. 36,834-2.

25. Godfrey, C.R.A. (Ed.), *Agrochemicals from Natural Products*, Marcel Dekker, New York, 1995.

26. Baxendale, I.R., Ley, S.V., Nesi, M., and Piutti, C., Total synthesis of the amaryllidaceae alkaloid (+)-plicamine and its unnatural enantiomer by using solid-supported reagents and scavengers in a multistep sequence of reactions, *Angew. Chem. Int. Ed. Engl.*, 41, 2194, 2002.

27. Ley, S.V., Nesi, M., and Piutti, C., Total synthesis of the amaryllidacea alkaloid (+)-plicamine using solid-supported reagents, *Tetrahedron*, 58, 6285, 2002.

28. (a) Salmond, W.G., Barta, M.A., and Havens, J.L., Allylic oxidation with 3,5-dimethylpyrazole, Chromium trioxide complex steroidal-δ–5-7-ketones, *J. Org. Chem.*, 43, 2057, 1978. (b) Danishefsky, S., Morris, J., Mullen, G., and Gammill, R., Total synthesis of dl-tazettine and 6a-epipretazettine: a formal synthesis of dl-pretazettine: some observations on the relationship of 6a-epipretazettine and tazettine, *J. Am. Chem. Soc.*, 104, 7591, 1982.

29. Baxendale, I.R. and Ley, S.V., Synthesis of the alkaloid natural products (+)-plicane and (–)-obliquine using polymer-supported reagents and scavengers, *Ind. Eng. Chem. Res.*, 44, 8588–8592, 2005.

30. Yoshida, M. et al., Potent and specific inhibition of mammalian histone deacetylase both *in vivo* and *in vitro* by trichostatin A, *J. Biol. Chem.*, 265, 17174, 1990.

31. Kouzarides, T., Acetylation: a regulatory modification to rival phosphorylation?, *EMBO J.*, 19, 1176, 2000.

32. Vickerstaff, E., Ladlow, M., Ley, S.V., and Warrington. B., Fully automated multi-step solution phase synthesis using polymer-supported reagents: preparation of histone deacetylase inhibitors, *Org. Biomol. Chem.*, 1, 2419, 2003.

33. Lee, C.K.Y., Holmes, A.B., Ley, S.V., McConvey, I.F., Al-Duri, B., Leeke, G.A., Santos, R.C.D., and Seville, J.P.K., Efficient batch and continuous flow Suzuki cross coupling reactions under mild conditions, catalyzed by polyurea-encapsulated palladium(II)acetate and tetra-*n*-butylammonium salts, *J. Chem. Soc., Chem. Commun.*, 2175, 2005.

34. Baxendale, I.R. and Ley, S.V., Polymer-supported reagents for multi-step organic synthesis: application to the synthesis of sildenafil, *Bioorg. Med. Chem. Lett.*, 10, 1983, 2000.

35. Bapna, A., Ladlow, M., Vickerstaff, E., Warrington, B.H., Fan, T-P., and Ley, S.V., Polymer-assisted multistep solution phase synthesis and biological screening of histone deacyclase inhibitors, *Org. Biomol. Chem.*, 2, 611, 2004.

36. (a) Allan, R.D. et al., *Neurochem.*, 34, 652, 1980. (b) Kusama, T. et al., Pharmacology of GABA ρ-1 and GABA α/β receptors expressed in xenopus oocytes and cos cells, *Br. J. Pharmacol.*, 109, 200, 1993. (c) Kusama, T. et al., GABA ρ-2 receptor pharmacological profile, GABA recognition site similarities to ρ-1, *Eur. J. Pharmacol.*, 245, 83, 1993.

37. Baxendale, I.R., Ernst, M., Krahnert, W.-R., and Ley, S.V., Application of polymer-supported enzymes and reagents in the synthesis of γ–aminobutyric acid (GABA) analogues, *Synlett*, 1641, 2002.

38. Baxendale, I.R., Lee, A.-L., and Ley, S.V., A polymer-supported iridium catalyst for the stereoselective isomerisation of double bonds, *Synlett*, 516, 2002.

39. Lee, A.-L. and S.V. Ley, The synthesis of the anti-malarial natural product polysyphorin and analogues using polymer-supported reagents and scavengers, *Org. Biomol. Chem.*, 1, 3957, 2003.

40. (a) Chapman, O.L. et al., *J. Am. Chem. Soc.*, 6696, 1971. (b) Matsumoto, M. and Kuroda, K., Transition metal(II) Schiff-base complexes catalyzed oxidation of trans-2-(1-propenyl)-4,5-methylenedioxyphenol to carpanone by molecular-oxygen, *Tetrahedron Lett.*, 22, 4437, 1981. (c) Nishiyama, A., et al., Anodic-oxidation of some propenylphenols, Synthesis of physiologically active neolignans, *Chem. Pharm. Bull.*, 31, 2834, 1983. (d) Iyer, M.R. and Trivedi, G.K., Silver(I) oxide catalyzed oxidation of ortho-allylphenols and ortho-(1-propenyl)phenols, *Bull. Chem. Soc. Jpn.*, 65, 1662, 1992.

41. Seymour, E. and Fréchet, J.M.J., Separation of *cis*-diols from isomeric *cis*-trans mixtures by selective coupling to a regenerable solid support, *Tetrahedron Lett.*, 3669, 1976.

42. Deegan, T.L. Jr. et al., Non-acidic cleavage of Wang-derived ethers from solid support: utilization of a mixed-bed scavenger for DDQ, *Tetrahedron Lett.*, 38, 4973, 1997.

43. Habermann, J., Ley, S.V., and Smits, R., Three-step synthesis of an array of substituted benzofurans using polymer-supported reagents, *J. Chem. Soc., Perkin Trans.* 1, 2421, 1999.

44. Habermann, J., Ley, S.V., Scicinski, J.S., Scott, J.S., Smits, R., and Thomas, A.W., Clean synthesis of α-bromo ketones and their utilization in the synthesis of 2-alkoxy-2,3-dihydro-2-aryl-1,4-benzodioxanes, 2-amino-4-aryl-1,3-thiazoles and piperidino-2-amino-1,3-thiazoles using polymer-supported reagents, *J. Chem. Soc., Perkin Trans.* 1, 2425, 1999.

45. (a) Nicolaou, K.C., Pfefferkorn, J.A., Roecker, A.J., Cao, G.-Q., Barluenga, S., Mitchell, H.J., Natural product-like combinatorial libraries based on privileged structures. 1. General principles and solid-phase synthesis of benzopyrans, *J. Am. Chem. Soc.*, 122, 9939, 2000. (b) Nicolaou, K.C., Pfefferkorn, J.A., Mitchell, H.J., Roecker, A.J., Barluenga, S., Cao, G.-Q., Affleck, R.L., Lillig, J.E., Natural product-like combinatorial libraries based on privileged structures. 2. Construction of a 10,000-membered benzopyran library by directed split-and-pool chemistry using Nanokans and optical encoding, *J. Am. Chem. Soc.*, 122, 9954, 2000. (c) Nicolaou, K.C., Pfefferkorn, J.A., Barluenga, S., Mitchell, H.J., Roecker, A.J., and Cao, G.-Q., Natural product-like combinatorial libraries based on privileged structures. 3. The "libraries from libraries" principle for diversity enhancement of benzopyran libraries, *J. Am. Chem. Soc.*, 122, 9968, 2000.

46. Nicolaou, K.C., Roecker, A.J., Pfefferkorn, J.A., and Cao, G.-Q., A novel strategy for the solid-phase synthesis of substituted indolines, *J. Am. Chem. Soc.*, 122, 2966, 2000.

47. (a) Kirschning, A., Jesberger, M., and Monenschein, H., Application of polymer-supported electrophilic reagents for the 1,2-functionalization of glycals, *Tetrahedron Lett.*, 40, 8999, 1999. (b) Monenschein, H. et al., Reactions of alkenes, alkynes, and alkoxyallenes with new polymer-supported electrophilic reagents, *Org. Lett.*, 1, 2101, 1999.

48. Kirschning, A., Jesberger, M., and Schönberger, A., The first polymer-assisted solution-phase synthesis of deoxyglycosides, *Org. Lett.*, 3, 3623, 2001.

49. Jaunzems, J. et al., Solid-phase-assisted solution-phase synthesis with minimum purification-preparation of 2-deoxyglycoconjugates from thioglycosides, *Angew. Chem. Int. Ed. Engl.*, 42, 1166, 2003 and references therein.

50. (a) Kirschning, A., Monenschein, H., and Schmeck, C., Stable polymer-bound iodine azide, *Angew. Chem., Int. Ed. Engl.*, 38, 2594, 1999. (b) Kirschning, A., Monenschein, H., and Schmeck C., Stabiles festphasengebundenes iodazid, *Angew. Chem.*, 111, 2720, 1999. (c) Sourkouni-Argirusi, G. and Kirschning, A., A new polymer-attached reagent for the oxidation of primary and secondary alcohols, *Org. Lett.*, 2, 3781, 2000.

51. Field, R.A., Iodine and its interhalogen compounds: versatile reagents in carbohydrate chemistry IX: a mild and selective deprotection of *tert*-butyldimethylsilyl (TBDMS) ethers in the presence of various protecting groups using iodine monobromide, *Synlett*, 3, 311, 1999.

52. MacCoss, R.N., Brennan, P.E., and Ley, S.V., Synthesis of carbohydrate derivatives using solid-phase work-up and scavenging techniques, *Org. Biomol. Chem.*, 1, 2029, 2003.

53. (a) Altaha, R. et al., Epothilones: a novel class of non-taxane microtubule-stabilizing agents, *Curr. Pharm. Des.*, 8, 1707, 2002. (b) Borzilleri, R.M. and Vite, G.D., Epothilones: new tubulin polymerization agents in preclinical and clinical development, *Drug Future*, 27, 1149, 2002. (c) Florsheimer, A. and Altmann, K.H., Epothilones and their analogues, a new class of promising microtubule inhibitors, *Expert Opin. Ther. Pat.*, 11, 951, 2001. (d) Nicolaou, K.C., Roschangar, F., and Vourloumis, D., Chemical biology of epothilones, *Angew. Chem., Int. Ed. Engl.*, 37, 2015, 1998. (e) Höfle, G. et al., Semisynthesis and degradation of the tubulin inhibitors epothilone and tubulysin, *Pure Appl. Chem.*, 75, 167, 2003.

54. Storer, R.I., Takemoto, T., Jackson, P.S., and Ley, S.V., A total synthesis of epothilones using solid-supported reagents and scavengers, *Angew. Chem., Int. Ed. Engl.*, 42, 2521, 2003.

55. Storer, R.I., Takemoto, T., Jackson, P.S., Brown, D.S., Baxendale, I.R., and Ley S.V., Multi-step application of immobilized reagents and scavengers: a total synthesis of epothilone C, *Chem. Eur. J.*, 10, 2529, 2004.

56. (a) Mulzer, J., Mantoulidis, A., and Ohler, E., Total syntheses of epothilones B and D, *J. Org. Chem.*, 65, 7456, 2000. (b) Taylor, R.E. and Chen, Y., Total synthesis of epothilones B and D, *Org. Lett.*, 3, 2221, 2001.

57. (a) Kiyooka, S., Development of chiral lewis acid, oxazaborolidinone, promoted asymmetric aldol reaction, *J. Synth. Org. Chem. Jpn.*, 55, 313, 1997. (b) Kiyooka, S., Development of a chiral Lewis acid-promoted asymmetric aldol reaction using oxazaborolidinone, *Rev. Heteroat. Chem.*, 17, 245, 1997.

58. Takemoto, T., Yasuda, K., and Ley, S.V., Solid-supported reagents for the oxidation of aldelydes to carboxylic acids, *Synlett*, 1555, 2001.

7 Carbohydrate-Derived Small-Molecule Libraries

Armen M. Boldi

CONTENTS

7.1 INTRODUCTION

Stereochemically and functionally rich carbohydrates are abundantly present in all living organisms. These natural products represent an inexpensive class of molecules and provide facile access to synthetically useful starting materials. For example, monosaccharides are "readily available, chiral, conformationally rigid, and highly functionalized, displaying the five hydroxyl groups in a well-defined, three-dimensional arrangement as vectors for the introduction of diverse side chains."[1] For many years, carbohydrates have been critical chiral synthons in an array of transformations.[2]

In the arena of drug discovery, carbohydrates, carbohydrate-based mimetics, and carbohydrate-derived molecules are receiving greater attention and interest. *Burger's Medicinal Chemistry and Drug Discovery* has reviews of carbohydrate-based drugs that have made it to market, therapeutics in development, and carbohydrate agents in research.[3] In the last decade, new therapeutics based on carbohydrates have been introduced to treat diabetes, diseases of the gastrointestinal system, diseases of the central nervous system, infection, thrombosis, inflammation, and enzyme replacement. Several low molecular-weight therapeutics are shown in Table 7.1. Two azasugar analogs, voglibose and miglitol, are α-glucosidase inhibitors that lower blood glucose levels similar to the approved natural product drug, acarbose. Structurally similar to miglitol, miglustat is a recently approved treatment for type 1 Gaucher's disease, a metabolic disorder.[4] Sugar sulfamates such as topiramate are also known anticonvulsants and antiepileptics.[5] Oseltamivir and zanamivir are neuraminidase inhibitors used for the treatment of influenza virus infection.[6] Interestingly, an enantioselective synthesis of 4320 encoded and spatially segregated dihydropyrancarboxamides explored the diversity around the zanamivir core (Figure 7.1).[7]

TABLE 7.1
Representative Recently Approved Small-Molecule Carbohydrate-Based and Carbohydrate-Mimetic Drugs

Generic Name (Brand Name)	Structure	Indication	Company
Voglibose (Basen, Glustat®)		Diabetes	Takeda, Abbott
Miglitol (Glyset®)		Diabetes	Bayer
Miglustat (Zavesca®)		Type 1 Gaucher's disease (metabolic disorder)	Pfizer, Actelion
Topiramate (Topamax®)		Anticonvulsant, antiepileptic	Ortho-McNeil, Johnson & Johnson
Zanamivir (Relenza®)		Antiviral	GlaxoSmithKline

TABLE 7.1 (CONTINUED)

Representative Recently Approved Small-Molecule Carbohydrate-Based and Carbohydrate-Mimetic Drugs

Generic Name (Brand Name)	Structure	Indication	Company
Oseltamivir (Tamiflu®)	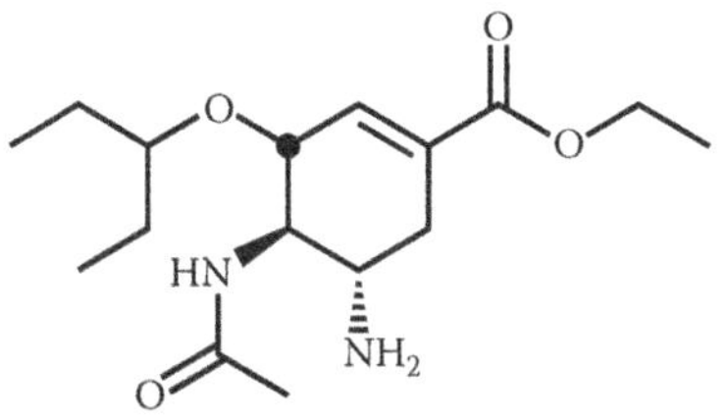	Antiviral	Hoffmann-La Roche, Gilead

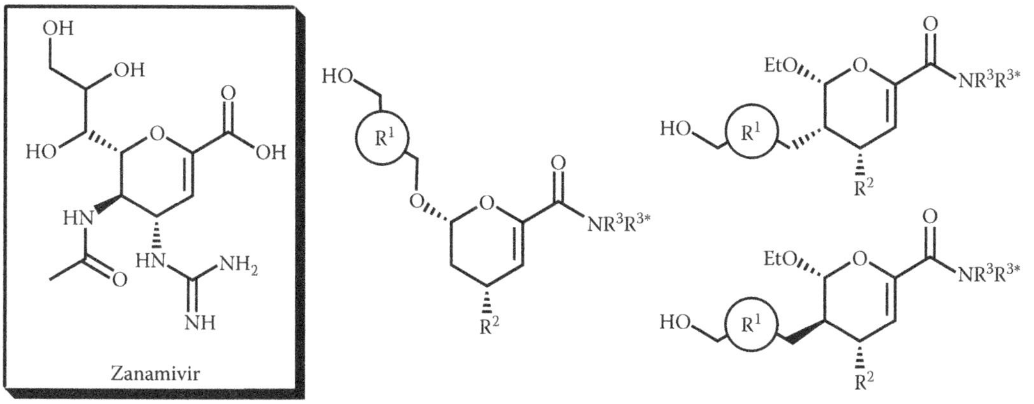

FIGURE 7.1 Zanamivir and related natural product-like libraries.

As therapeutic agents, oligosaccharides are an emerging possible treatment for various diseases, resulting in the rapid growth of chemical glycobiology.[8] For example, sugar amino acids, found naturally as sialic acids or glycosaminuronic acids, are the focus of much work in designing carbohydrate-based mimetics.[9] Since Frechét's first report of the solid-phase synthesis of oligosaccharides over 35 years ago,[10] the field of oligosaccharide synthesis has advanced tremendously, and combinatorial approaches have been comprehensively reviewed elsewhere (see Chapter 11, Subsection 11.2.1).[11] In Chapter 8, Sofia provides useful perspectives on how to use sugars in the development of novel therapeutic agents.

7.2 CARBOHYDRATES AS PRIVILEGED PLATFORMS FOR DRUG DISCOVERY

Hirschmann, Nicolaou, Smith, and coworkers introduced the use of monosaccharides as "privileged structures"[12] to modulate receptor and receptor subtype affinities.[13] The β-D-glucose scaffold serves as a nonpeptidal peptidomimetic of the hormone somatostatin (SRIF). Over the past decade, a number of analogs have been synthesized and tested to explore the structure–activity relationships of differential substitution of the sugar core.[14,15] In collaboration with Merck researchers, Hirschmann and coworkers discovered both somatostatin and NK-1 antagonists (Figure 7.2). A small

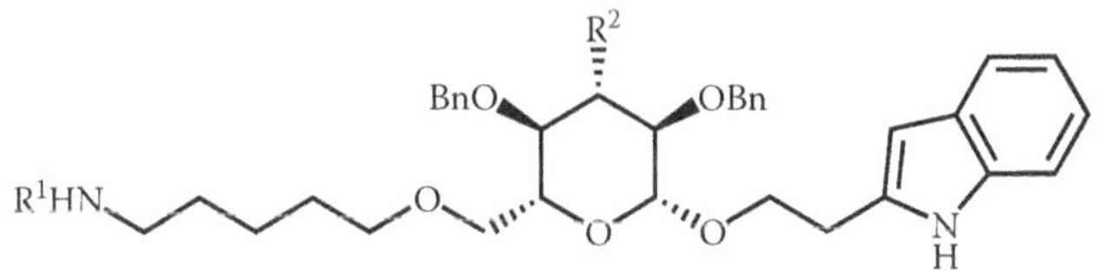

Somatostatin antagonist; R^1 = H; R^2 = H
NK-1 receptor antagonist; R^1 = Ac; R^2 = –OBn

FIGURE 7.2 β-D-glucose scaffold nonpeptidal peptidomimetics.

solution-phase library of monosaccharides was also prepared to demonstrate the feasibility of larger arrays of derivatives on solid phase.[1]

7.2.1 MONOSACCHARIDE LIBRARIES

Despite these reported biological activities and the recent advances in oligosaccharide synthesis,[11] small-molecule libraries derived from sugars have received surprisingly limited attention. In 1996, Lansbury produced a solution-phase library of monosaccharide derivatives and suggested that solid-phase synthesis would "access the tremendous structural diversity of carbohydrates without the synthetic difficulties and uncertain bioavailabilities associated with natural carbohydrates."[16] Soon thereafter, Sofia and coworkers reported the first carbohydrate-based small-molecule solid-phase libraries — **3** and **6** — that mapped pharmacophoric space in three-dimensions (Scheme 7.1).[17] Two related libraries were generated from D-glucosamine and β-methyl-glucopyranoside. The strategy took advantage of the stereochemical richness of carbohydrates in exploring various interactions with biological targets. More recently, Murphy et al. reported the synthesis of carbohydrate-based peptidomimetics that inhibited HIV-1 protease.[18]

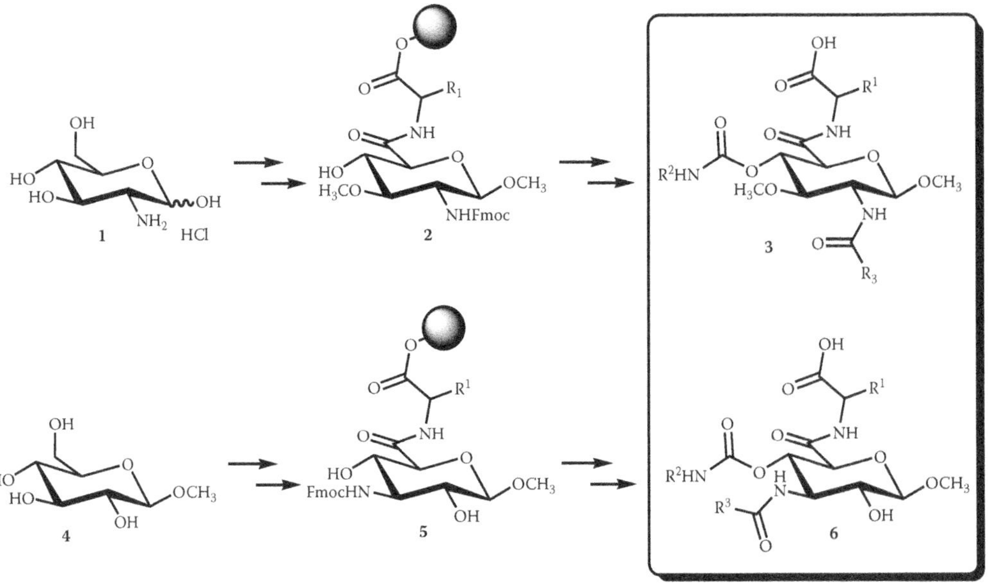

SCHEME 7.1 Solid-phase synthesis of pharmacophore mapping libraries from D-glucosamine and β-methyl-glucopyranoside.

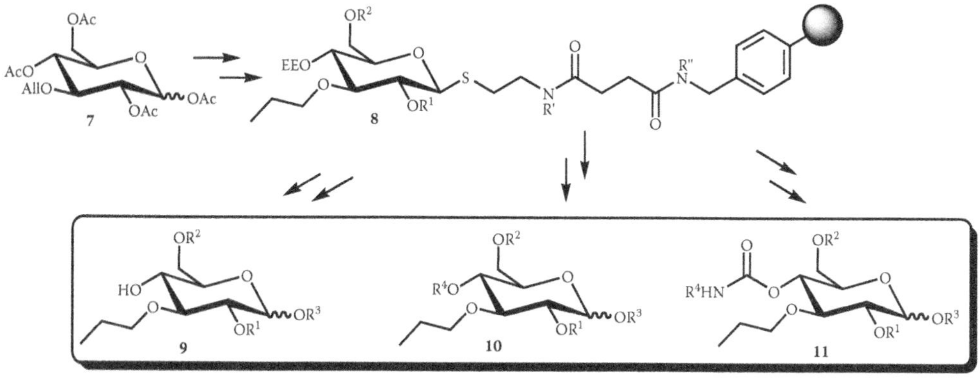

SCHEME 7.2 Solid-phase synthesis of orthogonally functionalized D-glucose derivatives.

Other early examples of solid-phase combinatorial synthesis of libraries using carbohydrate scaffolds were reported by Kunz and coworkers. Starting from D-glucose derivative **7**, a functionalized sugar intermediate was loaded onto the aminomethyl polystyrene resin to give **8** (Scheme 7.2). Three arrays **9**, **10**, and **11** were prepared with diversity displayed in different positions.[19] Similarly, starting from peracetylated D-galactose **12**, all five positions of the monosaccharide were functionalized on solid support to give compound types **15** to **19** (Scheme 7.3).[20] Later, Kunz and coworkers reported a method for derivatizing all five positions of a D-glucose derivative.[21] A related solution-phase library of $\alpha_4\beta_1$ integrin antagonists **21** based on β-D-mannose **20** was also reported by Kessler (Scheme 7.4).[22]

Aminoglycoside antibiotics recognize RNA, specifically binding to 16S ribosomal RNA and interfering with bacterial protein biosynthesis. Wong and coworkers reported the use of the monosaccharide, *N*-acetylglucosamine (**22**), to prepare an array of RNA binders **23** (Scheme 7.5).[23] They identified 1,2- and 1,3-hydroxyamines as a possible key binding motif for RNA recognition. All library members preserved this structural bias for the recognition of 16S ribosomal RNA, and diversity was introduced at the α-anomeric position and the 2-position.

7.2.2 Bicyclics from Glucose, Mannose, and Galactose

Nonaromatic polycyclic motifs represent a subclass of natural product-based libraries that can be prepared from carbohydrates.[24] Schreiber recently described the synthesis and screening of carbohydrate-derived libraries and demonstrated that stereochemistry and conformational restriction effect cellular response in biological screening (Scheme 7.6).[25] Glucose, mannose, and galactose derivatives **25** on solid support were functionalized, cleaved off of solid support to give *bis*-olefins **26**, and screened. Alternatively, the functionalized monosaccharides containing two olefin moieties **25** were treated with Grubb's catalyst for ring-closing metathesis (RCM) to give bicyclic compounds **28** upon cleavage off solid support. A total of 12 distinct stereochemical patterns resulted. Scheme 7.6 summarizes the synthetic approach. For simplicity, the stereochemistry has been omitted.

Another example of a conformationally restricted carbohydrate-derived library was reported by Overkleeft and coworkers.[26] They prepared carboxylic acid **30** in eight steps from D-mannitol (**29**) derived from D-mannose (Scheme 7.7). After coupling **30** onto the Rink amide linker, the hydroxyl and azide, upon conversion into carbamates and amides, respectively, furnished *bis*-olefin **32**. The oxacycle core-containing compounds **33** were generated upon RCM cleavage off solid-support.

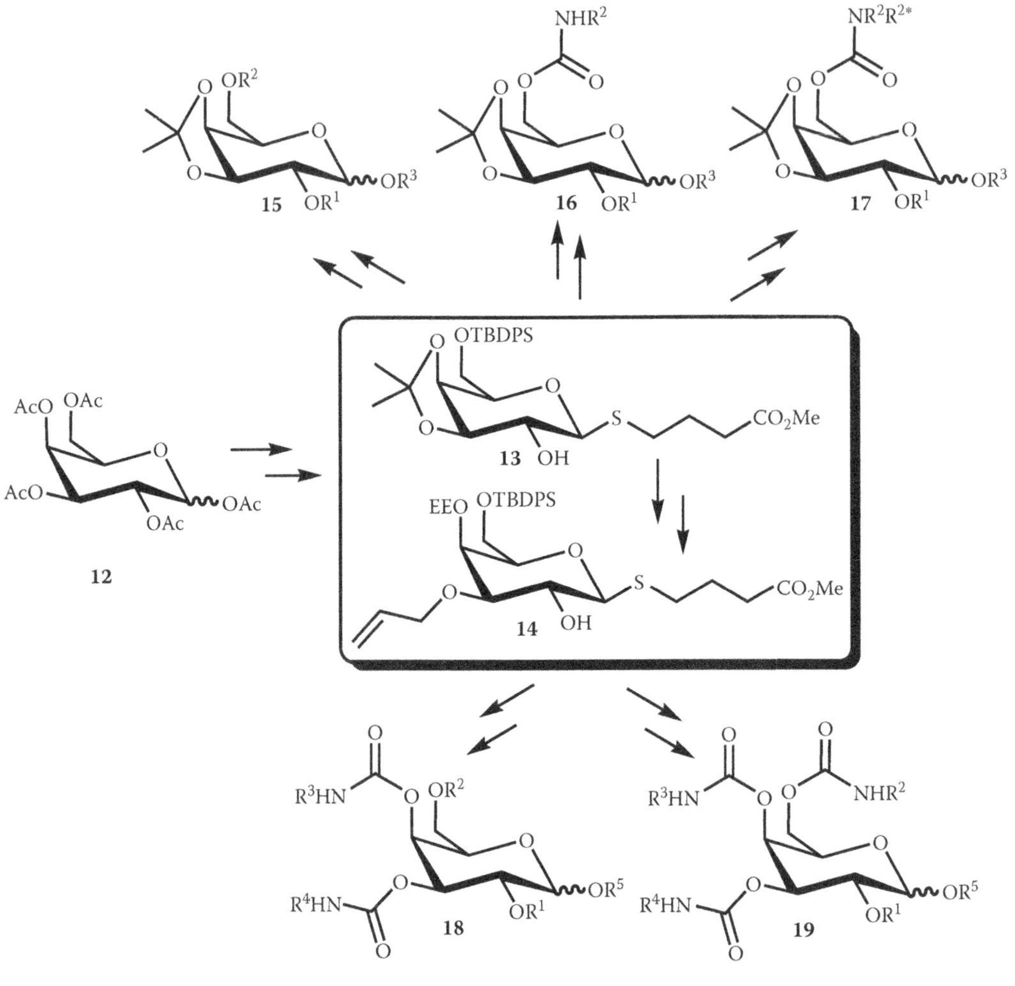

SCHEME 7.3 Solid-phase synthesis of orthogonally functionalized D-galactose derivatives.

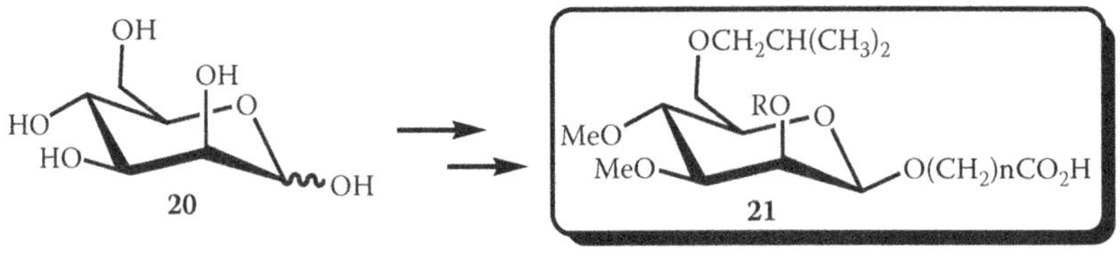

SCHEME 7.4 Solution-phase synthesis of $\alpha_4\beta_1$ integrin antagonists based on β-D-mannose derivatives.

7.2.3 TRICYCLICS FROM GLYCAL TEMPLATES

Even more structural complexity is obtained from carbohydrate starting materials as illustrated by the following two examples. Common to the iridoid family of natural products (Figure 7.3) is the tricyclic perhydrofuropyran core. Häner and coworkers accessed this tricyclic core by converting D-(−)-ribose (**34**) to dihydrofuranoside **35** in five steps (Scheme 7.8). Upon esterifcation of the free hydroxyl with acrylic acids **36**, esters **37** were heated in high-boiling aromatic solvents to promote the intramolecular hetero-Diels–Alder reaction and to give eight tricyclic derivatives **38** with the expected *endo* selectivity.

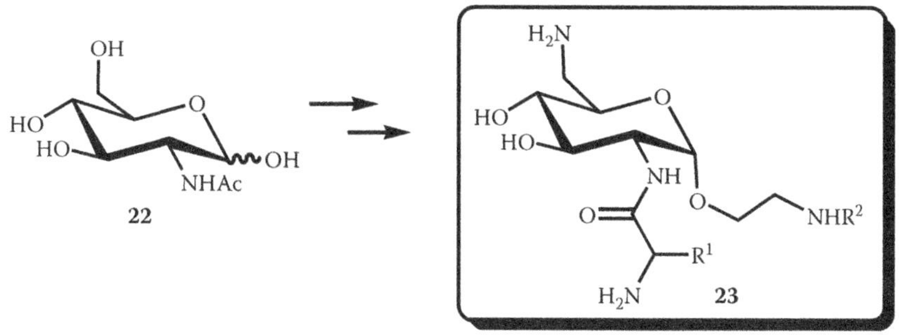

SCHEME 7.5 Synthesis of RNA binders.

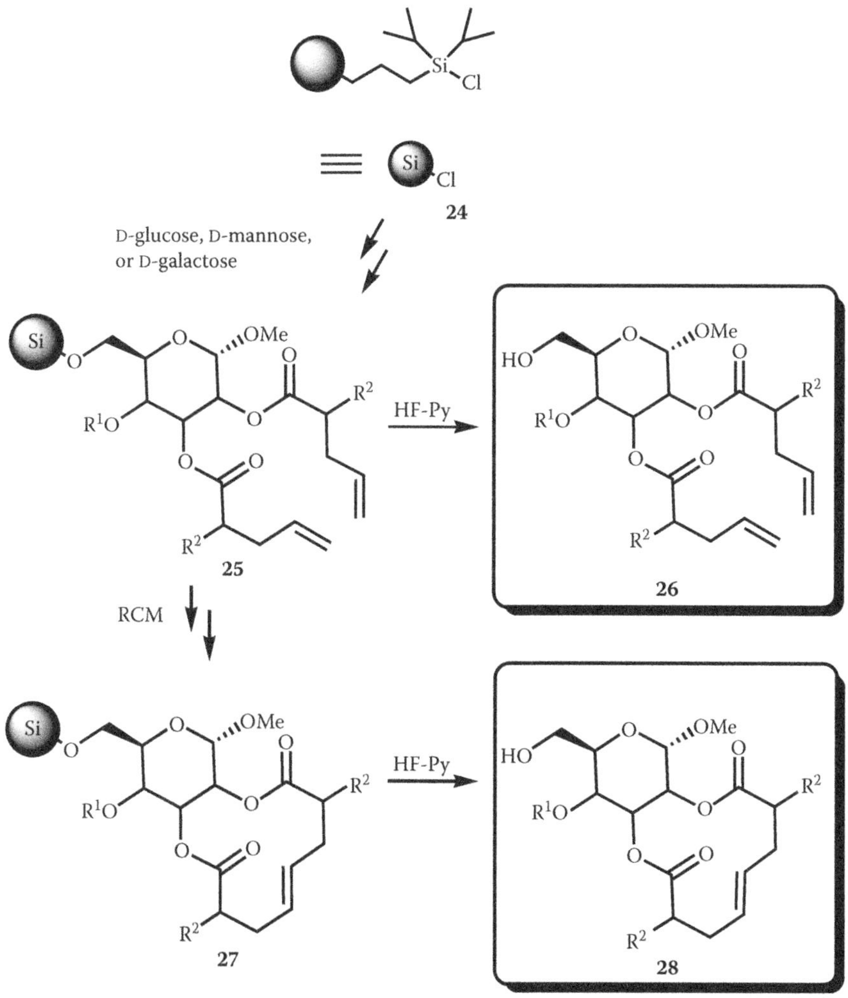

SCHEME 7.6 Synthesis of monosaccharides and medium-sized [10.4.0] bicyclic compounds.

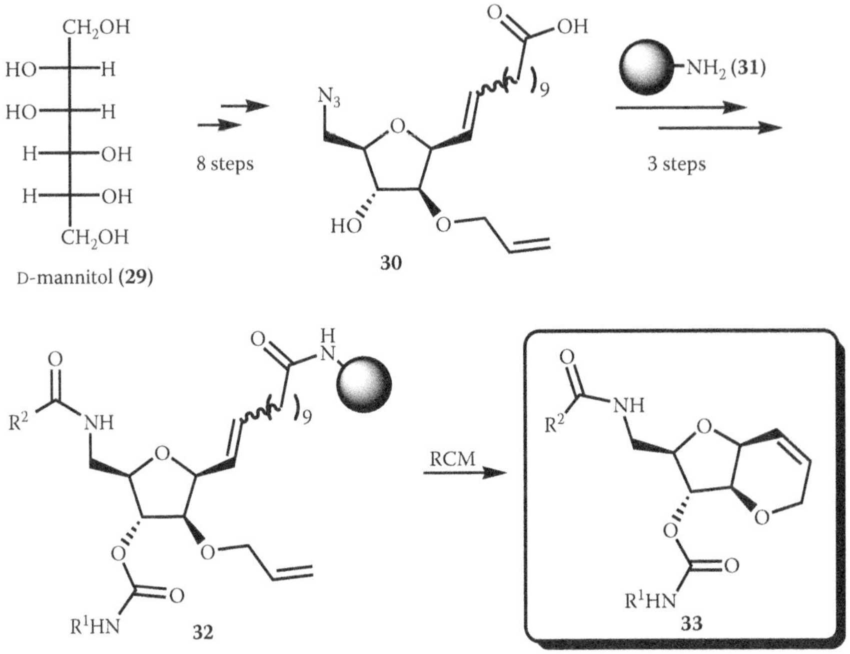

SCHEME 7.7 Synthesis of mannitol-derived oxacycles.

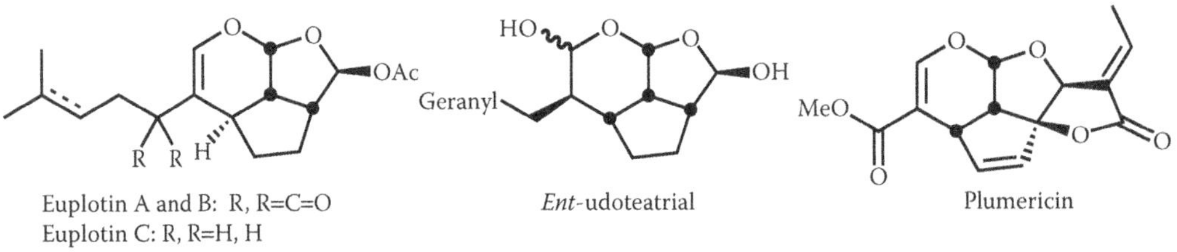

Euplotin A and B: R, R=C=O
Euplotin C: R, R=H, H

Ent-udoteatrial

Plumericin

FIGURE 7.3 Natural compounds belonging to the iridoid family.

To introduce additional diversity, the 3-nitrophenyl and 4-nitrophenyl derivatives in the R^1 position were further functionalized. For example, the nitro group and enol ether of 3-nitrophenyl **39** were reduced, and the resulting aniline was acylated to give **41** (Scheme 7.9). Alternatively, selective reduction of the nitro group worked most effectively in the presence of ammonium formate to give compounds **42**.

Furthermore, Schreiber and coworkers reported the split-pool synthesis of tricyclic libraries from a glycal template.[27] In their diversity-oriented synthesis approach,[28] each synthetic step introduced structural and stereochemical complexity. Starting from 3,4,6-tri-*O*-acetyl-D-glucal (**43**) prepared from D-glucose, (*R*) or (*S*) propargylic alcohols **44** were coupled using the Ferrier reaction to furnish propargylic ethers **45** (Scheme 7.10). After loading alcohols **45** onto solid-support **24**, the free hydroxyls were converted to carbamates **47** with isocyanates. Deprotection, formation of the triflate, and amine addition introduced the second site of diversity and gave (*R*)-**48** and (*S*)-**48**. At this point, diversity was introduced to the terminal alkyne of the (*R*)-**48** series, and the Pauson–Khand reaction was performed on (*S*)-**48** (Scheme 7.11). Subsequent to these transformations, the Pauson–Khand reaction was performed on **49**, and alkyl thiols were added by Michael addition to **50**. Reduction of the ketone in each series followed by acylation introduced the fourth site of diversity and furnished **53** and **54**.

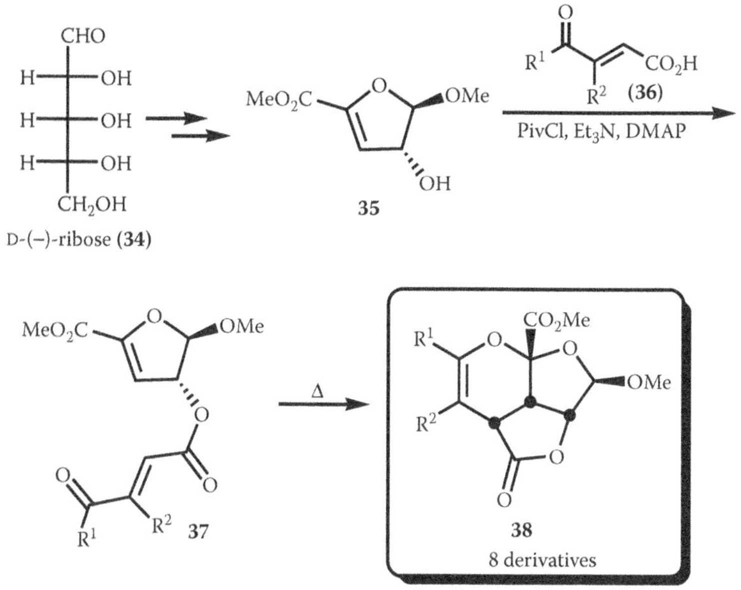

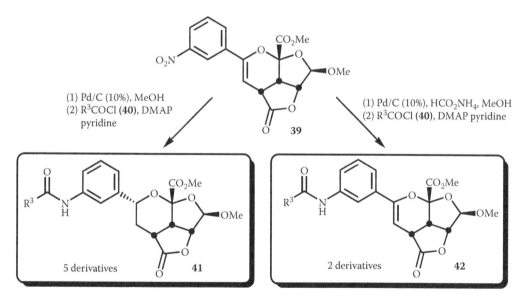

SCHEME 7.8 Synthesis of a tricyclic perhydrofuropyran core derived from D-(−)-ribose.

SCHEME 7.9 Functionalization of a tricyclic perhydrofuropyran core derived from D-(−)-ribose.

7.2.4 SPIROCYCLIC PYRANOSE LIBRARIES

In the universe of biologically active natural products, the spirocyclic moiety is a unique and key structural motif. For example, there are over 300 spiroketal natural products[29] including the avermectins and milbemycins, which demonstrate a range of antiparasitic activity with low toxicity (Figure 7.4).[30] A semisynthetic derivative, ivermectin, is used in animals to treat parasite infestations. The spongistatins exhibit anticancer activity by disrupting tubulin polymerization. Okadaic acid is a cytotoxin found in marine sponges.

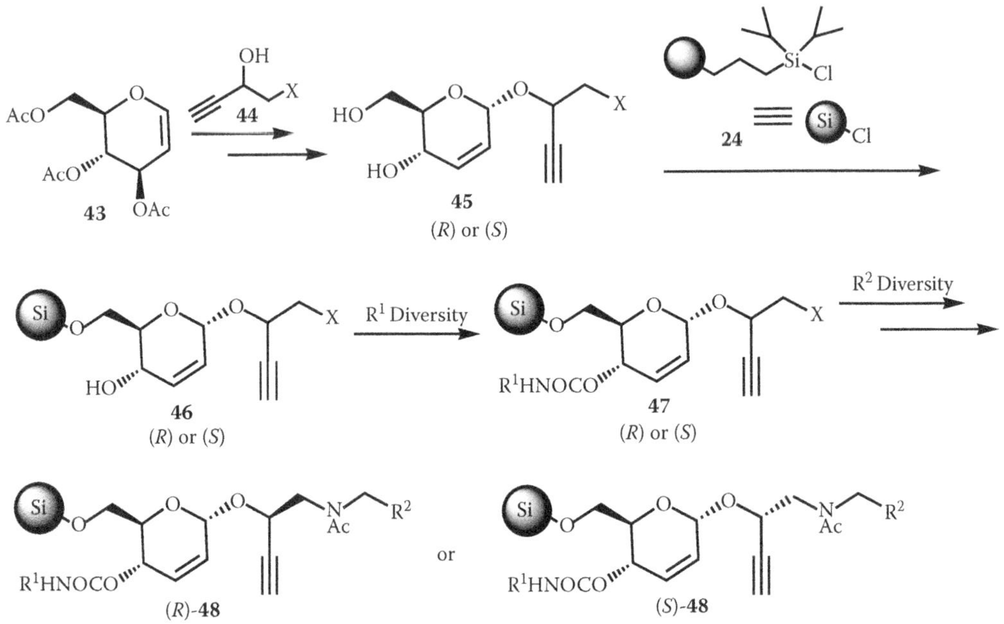

SCHEME 7.10 Synthesis of Pauson–Khand substrates from glycal template.

There have been several recent solid-phase and solution-phase libraries reported that contain the spiroketal moiety (Figure 7.5). These libraries introduce additional structural complexity and new interaction modes of carbohydrate-like molecules with biological targets. Ley and coworkers reported the solid-supported synthesis of spiroketals.[31] Using stereoselective aldol reactions of boron enolates, the Waldmann and Paterson groups reported the solid-phase synthesis of 6,6-spiroketals.[32] The 1,7-dioxaspiro [5.5]undecane spiroketal served as the core template for Porco's three-point diversity library to explore protein–protein interactions.[33]

Walters and coworkers prepared and functionalized a 6,6-disubsituted spirotetrahydropyran 3,4-epoxide **57** to furnish various spirocyclic compounds (Scheme 7.12).[34] The spirocyclic ring system **56** was synthesized using a ring-closing metathesis reaction of an intermediate prepared from Boc-4-piperidone (**55**). After Boc-deprotection and acylation of the piperidine nitrogen, epoxidation of the endocyclic olefin of **56** gave epoxide **57**. Regioselective addition of amines **58** to epoxide **57**, under nonchelating conditions consistent with the Fürst–Plattner rule and steric considerations, gave predominantly one regioisomer of amino alcohols **59**.

7.3 CARBOHYDRATE-DERIVED NATURAL PRODUCT LEAD DISCOVERY LIBRARIES

In the context of an ongoing combinatorial chemistry development program at Discovery Partners International, carbohydrates[35] have been useful for the construction of small-molecule libraries. Various carbohydrates have served as starting materials for the preparation of natural product-like arrays (Figure 7.6). Using diacetone-D-glucose prepared from D-glucose, isosorbide prepared from D-glucitol (sorbitol), and isomannide prepared from D-mannitol, libraries of furanoses, dihydroxy-tetrahydrofurans, and hexahydrofurofurans were prepared, respectively.

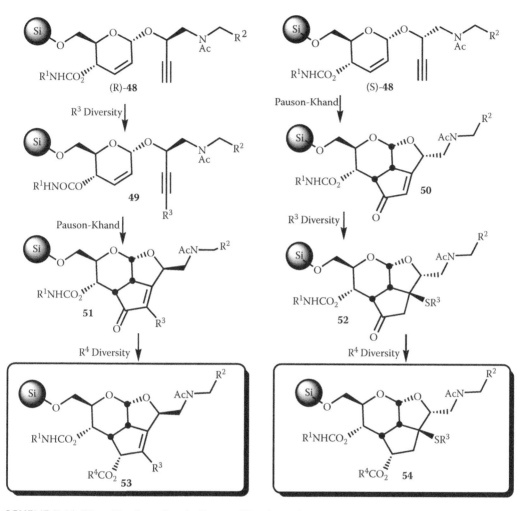

SCHEME 7.11 Diversification using the Pauson–Khand reaction.

7.3.1 FURANOSES FROM DIACETONE D-GLUCOSE

The tetrahydrofuran ring system is a particularly interesting chemotype for screening libraries. Furanoses show promise as filaricides,[36] and the furanose derivative amiprilose was reported to provide some effectiveness in the treatment of rheumatoid arthritis.[37] A recent report described furanose sugar amino acids as library scaffolds.[38] In 2002, we reported the synthesis of furanose libraries.[39] Diacetone D-glucose (**60**) was treated with powdered potassium hydroxide in DMSO and alkyl halides to furnish alkyl ethers (Scheme 7.13). Subsequent chemoselective hydrolysis of the primary isopropylidines with 70% acetic acid in water gave diols **62**. Finally, sodium periodate adsorbed on silica gel cleaved diols **62** to provide aldehydes **63**.

Amido-, urea-, and aminofuranoses were prepared by first treating alkylated furanose aldehydes **63** with primary amines **64** in the presence of sodium triacetoxyborohydride to give secondary amines **65** (Scheme 7.14). Amberlite® IRA-743 was used to remove borates from the reductive amination reactions. A polymer-supported benzaldehyde was used to scavenge out excess amine. Subsequent acylation with acid chlorides and isocyanates gave amidofuranoses **67a** and ureafuranoses **67b**, respectively. Dowex SBR LC NG OH form anion-exchange resin and tris(2-aminoethyl)amine resin were used to scavenge excess acid chloride and isocyanate, respectively.

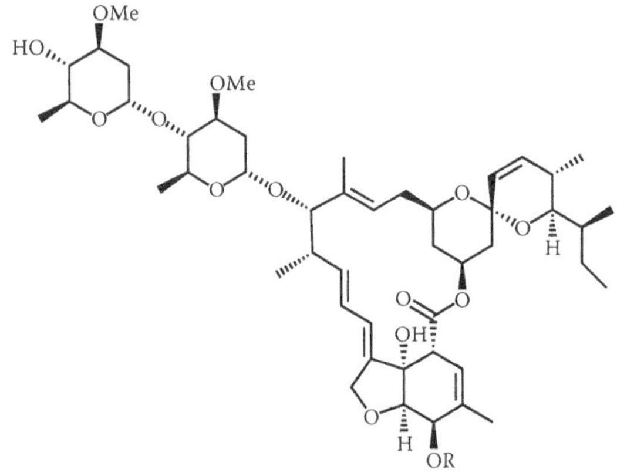

Avermectin A_{1a} (R=Me)
Avermectin B_{1a} (R=H)

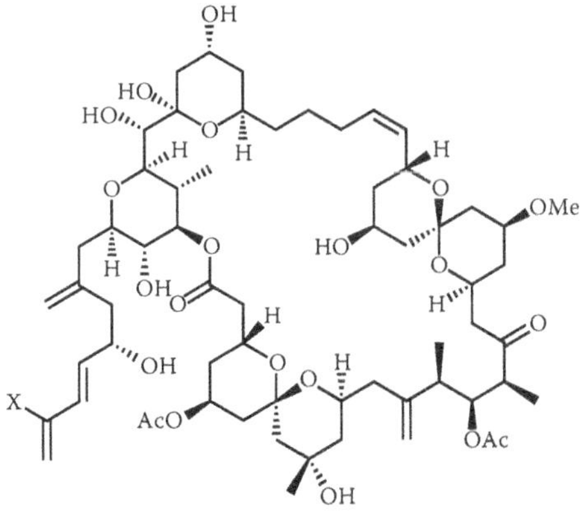

Spongistatin 1 (X=Cl)
Spongistatin 2 (X=H)

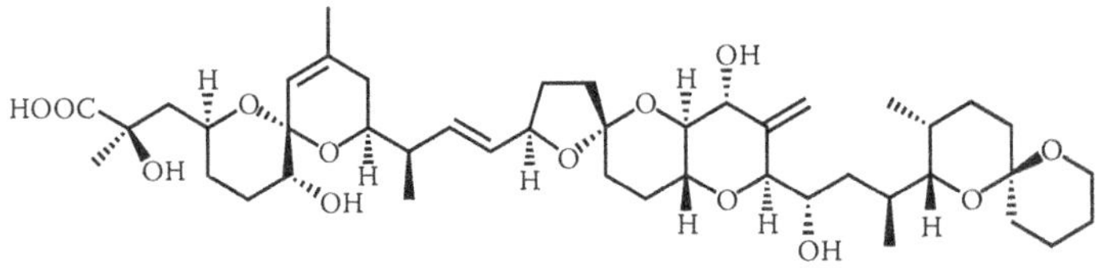

Okadaic acid

FIGURE 7.4 Spirocyclic natural products.

Furthermore, reductive amination of furanose aldehydes **63** with secondary amines **68** yielded tertiary amines **69**. Here, after removal of borates with Amberlite® IRA-743, polymer-supported isocyanate scavenged out excess amine. The resulting acetonides **69** were treated with alcohols **70** in the presence of acid to yield mixed acetals **71**. A final treatment with polymer-supported piperidine removed trace amounts of acid and ensured long-term stability of acetals **71**. These library syntheses highlighted the use of functionalized scavenger resins to purify intermediates and products.

7.3.2 Dihydroxytetrahydrofurans from Isosorbide

A library of dihydroxytetrahydrofurans was readily prepared from isosorbide.[40] Epoxide **73** was produced in two steps from isosorbide (**72**).[41] Subsequently, **73** was heated with secondary amines **68** to afford the corresponding amino alcohols **74** (Scheme 7.15). PS-Isocyanate resin was used to remove excess secondary amine **68**. Treatment of alcohols **74** with isocyanates **75** gave the desired carbamates;[42] PS-Trisamine resin removed excess isocyanate. Aqueous, acid-promoted hydrolysis of acetonides gave diols **76**.

7.3.3 Hexahydrofurofurans from Isomannide

Isomannide (**80**) was the core for a hexahydrofurofuran library.[43] Primary amines were loaded onto solid-support by reductive amination and acylated with bromoacetic acid to give bromides **79** (Scheme 7.16). Alkylation of bromides **79** on solid-support with isomannide (**80**) gave the solid-supported alcohols **81**. A Mitsunobu reaction with phthalimide (**82**) proceeded to furnish amines **83** in excellent yield and purity after removal of the protecting group.[44] Support-bound primary amines **83** were converted to secondary amines by stepwise imine formation with aldehydes **84** and reduction with sodium borohydride.[45] The hindered secondary amines **85** were acylated with acid chlorides, sulfonyl chlorides, isocyanates, and isothiocyanates to yield **87** after cleavage from solid-support.

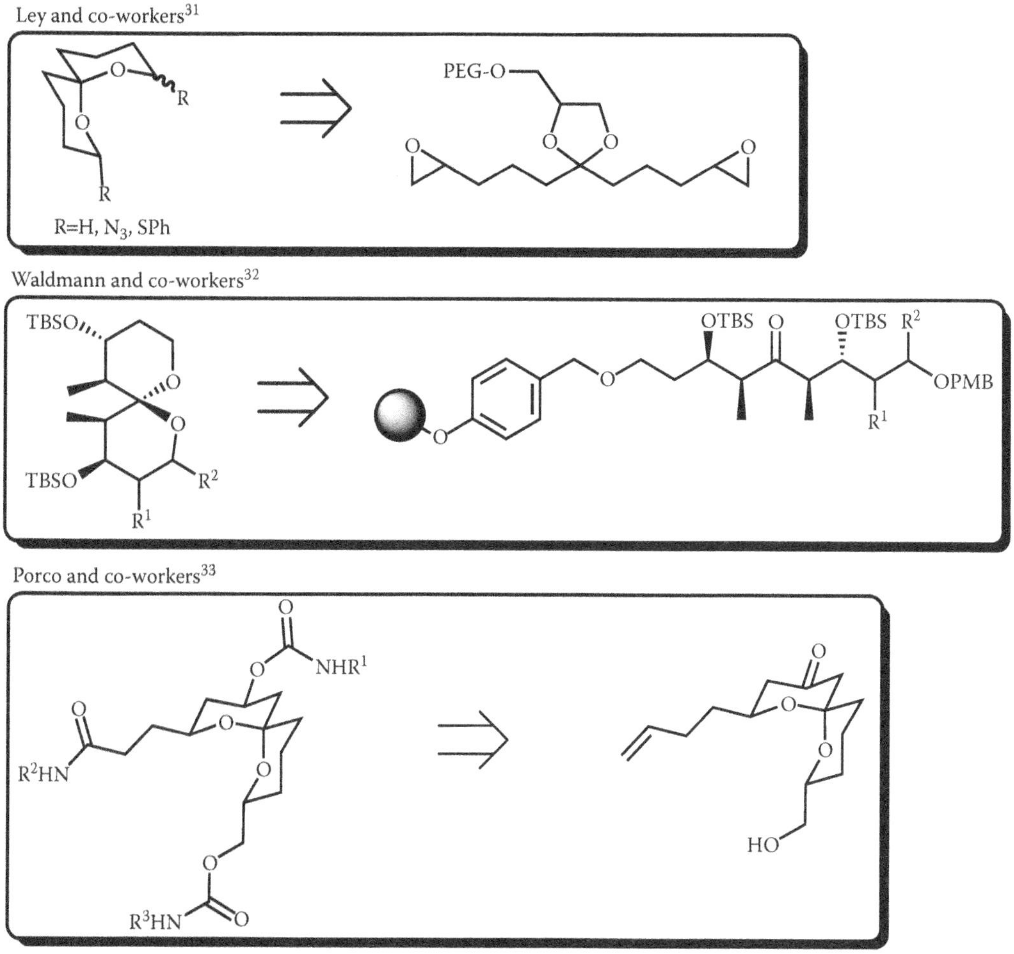
Ley and co-workers31
R
R
R=H, N3, SPh
PEG-O
Waldmann and co-workers32
TBSO
TBSO
R2
R1
OTBS
OTBS
R2
OPMB
R1
Porco and co-workers33
NHR1
R2HN
R3HN
HO

FIGURE 7.5 Spirocyclic pyranose-based natural product-like libraries.

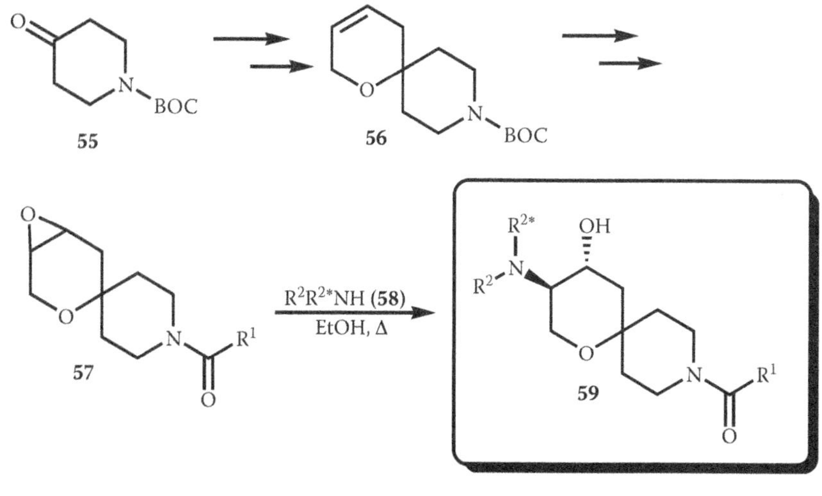
BOC
55
56
BOC
57
R1
R2R2*NH (58)
EtOH, Δ
R2*
OH
R2
59
R1

SCHEME 7.12 Spirocyclic 1-oxa-9-aza-spiro[5.5]undecane scaffold library.

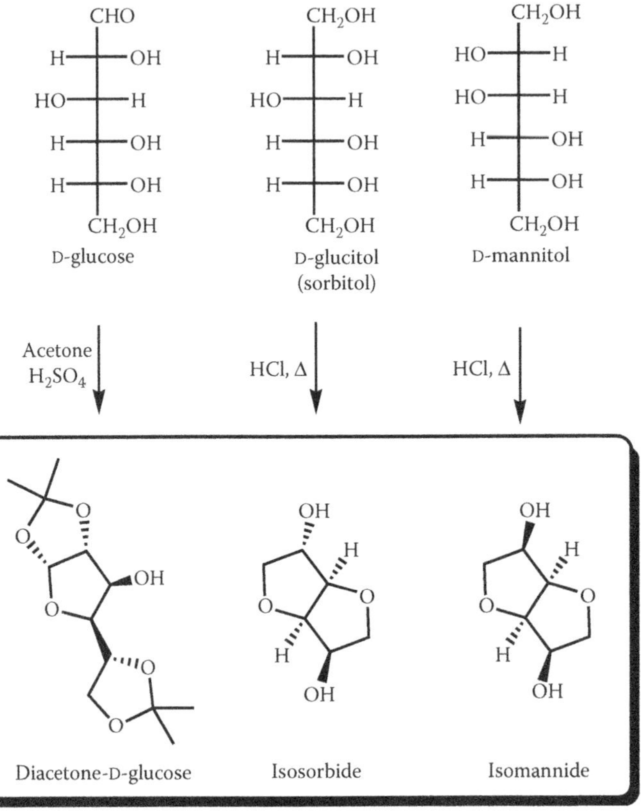

FIGURE 7.6 Carbohydrate starting materials used for the synthesis of natural-product-like libraries.

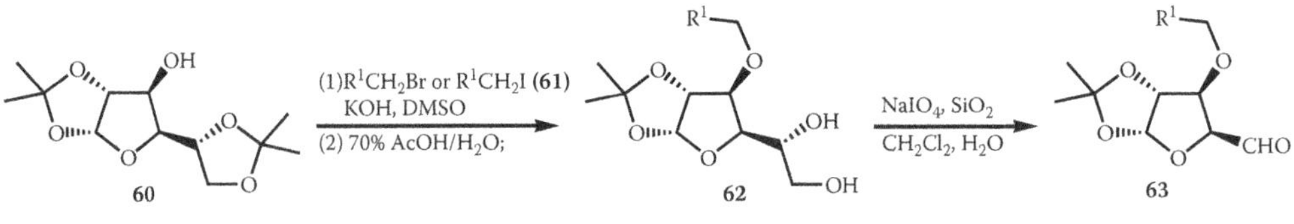

SCHEME 7.13 Preparation of aldehyde scaffolds from diacetone-D-glucose.

7.4 SUMMARY

The examples and approaches described in this chapter to generate natural product-based libraries illustrate the utility of carbohydrates for generating interesting chemotypes to explore biological activity. These pharmacologically rich templates are ideally suited for high-throughput synthesis and combinatorial chemistry. Through continued efforts to use carbohydrates as scaffolds for the generation of compound libraries, our understanding of the structure–activity relationships of small-molecule interactions with their biological targets will increase. Furthermore, new pharmaceutical products will result from the application of this approach to the discovery of therapeutics.

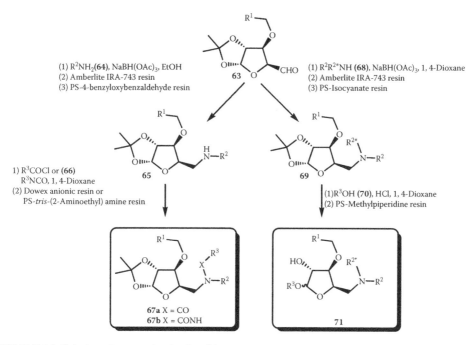

SCHEME 7.14 Solution-phase synthesis of amido-, urea-, and aminofuranoses.

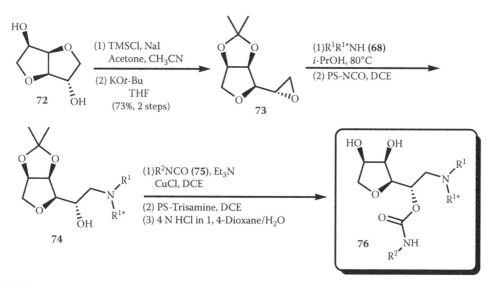

SCHEME 7.15 Solution-phase synthesis of dihydroxytetrahydrofurans from isosorbide.

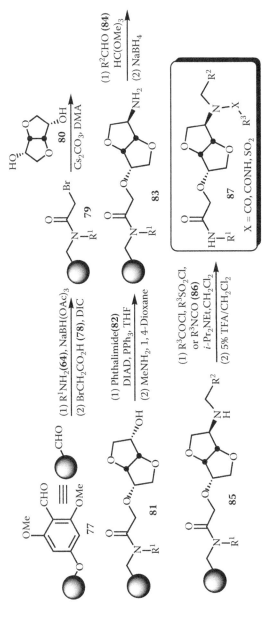

SCHEME 7.16 Solid-phase synthesis of hexahydrofurofurans from isomannide.

REFERENCES

1. Hirschmann, R., Ducry, L., and Smith, A.B., III, Development of an efficient, regio-, and stereoselective route to libraries based on the β-D-glucose scaffold, *J. Org. Chem.*, 65, 8307, 2000.
2. (a) Witczak, Z.J. and Tatsuta, K. (Eds.), Carbohydrate synthons in natural products chemistry: synthesis, functionalization, and applications, American Chemical Society, Washington, D.C., 2003. (b) Hollingsworth, R.I. and Wang, G., Toward a carbohydrate-based chemistry: progress in the development of general-purpose chiral synthons from carbohydrates, *Chem. Rev.*, 100, 4267, 2000.
3. Musser, J.H., Carbohydrate-based therapeutics, in *Burger's Medicinal Chemistry and Drug Discovery*, Vol. 2, Abraham, D.J., Ed., John Wiley & Sons, New York, 2003, chap. 7.
4. Hegde, S. and Carter, J., To market, to market-2003, *Annu. Rep. Med. Chem.*, 39, 337, 2004.
5. Maryanoff, B.E. et al., Structure-activity studies on anticonvulsant sugar sulfamates related to topiramate. Enhanced potency with cyclic sulfate derivatives, *J. Med. Chem.*, 41, 1315, 1998.
6. Oxford, J.S., Zanamivir (Glaxo Wellcome), *IDrugs*, 3, 447, 2000.
7. Stavenger, R.A. and Schreiber, S.L., Asymmetric catalysis in diversity-oriented organic synthesis: enantioselective synthesis of 4320 encoded and spatially segregated dihydropyrancarboxamides, *Angew. Chem., Int. Ed. Engl.*, 40, 3417, 2001.
8. (a) Dube, D.H. and Bertozzi, C.R., Glycans in cancer and inflammation — potential for therapeutics and diagnostics, *Nat. Rev. Drug Discovery*, 4, 477, 2005. (b) Shriver, Z., Raguram, S., and Sasisekharan, R., Glycomics: a pathway to a class of new and improved therapeutics, *Nat. Rev. Drug Discovery*, 3, 863, 2004. (c) Bertozzi, C.R. and Kiessling, L.L., Chemical glycobiology, *Science*, 291, 2357, 2001. (d) Dwek, R.A., Glycobiology: toward understanding the function of sugars, *Chem. Rev.*, 96, 683, 1996.
9. Gruner, S.A.W. et al., Carbohydrate-based mimetics in drug design: sugar amino acids and carbohydrate scaffolds, *Chem. Rev.*, 102, 491, 2002.
10. Frechét, J.M. and Schuerch, C., Solid-phase synthesis of oligosaccharides. I. Preparation of the solid-support. Poly [p-1-propen-3-ol-1-yl)styrene], *J. Am. Chem. Soc.*, 93, 492, 1971.
11. (a) Hölemann, A. and Seeberger, P.H., Carbohydrate diversity: synthesis of glycoconjugates and complex carbohydrates, *Curr. Opin. Biotech.*, 15, 615, 2004. (b) Ohnsmann, J., Madalinksi, M., and Kunz, H., Carbohydrates as polyfunctional scaffolds in combinatorial synthesis, *Chimica Oggi*, 20, 2005. (c) Marcaurelle, L.A. and Seeberger, P.H., Combinatorial carbohydrate chemistry, *Curr. Opin. Chem. Biol.*, 6, 289, 2002. (d) Seeberger, P.H., Ed., *Solid-Support Oligosaccharide Synthesis and Combinatorial Carbohydrate Libraries*, John Wiley & Sons, New York, 2001. (e) Nishimura, S., Combinatorial syntheses of sugar derivatives, *Curr. Opin. Chem. Biol.*, 5, 325, 2001. (f) Seeberger, P.H. and Haase, W.-C., Solid-phase oligosaccharide synthesis and combinatorial carbohydrate libraries, *Chem. Rev.*, 100, 4349, 2000. (g) Schweizer, F. and Hindsgaul, O., Combinatorial synthesis of carbohydrates, *Curr. Opin. Chem. Biol.*, 3, 291, 1999. (h) Sofia, M.J. and Silva, D.J., Recent developments in solid-and solution-phase methods for generating carbohydrate libraries, *Curr. Opin. Drug Discovery Dev.*, 2, 365, 1999. (i) Kahne, D., Combinatorial approaches to carbohydrates, *Curr. Opin. Chem. Biol.*, 1, 130, 1997.
12. Evans, B.E. et al., Methods for drug discovery: development of potent, selective, orally effective cholecystokinin antagonists, *J. Med. Chem.*, 31, 2235, 1988.
13. (a) Hirschmann, R. et al., De novo design and synthesis of somatostatin non-peptide peptidomimetics utilizing β-D-glucose as a novel scaffolding, *J. Am. Chem. Soc.*, 115, 12550, 1993. (b) Hirschmann, R. et al., Synthesis of potent cyclic hexapeptide NK-1 antagonists. Use of a mini-library in transforming a peptidal somatostatin receptor ligand into an NK-1 receptor ligand via a polyvalent peptidomimetic, *J. Med. Chem.*, 39, 2441, 1996. (c) Hirschmann, R. et al., Modulation of receptor and receptor subtype affinities using diastereomeric and enantiomeric monosaccharide scaffolds as a means to structural and biological diversity: a new route to ether synthesis, *J. Med. Chem.*, 41, 1382, 1998.
14. Prasad, V. et al., Effects of heterocyclic aromatic substituents on binding affinities at two distinct sites of somatostain receptors. Correlation with the electrostatic potential of the substituents, *J. Med. Chem.*, 46, 1858, 2003.
15. For other examples, see (a) Papageorgiou, C. et al., *Bioorg. Med Chem. Lett.*, 2, 135, 1992. (b) Nicolaou, K.C., Trujillo, J.I., and Chibale, K., *Tetrahedron*, 53, 8751, 1997. (c) Diguarher, T.L. et al., *Bioorg. Med. Chem. Lett.*, 6, 1983, 1996.

16. McDevitt, J.P. and Lansbury, P.T., Jr., Glycosamino acids: new building blocks for combinatorial synthesis, *J. Am. Chem. Soc.,* 118, 3818, 1996.

17. Sofia, M.J. et al., Carbohydrate-based small-molecule scaffolds for the construction of universal pharmacophore mapping libraries, *J. Org. Chem.,* 63, 2802, 1998.

18. (a) Murphy, P.V. et al., Synthesis of novel HIV-1 protease inhibitors based on carbohydrate scaffolds, *Tetrahedron,* 59, 2259, 2003. (b) Chery, F. et al., Synthesis of peptidomimetics based on iminosugar and β-D-glucopyranoside scaffolds and inhibition of HIV-protease, *Tetrahedron,* 60, 6597, 2004.

19. Wunberg, T. et al., Carbohydrates as multifunctional chiral scaffolds in combinatorial synthesis, *Angew. Chem., Int. Ed. Engl.,* 37, 2503, 1998.

20. Kallus, C. et al., *Tetrahedron Lett.,* 40, 7783, 1999.

21. (a) Opatz, T. et al., D-Glucose as a pentavalent chiral scaffold, *Eur. J. Org. Chem.,* 1527, 2003. (b) Opatz, T. et al., *Carbohydrate Res.,* 337, 2089, 2002.

22. Boer, J. et al., Design, synthesis, and biological evaluation of $\alpha_4\beta_1$ integrin antagonists based on β-D-mannose as rigid scaffold, *Angew. Chem., Int. Ed. Engl.,* 40, 3870, 2001.

23. Wong, C.-H. et al., A library approach to the discovery of small molecules that recognize RNA: use of a 1,3-hydroxyamine motif as core, *J. Am. Chem. Soc.,* 120, 8319, 1998.

24. Messer, R., Fuhrer, C.A., and Häner, R., Natural product-like libraries based on non-aromatic, polycyclic motifs, *Curr. Opin. Chem. Biol.,* 9, 259, 2005.

25. Kim, Y.-K. et al., Relationship of stereochemical and skeletal diversity of small molecules to cellular measurement space, *J. Am. Chem. Soc.,* 126, 14740, 2004.

26. Timmer, M.S.M. et al., The use of a mannitol-derived fused oxacycle as a combinatorial scaffold, *J. Org. Chem.,* 68, 9406, 2003.

27. (a) Kubota, H., Lim, J., Depew, K.M., and Schreiber, S.L., Pathway development and pilot library realization in diversity-oriented synthesis: exploring Ferrier and Pauson–Khand reactions on a glycal template, *Chem. Biol.,* 9, 265, 2002. (b) Hotha, S. and Tripathi, A., Diversity oriented synthesis of tricyclic compounds from glycals using the Ferrier and the Pauson-Khand reactions, *J. Comb. Chem.,* 7, 968, 2005.

28. (a) Burke, M.D., Berger, E.M., and Schreiber, S.L., A synthesis strategy yielding skeletally diverse small molecules combinatorially, *J. Am. Chem. Soc.,* 126, 14095, 2004. (b) Burke, M.D. and Schreiber, S.L., A planning strategy for diversity-oriented synthesis, *Angew. Chem., Int. Ed. Engl.,* 43, 46, 2004. (c) Schreiber S.L., Target-oriented and diversity-oriented organic synthesis in drug discovery, *Science,* 287, 1964, 2000.

29. Buckingham, J., *Dictionary of Natural Products,* Taylor and Francis/CRC Press, London, 2005 (CD-ROM).

30. Omura, S., Ed., *Macrolide Antibiotics: Chemistry, Biology, and Practice,* 2nd ed., Academic Press, San Diego, CA, 2002.

31. Haag, R. et al., New polyethylene glycol polymers as ketal protecting groups — a polymer supported approach to symmetrically substituted spiroketals, *Synth. Commun.,* 31, 2965, 2001.

32. (a) Barun, O., Sommer, S., and Waldmann, H., Asymmetric solid-phase synthesis of 6,6-spiroketals, *Angew. Chem., Int. Ed. Engl.,* 43, 3195, 2004. (b) Paterson, I., Gottschling, D., and Menche, D., Towards the combinatorial synthesis of spongistatin fragment libraries by using asymmetric aldol reactions on solid-support, *Chem. Commun.,* 3568, 2005.

33. Kulkarni, B.A. et al., Combinatorial synthesis of natural product-like molecules using a first-generation spiroketal scaffold, *J. Comb. Chem.,* 4, 56, 2002.

34. Walters, M.A. et al., Templates for exploratory library preparation. Derivatization of a functionalized spirocyclic 3,6-dihydro-2H-pyran formed by ring-closing metathesis reaction, *J. Comb. Chem.,* 4, 125, 2002.

35. (a) Lichtenthaler, F.W., Enantiopure building blocks from sugars and their utilization in natural product synthesis, in *Modern Synthetic Methods,* Scheffold, R., Ed., VCH, New York, 1992, pp. 273–376. (b) Hollingsworth, R.I. and Wang, G., *Chem. Rev.,* 100, 4267, 2000.

36. Chauhan, P.M.S., *Drugs Future,* 25, 481, 2000.

37. Prous, J.R., Ed., *Drugs Future,* 25, 402, 2000.

38. Edwards, A.A. et al., Tetrahydrofuran-based amino acids as library scaffolds, *J. Comb. Chem.,* 6, 230, 2004.

39. (a) Krueger, E.B. et al., Solution-phase library synthesis of furanoses, *J. Comb. Chem.*, 4, 229, 2002. (b) Boldi, A.M. et al., Aminofuranose Compounds, U.S. Patent 6,794,497, September 21, 2004.

40. (a) Boldi, A.M., Building discovery libraries from natural products, *ACS Prospectives Conference on Combinatorial Chemistry: New Methods, New Discoveries*, September 21–24, 2003. (b) Krueger, E.B., Hu, C., and Boldi, A.M., unpublished results.

41. Ejjiya, S., Saluzzo, C., and Amouroux, R., *O4,O5*-Isopropylidene-1,2:3,6-dianhydro-D-glucitol from isosorbide, *Org. Synth.*, 77, 91, 1999.

42. (a) Duggan, M.E. and Imagire, J.S., Copper(I) chloride catalyzed addition of alcohols to alkyl isocyanates, A mild and expedient method for alkyl carbamate formation, *Synthesis,* 131, 1989. (b) Ohe, K. et al., A facile preparation of 4-alkylidene-3-tosyloxazolidin-2-ones from propargylic alcohols and *p*-toluenesulfonyl isocyanate using a cuprous iodide/triethylamine catalyst, *J. Org. Chem.*, 56, 2267, 1991.

43. (a) Boldi, A.M., Libraries containing natural product structural motifs, *CHI 6th Annual Conference on High-Throughput Organic Synthesis*, San Diego, CA, February 14–16, 2001. (b) Boldi, A.M., Libraries containing natural product Structural Motifs, *CHI 5th Annual Conference on Drug Discovery Japan*, Tokyo, Japan, March 27–29, 2001. (c) Eissa, H.O. and Boldi, A.M., unpublished results.

44. Adams, J.H. et al., A reinvestigation of the preparation, properties, and applications of aminomethyl and 4-methylbenzhydrylamine polystyrene resins, *J. Org. Chem.*, 63, 3706, 1998.

45. Szardenings, A.K. et al., A reductive alkylation procedure applicable to both solution- and solid-phase syntheses of secondary amines, *J. Org. Chem.*, 61, 6720, 1996.

8 In Search of Novel Antibiotics Using a Natural Product Template Approach

Michael J. Sofia

CONTENTS

8.1 INTRODUCTION

Natural products have played a prominent role in the history of drug discovery (Figure 8.1). The uniqueness of natural products and the rationale for their exceptional impact has been widely cited.[1,2] As biosynthetically derived secondary metabolites, natural products inherently contain structural characteristics that are recognized by proteins. The fact that many natural products express biological activity in their natural environment supports a view that structural characteristics common to many natural products favor interaction with important regulatory biomolecules such as proteins, oligonucleotides, and oligosaccharides.[3] Analyses also have shown that natural products contain many of the characteristics found in marketed synthetic drugs. These common drug characteristics include many ring systems, physicochemical characteristics, and constellations of pharmacophoric groups.[4–6] However, natural products are unique when compared to biologically active synthetic compounds. Natural products are distinguished by a larger diversity of ring systems, and ring systems of more complex architecture. In addition, natural products are uniquely characterized by their high stereochemical complexities and functional group density. It is therefore the combination of inherent biological activity, drug-like characteristics, and distinctive chemical architecture that makes natural products valuable as molecular templates when searching for novel biologically active medicinal agents through the implementation of chemical library approaches.

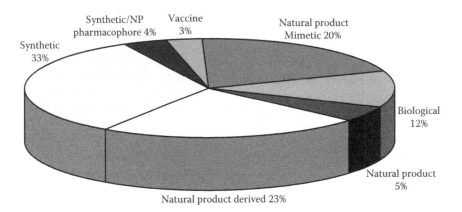

FIGURE 8.1 Sources of drugs from 1981 to 2002.

Over the last 20 years, natural products have had a tremendous impact on the field of infectious disease drug discovery by contributing to the development of approximately 80% of the therapeutically used agents.[1,7] Particularly striking is the impact on the discovery of novel antibacterial agents, where over 50 new drugs have made it to market. These natural product or natural product-derived agents have been developed against three major mechanisms of action by which clinically used antibacterial agents act: inhibition of cell wall biosynthesis, inhibition of protein synthesis, and inhibition of cell wall function (Figure 8.2). Natural product inhibitors of cell wall biosynthesis include β-lactams (penicillins and cephalosporins), and glycopeptides (vancomycin and teicoplanin). Macrolides (erythromycin and clarithromycin), aminoglycosides (gentamicin and kanamycin), and the tetracyclines (tetracycline and chlortetracycline) exemplify natural product inhibitors of bacterial protein synthesis. Peptide antibiotics such as gramicidin and bacitracin inhibit cell wall function. The inherent success of using natural product approaches for the discovery of new antibacterial agents makes it fertile ground for continued exploration. In addition, the rise of multidrug resistant pathogenic bacteria has made new antibacterial agents a medical imperative.[8–10]

In pursuing the discovery of novel medicinal agents using natural product templates, two strategies can be considered.[3,11–14] These strategies are a hypothesis-driven and a broad-screening (or serendipitous) strategy. The hypothesis-driven approach relies on the identification of a natural product or derivative that shows either biological activity in a therapeutic area of interest or whose biological mode of action is mechanistically understood. Understanding of a natural product's fundamental biological mode of action against a particular class of protein or other biomolecule suggests its use as a molecular template to probe for activity against any biological phenomenon in which a related biomolecular target is involved. In this approach biological screening is focused against a particular target or pathway, and available structure–activity relationships (SARs) direct the chemical diversity strategy around the template.

The broad-screening or serendipitous approach takes advantage of a natural product's unique structural and physicochemical characteristics to probe for novel biological activity. In this approach, no preexisting knowledge of biological activity associated with the natural product is required, and a strategy for introducing maximal chemical diversity is the primary focus of the chemical effort. The serendipitous approach requires access to significant screening capabilities that can expose the chemical library to many different targets, thus increasing the probability of success.

Because the hypothesis-driven approach takes advantage of existing knowledge about the natural product and potential biological target opportunities, the chance of success is significantly improved compared to the serendipitous approach. Therefore, the demands on large and expensive screening campaigns to obtain value from the template-derived library are substantially reduced. Consequently, most reports describing the use of natural product templates as platforms for novel

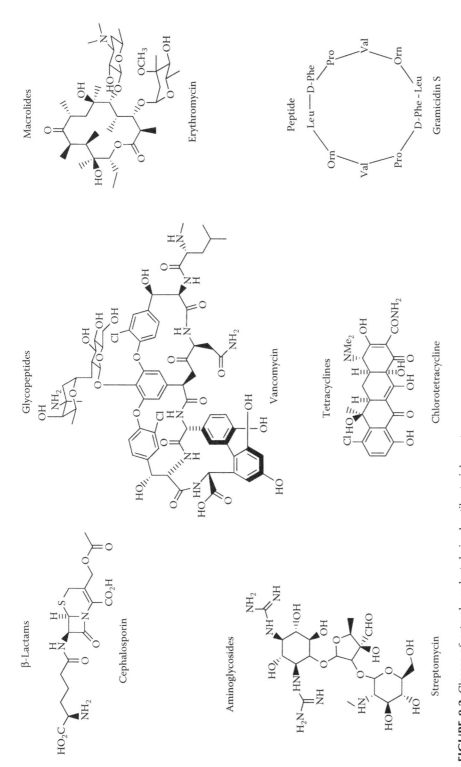

FIGURE 8.2 Classes of natural product-derived antibacterial agents.

molecular diversity have focused on hypothesis-driven strategies that allow focused screening and directed chemical diversity approaches.[15–18]

A natural product template can be characterized as the natural product itself or as its derivative. It can also be that portion of the natural product which conveys biological activity or is the key pharmacophoric unit. Producing a library of structurally diverse compounds starting with a natural product template can proceed with varying degrees of complexity, depending on the sophistication of the chemical modifications. In many cases the natural product is modified in such a way as to allow selective introduction of chemical diversity into one or more chemically accessible sites. This is usually accomplished by either taking advantage of selective functional group reactivity or by protecting the template in such a fashion to allow selective chemical modification. Introduction of more complex structural modifications may require multistep synthesis, in which diversity is introduced by utilizing a synthetic strategy that constructs the template using a diverse set of reagent building blocks. Solid-phase or solution-phase methods have been used for natural product-based library synthesis, with solid-phase methods being preferred for the development of larger libraries and libraries requiring multiple synthetic steps.[11–19]

Irrespective of the strategy employed, two major challenges exist when exploring chemical diversity using natural product templates. These challenges are access to template material and synthetic accessibility for introducing chemical diversity. The template can be obtained through several different approaches: Isolation from the natural source (microbe, plant, marine organism), total synthesis, and semisynthesis. The accessibility of the natural source and the abundance of the parent natural product will determine the viability of both the isolation and semisynthetic approaches. The complexity of the natural product template will dictate the feasibility of template total synthesis. Also, the extent and diversity of the desired chemical modifications will influence the preferred route of access to template material. Introduction of chemical diversity that requires significant structural modifications will favor template construction by total synthetic or semisynthetic means. A diversity strategy that requires substituent variation of an existing reactive functional group is more easily accomplished if the template can be obtained through isolation, especially if the template is structurally complex.

This chapter will present several examples of the use of natural product template strategies to search for novel antimicrobial agents. These examples will highlight the use of hypothesis-driven strategies. They will also demonstrate varied levels of synthetic complexity and highlight several different approaches to template accessibility.

8.2 MOENOMYCIN (FLAVOMYCIN®)

8.2.1 Background

Inhibition of bacterial cell wall biosynthesis is a well-validated strategy for the development of clinically effective antibacterial agents.[20] The cell wall is composed of cross-linked peptidoglycan that results from the polymerization of a monomeric lipid II intermediate (N-acetylglucosamine-β-1,4-MurNAc-pentapeptide-pyrophosphoryl-undecaprenol). The lipid II intermediate is synthesized on the cytoplasmic side of the cell membrane, and polymerization and cross-linking steps occur on the periplasmic side to give the fully elaborated peptidoglycan cell wall (Figure 8.3). To elaborate the peptidoglycan structure, two enzymatic steps are needed. Transglycosylation links the disaccharide portions of lipid II to give the glycan backbone, and a transpeptidization step catalyzes the cross-linking of the peptide chains. Both the transglycosylation and transpeptidization activities reside in a class of proteins known as the penicillin-binding proteins (PBPs).[20] Several clinically used drug classes are known to inhibit bacterial cell wall biosynthesis. The penicillins and cephalosporins target the transpeptidase activity of PBPs. Glycopeptides such as vancomycin and teicoplanin also act on the transpeptidase activity of PBPs. However, there are no known clinically useful agents that act on the other enzymes involved in bacterial cell wall biosynthesis.[21]

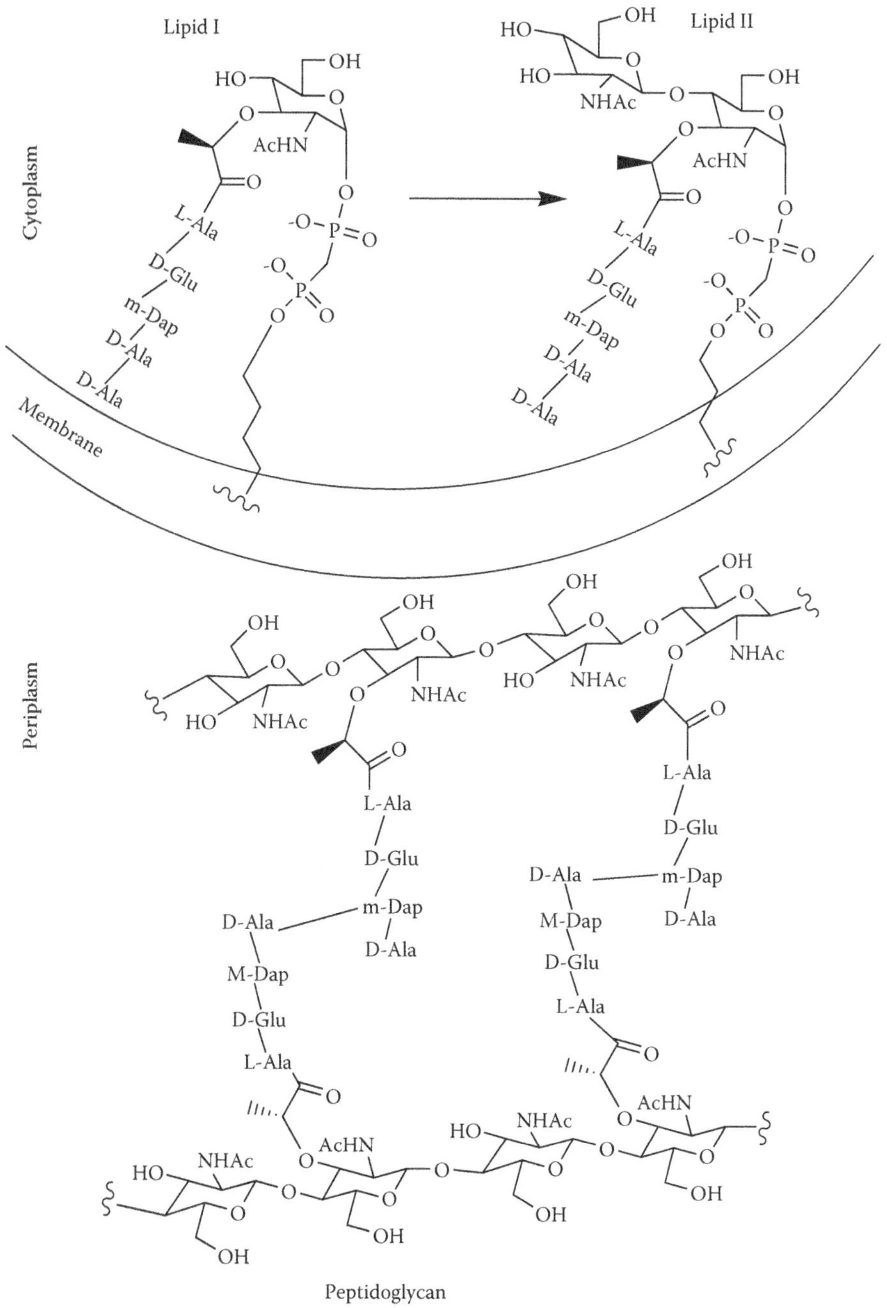

FIGURE 8.3 Bacterial cell wall structure.

The hypothesis that an inhibitor of bacterial cell wall biosynthesis targeting a critical yet unexplored enzyme could result in a novel approach for the development of new antibacterial agents led us to investigate the natural product moenomycin as a template around which to develop a lead generation library effort. Moenomycin A is a pentasaccharide containing a long C-25 lipid, moenocinol, attached to the reducing sugar of the pentasaccharide through a phosphoglycerate unit (Figure 8.4, **1**).[22–25] Isolated from *Streptomyces*, Moenomycin A is a potent inhibitor of the transglycosylase activity of peptidoglycan biosynthesis (MIC [*S. aureus*] 0.05 µg/ml). It has a broad spectrum of

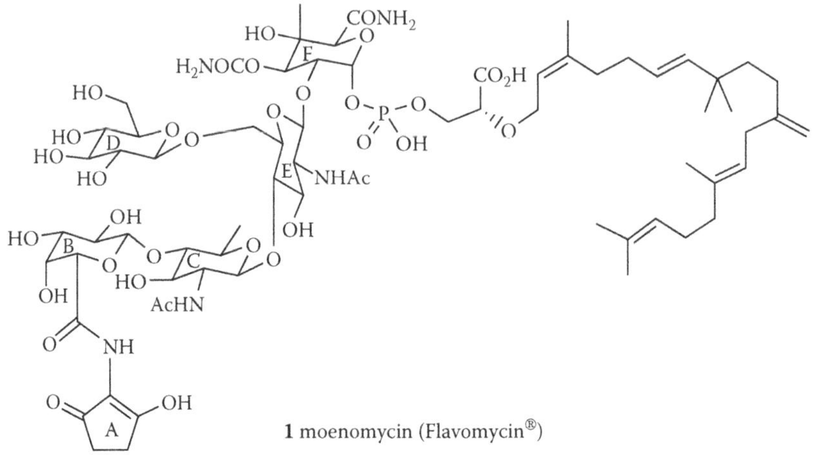

FIGURE 8.4 Moenomycin (Flavomycin®).

activity against Gram-positive microorganisms and is active in mouse models of infection. However, because moenomycin A is poorly absorbed, it is not used as a human clinical agent but as an animal feed growth promoter under the trademark Flavomycin®.[26,27]

Degradation studies and chemical modification have shown that although trisaccharide analogs (Figure 8.5, **2**) retain substantial *in vitro* transglycosylase inhibitory activity in *Escherichia coli*, they are 50-fold less active as antibacterial agents against *Staphylococcus aureus*.[26] In contrast, the disaccharide degradation product **3** does not retain any antibacterial activity, but does retain transglycosylase inhibitory activity *in vitro*.[26] Because the disaccharides possess intrinsic transglycosylase activity and are synthetically more accessible than either moenomycin itself or the trisaccharide analogs, we chose the moenomycin disaccharide as a template on which to build a library that would explore structural and functional group modifications in the hope of identifying analogs that would possess both intrinsic transglycosylase inhibitory activity and antibacterial activity in Gram-positive microbes.

8.2.2 SYNTHETIC STRATEGY

To address the synthetic challenge of building a library of moenomycin disaccharide analogs, we chose a solid-phase strategy that would allow us to investigate a number of structural and functional group modifications combinatorially. Work by Welzel had shown that certain structural features and functionality were important in the biological activity of moenomycin.[28] However, little SAR work was available to guide the choice of the desired modifications or provide any insight into the interrelationship between the functional groups or structural units.[26,28–37] For our SAR work, we decided to focus our attention on three key structural elements: the C-3 carbamate, the C-2′ amide, and the phosphoglycerate moiety at C-1 (Figure 8.5). Individually, each of these groups was shown to be important for activity.[26,28–37] In addition, we chose to investigate how varying the nature of each sugar unit of the disaccharide would affect overall biological activity. Little work had been done to study if variation of the sugars had any affect on antibacterial activity and whether less complex sugars could be substituted for the unusual F-sugar of moenomycin.

To execute a solid-phase library synthesis strategy, we envisioned constructing a set of disaccharide templates that would support introduction of chemical diversity into the C-1, C-3, and C-2′ positions (Figure 8.6).[38] This strategy would require either preconstruction of the disaccharide template and attachment to a solid support or construction of the disaccharide on the solid support through glycosylation of a solid-phase-bound monosaccharide building block. Once the appropriately protected solid-phase-bound disaccharides were available, introduction of diversity at the C-3

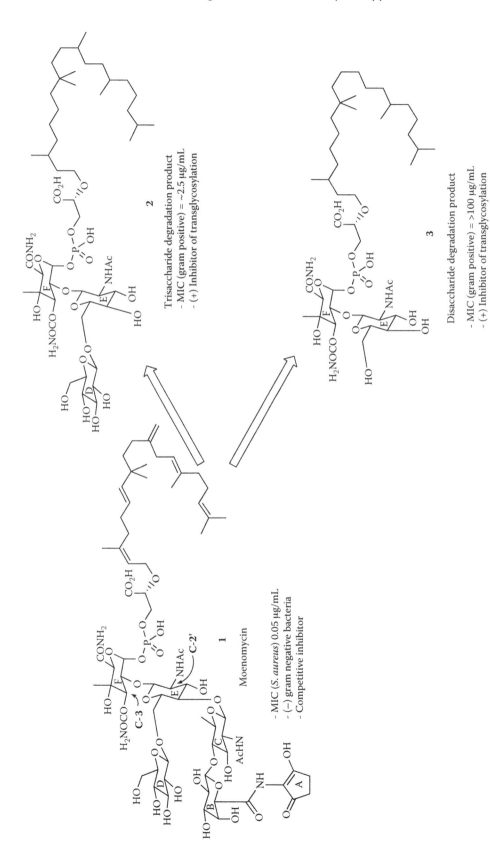

FIGURE 8.5 Degradation of moenomycin (Flavomycin®).

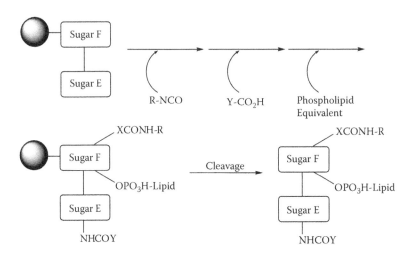

FIGURE 8.6 Disaccharide solid-phase synthesis strategy.

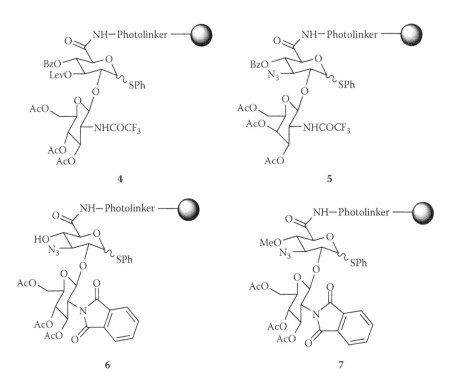

FIGURE 8.7 Disaccharide templates attached to solid support.

and C-2′ positions could proceed by revealing the required functional groups and subsequent derivatization. The final diversity-introducing synthetic step would require attachment of various phosphogylcerate lipid units. The key to this strategy is preparation of disaccharide templates in which functional groups are differentially protected to enable selective site modifications when required.

To execute this strategy, four disaccharide templates **4–7** were constructed (Figure 8.7).[38] Each of the disaccharides was prepared with a β-linkage similar to the linkage configuration of the E-F disaccharide of moenomycin. The C-2′ amino group of each disaccharide was protected with either a trifluoroacetamido or phthalimido base-labile protecting group. These protecting groups could be easily removed to enable further C-2′ amine derivatizations. To explore chemical diversity at the C-3 position of the reducing sugar (F-sugar of moenomycin), a 3-O-levulinate protecting group was employed. In the case where a 3-amino group was desired, an azido group was used as a masked amine. The levulinate protecting group could be removed without removal of other ester protecting groups, thus enabling selective C-3 hydroxyl derivatization. Each disaccharide was also designed to contain an anomeric (C-1) thiophenyl group as a masked anomeric hydroxyl moiety. Once derivatization of the C-2′ and C-3 positions was accomplished, revealing the anomeric hydroxyl group would set the stage for introduction of the phosphogylcerate lipid unit.

To access these disaccharide templates, the synthesis of novel differentially protected monosaccharide building blocks was required. Particularly challenging was the preparation of the unique monosaccharides representing the F-sugar (reducing sugar) of moenomycin disaccharide. The syntheses of these sugars demanded a sophisticated protecting-group scheme that was compatible with a glycosylation step needed to introduce the second sugar, afforded a site for stable attachment to the resin, and supported differential and selective functionalization at the C-3 and C-1 sites. These requirements resulted in building-block syntheses that were multistep and complex, and significant resources were required to develop and prepare sufficient quantities for library synthesis.[39]

Two approaches were used to construct the disaccharide templates — solid-phase glycosylation of a resin-bound monosaccharide and solution-phase construction of the disaccharide followed by linkage to the solid support. Scheme 8.1 demonstrates the application of solid-phase glycosylation to the construction of a disaccharide template.[40,41] This approach used glycosylsulfoxide donor technology and afforded the desired disaccharide **10** in 90 to 95% yield (Scheme 8.1).[40] However, with other donor–acceptor combinations, we were not able to achieve the high-yield glycosylation necessary to eliminate the presence of monosaccharide by-products. Therefore, we prepared disaccharides **5**, **6**, and **7** in solution using standard glycosyl bromide chemistry. Each disaccharide was attached to aminoethyl-photolinker AM resin by amide bond formation of either the monosaccharide building block or the preconstructed disaccharide template. The aminoethyl-photolinker AM resin was chosen because earlier studies had shown that with Rink amide resin the fully elaborated disaccharide would not withstand the acid treatment needed to liberate the disaccharide from the solid support.[38]

As shown in Scheme 8.2, template **4** allowed preparation of carbamates at the C-3 position by selective removal of the levulinate protecting group followed by reaction with isocyanates. The three templates **5**, **6**, and **7** supported introduction of diversity at both the C-3 and C-2′ positions by removal of the base-labile protecting group at C-2′ and subsequent derivatizations to form amides (Scheme 8.3). Reduction of the C-3 azido group to a primary amine and reaction with isocyanates

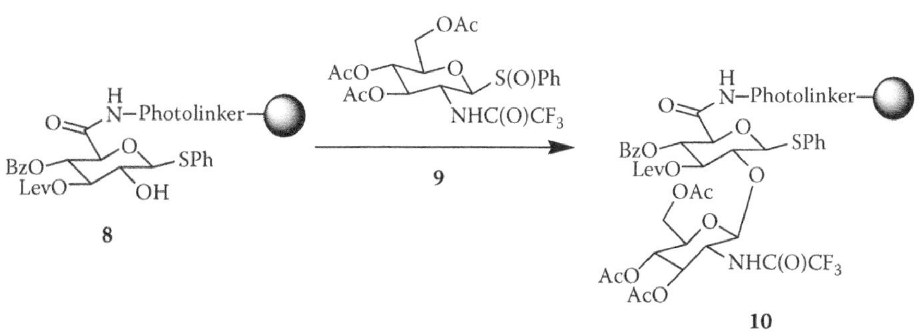

SCHEME 8.1 Solid-phase glycosylation to form β-linked disaccharides.

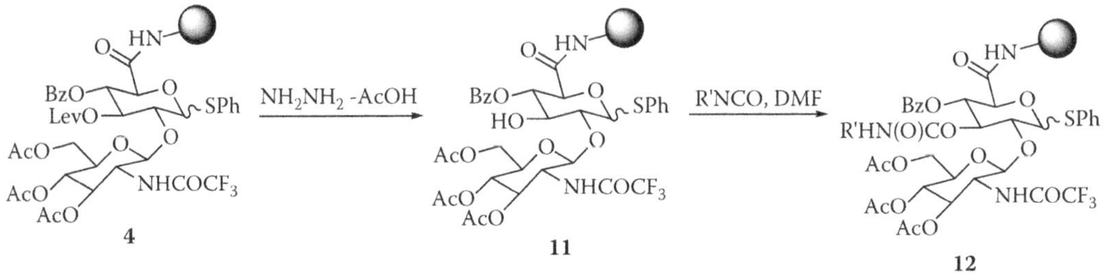

SCHEME 8.2 Selective derivatization of a disaccharide C-3 hydroxyl group on solid phase.

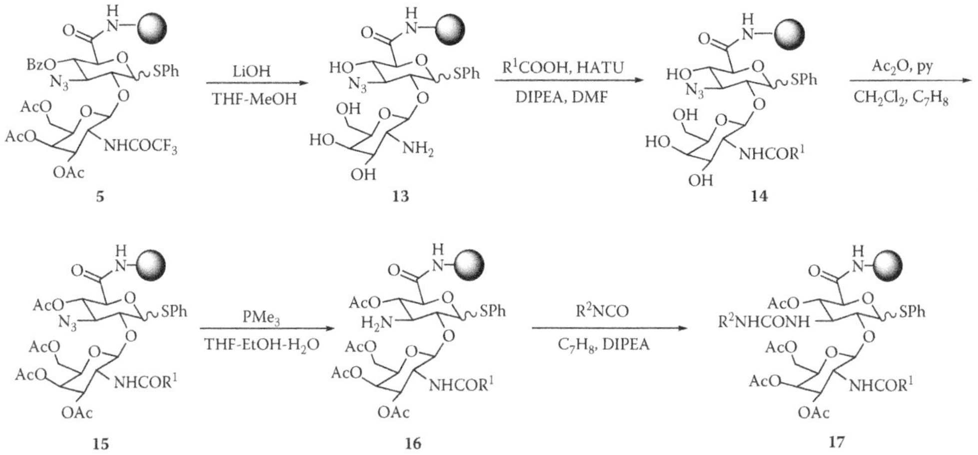

SCHEME 8.3 Solid-phase synthesis strategy for introducing diversity at the C-3 and C-2′ positions of a disaccharide template.

or acids provided C-3 urea and amide derivatives. The final diversity step required that the masked anomeric hydroxyl group be liberated by treating the disaccharides with $Hg(OCOCF_3)_2$ (Scheme 8.4). Introduction of the phosphoglycerate lipid units employed phosphoramidite chemistry (Scheme 8.4). A set of custom-synthesized cyanoethyl phosphoramidite building blocks (Figure 8.8), upon oxidation and global base deprotection, provided the fully elaborated resin-bound disaccharide analogs;[38] photolytic release of the disaccharide from the resin provided the target carboxamide disaccharides **21**. A library of 1300 discrete disaccharide analogs was prepared using 13 acids, 24 isocyanates, and 4 acids to generate C-3 amides and 8 lipid phosphoramidites. The library was generated using the directed-sorting mix-and-split technology with radiofrequency tags developed by IRORI.[38,42]

8.2.3 Biological Activity

The library of moenomycin disaccharide analogs was screened *in vitro* in a bacterial cell wall inhibition assay and screened against a panel of both Gram-positive and Gram-negative microbes using an agar lawn assay approach. This initial screening campaign identified disaccharides having both cell wall inhibition and antibacterial activity. These compounds arose primarily from the C-3 amino functionalized templates **5**, **6**, and **7**. Each of the active compounds contained hydrophobic aromatic groups appended to the C-3 and C-2′ centers and displayed an array of unique moieties replacing the moenomycin phosphoglycerate lipid unit. These analogs were structurally quite distinct from the moenomcyin disaccharide degradation product and

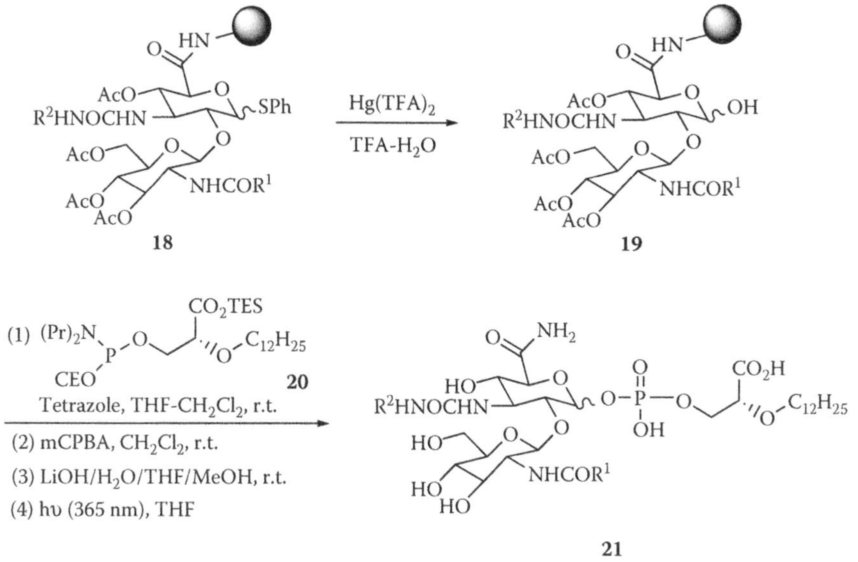

SCHEME 8.4 Solid-phase synthesis methodology to attach a phosphoglycerate lipid unit to an elaborated disaccharide and cleavage from solid support.

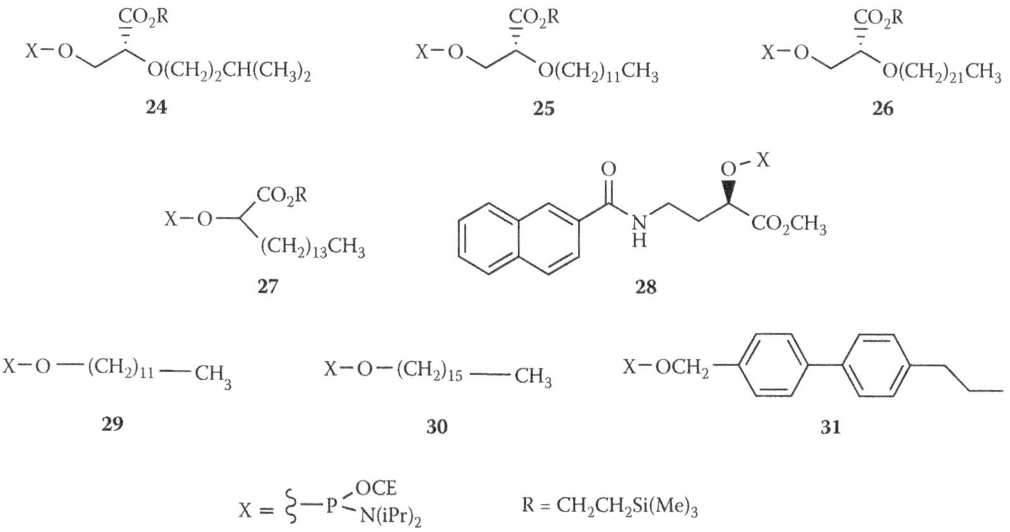

FIGURE 8.8 Lipid building blocks.

demonstrated antibacterial activity not observed in either the moenomycin disaccharide or moenomycin itself. Analogs **30, 31,** and **33** (Figure 8.9) were shown to inhibit cell wall biosynthesis with IC_{50} values in the 8 to 10 µg/ml range and to be active against a panel of clinically relevant antibiotic-sensitive and resistant Gram-positive bacteria with MIC values in the 3 to 25 µg/ml range (Table 8.1). The disaccharide analogs were also active on *Ent. faecium* ATTC49624, a strain that is resistant to moenomycin. Although several of these analogs were less potent (2- to 16-fold) than vancomycin or vancomycin-sensitive Gram-positive bacteria, they maintained their potency on vancomycin-resistant strains. When evaluated against the NIH3T3, HL-60, and

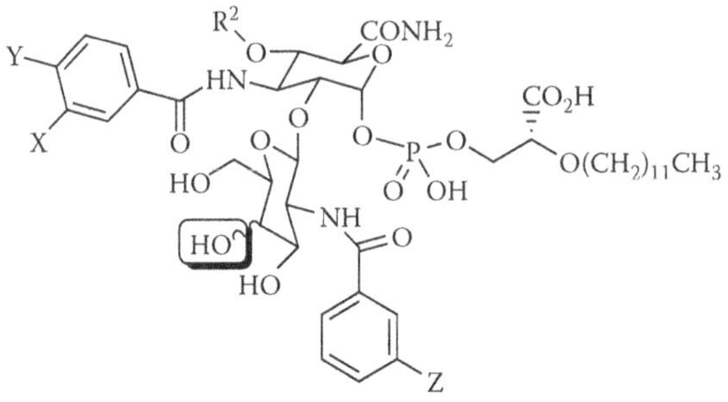

The table below the structure:

	X	Y	Z	R^2	OH
30	H	OCF_3	CF_3	H	Axial
31	CF_3	Cl	CF_3	H	Equatorial
32	CF_3	Cl	H	CH_3	Equatorial

FIGURE 8.9 Active moenomycin disaccharide analogs with the moenomycin phosphoglycerate unit.

HBL-100 mammalian cell lines, none of the disaccharides exhibited cytotoxicities that were relevant to their antimicrobial activity.[38,43]

The SAR generated from the disaccharide analog library clearly identified active compounds with structural features unanticipated from available literature data on moenomycin degradation products and the small number of known analogs. For the first time we were able to identify simple lipid replacements that eliminated the need for the C-25 moenomycin lipid (Figure 8.10, Table 8.2). In addition, we were able to identify a less complex sugar to replace the unusual and difficult-to-synthesize moenomycin F-sugar (Figure 8.9, Figure 8.10). The identification of aromatic hydrophobic units at C-3 and C-2′ sites was also completely unanticipated based on historic SAR data.[38,43]

In addition to demonstrating a unique spectrum of antibacterial activity, the disaccharide analogs also demonstrated antibacterial characteristics that differ from moenomycin itself.[43] Analysis of analogs **30**, **31**, and **32** showed that these compounds were substantially more bactericidal than moenomycin. They demonstrated 4 to 6 log units of killing when Gram-positive bacteria were treated with 4 to 8 times their MIC as compared to 1 log unit of killing when treated with 100 times the MIC of moenomycin (Figure 8.11).

Experimental evidence also supported the hypothesis that the disaccharide analogs inhibit bacterial growth by inhibiting cell wall biosynthesis. They demonstrated bactericidal activity only in actively growing cells, and cells pretreated with protein synthesis inhibitors were resistant to the effects of the disaccharide analogs. In addition, they inhibited incorporation of lysine into cell wall material in intact *Ent. faecalis*.[43]

The development of a library of moenomycin disaccharides led to a series of compounds that were not only potent but also unique in their antibacterial activities. It is accurate to say that without applying a combinatorial library approach, the probability of identifying these novel and potent agents would have been very low. The effort to prepare each library analog via traditional single compound synthesis would have taken many man-years. The library approach allowed us to obtain significant SAR and identify the key actives within one screening campaign.

TABLE 8.1
Antibacterial Activity of Moenomycin Disaccharide Analogs 30 to 32

	PGP IC$_{50}$ (µg/ml) X ± SE	MIC (µg/ml)										
		Sensitive Strains					Resistant Strains					
Compound		E. faecalis ATCC 29212	E. faecium ATCC 49624	S. aureus ATCC 29213	S. epi ATCC 12228	E. coli BAS849[a]	S. aureus ATCC 43300[b]	E. faecium CL4931 (VanA)[c]	E. faecalis CL5244 (VanB)[d]	E. faecalis CL4877 (Van B)[d]	E. faecalis ATCC 51575[d]	E. faecium ATCC 51559[c]
30	15.2	6.25	6.25	6.25	6.25	> 25	6.25	6.25	6.25	3.13	25	12.5
31	10.6	6.25	6.25	6.25–12.5	6.25–12.5	12.5	6.25	6.25	3.13	3.13	12.5	12.5
32	6.9	12.5	12.5	12.5	12.5	25	25	25	12.5	6.25	25	12.5
Moenomycin	0.025 ± 0.014	0.078	> 200	0.05	0.025	0.025	0.062	0.39	0.062	0.062	0.062	1.56
Vancomycin	5.2 ± 0.08	6.25	0.78	3.13	3.13	0.78	3.13	> 125	15.6	> 125	> 125	> 125

a Super-sensitive permeability mutant of *E. coli.*
b Methicillin-resistant *S. aureus.*
c Vancomycin-resistant *E. faecium.*
d Vancomycin-resistant *E. faecalis.*

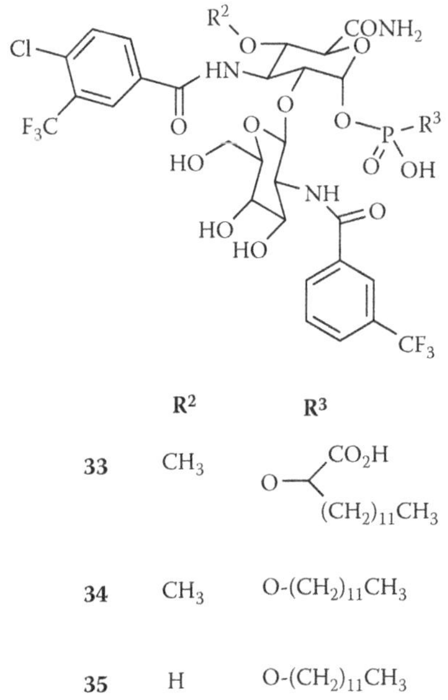

	R^2	R^3
33	CH_3	(structure: O-CH with CO_2H and $(CH_2)_{11}CH_3$)
34	CH_3	$O\text{-}(CH_2)_{11}CH_3$
35	H	$O\text{-}(CH_2)_{11}CH_3$

FIGURE 8.10 Active moenomycin disaccharide analogs with phosphogylcerate lipid unit replacements.

8.3 ANISOMYCIN

8.3.1 BACKGROUND

Inhibition of protein synthesis through action on prokaryotic ribosomes is known to be a viable strategy for developing clinically useful antibiotics.[21] Macrolides (erythromycin, clarithromycin), aminoglycosides (gentamicin, kanamycin), and oxazolidinones (linezolid) are all examples of classes of antibiotics that function by inhibiting protein synthesis through action on ribosomes.[21,44] (−)-Anisomycin **36** is a small-molecular-weight natural product isolated from *Streptomyces griseolus* and *Streptomyces roseochromogenes* (Figure 8.12).[45] Although it is used clinically to treat amoebic dysentery and trichomonas vaginitis and is used as a fungicide to eradicate bean mildew, anisomycin does not show antibacterial activity.[46–48] (−)-Anisomycin functions by inhibiting protein synthesis through inhibition of peptide chain elongation on 60S eukaryotic ribosomes.[49–51] Although anisomycin does not possess antibacterial activity, its mechanism of action makes it a candidate as a natural product template to explore the identification of novel antimicrobial agents. Therefore, we were interested in developing a library using anisomycin as a molecular template and screening it for antibacterial and antifungal activity.

To guide our library strategy, we looked to the available SAR data around anisomycin. However, literature analysis showed that little SAR work had been reported. Most of the chemical work focused on the total synthesis of anisomycin itself and minor modifications to the benzyl moiety (Figure 8.13).[52–59] Halogen substitution on the phenyl ring resulted in analogs that retained antifungal activity but did not induce antibacterial activity. Modifications to the benzyl methylene eliminated the antifungal activity and did not produce antibacterial properties. The limited SAR information was not useful in guiding our diversity strategy and, therefore, we investigated an approach that would allow introduction of broad chemical diversity at several different sites on the anisomycin template. This strategy also dictated the use of a solid-phase approach.

TABLE 8.2
Antibacterial Activity of Moenomycin Disaccharide Analogs 33 to 35

MIC (μg/ml)

Compound	PGP IC$_{50}$ (μg/ml) X ± SE	Sensitive Strains						Resistant Strains				
		E. faecalis ATCC 29212	E. faecium ATCC 49624	S. aureus ATCC 29213	S. epi ATCC 12228	E. coli BAS849[a]	S. aureus ATCC 43300[b]	E. faecium CL4931 (VanA)[c]	E. faecalis CL5244 (VanB)[d]	E. faecalis CL4877 (Van B)[d]	E. faecalis ATCC 51575[d]	E. faecium ATCC 51559[c]
33	9.2 (n = 2)	3.12	3.12	6.25	6.25	> 25	3.12	6.25	6.25	3.12	6.25	3.12
34	8.2 (n = 2)	6.25	6.25	6.25	6.25	12.5	6.25	6.25	6.25	3.12	6.25	6.25
35	15.4 ± 3.2	3.12	12.5	6.25	12.5	12.5	6.25	3.13	3.13	3.13	12.5	12.5
Moenomycin	0.025 ± 0.014	0.078	> 200	0.05	0.025	0.025	0.062	0.39	0.062	0.062	0.062	1.56
Vancomycin	5.2 ± 0.08	6.25	0.78	3.13	3.13	0.78	3.13	> 125	15.6	> 125	> 125	> 125

[a] Super-sensitive permeability mutant of E. coli.
[b] Methicillin-resistant S. aureus.
[c] Vancomycin-resistant E. faecium.
[d] Vancomycin-resistant E. faecalis.

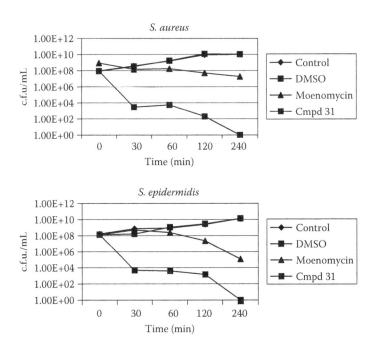

FIGURE 8.11 Bactericidal effect of moenomycin and disaccharide analog **31** on Gram-positive bacteria (*S. aureus* and *S. epidermidis*). Moenomycin was added to 100 times MIC and disaccharide analog **31** added to 8 times MIC. C.f.u.'s were determined by plating serial dilutions at various times following drug addition.

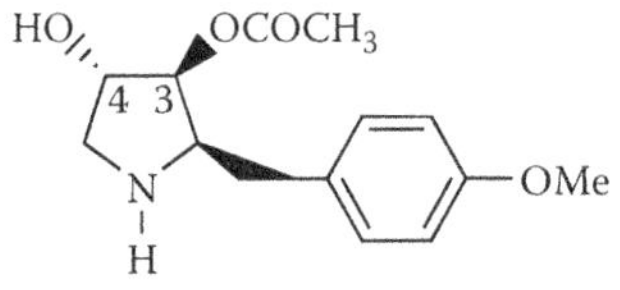

36 (–)-anisomycin

FIGURE 8.12 (–)-Anisomycin.

8.3.2 SYNTHETIC STRATEGY

Anisomycin has three sites that support rapid introduction of chemical diversity — the free hydroxyl group, the nitrogen of the pyrrolidine nucleus, and the hydroxyl group that is masked as an acetate moiety (Figure 8.13). A solid-phase synthesis strategy required that we be able to attach the anisomycin template to the solid support and introduce chemical diversity at multiple sites. We chose to introduce diversity at the nitrogen of the pyrrolidine nucleus and at the C-3 oxygen functionality that in anisomycin exists as an acetate group. This was accomplished by developing a functionalized template **37** that allowed us to attach the template to solid support via the existing anisomycin C-4 free hydroxyl group, protect the nitrogen function with an acid-labile Fmoc group, and use the existing acetate function as a built-in base-labile protecting group. This functionalized template was easily prepared directly from anisomycin by a one-step protection of the amine functionality.[60]

To support preparation of a diverse anisomycin analog library, sufficient quantities of function-alized template **37** were required. To prepare this material, significant quantities of anisomycin were needed. However, commercial supplies of anisomycin were found to be limited and expensive ($1000/g). Therefore, we undertook a process of in-house fermentation, isolation, and purification

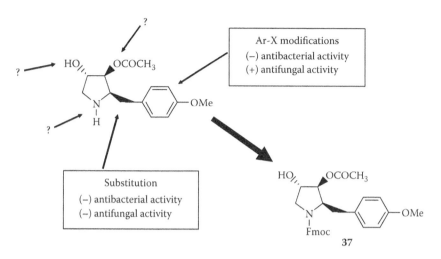

FIGURE 8.13 Known anisomycin structure–activity relationships.

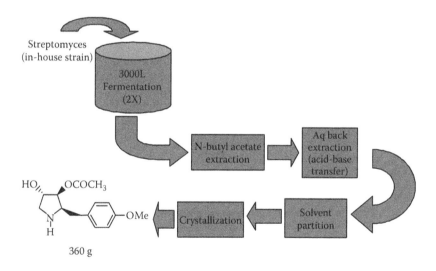

FIGURE 8.14 Anisomycin fermentation and isolation process.

of anisomycin (Figure 8.14).[60] Two 3000-l fermentations of an in-house *Streptomyces* strain were processed through an *n*-butyl acetate extraction and subsequently back-extracted using aqueous acid base transfer. Solvent partitioning followed by crystallization provided 360 g of anisomycin. Although this process required special capabilities such as large fermentation capacity and special expertise, it proved to be much more efficient than obtaining anisomycin by the reported total syntheses. A total synthesis of anisomycin as reported in the literature would require 8 to 12 synthetic steps and would provide anisomycin in 23 to 44% overall yield.[52–59]

The method used for solid-phase library synthesis of anisomycin analogs was straightforward. Template derivative **37** was attached to Merrifield resin equipped with a dihydropyran linker (DHP HM resin) using a modification of the Ellman method (Scheme 8.5).[61] The Fmoc group of intermediate **39** was then removed under standard conditions to reveal the secondary amine of the pyrrolidine. The first diversity element was introduced at the pyrrolidine nitrogen using a wide range of reagents that provided a diverse set of functional groups. These reagents included acid chlorides, chloroformates, isocyanates, carbamoyl chlorides, aromatic sulfonyl chlorides, and

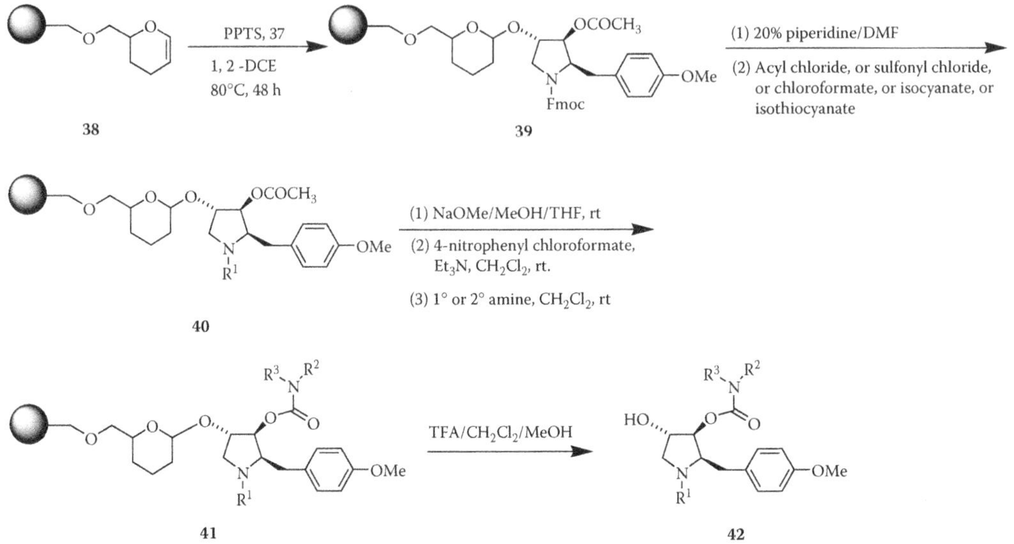

SCHEME 8.5 Solid-phase synthesis methodology for preparing a library of anisomycin analogs.

isothiocyanates. Introduction of the second diversity element at the C-3 hydroxyl group proceeded by base removal of the acetate protecting group and then preparation of carbamates by treatment with 4-nitrophenyl chloroformate and a primary or secondary amine. Acid treatment provided the desired anisomycin analogs having structural diversity at the N-1 and C-3 positions. Using this methodology, a library of over 10,000 compounds was prepared with 100 R^1 reagents and 100 R^2 reagents. The products were prepared as discretes with a submission rate of 80% based on the criteria of 70% purity by HPLC and identity by MS.

8.3.3 BIOLOGICAL ACTIVITY

The library was tested in both a bacterial and a fungal screen. An *S. aureus* antibacterial screen measured the growth inhibition of liquid cultures via optical densitometry when exposed to the individual library compounds. Growth inhibition against the fungal target *Candida albicans* (efflux-deficient strain) was measured by fluorescence of the growth indicator alamarBlue™. The hit rate for inhibiting *S. aureus* at 50% was 0.25%. The hit rate for antifungal activity at 50% inhibition was 0.47%.

Examples of compounds demonstrating antibacterial activity are shown in Figure 8.15. Initial screening data demonstrated that compounds **43**, **44**, and **45** (16 to 32 µg/ml) showed comparable antibacterial activity to the known clinical agents ceftazidine and aztreonam (Table 8.3). In contrast, anisomycin itself did not show any antibacterial activity. Evaluation of these compounds in a biochemical screen of bacterial cell wall inhibition (MurPath and MraY/MurG) did not identify any compounds of significant activity, thus eliminating cell wall inhibition as a potential mechanism of action. This is the first time that analogs of anisomycin have demonstrated antibacterial activity, validating our hypothesis-driven natural product template strategy.

The observation that the anisomycin analogs showed substantially reduced cytotoxicity compared to anisomycin itself was of particular significance (Table 8.3). When tested against HEK 293 cells, anisomycin showed a cytotoxicity IC$_{50}$ of 0.02 µM. In contrast, the analogs showed IC$_{50}$ values in the 2 to 10 µM range — a 100- to 500-fold reduction in cytotoxicity. This reduced cytotoxicity makes these analogs more viable as leads for further development.

In addition to identifying novel antibacterial activity, a number of library compounds also showed antifungal properties (Figure 8.16, Table 8.4). When tested at a concentration of 15.5 µM,

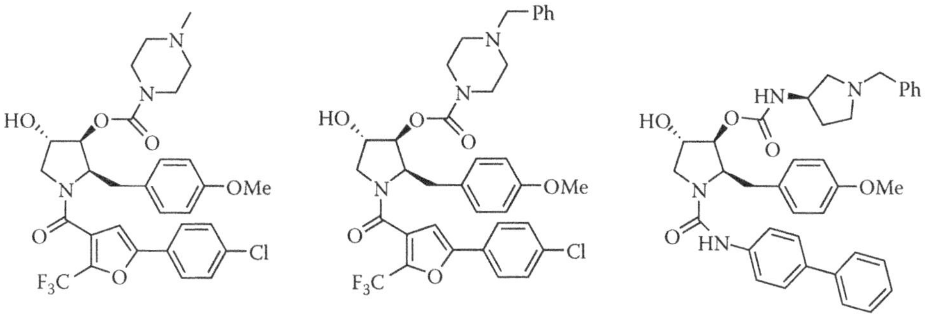

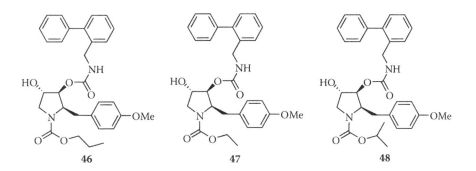

FIGURE 8.15 Anisomycin analogs demonstrating antibacterial activity.

TABLE 8.3
Antibacterial Activity of
Anisomycin Analogs 43 to 45

Compound	S. aureus MIC (μg/ml)	Cytotoxicity HEK 293 IC$_{50}$ (μM)
43	32	10
44	16	8
45	16	2
Ceftazidime	16	NT
Aztreonam	32	98
Linezolid	2	> 100
Anisomycin	> 32	0.02

FIGURE 8.16 Anisomycin analogs demonstrating antifungal activity.

several analogs inhibited fungal growth by 82 to 96%. This is compared to several known antifungal agents such as sordarin, fluconazole, and anisomycin that showed inhibition between 85 and 99%. Again, these novel analogs demonstrated substantial reduction in cytotoxic activity (Table 8.4) when compared to the clinical agent anisomycin. This is the first time that substantial antifungal SAR has been developed for the anisomycin class of agents.

TABLE 8.4
Antifungal Activity of Anisomycin
Analogs 46 to 48

Compound	*C. albicans* % Inhibition at 15.5 μM	Cytotoxicity HEK 293 IC_{50} (μM)
46	95	9
47	83	22
48	82	11
Anisomycin	99	0.02
Fluconazole	96	> 100
Sordarin	85	> 100

The identification of antibacterial agents resulting from a library of anisomycin analogs validates the hypothesis that a natural product template with no antibacterial activity, yet possessing a biological mechanism of action known to be relevant to the discovery of antibiotics, can result in the identification of novel antibacterial agents. Clearly, this strategy can be applied to other therapeutic areas in which a natural product possesses a relevant biological mechanism of action.

8.4 CONCLUSION

Natural products have played a major role in the field of antibiotics. Most of their impact has resulted directly from the natural product itself or from single-compound analog synthesis. However, the use of natural-product-template-driven library approaches has the potential to identify novel agents with a new spectrum of activity. The identification of novel antibiotics from the natural products moenomycin and anisomycin demonstrates the viability of the natural product template strategy. These examples also highlight some of the strategies and challenges associated with this approach.

ACKNOWLEDGMENTS

I would like to acknowledge all those who contributed to the work described in this chapter. For the moenomycin project the contributors include Nigel Allanson, Nicole Hatzenbuhler, Rakesh Jain, Ramesh Kakarla, Natan Kogan, Rui Liang, Dashan Liu, Domingos Silva, Huimig Wang, David Gange, Jan Anderson, Anna Chem, Feng Chi, Richard Dulina, Buwen Huang, Muthoni Kamau, Chunguang Wang, Eugene Baizman, Arthur Branstrom, Robert Goldman, Clifford Longley, Sunita Midha, and Helena Axelrod. For the anisomycin project the contributors include Shuhoa Shi, Samuel Gerritz, Ramesh Padmanabha, Kim Esposito, John Herbst, Wenying Li, Henry Wong, Yue Shu, and Kim Lam.

REFERENCES

1. Butler, M.S., The role of natural product chemistry in drug discovery, *J. Nat. Prod.*, 67, 2141, 2004.
2. Newman, D.J., Cragg, G.M., and Snader, K.M., Natural products as sources of new drugs over the period 1981–2002, *J. Nat. Prod.*, 66, 1022, 2003.
3. Breinbauer, R., Vetter, I.R., and Waldmann, H., From protein domains to drug candidates — natural products as guiding principles in the design and synthesis of compound libraries, *Angew. Chem., Int. Ed. Engl.*, 41, 2879, 2002.

4. Feher, M. and Schmidt, J.M., Property distributions: differences between drugs, natural products, and molecules from combinatorial chemistry, *J. Chem. Inf. Comput. Sci.*, 43, 218, 2003.

5. Bajorath, J., Chemoinformatics methods for systematic comparison of molecules from natural and synthetic sources and design of hybrid libraries, *Mol. Diversity*, 5, 305, 2000.

6. Lee, M-L. and Schneider, G., Scaffold architecture and pharmacophoric properties of natural products and trade drugs: application in the design of natural product-based combinatorial libraries, *J. Comb. Chem*, 3, 284, 2001.

7. Overbye, K.M. and Barrett, J.F., Antibiotics: where did we go wrong?, *Drug Discovery Today*, 10, 45, 2005.

8. Barker, K.F., Antibiotic resistance: a current perspective, *Br. J. Clin. Pharmacol.*, 48, 109, 1999.

9. Cohen, M.L., Changing patterns of infectious diseases, *Nature*, 406, 762, 2000.

10. Guerrant, R.L. and Glackwood, B.L., Threats to a global survival: the growing crisis of infectious diseases — our unfinished agenda, *Clin. Infect. Dis.*, 28, 966, 1999.

11. Ganesan, A., Natural products as a hunting ground for combinatorial chemistry, *Curr. Opin. Biotechnol.*, 15, 584, 2004.

12. Waldmann, H., At the crossroads of chemistry and biology, *Bioorg. Med. Chem.*, 11, 3045, 2003.

13. Boldi, A.M., Libraries from natural product-like scaffolds, *Curr. Opin. Chem. Biol.,* 8, 281, 2004.

14. Tietze, L.F., Bell, H.P., and Chandrasekhar, S., Natural product hybrids as new leads for drug discovery, *Angew. Chem., Int. Ed. Engl.,* 42, 3996, 2003.

15. Hall, D.G., Manku, S., and Wang, F., Solution- and solid-phase strategies for the design, synthesis, and screening of libraries based on natural product templates: a comprehensive survey, *J. Comb. Chem.,* 3, 125, 2001.

16. Abel, U. et al., Modern methods to produce natural-product libraries, *Curr. Opin. Chem. Biol.*, 6, 453, 2002.

17. Ortholand, J-Y. and Ganesan, A., Natural products and combinatorial chemistry: back to the future, *Curr. Opin. Chem. Biol.,* 8, 271, 2004.

18. Nielsen, J., Combinatorial synthesis of natural products, *Curr. Opin. Chem Biol.*, 6, 297, 2002.

19. Ganesan, A., Recent developments in combinatorial organic synthesis, *Drug Discovery Today*, 7, 47, 2002.

20. Goldman, R.C. and Branstrom, A., Targeting cell wall synthesis and assembly in microbes: similarities and contrasts between bacteria and fungi, *Curr. Pharm. Design*, 5, 473, 1999.

21. Lancini, G., Parenti, F., and Gallo, G.G., *Antibiotics a Multidisciplinary Approach*, Plenum Press, New York, 1995, chap. 3, 5.

22. Wasielewski, W.V. et al., Moenomycin, a new antibiotic III. Biological properties, *Antimicrob. Agents Chemother.*, 743, 1965.

23. Donnerstag, A. et al., A structurally and biogenetically interesting moenomycin antibiotic, *Tetrahedron*, 51, 1931, 1995.

24. Kurz, M., Guba, W., and Vertesy, L., Three-dimensional structure of moenomycin A: a potent inhibitor of penicillin-binding protein 1b, *Eur. J. Biochem.*, 252, 500, 1998.

25. Scherkenbeck, J. et al., *Tetrahedron*, 49, 3091, 1993.

26. El-Abadia, N. et al., Moenomycin A: the role of the methyl group in the moenuronamide unit and a general discussion of structure-activity relationships, *Tetrahedron*, 55, 699, 1999.

27. Huber, G., Moenomycin and related phosphorus-containing antibiotics, in *Antibiotics,* Vol 5, Part 1, Hahn, F.E., Ed., Springer-Verlag, New York, 1979, p. 135.

28. Welzel, P. et al., Moenomycin A: minimal structural requirements for biological activity, *Tetrahedron*, 43, 585, 1997.

29. Welzel, P. et al., Stepwise degradation of moenomycin A, *Carbohydr. Res.,* 126, C1, 1984.

30. Metten, K.-H. et al., The first enzymatic degradation products of the antibiotic moenomycin A, *Tetrahedron*, 48, 8401, 1992.

31. Fehlhaber, H.-W. et al., Moenomycin A: a structural revision and new structure-activity relations, *Tetrahedron*, 46, 1557, 1990.

32. Hessler-Klintz, M. et al., The first moenomycin antibiotic without the methyl-branched uronic acid constituent. Unexpected structure-activity relations, *Tetrahedron*, 49, 7667, 1993.

33. Moller, U. et al., Moenomycin A — structure-activity relations synthesis of the D-galacturonamide analogue of the smallest antibiotically active degradation product of moenomycin, *Tetrahedron*, 49, 1653, 1993.

34. Marzian, S.M. et al., Moenomycin A: reactions at the lipid part, new structure-activity relations, *Tetrahedron*, 50, 5299, 1994.

35. Lunig, J., Markus, A., and Welzel, P., Moenomycin-type transglycoslyase inhibitors: inhibiting activity vs. topology around the phosphoric acid diester group, *Tetrahedron Lett.*, 35, 1859, 1994.

36. Hauer, M. et al., Structural analogues of the antibiotic moenomycin A with a D-glucose derived unit F, *Tetrahedron*, 50, 2029, 1994.

37. Riedel, S. et al., Synthesis of transglycosylase inhibiting properties of a disaccharide analogue of moenomycin A lacking substitution at C-4 of unit F, *Tetrahedron*, 55, 1921, 1999.

38. Sofia, M.J. et al., Discovery of novel disaccharide antibacterial agents using a combinatorial library approach, *J. Med. Chem.*, 42, 3193, 1999.

39. Ghosh, M., Kakarla, R., and Sofia, M.J., Diastereofacial selection in the 1,2-addition of MeMgX and MeLi to 4-oxo sugar: efficient synthesis of 4-C-methyl-1-S-β-D-gluco- and galactopyranoside building blocks of moenomycin, *Tetrahedron Lett.*, 40, 4511, 1999.

40. Yan, L. et al., Glycosylation on the Merrifield resin using anomeric sulfoxides, *J. Am. Chem. Soc.*, 116, 6953, 1994.

41. Silva, D. et al., Stereospecific solution- and solid-phase glycosylations. Synthesis of β-linked saccharides and construction of disaccharide libraries using phenylsulfenyl 2-deoxy-2-trifluoroacetamido glycopyranosides as glycosyl donors, *J. Org. Chem.*, 64, 5926, 1999.

42. Nicolaou, K.C. et al., Radio frequency encoded combinatorial chemistry, *Angew. Chem., Int. Ed. Engl.*, 34, 2289, 1995.

43. Baizman, E.R. et al., Antibacterial activity of synthetic analogues based on the disaccharide structure of moenomycin, an inhibitor of bacterial transglycosyulase, *Microbiology*, 146, 3129, 2000.

44. Bozdogan, B. and Appelbaum, P.C., Oxazolidinones: activity, mode of action, and mechanism of resistance, *Int. J. Antimicrob. Agents*, 23, 113, 2004.

45. Sobin, B.A. and Tanner, F.W., Anisomycin: a new antiprotozoan antibiotic, *J. Am. Chem. Soc.*, 76, 4053, 1954.

46. Tanner, F.W., Jr., Sobin, B.A., and Gardocki, J.F., Some chemical and biological properties of anisomycin, *Antibiot. Ann.*, 2, 809, 1955.

47. Lynch, J., Holley, E.C., and Salmirs, A.M., Effect of anisomycin on the growth of *Trichomonas vaginalis*, *Antibiot. Chemother.*, 5, 300, 1955.

48. Windholz, W., *Merck Index*, 10th ed., Merck, Rahway, NJ, 98, 1983.

49. Grollman, A.P. and Walsh, M., Inhibitors of protein biosynthesis. II. Mode of action of anisomycin, *J. Biol. Chem.*, 242, 3226, 1967.

50. Kageyama, A. et al., Comparison of the apoptosis-inducing abilities of various protein synthesis inhibitors of U937 cells, *Biosci. Biotechnol. Biochem.*, 66, 835, 2002.

51. Hanson, J.L., Moore, P.B., and Steitz, T.A., Structure of five antibiotics bound at the peptidyl transferase center of the large ribosomal subunit, *J. Mol. Biol.*, 330, 1061, 2003.

52. Delair, P. et al., Formal total synthesis of enantiopure (−)-anisomycin from streptomyces, *J. Org. Chem.*, 64, 1383, 1999.

53. Shi, Z. and Li, G., Enantioselective synthesis of (−)-anisomycin and its proprionate derivative (3097-B1), *Tetrahedron*, 6, 2907, 1995.

54. Buchanan, J.G. et al., Synthesis of chiral pyrrolidines from carbohydrates, *ACS Symp. Ser.*, 386, 107, 1989.

55. Buchanan, J.G. et al., A new chiral synthesis of (−)-anisomycin and its demethoxy analogs, *J. Chem. Soc. Chem. Commun.*, 9, 486, 1983.

56. Felner, I. and Schenker, K., Total synthesis of the antibiotic anisomycin, *Helv. Chem. Acta*, 53, 754, 1970.

57. Wong, C.M., Buccini, J., and TeRaa, J., Synthesis of optically active deacetyl anisomycin, *Can. J. Chem.*, 46, 3091, 1968.

58. Hall, S.S., Loebenberg, D., and Schumacher, D.P., Structure-activity relationships of synthetic antibiotic analogs of anisomycin, *J. Med. Chem.*, 26, 469, 1983.

59. Rosser E.M. et al., Synthetic anisomycin analogues activating the JNK/SAPK1 and p38/SAPK2 pathways, *Org. Biomol. Chem.*, 2, 142, 2004.
60. Shi, S. et al., Solid-phase synthesis and anti-infective activity of a combinatorial library based on the natural product anisomycin, *Bioorg. Med. Chem. Lett.*, submitted for publication.
61. Thompson, L.A. and Ellman, J.A., Straight forward and general method for coupling alcohols to solid supports, *Tetrahedron Lett.*, 35, 9333, 1994.

9 Synthetic Libraries of Fungal Natural Products

Michael C. Pirrung, Zhitao Li, and Hao Liu

CONTENTS

9.1 NATURAL PRODUCTS IN COMBINATORIAL CHEMISTRY

There is a significant ongoing need to add sophistication to the methods used for creating diversity in combinatorial chemistry. One approach is to use as a starting point the complex natural products that have for so long been the lead structures in drug development. The combinatorial synthesis of natural products was initially a relatively unexplored area, but the publication of this volume is testimony to the increasing interest, particularly among academic chemists, in combining a traditional fascination for natural products with modern combinatorial technologies. A virtue that we are seeing in this endeavor is that it entails the adaptation and development of ever so many types of organic reactions to combinatorial and, often, solid-phase synthesis.[1] Too often, however, this has involved using the natural product as a "template" and only changing substituents around its periphery with acylation and alkylation reactions (C-heteroatom bond formation).[2] Although such molecules certainly may exhibit enhanced biological activities, their preparation does not provide much challenge to current chemical synthesis technologies. Methods to assemble modified core structures through C-C bond formation in a combinatorial synthesis have been much less known. Development of such technologies to provide access to complex and novel molecules from combinatorial synthesis is therefore an ongoing challenge.

A long-held rationale for natural products synthesis is that, in the course of planning and executing a complex multistep synthesis, incompatibilities between structures, synthetic methodologies, and functionalities arise that reveal the gaps in current synthetic technology. These could not have been envisioned by the developers of reaction methodology, and the solution of such problems involves developing novel chemistry that is by definition useful for the task of synthesizing desirable targets. The planning and execution of the preparation of a molecule as complex as a Taxol® is worthwhile because it tests the limits of our rationality in predicting chemical reactivity. When it falls short, synthetic chemists are forced to reevaluate their design principles and develop new models for chemical reactivity and new synthetic methodology. Particularly in the context of multistep synthetic procedures, even when each individual step has strong precedence, incompatibilities of functionalities often prevent success when conventional methods are applied in a more complex context. Failure necessitates the development of totally new methodologies in synthesis, which by definition are useful because they address a mode of reactivity that is currently unavailable. In some cases, failure may prompt the use of an existing alternative methodology, demonstrating for it a new application that does not interfere with an incompatible group or reagent.

As stated in the preceding text, this thinking also provides the rationale for *combinatorial natural product synthesis*. This endeavor rehearses a possible model for future drug development, wherein natural product leads discovered by either traditional or *avant garde*[3] means are placed into combinatorial chemistry programs for further development and optimization. This paradigm is important because over 60% of approved drugs and pre-NDA candidates are of natural origin, defined as original natural products, products derived semisynthetically from them, or synthetic products based on natural product models.[4] For a given natural product drug lead, the ability to synthesize a derivative *library* for evaluation would greatly increase the efficiency of drug development compared to conventional medicinal chemistry. The "combinatorialization" of a natural product represents a significant challenge for synthetic chemistry.

In conventional organic synthesis, great value is placed on convergence because of the overall increase in efficiency and shortening of the linear reaction sequence. In combinatorial organic synthesis, it is crucial to design a synthesis that is *modular* (but not necessarily convergent) so that groups of modules (ideally, commercially available compounds) can be collected and combined at the library preparation stage. If each module requires a multistep synthesis to prepare, far more time and effort is spent on preparing building blocks than on preparing the library.

Demethylasterriquinone B1 (DAQ B1), one of the targets of interest here (Figure 9.1), is inherently modular, so that it is readily possible to design a modular synthesis route to it. However, it is not necessarily true that any synthesis of a modular molecule must itself be modular; one reported synthesis of DAQ B1 is not.[5] Interestingly, it is also not true that nonmodular targets must be assembled by nonmodular syntheses. We provide as another example the illudins, targets that present no obvious structural modules. Yet by exploiting chemistry developed by Padwa,[6] the illudins can be readily assembled by a modular synthesis from variable building blocks.

The illudin and the asterriquinone families are both produced by fungi, the former by the jack-o'-lantern mushroom *Omphalotus illudens*, and the latter by several filamentous fungi including *Aspergillus terreus* and *Pseudomassaria sp*. The biological activities of the illudins are primarily related to cancer,[7] whereas the biological activities of the asterriquinones have been studied in many

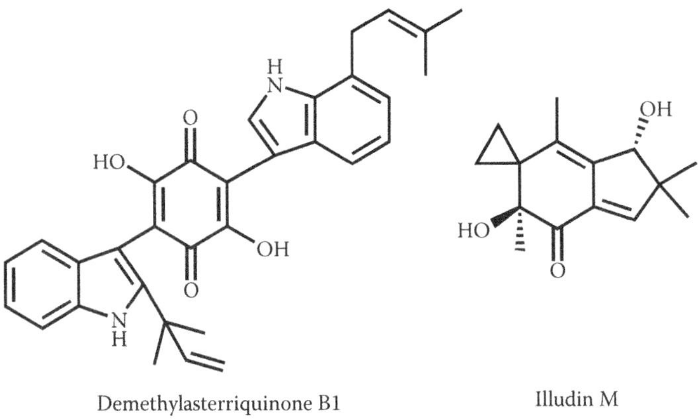

Demethylasterriquinone B1   Illudin M

FIGURE 9.1 Fungal natural products serving as a basis for libraries described.

fields, including infectious disease, cancer, and diabetes.[8] Thus, there is good reason to believe that biologically active agents might be found among libraries of derivatives of these two natural products.

Indeed, the idea that molecules derived from any natural source might have special capabilities in modulating biological processes in any particular target organism of interest might be called the central dogma of chemical genetics. This novel phraseology recalls the well-worn central dogma of molecular biology, which is "DNA makes RNA makes protein." The seemingly inviolable rules of this original central dogma have for decades formed a framework for biological research. Although newly coined, the central dogma of chemical genetics has provided a framework for natural products research since Fleming discovered penicillin. The dogma may be expressed as "molecules useful to microbes, or trees, or sponges may be useful to *H. sapiens.*" Readers are invited to develop their own formulations for this new central dogma that rival the supremely euphonious "DNA makes RNA makes protein."

Although decades of natural products research demonstrates the faith that scientists have held in the central dogma of chemical genetics, one can reasonably ask why it should be true. There is no doubt that because microbes engage in chemical warfare with one another for precious resources of space and nutrients, it is rational to seek novel antimicrobial agents from the microbial world. But the idea that a product from a sponge might yield a useful therapy for mammalian diseases is quite a different matter. However, this new central dogma answers such questions with fundamental principles. In the same way that the genetic code is broadly preserved throughout living beings, biological mechanisms and signaling pathways from the simplest to the most complex organisms bear striking similarities. Examples include nucleic acid replication and protein phosphorylation. Small molecules (also called secondary metabolites, because their production does not obviously address one of the basic functions of life) may be made in one organism to interact with a specific macromolecule (such as an enzyme or receptor) in its own environment. That very small molecule might be useful to another organism in a different ecological niche that interacts with a macromolecule that is different from the molecule's original target but happens (for evolutionary reasons, *inter alia*) to be similar to it.

These considerations have been so powerful in motivating research in natural products isolation that it is somewhat surprising in retrospect that they did not gain traction in combinatorial chemistry more quickly. After all, both fields have been viewed as valid approaches to gaining new molecules with useful activities. The publication of this volume is a reflection of this dawning realization, and one can expect the melding of natural products synthesis and combinatorial chemistry to continue. The two examples provided in the following text summarize some contributions of our laboratory to this thinking.

9.2 SYNTHESIS OF AN ASTERRIQUINONE LIBRARY

9.2.1 PREPARATORY CHEMISTRY STUDIES

Although DAQ B1 itself can be retrosynthetically split into three simple modules, the quinone ring and two indoles, it represents a more challenging and complex molecular target than those in many libraries that have been prepared by combinatorial methods. The three modules are joined by C-C bonds, and C-C bond formation is always more difficult than formation of C-heteroatom bonds, especially in solid-phase synthesis. For that reason, our work on a DAQ B1–based combinatorial library focused on parallel, solution-phase synthesis. Previous research from our laboratory on the total synthesis of DAQ B1 provided the tools to achieve this goal. We developed a Brønsted acid-catalyzed method to add an indole to dichlorobenzoquinone to produce an indolylbenzoquinone,[9] and this method was used in our first-generation total synthesis.[10] However, the second indole was added by a palladium-catalyzed Stille coupling reaction, which requires an indolylstannane as starting material. The indolylstannane was prepared in a multistep route; it was not stable, and it must be used *in situ* in the total synthesis, which would be quite inconvenient in combinatorial synthesis. The coupling reaction also requires strict control of reaction conditions, which could be difficult. An ideal coupling reaction should use mild reaction conditions and employ indoles directly as building blocks.

While investigating an improved synthesis that enabled us to prepare a small library of methylated derivatives of the natural product, we developed a Lewis acid-catalyzed method that consisted of adding an indole to an indolylbenzoquinone to produce a *bis*-indolylbenzoquinone.[11] This discovery enabled a fully streamlined synthetic route to be performed involving three stages: sequential addition of the indoles, followed by hydrolysis of the dichloroquinones to the dihydroxyquinones, which were readily adapted to library synthesis (Figure 9.2). This route uses chlorides as electron-deficient hydroxyl equivalents to enhance (along with the acids) the electrophilicity of the quinone toward the weakly nucleophilic indoles. Note that the first stage requires the greatest chemoselectivity, because the product **2{n}** is also a quinone and could, in principle, undergo addition of a further indole molecule to provide a symmetrical *bis*-indolylbenzoquinone ($R^1 = R^2$). Although such products are occasionally observed, they can largely be suppressed by careful attention to reaction stoichiometry and sequential addition of the indole (generating an intermediate, stable hydroquinone) followed by the oxidant. However, it also appears that protic acids are simply poor catalysts for the addition of indoles to indolylbenzoquinones. Although in some ways disappointing as we attempted to achieve the addition of the second indole with protons, this turned out to be an advantage in controlling the addition of the first indole.

In the second stage, there is no need to exclude oxidant during the addition of the indole and, indeed, its presence offers an advantage. A side reaction sometimes encountered in the nucleophilic addition of indoles to these dichloroquinones is reduction. The mechanism of this process is unknown, but it is certainly problematic because the hydroquinones are unreactive and either contaminate the product or are oxidized upon workup to return starting material. Inclusion of the oxidant during the addition step enables "recycling" of the *in situ* generated hydroquinone to the quinone, which can then undergo addition.

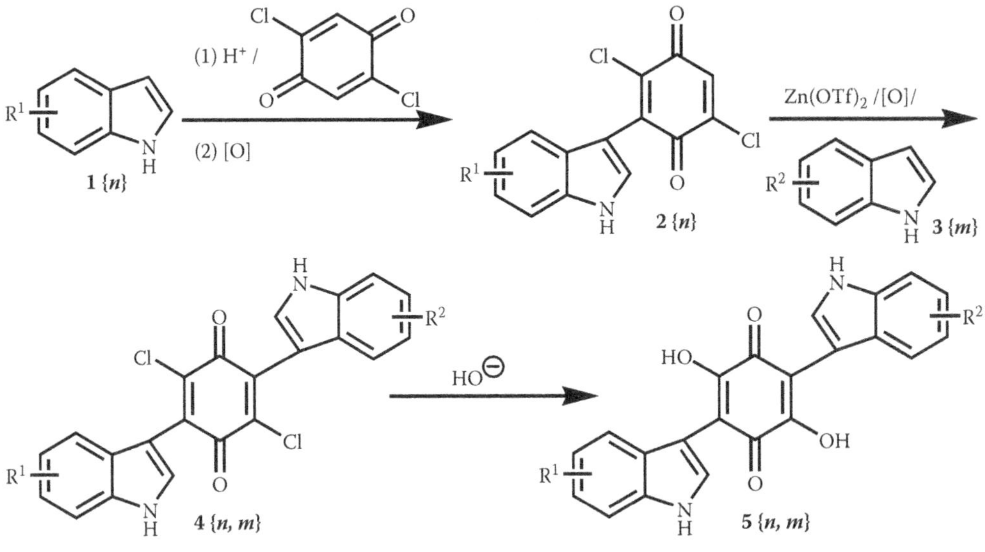

FIGURE 9.2 Three-step synthetic route to *bis*-indolylquinones from two indoles and dichlorobenzoquinone.

FIGURE 9.3 High-throughput purification of coupling products on the Quest synthesizer, using reaction vessels for plug filtration.

9.2.2 PREPARATORY PURIFICATION STUDIES

Although the chemistry for library synthesis seemed secure and was readily adapted to a Quest 210 SLN solution-phase synthesizer, attention was required to the workup procedures. For the Lewis acid-catalyzed coupling reaction, our conventional workup method was silica gel chromatography. The problem with this method is practicality when reactions are performed 20 at a time. Because the impurities in these reaction mixtures are the starting materials (indoles and indolylquinones), which are much less polar than the product, it is possible to purify the product by filtration through a short plug of silica gel. Figure 9.3 demonstrates the use of a silica gel plug in this case. A 10-mL Quest reaction vessel half-filled with silica gel was used as the column. A small amount of silica gel loaded with the crude product was added to the top of this column. The column was washed successively with 20% ethyl acetate in hexanes and 40% ethyl acetate in hexanes. The less polar eluent (20% ethyl acetate in hexanes) gave starting materials (indolylquinones) as the first band, and 40% ethyl acetate in hexanes gave the product as the second band. Owing to the different colors of the starting material and product bands, they can be easily recognized on the column. Different bands were collected and subjected to evaporation to give the product, generally in solid form. These small columns were installed on the Quest synthesizer and driven by its high-pressure nitrogen gas. The Quest accepts 20 reaction vessels, so 20 reactions could be worked up in 1 chromatographic run. This purification procedure is much more efficient than the traditional method.

An example of a single coupling process demonstrates the convenience of this protocol (Figure 9.4). Indolylquinone **2{1}** reacts with indole **3{1}** to give *bis*-indolylquinone **4{1,1}**. The

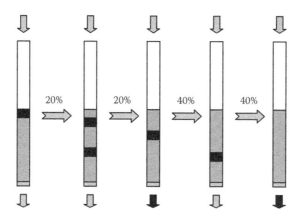

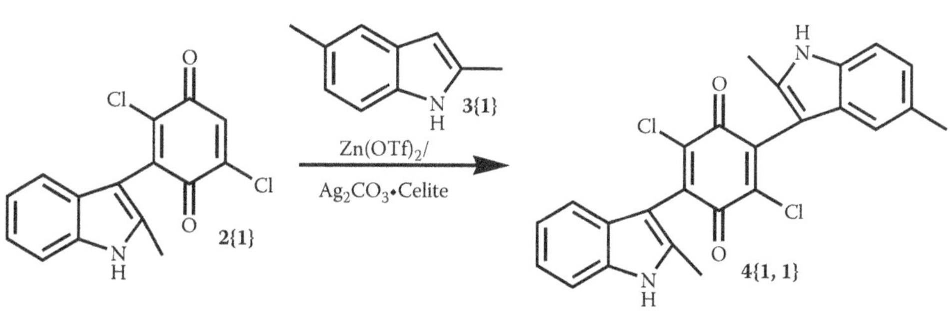

FIGURE 9.4 Model reaction between 2,5-dimethylindole and indolylquinone **2{1}**.

TABLE 9.1
Combinatorial Reactions between Indoles and Indolylquinone 2{1}

Entry	Indole	Yield (%) Purity (%)	Entry	Indole	Yield (%) Purity (%)
1	Indole	89 50[a] (29)	11	6-Benzyloxyindole	76 77[a] (10)
2	1-Methylindole	88 73[a] (12)	12	5-Methylindole	82 78[a] (17)
3	2-Methylindole	89 85	13	7-Chloroindole	19 40[a] (45)
4	2-Phenylindole	86 91	14	5-Iodoindole	49 88
5	4-Methoxyindole	0 —	15	5-Hydroxyindole	0 —
6	4-Benzyloxyindole	76 63	16	5-Chloroindole	51 90
7	5-Fluoroindole	74 84	17	6-Chloroindole	63 84
8	5-Methoxyindole	81 86	18	2-Isopropylindole	86 83
9	2-Methyl-5-chloroindole	100 81	19	6-Fluoroindole	61 68
10	2-Methyl-5-methoxyindole	67 91[a] (9)	20	5,6-Methylene-dioxyindole	44 73

[a] The hydroquinone was also present at the level indicated in parentheses.

reaction was performed on the Quest in a 5-mL reaction vessel equipped with a frit. A mixture of indolylquinone **2{1}**, indole **3{1}**, zinc triflate, and silver carbonate on Celite was heated at reflux in THF overnight. TLC showed that **2{1}** was completely consumed, and a new spot appeared with a lower R$_f$ value. The reaction mixture was filtered through the frit and purified by the high-throughput method described in the preceding text to give the desired product **4{1,1}** in 92% yield.

On the basis of this result, rehearsal reactions of **2{1}** were performed with 20 indoles. The products were subjected to LC-MS to determine their purity and identity. The results are summarized in Table 9.1. Indoles bearing electron-donating groups give the best yields (>70%), whereas indoles bearing halides give lower yields (19 to 74%). Two 4-substituted indoles, 4-benzyloxyindole and 4-methoxyindole give quite divergent results, 76% and no product (respectively), which is very difficult to explain. The crude purity of the majority of products is good (above 80%), whereas that of a minority of products is poor. Six products showed the hydroquinone, as identified by mass spectrometry, to be the major impurity. These compounds have structures very similar to **4{1,m}** as well as polarities very similar to the desired benzoquinone products, so that it was impossible to separate them from **4{1,m}** using such short columns. Given the inclusion of the silver carbonate oxidant in these reactions, the presence of the hydroquinone is difficult to explain.

Rehearsal reactions of 20 indolylquinones, available via our previously described method,[9] were also performed with indole. The results are summarized in Table 9.2. Most of the reactions give an acceptable yield (36 to 100%). Even though yields varied, no obvious relationship between the structure of the indolylquinone and reaction yield could be established at this stage.

An important consideration in planning of the actual library synthesis was the order in which the two indoles would be introduced. For a given *bis*-indolylquinone target, there are two possibilities. One indole reacts with dichloroquinone via an acid-catalyzed reaction in the first step to

TABLE 9.2
Combinatorial Reactions between Indole and Indolylquinones 3{m} Derived from the Named Indole

Entry	Indolylquinone	Yield (%)	Entry	Indolylquinone	Yield (%)
1	Indole	45	11	2-Ethylindole	70
2	1-Methylindole	91	12	5-Methylindole	80
3	6-Fluoroindole	36	13	7-Methylindole	60
4	2-Phenylindole	82	14	6-Methylindole	73
5	4-Methoxyindole	64	15	2-Isopropylindole	86
6	5-Fluoroindole	70	16	4-Fluoroindole	100
7	2-Cyclopropylindole	80	17	6-Chloroindole	49
8	5-Methoxyindole	75	18	2,5-Dimethylindole	48
9	2-Methyl-5-chloroindole	100	19	7-tert-Butylindole	70
10	2-Methyl-5-methoxyindole	80	20	2-(1-Methyl-cyclopropyl)indole	32

give the indolylquinone, which is used as one building block in parallel synthesis. The other indole is used as another building block to react with the indolylquinone via Lewis acid-catalyzed reaction in the second step. Based on the preliminary study of the reactivity of indoles and indolylquinones, the following guides were established to choose which indole should be introduced in the first step and which should be introduced in the second step:

- Indoles that have given good yields in Lewis acid-catalyzed coupling reactions, such as those with electron-donating groups, should be introduced in the second step via reaction with the indolylquinones.
- Indoles that have given poor reactivity in Lewis acid-catalyzed coupling reactions, such as those with electron-withdrawing groups or 4-substituents, should be introduced in the first step via acid-catalyzed reaction with dichlorobenzoquinone because it is more tolerant of substitution.

A final issue is how exactly to conduct 20 reactions at once. There are two obvious ways to simplify reaction setup, which are to use 1 indolylquinone to react with 20 different indoles or to use 1 indole to react with 20 different indolylquinones. These two methods can result in the same overall library, but they are quite different in the actual experimental manipulations, and each has advantages and disadvantages. To use one indolylquinone to react with different indoles has several advantages.

First, the reaction is easier to monitor via TLC. In the Lewis acid-catalyzed coupling reaction, one of the starting materials (the indolylquinone) and the products (the *bis*-indolylquinone) are colored compounds with different colors and R_f values, which makes reaction monitoring quite convenient. The other materials, the indoles, are much less polar than the indolylquinone and *bis*-indolylquinone, and will not interfere with them on TLC. For typical reactions, there are two spots on TLC; spots with higher R_f values are the indolylquinones, whereas spots with lower R_f values are products. When a reaction shows only one spot on TLC, it is easy to tell by comparison with the other lanes if the reaction contains starting material and no product is produced, or if the only spot is product and no starting material remains, because the same indolylquinone was used in all reactions. Second, the reaction is easier to work up. When purifying the product on the short column, the common starting material in all reactions will be eluted at the same time with the same solvent, whereas the different products will remain on the column to be eluted later. Third, the indolylquinone is easier to collect and recover. After the short plug column, all of the indolylquinone fractions can be combined and purified, enabling the indolylquinone to be recycled. This consideration is especially important when the source of the indolylquinone is limited.

The disadvantage of this method is also a very practical one. Because most of the reactions are run at a modest scale, such as 0.1 mmol or 0.05 mmol, only about 30 mg of indolylquinone and less than 20 mg of indole are needed for each reaction. Whereas all of the indolylquinones are crystalline and easy to handle, many indoles are liquids or oils or even sticky solids. It is inconvenient to measure such a small amount of oils, given the number of reactions that must be run. This might be the main disadvantage of this method, and it is a large one.

Another strategy is to allow one indole to react with different indolylquinones, as was done in testing the reactivity of different indolylquinones (Table 9.2). This method overcomes the disadvantages of the first method. Because all the indolylquinones are crystalline, 0.1 mmol of each compound can be weighed easily, and the indoles can be weighed more conveniently at the 2-mmol level (the total amount needed for 20 reactions). Another advantage is the possibility of recovering and recycling the excess indole used in the reaction, when it is valuable. The disadvantage of this method is inconvenience in the workup, because all the starting materials have different R_f values. When there are two spots, the less polar spot should be the starting material, whereas if there is only one spot, it may be difficult to know whether it is product or unreacted starting material. Purification of the product will also be more difficult. Different indolylquinones often have different R_f values, some of them quite wide-ranging. When running the short column with a single solvent, some indolylquinones will be fully eluted whereas others will move only a little. This situation requires monitoring each elution individually and using a different solvent gradient for each column. The other disadvantage compared to the first strategy is the difficulty in recovering the indolylquinones. Generally, because the indolylquinones are all synthesized whereas some of the indoles are commercial, it is more important to recover the former.

After evaluating both methods in model experiments, the first method was initially chosen, considering its greater convenience in the workup. To overcome its major shortcoming, the inconvenience in weighing indoles, all the indoles were converted to 0.1-M THF solutions. One indolylquinone was used in reactions with 10 different indoles, with each indole added as a 0.1-M solution in THF. This method worked well initially, but soon we discovered that some indoles are not stable in solution. After storage for some time, they decompose and introduce impurities into the reactions. The yield of the coupling reaction was also lower owing to this decomposition, reducing the titer of the indoles. Finally, we developed a compromise between these two methods. It involves the use of one indolylquinone to react with five different indoles; simultaneously, another indolylquinone reacts with the same five indoles, and so on. To run 20 reactions simultaneously, 4 indolylquinones are used and each indole is used 4 times. The amount of each indole required is 0.4 mmol, which can be weighed easily and accurately.

9.2.3 LIBRARY PRODUCTION

The synthesis of the two building-block sets was done via conventional methods, and the compounds were purified by chromatography. These building blocks include 21 indolylquinones ($2\{1\}$–$2\{21\}$), 17 of which were known compounds from our earlier work, while the other 4 were new (Figure 9.5), and 35 indoles ($3\{1\}$–$3\{35\}$), 26 of which were commercially available whereas the other 9 were synthesized (Figure 9.6). Indolylquinones were made from the corresponding indoles and dichloroquinone by using our Brønsted acid-catalyzed coupling reaction. Sulfuric acid or acetic acid was used, based on the nature of the indoles. The library is formally 21 × 35, or 735, but the fact that some indole substitutions are present in both building-block sets makes some combinations degenerate, leading to 535 unique combinations.

The final step of the parallel synthesis of DAQ analogs is the hydrolysis of dichloroquinone $4\{n,m\}$ to dihydroxyquinone $5\{n,m\}$. These reaction products have traditionally been purified (including in our laboratory) by chromatography on oxalic acid precoated silica gel. This would

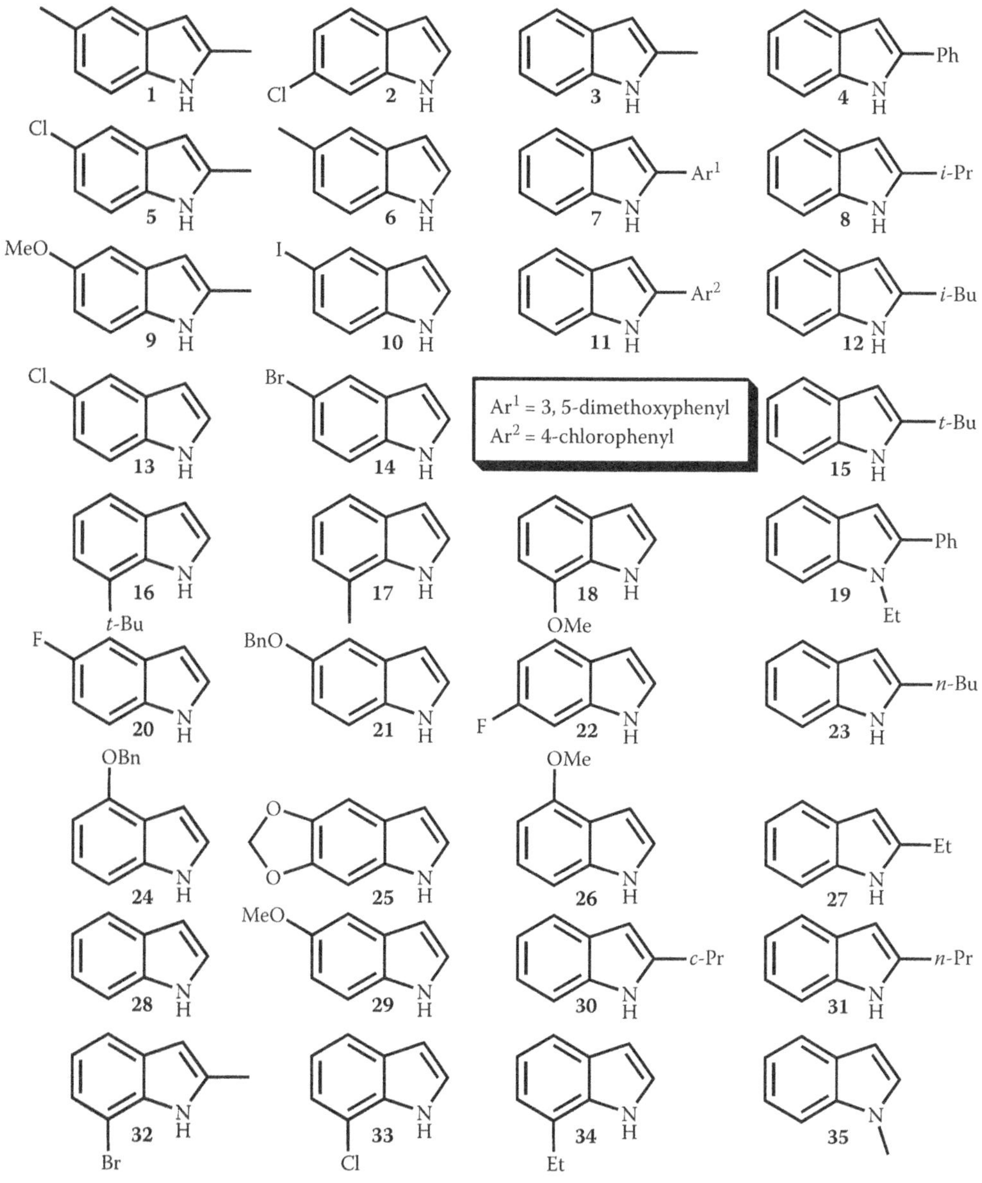

FIGURE 9.5 Indoles **3**{*m*} used in combinatorial synthesis.

clearly be unacceptable for high throughput. To avoid chromatography, we first tested purification by crystallization on the synthesizer. To a refluxing solution of dichloroquinone **4**{*n,m*} in methanol was added NaOH solution, and the mixture was heated at reflux for 30 min. After cooling to room temperature and acidification with sulfuric acid, a precipitate formed and was retained in the reaction vessel after filtration. This crude product was then dissolved in hot toluene. Hexane was added, and the mixture was cooled to room temperature. After filtration, the purified product was retained in the reaction vessel. The final product was collected by dissolution in ethyl acetate, followed by elution and evaporation of this solution. Most reactions gave good yields and high purities of products, and LC-MS gave the expected molecular ion peaks.

We also developed a protocol that uses ion-exchange chromatography. This method has been used for the purification of benzoic acids,[12] and because dihydroxybenzoquinones have pK_as similar

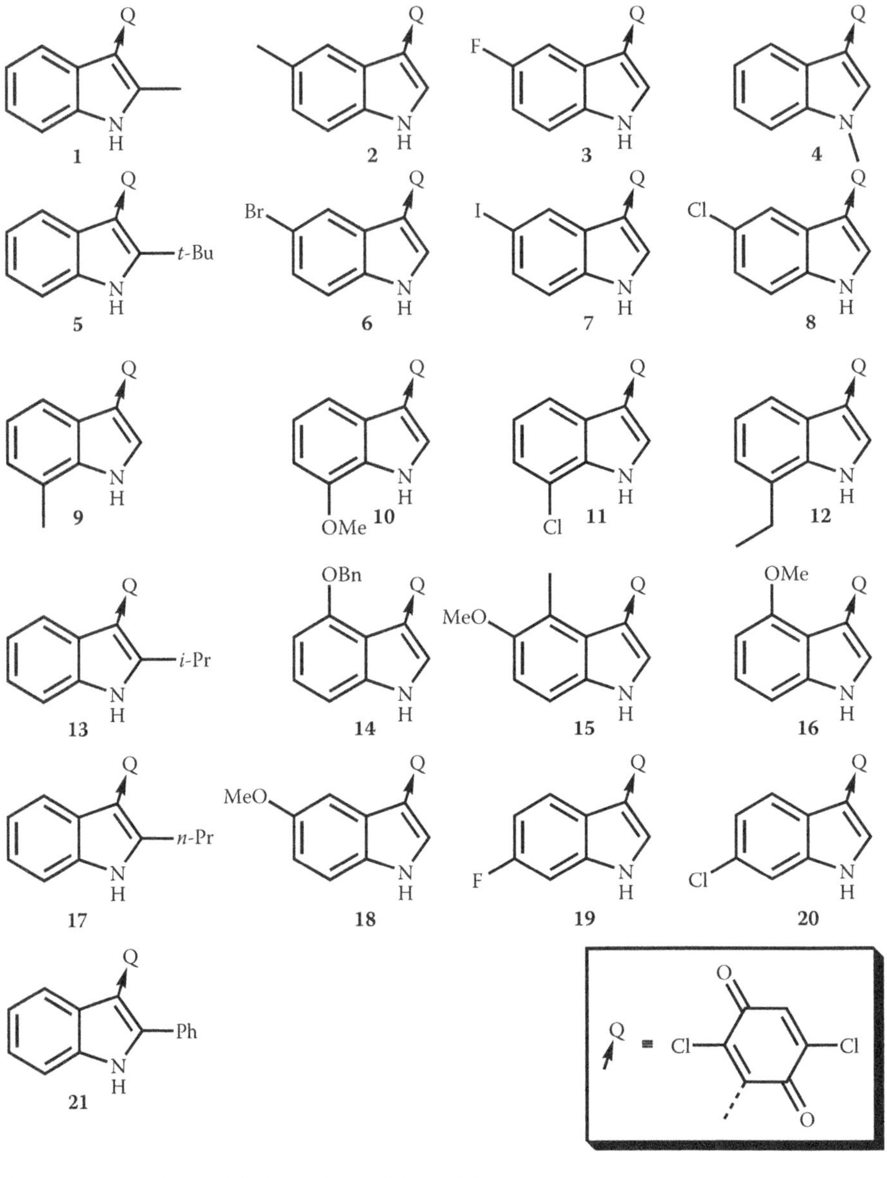

FIGURE 9.6 Indolylquinones 2{*n*} used in combinatorial synthesis.

to carboxylic acids, it should also work with **5{*n,m*}**. Dowex 1 × 8-400 (formate form) was used for the purification. The crude product was dissolved in methanol and treated with triethylamine, followed by addition of the ion-exchange resin and agitation for 30 min. This loaded the product onto the resin, which was then washed with DMF, THF, methanol, and dichloromethane. Treatment with formic acid in dichloromethane solution released the product, which after evaporation showed purity similar to that of the crystallization method.

The results of the hydrolysis reactions were quite dependent on the nature of the indole groups of the *bis*-indolyldichloroquinones. Although it is very difficult to explain chemically why this should be the case, the following empirical rules were developed: *bis*-indolylquinones bearing at least one 2-alkyl-indole are more likely to give good results; *bis*-indolylquinones bearing one 5-alkyl-, 5-alkoxy-, 5-fluoro-, or 6-fluoro-indole, or two 2-unsubstituted indoles are more likely to give poor results.

Of the 535 possible *bis*-indolylquinones that can be prepared from 21 indolylquinones and 35 indoles, 424 (79%) were successfully prepared via solution-phase synthesis. The yields of **4{*n,m*}** were 6 to 100%, averaging 60%, and their purities were 50 to 100%, averaging 80%. Of these 424 dichlorobenzoquinones **4{*n,m*}**, 72% were produced in purities > 80% and in amounts > 10 mg. These 305 compounds were chosen for hydrolysis to give 269 (88%) of the final products **5{*n,m*}**. Their yields were 14 to 100%, averaging 50%, and their purities were 29 to 96%, averaging 70%.

9.3 SYNTHESIS OF AN ILLUDIN LIBRARY

9.3.1 PREPARATORY CHEMISTRY STUDIES

Unlike the situation with the asterriquinones, a modular synthesis of illudins had already been developed through the efforts of Padwa and Kinder.[6] The dipolar cycloaddition of a carbonyl ylide derived from a diazo-β-diketone with a cyclopentenone forms two C-C bonds and establishes the ring skeleton (Figure 9.7) in the key intermediate. The final carbon was added by a methyl Grignard addition to the more reactive ketone. Oxidation states and functional groups were adjusted to provide dehydroilludin M, and it was converted to illudin M.

The key intermediate in this synthesis is a good general structure for a combinatorial synthesis because it is created from multiple modules. This enables diversity to be incorporated into the library at the stage of the preparation of the "core," rather than simply decorating the core with substituents. This strategy allows for changes in the core itself (to a limited extent; see the following text) as a natural outcome of the synthesis, which differs from many combinatorial natural product synthesis strategies that modify side chains on a common core.

The fact that this route was extant did not mean that it was ready to be applied in parallel synthesis. We envisioned executing it using three modules: an enone, a diazo-β-diketone, and an organometallic reagent. The last named must selectively react with the two carbonyls in the key intermediate, modifying the carbonyl bearing the α-ether grouping while preserving other carbonyl, which must promote the base-catalyzed oxygen bridge opening. Rehearsal reactions were conducted with several possible classes of compounds for each module.

The first class investigated was the enones. When cyclopentenone is substituted for **7{*2*}**, the dipolar cycloaddition reaction proceeds even better, resulting in a 75% yield and 6:1 ratio of *exo:endo* products vs. the original 27% yield, *exo* product only; the *exo* adduct is defined as the

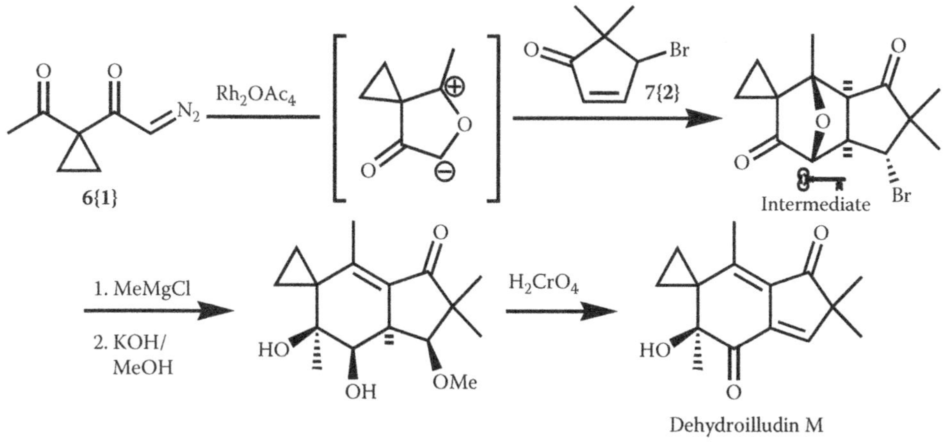

FIGURE 9.7 Padwa–Kinder synthesis of illudins.

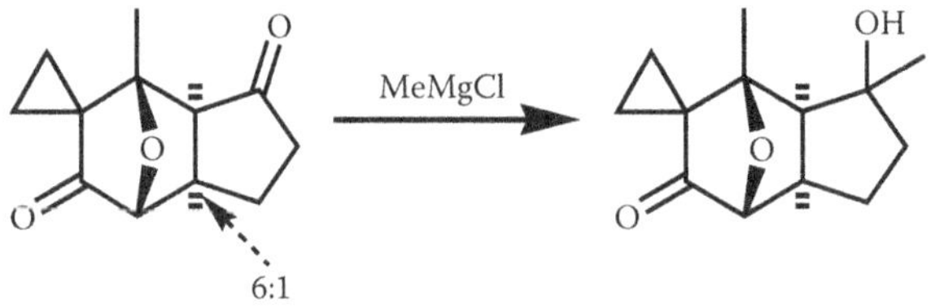

FIGURE 9.8 Unsubstituted cyclopentenone cycloadduct undergoes Grignard addition at the wrong site.

compound in which the enone carbonyl group is on the same face of the cyclohexane ring as the oxygen bridge. However, the resulting product undergoes Grignard reagent addition not at the desired carbonyl but at the carbonyl in the enone module (Figure 9.8). Evidently, the Grignard is less sensitive to the inductive electron-withdrawing effect of the ether oxygen than the steric environment around the carbonyls. The Grignard adds to the carbonyl that came from the enone module when there are no substituents α to it (CH and CH_2 groups in the cycloadduct), rather than the carbonyl that came from the diazo module (CH group and a spirocyclopropane α to it). In the original Padwa–Kinder synthesis, the steric environment was reversed (CH group and quaternary carbon α to the carbonyl from the enone module). Therefore, the dipolarophile module must be limited to enones bearing an α' quaternary carbon. Although such compounds are not that readily available in cyclic systems, we envisioned combinatorial generation of acyclic enone building blocks via aldol condensation of aromatic aldehydes with five commercially available methyl ketones bearing α quaternary carbons. The aldol reaction generally affords the desired acyclic enones as only the trans forms in high yield (>80%), and several variants were prepared using this method (Figure 9.9).

In preliminary studies of the cycloaddition of acyclic enones with **6{1}**, the catalyst used was Rh_2OAc_4 and the solvent was methylene chloride. Two equivalents of the enones were used to obtain higher yields based on the more precious diazo compounds. The yields of cycloadducts were 34 to 60%, with *exo:endo* ratios ranging from 0.92:1 to 3:1.

We then investigated conditions for the cycloaddition of **8{1}** with **6{1}**. The effects of solvents and catalysts were studied. Using 1,3,5-trimethoxybenzene as an internal standard, the yield of the reaction and *exo:endo* ratio could be determined directly by careful integration of the ^1H NMR of the crude reaction mixture. For solvent studies, Rh_2Oct_4 was used as the catalyst, and 1.5 equivalents of enone were used in the reaction. The results are listed in Table 9.3. The effect of solvents on the Rh (II)–mediated cyclization/cycloaddition reaction is quite subtle. Ethereal solvents generally

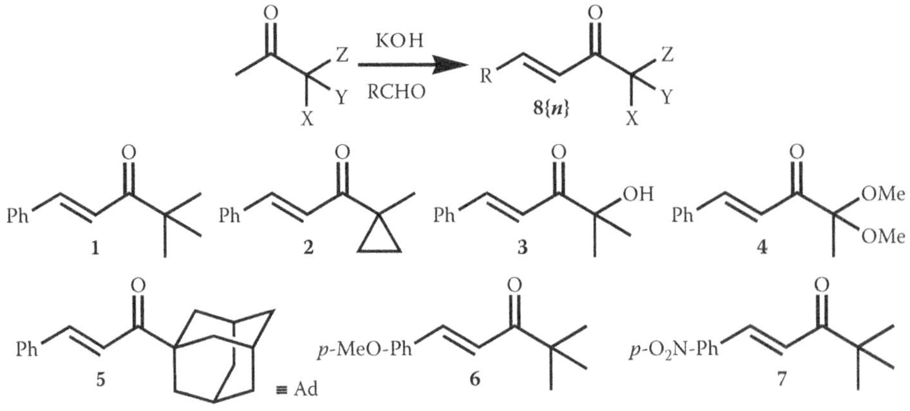

FIGURE 9.9 Preparation of acyclic enones.

TABLE 9.3
Solvent Effects on the Cycloaddition
of 6{1} with 8{1} Using Rh$_2$(Oct)$_4$ as
Catalyst

Solvent	Yield (%)	exo:endo
Diethyl ether	69	2.52:1
1,4-Dioxane	65	1.33:1
Tetrahydrofuran	54	1.88:1
Fluorobenzene	8	—
o-Xylene	40	1.78:1
Acetonitrile	44	1.74:1
Dichloromethane	38	0.83:1
Dimethylformamide	0	—
Cyclohexane	52	2.19:1
Nitromethane	0	—
Chloroform	45	0.81:1
tert-Butyl methyl ether	53	2.03:1
Ethyl acetate	0	—
Acetone	35	1.20:1

gave the best yields (53 to 69%). This is quite surprising, as ethereal solvents are rarely used for rhodium carbenoid reactions. We postulate that this is due to complexation of solvent to the Rh catalyst, which should slow the rate of rhodium carbenoid formation and, therefore, the rate of carbonyl ylide formation. If the rate of cycloaddition of the ylide to the enone is slow, it makes sense that slow generation of the ylide provides "just in time" delivery for the following reaction. If the ylide accumulates, side reaction pathways may compete with the desired process, leading to polar products that are not detected. The *exo:endo* ratio in the cycloadducts also varied with solvent in ways not readily systematized.

The effect of different rhodium catalysts on the yield of the reaction was also examined in ether, with results listed in Table 9.4. Rhodium catalysts with more polarizable (carbene-stabilizing) ligands gave higher yields, in accordance with the theory presented in the preceding text. Solubility of the catalyst in ether may also play a role. A chiral rhodium catalyst, Rh$_2$(5-R-MEPY)$_4$, gave a moderate yield, but the *ee* of the product was low (*ee* = 13.5%), as determined by chiral HPLC. If the 1,3-dipole were completely dissociated from the catalyst, the chiral ligand could not exert its influence on the subsequent reactions of the dipole, and there should be no *ee* at all.[13] The observed *ee*, though small, shows that the 1,3-dipole is not completely dissociated from the rhodium catalyst in the product-determining transition state.

The dipolar cycloaddition product of the adamantyl enone **8{5}** was selected for further study because its *exo/endo* stereoisomers were readily separable chromatographically. We wished to determine their reactivity independently so that we would know if it was important to address the stereoselectivity in the dipolar cycloaddition reaction. Methyl Grignard adds to the desired carbonyl of the *endo* adduct whose phenyl group is up (Figure 9.10) but fails to differentiate between the two carbonyls when the phenyl group is down, giving the double addition product. Using the addition product that could be obtained selectively, the subsequent reaction was examined. However, under a wide variety of basic forcing conditions, it was impossible to eliminate the oxygen bridge at the β-position of the enone. We hypothesize that this is because of the large phenyl group blocking access of the base to the hydrogen that is *cis* to it. In sum, acyclic enones had to be abandoned as building blocks because they suffer from difficulty in Grignard addition to the *exo* adduct or elimination of the *endo* Grignard adduct.

TABLE 9.4
Catalyst Effects on the
Cycloaddition of 6{1} with
8{1} Using Ether as Solvent

Catalyst	Yield (%)
$Rh_2(Oct)_4$	69
$Rh_2(Piv)_4$	54
$Rh_2(OCHO)_4$	47
$Rh_2(OAc)_4$	48
$Rh_2(O_2C\text{-}Ph\text{-}OMe^p)_4$	53
$Rh_2(O_2C\text{-}Ph\text{-}COOMe^p)_4$	32
$Rh_2(O_2C\text{-}Ph\text{-}Cl^m)_4$	55
$Rh_2(5\text{-}R\text{-}MEPY)_4$	56
$Rh_2(O_2CCCl_3)_4$	0
$Rh_2(tfa)_4$	0
$Rh_2(cap)_4$	0
$Cu(acac)_2$	56

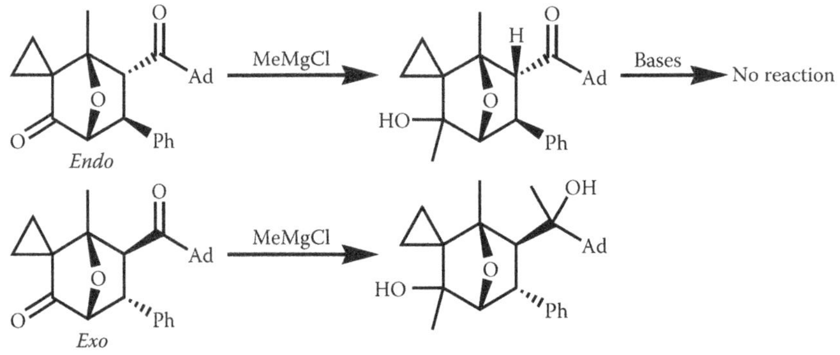

FIGURE 9.10 Transformations of cycloadducts of acyclic enones.

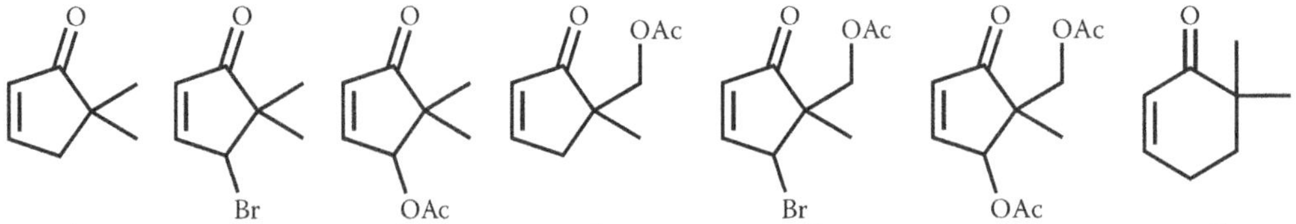

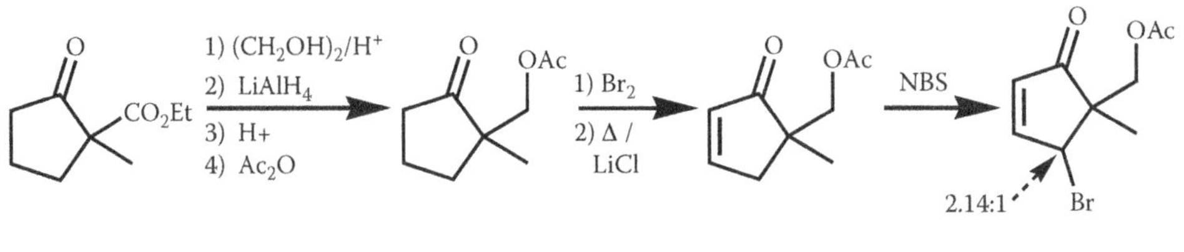

FIGURE 9.11 Cyclopentenone building blocks 7{p}.

TABLE 9.5
Reaction Sequence to Diazo-β-
Dicarbonyl Compounds

6{p}	Yield 1 (%)	Yield 2 (%)	Yield 3 (%)	Yield 4 (%)
2	59	98	70	62
3	45	80	70	89
4	75	84	70	63
5	43	53	76	47
6	68	87	60	81
7	86	83	44	68

A family of cyclic enones with α′-quaternary centers was then selected (Figure 9.11). These compounds are all known,[14] and a synthetic sequence that produces two of them is presented that gives an overview of the methods. This is a fairly involved procedure for preparation of building blocks for combinatorial synthesis, which is generally not recommended, as it can take longer to prepare building blocks than to prepare the library. One redeeming feature of this sequence is that intermediates can themselves be used as building blocks.

The required diazo-β-diketone compounds were prepared by a variation on the known method for preparation of **6{1}**. The dianion of methyl acetoacetate was alkylated by a series of alkyl halides to introduce diversity (Figure 9.12). The remainder of the reaction sequence to the diaz-oketone was according to the literature and resulted in six compounds (Table 9.5). It would have been far more efficient to introduce the diversity at the stage of **6{1}** by direct alkylation, and there was certainly reason to believe that the diazoketone would not interfere with an enolate alkylation reaction. Yet, it proved impossible to accomplish this transformation with even the most reactive alkylating agents such as MeI. This is another instance of the widely appreciated properties of β-ketoester dianions as nucleophiles in alkylation reactions. Again, this is a fairly involved procedure for preparation of building blocks, but no alternatives were forthcoming.

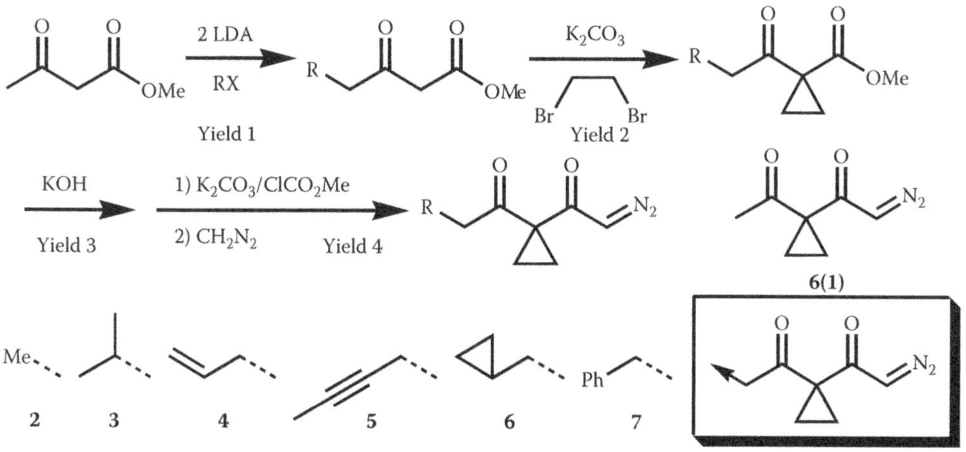

FIGURE 9.12 Preparation of diazo-β-dicarbonyl building blocks **6.**

A fast and easy method to purify products is important to parallel synthesis. Because excess enone is used in the cycloaddition step, a method to remove the unreacted enone at the end of the reaction is required. Many types of scavenger resins have been developed to remove impurities and unreacted starting materials in solution-phase synthesis (see Chapter 6).[15] Polymer-supported nucleophiles, such as amines and thiols, may undergo conjugate addition to the enone to bind it to the support, which permits removal of the enone by filtration. Both a polyamine resin and a thiophenol resin were tested in enone removal by solid-phase extraction. A mixture of compound 7{1} and its cycloadduct in dichloromethane was shaken at room temperature with a polyamine resin for 24 h. We recovered 100% of the adduct and 92% of the enone. As this resin is not reactive enough with the cyclopentenone, a thiophenol resin was examined. The resin was pretreated with a 0.7-M solution of n-Bu$_3$P in THF/H$_2$O (95:5) to reduce any disulfides that may have been present. Different bases and solvents were studied in the capture of enones 7, with analysis by gas chromatography using tridecane as an internal standard. Using a reaction time of 48 h in ethanol with diisopropylethylamine as the base, 72 to 99% of the enones were removed.

The organometallic addition step in the projected synthesis was then examined (Figure 9.13). With cycloadduct **A**, either methyl or vinyl Grignard adds selectively to the desired ketone. This suggested that a variety of Grignards might be used to introduce diversity. Methyl Grignard could also be added selectively to the desired ketone in cycloadduct **B** (including an acetate), provided only one equivalent of the Grignard reagent was used, and the temperature was kept below 0° C. However, extra care is required to add only stoichiometric amounts of Grignard reagents to reactions involving varying amounts of cycloadducts during parallel synthesis. Therefore, a Wittig reaction with methylenetriphenylphosphorane was investigated as a more practical choice for the library synthesis overall, with the large steric demand of the triphenylphosphine expected to promote selectivity between the carbonyl groups. In the reaction of cycloadduct **C** and excess Wittig reagent, two products are obtained, both of which involve the reaction of only the targeted carbonyl. Rather than attacking the second carbonyl after the first addition, the phosphorane acts as a base that promotes the elimination to open the oxygen bridge. This observation is very advantageous in library synthesis, because excess Wittig reagent will not mar the purity of the final product. The elimination of the oxygen bridge is completed by treatment with methanolic potassium hydroxide to form the secondary alcohol.

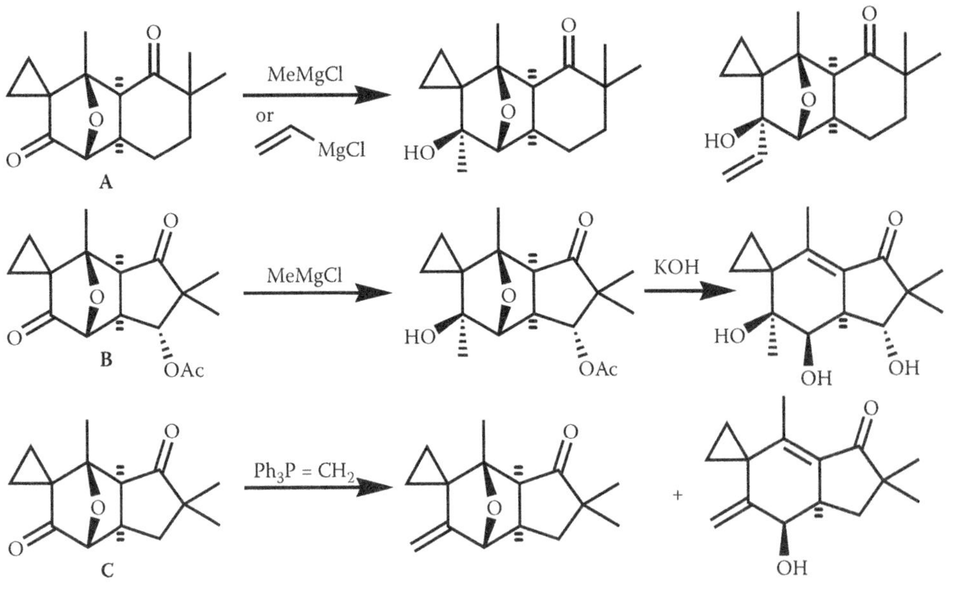

FIGURE 9.13 Nucleophilic addition to key intermediates and elimination of the oxido bridge.

9.3.2 LIBRARY PRODUCTION

The library was constructed as follows. In Quest reaction vessels, 0.30 mmol of 6{*q*} and 0.45 mmol of 7{*p*} were mixed in ether, and a small spatula of Rh_2Oct_4 was added to each vessel at room temperature. After a 6-h reaction time, the dipolar cycloaddition furnishes product 9{*p,q*}, along with some highly polar impurities and unreacted enones. The polar impurities can be easily removed by filtering the crude reaction mixture through a plug of silica gel using methylene chloride as eluent. After solvent exchange to ethanol, the excess enone can be removed by thiophenol resin in the presence of diisopropylethylamine. The yields of 9{*p,q*} were 46 to 100%, averaging 77%, and their purities were 65 to 100%, averaging 87%. Wittig reaction with two equivalents of methylenetriphenylphosphorane (assuming that the average yield in the previous steps was 70%) gives intermediate 10{*p,q*}, which is subsequently eliminated to give the final product 11{*p,q*}, using 10% methanolic potassium hydroxide at reflux. Only a small amount of water (50 µl) was used to quench the Wittig reaction to eliminate the need for an aqueous workup. Excess Wittig reagent does not compromise product purity, because it acts only as a base to promote the desired elimination reaction. Final products were filtered through a plug of silica gel on the synthesizer using methylene chloride as eluent, and then a 20 cm × 1.5 cm silica gel column using ethyl acetate:hexane = 1:3 as eluent removed the triphenylphosphine oxide by-product from the Wittig reaction. Seven building blocks of 6 and seven building blocks of 7 were used, which is formally a 49-compound library but potentially includes 119 compounds if the stereoisomers in building blocks 7 and the *exo/endo* isomerism in the dipolar cycloaddition are taken into consideration. In the final stage, methanolic potassium hydroxide introduces functional group and stereochemical diversity via the bromide and acetate. The acetates are removed to give alcohols with retention of configuration, whereas the halide undergoes nucleophilic substitution for methoxide with inversion (Figure 9.14). All products were characterized by ^{1}H NMR, FID GC, and GC-MS. The 11{*p,q*} were produced in 3 to 18 mg amounts.

9.4 CONCLUSION

The foregoing research provides a study in contrasts of our capabilities to apply combinatorial techniques to natural products. For the asterriquinones, accessing a larger library was made possible by the fact that many of the key indole building blocks are commercially available. For the illudinoids, a significant investment in synthesis of the building blocks was required, limiting the size of the library. Future designs of natural product synthesis may incorporate the commercial availability of appropriate building blocks into the retrosynthetic planning. This has certainly been true of past efforts to develop computer programs that plan organic syntheses of single targets, such as LHASA and SECS.[16] Perhaps an integration of these classical chemoinformatic tools with some of the modern software for planning combinatorial synthesis will be required for combinatorial natural product synthesis to reach its full potential.

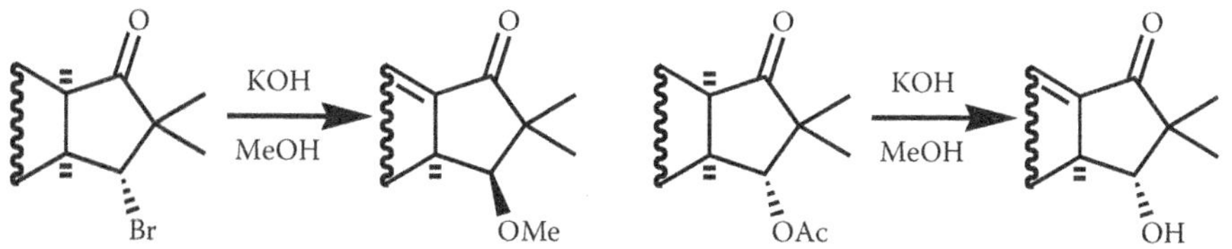

FIGURE 9.14 Stereochemical and structural diversity introduced through reactions at secondary sites in cyclopentenone building blocks.

REFERENCES

1. Nicolaou, K.C. et al., Synthesis of epothilones A and B in solid and solution phase, *Nature*, 387, 268, 1997.
2. Xiao, X.Y. et al., Solid-phase combinatorial synthesis using MicroKan reactors, RF tagging, and directed sorting, *Biotechnol. Bioeng.*, 71, 44, 2000.
3. Walsh, C.T., Combinatorial biosynthesis of antibiotics: challenges and opportunities, *ChemBioChem*, 3, 125, 2002.
4. (a) Cragg, G.M., Newman, D.J., and Snader, K.M., Natural products in drug discovery and development, *J. Nat. Prod.*, 60, 52, 1997. (b) Newman, D.J., Cragg, G.M., and Snader, K.M., Natural products as sources of new drugs over the period 1981–2002, *J. Nat. Prod.*, 66, 1022, 2003. (c) Mulzer, J. and Bohlmann, R., Ed., *The Role of Natural Products in Drug Discovery*, Springer-Verlag, Berlin, 2000.
5. Liu, K., Wood, H.B., and Jones, A.B., Total synthesis of asterriquinone B1, *Tetrahedron Lett.*, 40, 5119, 1999.
6. (a) Padwa, A., Sandanayake, V.P., and Curtis, E.A., Synthetic studies toward illudins and ptaquilosin: a highly convergent approach via the dipolar cycloaddition of carbonyl ylides, *J. Am. Chem. Soc.*, 116, 2667, 1994. (b) Kinder, F.R., and Bair, K.W., Total synthesis of (±)-illudin M, *J. Org. Chem.*, 59, 6965, 1994.
7. McMorris, T.C., Discovery and development of sesquiterpenoid derived hydroxymethylacylfulvene: a new anticancer drug, *Bioorg. Med. Chem.*, 7, 881, 1999.
8. (a) Kaji, A. et al., Relationship between the structure and cytotoxic activity of asterriquinone, an antitumor metabolite of *Aspergillus terreus*, and its alkyl ether derivatives, *Biol. Pharm. Bull.*, 21, 945, 1998. (b) Salituro, G.M., Pelaez, F., and Zhang, B.B., Discovery of a small molecule insulin receptor activator, *Recent Prog. Horm. Res.*, 56, 107, 2001.
9. Pirrung, M.C. et al., Synthesis of 2,5-dihydroxy-3-(indol-3-yl)-benzoquinones by acid-catalyzed condensation, *J. Org. Chem.*, 67, 8374, 2002.
10. Pirrung, M.C. et al., Total syntheses of demethylasterriquinone B1, an orally active insulin mimetic, and demethylasterriquinone A1, *J. Org. Chem.*, 67, 7919, 2002.
11. Pirrung, M.C. et al., Methyl scanning: total syntheses of demethylasterriquinone B1 and derivatives for identification of sites of interaction with and isolation of its receptors, *J. Am. Chem. Soc.*, 127, 4609, 2005.
12. Bookser, B.C. and Zhu, S., Solid phase extraction purification of carboxylic acid products from 96-well format solution phase synthesis with DOWEX 1x8-400 formate anion exchange resin, *J. Comb. Chem.*, 3, 205, 2001.
13. Hodgson, D.M. et al., Catalytic enantio selective intermolecular cycloadditions of 2-diazo-3,6-diketoester-derived carbonyl ylides with alkene dipolarophiles, *Proc. Natl. Acad. Sci. U.S.A.*, 101, 5450, 2004.
14. (a) Matsumoto, T. et al., Synthesis of 5,5-dimethyl-4-acetoxy-2-cyclopentenone and 5-methyl-t-5-acetoxymethyl-R-4-acetoxy-2-cyclopentenone intermediates for illudin M and S, *Bull. Chem. Soc. Jpn.*, 45, 1140, 1972. (b) Baigrie, L., Seiklay, H.R., and Tidwell, T.T., Stereospecific formation of enolates from reaction of unsymmetrical ketenes and organolithium reagents, *J. Am. Chem. Soc.*, 107, 5391, 1985.
15. Flynn, D.L., Devraj, R.V., and Parlow, J.J., Recent advances in polymer-assisted solution-phase chemical library synthesis and purification, *Curr. Opin. Drug Discovery Dev.*, 1, 41, 1998.
16. (a) Wipke, W.T. and Howe, W.J., SECS — simulation and evaluation of chemical synthesis: strategy and planning, *ACS Symp. Ser. (Comput.-Assisted Org. Synth., Symp. Cent. Meet. Am. Chem. Soc.)*, 61, 97, 1977. (b) Pensak, D.A. and Corey, E.J., LHASA — logic and heuristics applied to synthetic analysis, *ACS Symp. Ser. (Comput.-Assisted Org. Synth., Symp. Cent. Meet. Am. Chem. Soc.)*, 61, 1, 1977. (c) Corey, E.J., Computer-assisted analysis of complex synthetic problems, *Q. Rev. Chem. Soc.*, 25, 455, 1971.

10 Solid-Phase Combinatorial Synthesis Based on Natural Products

Takayuki Doi and Takashi Takahashi

CONTENTS

10.1 INTRODUCTION

We have been interested in a novel strategy for the efficient syntheses of natural products. During these studies, we were interested in the synthesis of a variety of their analogs utilizing combinatorial chemistry methods.[1–3] Here, we describe our recent studies of solid-phase combinatorial synthesis for a variety of naturally occurring skeletons.

10.2 ACTIVATED VITAMIN D$_3$ ANALOGS

1α, 25-Dihydroxyvitamin D$_3$ (Calcitriol) (**1**) is well known as the hormonally active form of vitamin D$_3$, whose physiological activities include regulation of cell differentiation and proliferation, intestinal calcium absorption, bone mobilization, and bone formation. Whereas most of these have been modified in either the A-ring or the side chain, a few derivatives were altered in the CD-ring, or in both the A-ring and the CD-ring side chain. These analogs were synthesized individually by various methods, many of which are not consistently convergent methods for analog synthesis.[4] We have developed two synthetic strategies for the high-speed combinatorial synthesis of vitamin D$_3$ analogs on solid phase.

10.2.1 11-HYDROXYVITAMIN D$_3$ USING A SILYL LINKER[5]

1α,11,25-Trihydroxyvitamin D$_3$ (**2**) was initially selected as the synthetic target (Scheme 10.1). The hydroxy group can be attached to a polymer support via a silyl linker at the 11-position, and the attached group should not affect the subsequent reactions. Two sequential carbon–carbon bond

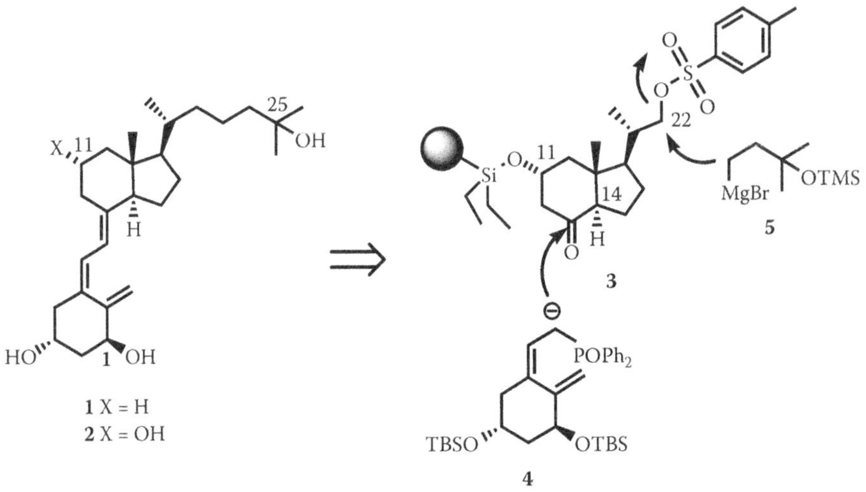

SCHEME 10.1 Strategy for the solid-phase synthesis of 11-hydroxyvitamin D_3 analogs.

formations on polymer support, Horner–Wadsworth–Emmons olefination of ketone **3** with **4,** and alkylation at the 22-position with **5** are involved. Because the Grignard reagent **5** reacts with the ketone in **3**, **3** was coupled with **4** to form a triene system followed by alkylation at the 22-position with **5**.

Polymer-supported ketone **3** was prepared from 11-hydroxyketone **7** and PS-diethylsilyl chloride resin **6** (Scheme 10.2). Triene formation was achieved using the Horner–Wadsworth–Emmons reaction of **3** with eight equivalents of phosphorane **4**. The disappearance of the IR absorption at 1714 cm[1] indicated that the reaction was complete. Next, alkylation of polymer-supported tosylate **8** with 30 equivalents of Grignard reagent **5** in the presence of CuBr•Me$_2$S provided **9**. The disappearance of the IR absorption at 1174 cm[1] indicated complete alkylation. Finally, cleavage from polymer support and concomitant removal of the TMS and the TBS groups with HF•pyridine at room temperature furnished **2** in a 61% overall yield from **3**. Neither isomerization of the triene or epimerization at the 14-position adjacent to the ketone was observed. Furthermore, in this rapid synthesis of 11-hydroxyvitamin D_3 analogs, the trialkylsilyl linker was stable under these carbon–carbon bond-forming reaction conditions, and was cleaved with weak acid without affecting the triene system.

10.2.2 Activated Vitamin D_3 Using a Traceless Sulfonate Linker[6]

In the next stage, we planned a combinatorial synthesis of activated vitamin D_3 analogs without the 11-hydroxy group. If the tosylate group is attached to a polymer support and is replaced with a Grignard reagent, the carbon–carbon bond formation previously described could be carried out on a polymer support. We designed **10** with an alkyoxyphenyl tosylate attached to a polymer support via a silyl linker (Scheme 10.3). The linker-containing polymer-supported **12** was prepared upon the reaction of chlorodiethylsilyl resin **6** with **11** (Scheme 10.4). The structure was determined by ^{13}C SR-MAS (swollen resin–magic angle spinning) NMR compared with the solution ^{13}C NMR of the authentic sample shown in Figure 10.1. The combinatorial library was synthesized using three CD ring derivatives, four A ring derivatives, and six Grignard reagents (Figure 10.2). Resin **12** (25 mg) was packed into 72 IRORI MicroKan reactors containing a radio-frequency (RF) tag (Figure 10.3).[7] After dividing these MicroKans into three flasks, each flask was treated with a CD ring derivative (Scheme 10.5). After the reaction was complete based on SR-MAS analysis, the MicroKans were combined and washed with solvent several times. Then, the MicroKans were split into four flasks and each flask was treated with an A ring derivative under Horner–Wadsworth–Emmons olefination

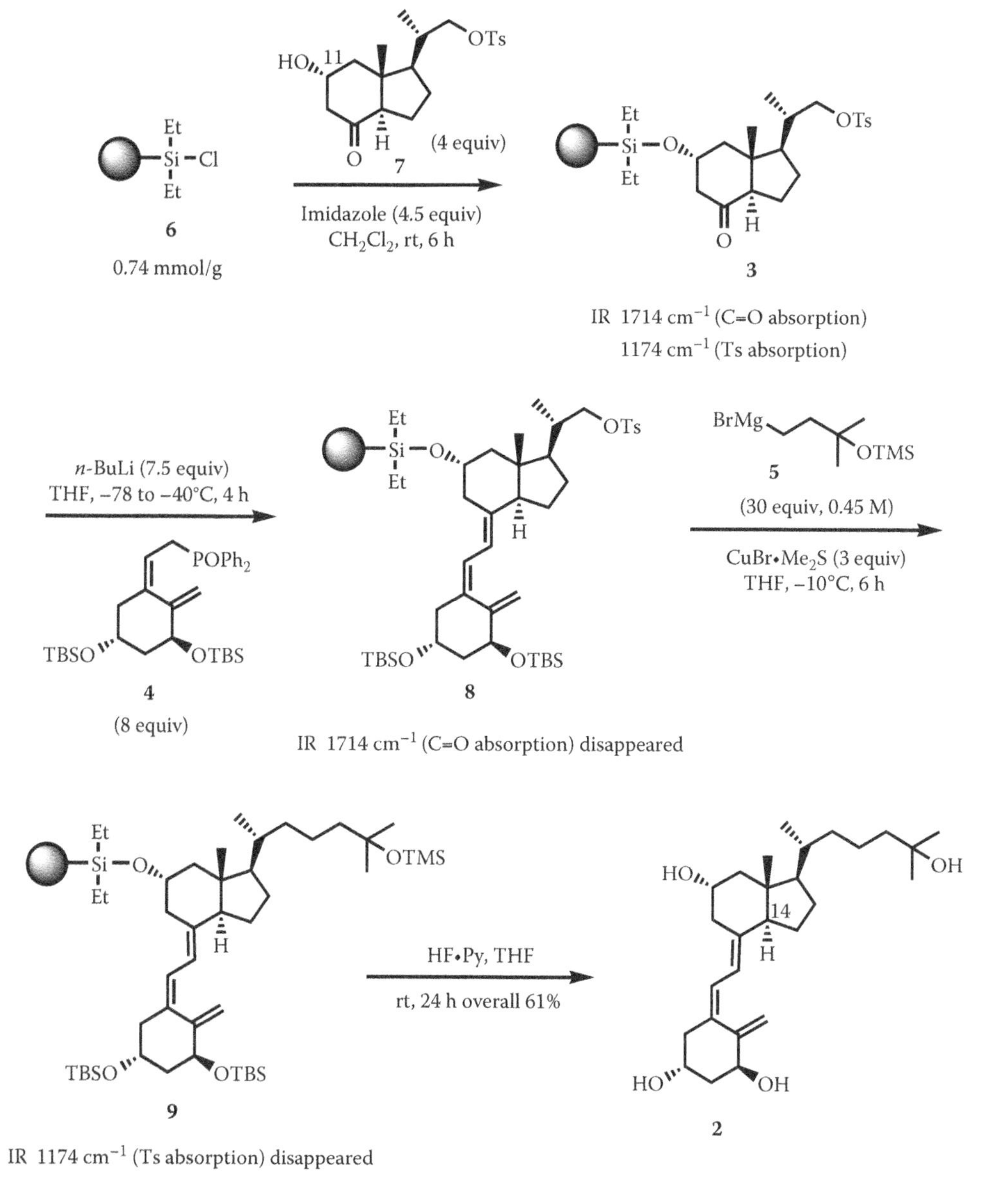

SCHEME 10.2 Solid-phase synthesis of 1,11,25-trihydroxyvitamin D_3 (2).

conditions. The MicroKans were combined and washed again. Each MicroKan was then placed in a Quest 210 (Figure 10.4) and treated with Grignard reagents in parallel to provide 72 products. Upon quenching with water and treatment with acid, activated vitamin D_3 derivatives were obtained. Simple silica-gel column chromatography purified this combinatorial library of activated vitamin D_3 analogs.

10.3 FUNCTIONALIZED TRISACCHARIDES[8]

The sulfonate traceless linker was also used in the combinatorial synthesis of trisaccharides. The linker is stable under acidic glycosylation conditions and is effective for the introduction of

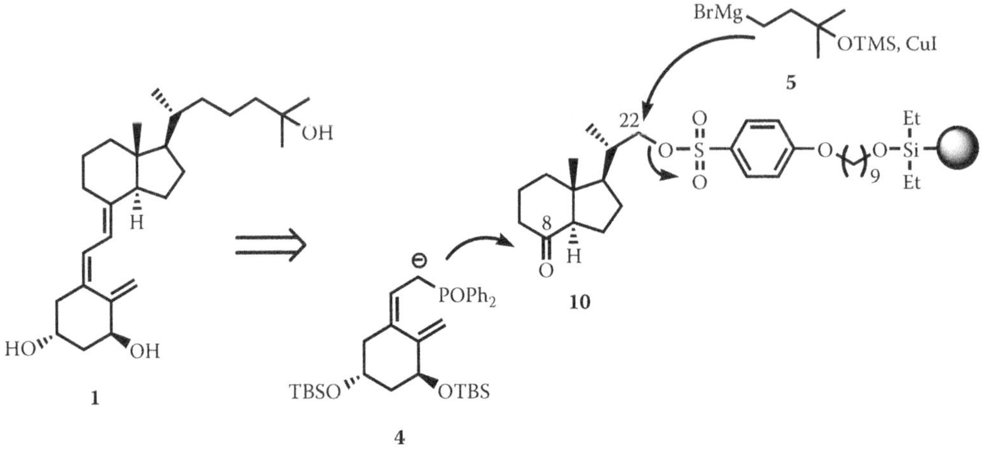

SCHEME 10.3 Strategy for the solid-phase synthesis of vitamin D$_3$ analogs using a sulfonate traceless linker.

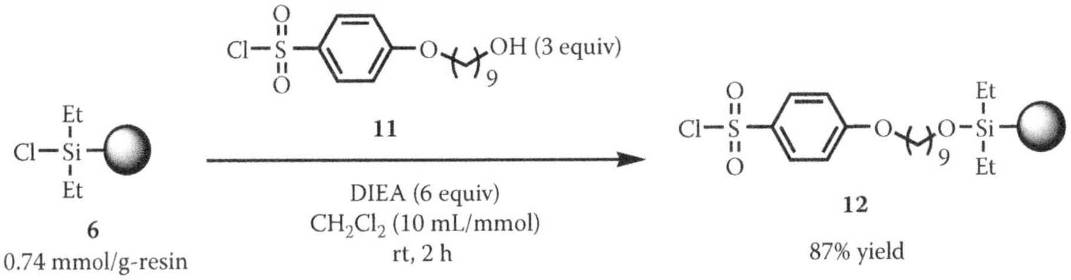

SCHEME 10.4 Preparation of polymer-supported 4-alkoxyphenylsulfonyl chloride.

functionality at the 6-position of a glycopyranose. The strategy included two-directional glycosylation (Scheme 10.6). The loading of **13** to polymer support can be carried out using palladium-catalyzed carbonylation of iodophenylsulfonate (reaction a), easily prepared from a 6-hydroxy free saccharide and 4-iodophenylsulfonyl chloride. We planned three patterns of the synthesis of polymer-supported trisaccharides:

- *Route 1* — deprotection of a silyl group of **15**, glycosylation with **16** (reaction b-1), deprotection of a silyl group in **16**, glycosylation with **18** (reaction c)
- *Route 2* — glycosidation of **15** with **17** (reaction b-2), deprotection of a silyl group in **17**, and glycosylation with **18** (reaction c)
- *Route 3* — the glycosidation of **15** with **17** (reaction b-2), the deprotection of a silyl group in **15**, and glycosylation with **18** (reaction c)

Finally, displacement of the phenylsulfonate linker with various nucleophiles (reaction d) will afford functionalized trisaccharides **19**, **20**, and **21**.[9,10]

Polymer-supported triglycine **14** was prepared by general solid-phase peptide synthesis using an Fmoc strategy (Scheme 10.7). Carbonylation of various iodophenylsulfonates **13**(*1–6*) with **14** was carried out in autoclaves. In route 1, removal of the TBS from three types of polymer-supported saccharides **15**(*1–3*), followed by glycosylation with two glycosyl donors **16**(*1–2*) provided six disaccharides **23** on polymer support (Scheme 10.8). In route 2, glycosidation of glycosyl fluoride **15**(*4*) with three glycosyl acceptors **17**(*1–3*) afforded three disaccharides **24** (Scheme 10.9). In route 3, glycosidation of two fluorides **15**(*5–6*) with a glycosyl acceptor **17**(*4*) provided two disaccharides **25** on polymer support (Scheme 10.9). The polymer-supported 11-member

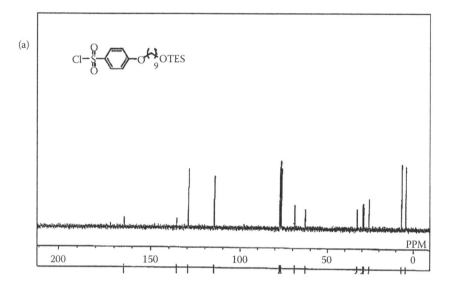

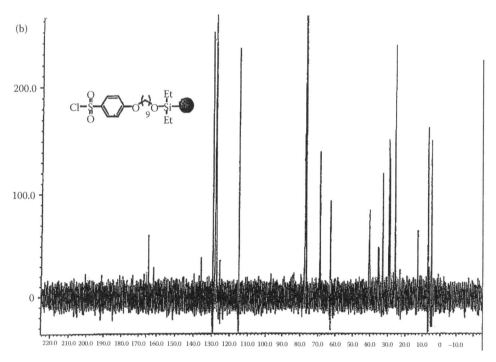

FIGURE 10.1 (a) ^{13}C NMR spectra of 4-[9-(triethylsilyloxy)nonyloxy]phenylsulfonyl chloride in solution (up), (b) ^{13}C SR-MAS NMR spectra of polymer-supported **12** (down).

disaccharides **23**, **24**, and **25** were mixed and treated with acid to remove the TBS protecting group (Scheme 10.10). The second glycosylation of hydroxy-free compounds of **23–25** with **18** was carried out in a single flask. Then, nucleophilic displacement of the phenylsulfonate linker with four nucleophiles, such as azide, iodide, acetate, and hydride, provided 44 trisaccharides **19–21** that were functionalized at the 6-position of one unit of the trisaccharides. Basic hydrolysis followed by hydrogenolysis will afford hydroxy-free trisaccharides.

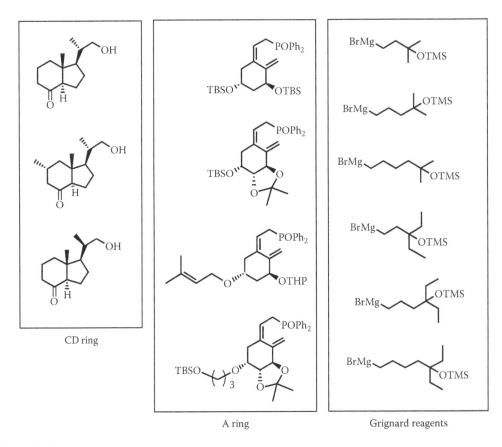

FIGURE 10.2 Building blocks for the combinatorial synthesis of vitamin D$_3$ analogs.

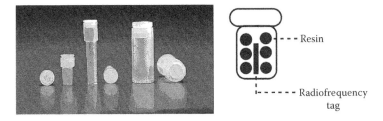

FIGURE 10.3 IRORI MicroKans (Discovery Partners International).

10.4 MACROSPHELIDE ANALOGS UTILIZING PALLADIUM-CATALYZED CARBONYLATION[11]

Macrosphelides (**26**), isolated from the culture medium of *Macrospaeropsis sp.* FO-5050 by the Omura group, exhibited adhesion of human leukemia HL-60 cells to human umbilical vein endothelial cells (HUVEC) in a dose-dependent fashion (Figure 10.5). To elucidate their structure–activity relationships and to find a lead for drug discovery, macrosphelide analogs were prepared by combinatorial synthesis (Scheme 10.11). As shown in the synthesis of trisaccharides on polymer support, carbonylation is equivalent to the formation of an activated ester with aryl iodides serving as masked activated esters.[12] We divided macrosphelide into three units, A, B,

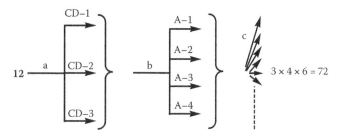

(a) Attachment to polymer-support
(b) Horner-Wadsworth-Emmons reaction
(c) Alkylation with six Grignard reagents in parallel

SCHEME 10.5 Combinatorial synthesis of vitamin D_3 analogs using a traceless linker.

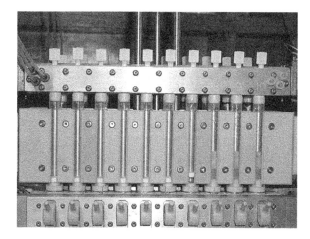

FIGURE 10.4 Quest 210 (Argonaut Technologies). Reaction can be carried out under nitrogen.

and C (Scheme 10.11). The hydroxy group of **28** (unit A) can be attached to the polymer support through tetrahydropyran (THP) linker **27**. Selective deprotection of R^1, esterification with acid **29** (unit B), and chemoselective carbonylation of vinyl iodide in **28** with alcohol **30**, containing a vinyl bromide moiety, can lead to **31**. The vinyl bromide **31** can undergo carbonylative macrolactonization after deprotection of R^2. A tedious protection–deprotection process is not necessary. Finally, acid hydrolysis would yield macrosphelide analogs **32** upon the cleavage from the polymer support.

We initially optimized the reaction conditions on solid phase through the total synthesis of macrosphelide A (Scheme 10.12). Upon loading 1-iodo-1-alken-3-ol **28**(1) onto PS-DHP resin (**27**) and removing the TBS protecting group, compound **33** was obtained. Esterification of polymer-supported alcohol **33** with **29**(2) using DIC-DMAP afforded polymer-supported ester **34**. The palladium-catalyzed carbonylative esterification of **34** with 1-bromo-1-alken-4-ol **30**(1) was investigated. Catalytic $PdCl_2(MeCN)_2$ under 30 atm of carbon monoxide provided **35** with the alkenyl bromide moiety intact. Removal of the MPM group using DDQ in the presence of sodium bicarbonate afforded **31**($1,2,1$). The palladium-catalyzed carbonylative macrolactonization was achieved using $Pd_2(dba)_3$-dppf at 80°C. Acid cleavage from the polymer support and removal of MEM group furnished macrosphelide A (**26a**) in a 38% overall yield over seven steps.

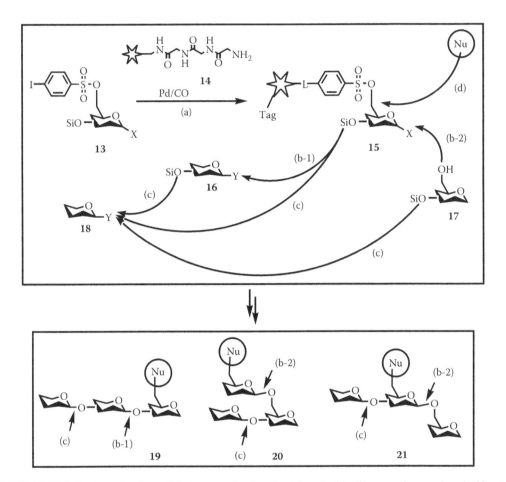

SCHEME 10.6 Strategy for the solid-phase synthesis of a trisaccharide library using a phenylsulfonate traceless linker.

A combinatorial synthesis of macrosphelide analogs was carried out using a split-and-mix method using RF tags.[8] Based upon 4 A units, 4 B units, and 8 C units, a 128-member library was possible (Scheme 10.11 and Scheme 10.13). The crude products obtained by cleavage from the polymer support were purified by automated reversed-phase preparative HPLC gave 122 products (0.2 to 1.5 mg from 30 mg of DHP resin **27**) and included macrosphelides A, C, E, and F.

10.5 CYCLIC DEPSIPEPTIDES[13]

Aurilide (**37**) is a 26-membered cyclic depsipeptide (see Chapter 11, Subsection 11.5.6) that includes 3 *N*-methyl amino acids, isoleucic acid, and an aliphatic acid (Scheme 10.14). To elucidate structure–activity relationships, solid-phase peptide synthesis and solution-phase macrolactamization were planned. The linear precursor **38** was selected based on Yamada's previously reported approach.[14] The linear cyclization precursor **38** can be synthesized from solid-supported tetrapeptide **39** in two ways. In method A, isoleucic acid **40**, aliphatic acid **41**, and *N*-methylalanine **42** are coupled sequentially. An efficient coupling method for ester formation on polymer support is required. In method B, fragment **43** is prepared in advance to avoid ester formation on polymer

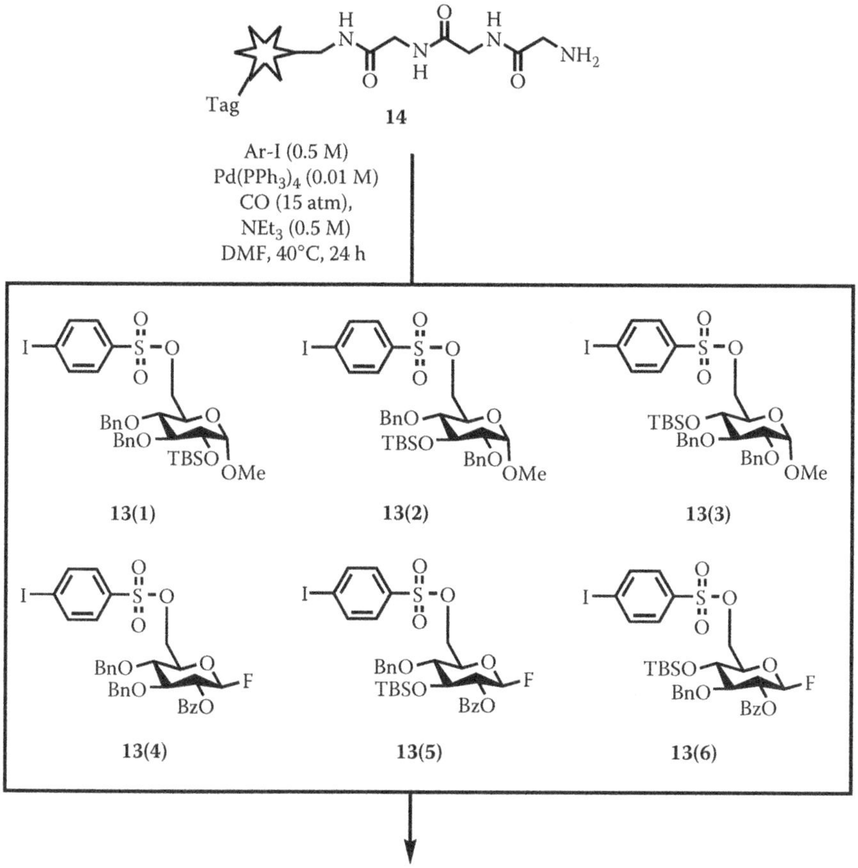

SCHEME 10.7 Loading of 6-phenylsulfonate glycosides **13** by palladium-catalyzed carbonylation onto solid support.

support, and **43** is coupled onto **39**. The tetrapeptide **39** is prepared by standard solid-phase peptide synthetic methods.

We selected a Multipin system for the parallel synthesis of polymer-supported tetrapeptide **39** and all its epimers (Scheme 10.15).[15] Polymer-supported Fmoc D- and L-valine **44** were prepared from amino PS-crown with two types of linkers, HMP (4-hydroxymethylphenoxyace-tamide) and trityl using standard methods. After removal of the Fmoc group with piperidine-CH$_2$Cl$_2$, Fmoc amino acids were coupled using three coupling reagents (DIC-HOBT, HBTU [2-(1H-benzotriazol-1-yl)-1,1,3,3-tetramethyluronium hexafluorophosphate], TFFH [tetramethylflu-oroformamidinium hexafluorophosphate]). These processes were repeated for all combinations of Fmoc sarcosine (**45**), Fmoc D- and L-N-methylleucines (**46**), and Fmoc D- and L-N-methylva-lines (**47**). Therefore, six sets (two linkers and three coupling reagents) of eight compounds (D- and L- combination) were prepared in parallel using the Multipin system shown in Figure 10.6. After removal of the Fmoc group, acid cleavage was carried out using 25% TFA-CH$_2$Cl$_2$ (HMP linker) or 1% TFA-CH$_2$Cl$_2$ (trityl linker), respectively. The crude products **48** were analyzed by LC-MS, and the purity was calculated based upon the UV peak areas at 214 nm. All diastereomers separated well by HPLC (Table 10.1). Acid cleavage conditions were crucial in this synthesis because the tetrapeptide was easily hydrolyzed between the two N-methylamino acids, N-MeLeu

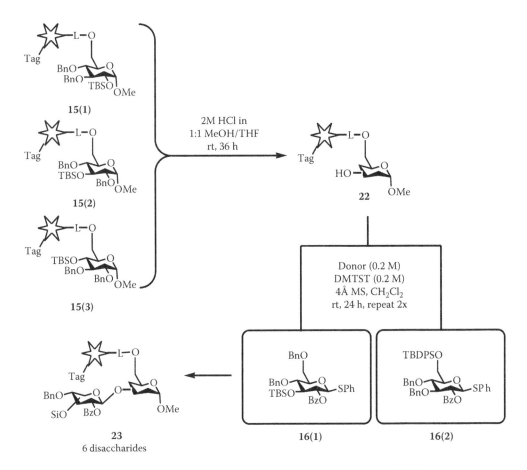

SCHEME 10.8 The first glycosylation of polymer-supported glycosyl acceptors **15**(*1–3*).

and Sar, under the 25% TFA-CH$_2$Cl$_2$ HMP linker cleavage conditions (30 to 66% purities, data not shown). Therefore, a trityl linker that requires milder cleavage conditions (1% TFA-CH$_2$Cl$_2$) was preferred. Only the desired diastereomer was observed. Therefore, no epimerization occurred during the coupling reactions. After reaction optimization, DIC-HOBT gave better results than that HBTU and TFFH (Table 10.1).

Method A was used to synthesize analogs containing a simple alkyl chain (Scheme 10.16). After treatment with 20% piperidine, solid-supported tetrapeptide **39** was coupled with D- and L-*allo*-isoleucic acids (**40**), then coupled with *O*-Fmoc 7-hydroxy-2-heptenoic acid (**49**) using DIC-DMAP. After removal of the Fmoc group, the free hydroxy group was coupled with Fmoc D- and L-*N*-methylalanines (**42**) using 1-(2-mesitylenesulfonyl)-3-nitro-1,2,4-triazole (MSNT), effectively used for the polymer-supported ester formation.[16] Acid cleavage gave linear precursors **50** in quantitative yields with 54 to 88% purities (UV at 214 nm). Macrolactamization of crude **50** was carried out using the reported conditions, EDCI-HOAt in 10% DMF–CH$_2$Cl$_2$.[14] The yield and purity of the crude products **51** are shown in Scheme 10.16. In the synthesis of analogs containing a simple aliphatic system, sequential coupling was achieved using DIC-DMAP for 2-alkenoic acid and MSNT for ester formation.

In the synthesis of aurilide (**37**), preparation of the aliphatic acid moiety was relatively time consuming. So, the ester linkage was constructed in solution, and acid **43** was coupled with polymer-supported tetrapeptide **39** using DIC-HOBt (Scheme 10.17). The reaction proceeded slowly, but

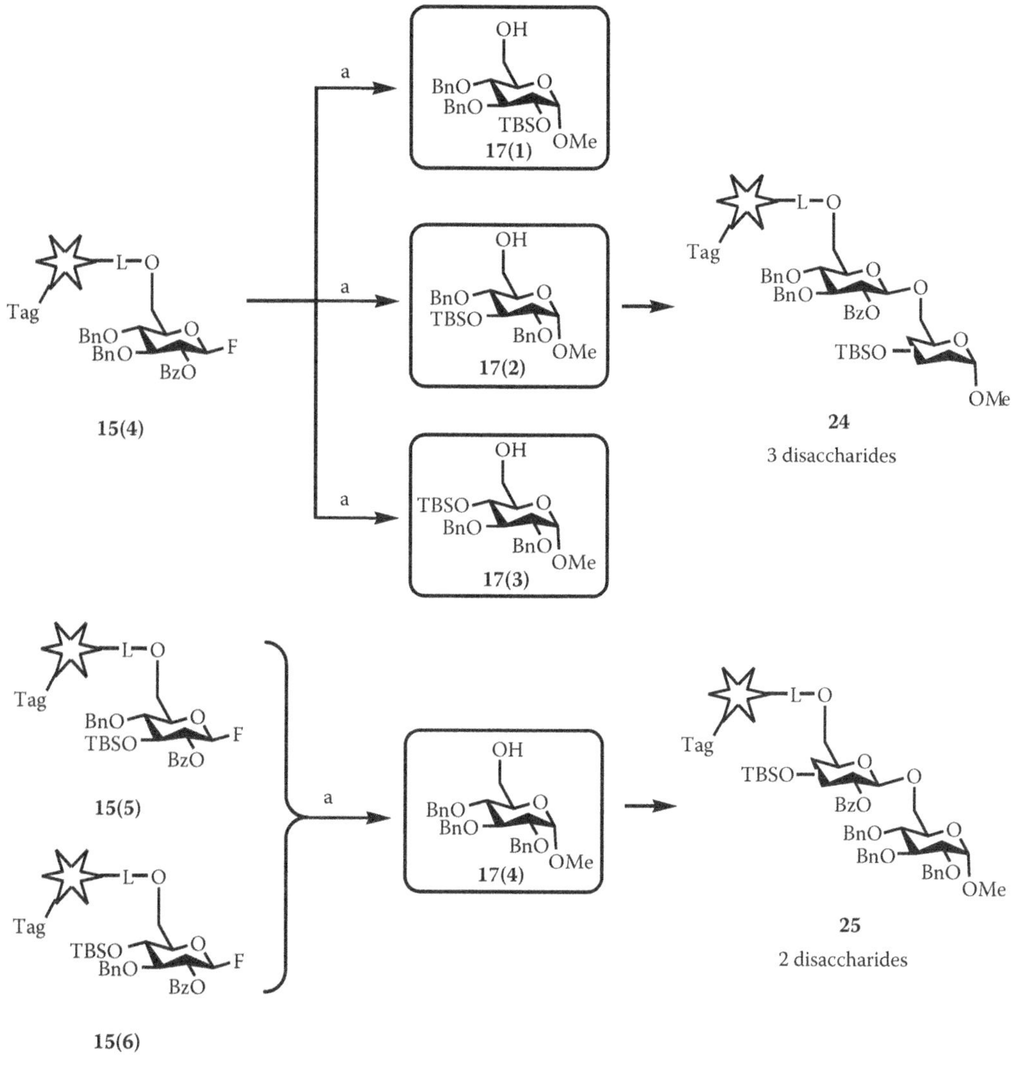

(a) Glycosylation conditions: Acceptor 0.2 M, Cp$_2$Zr(OTf)$_2$ 0.2 M, 4Å MS, CH$_2$Cl$_2$, rt, 24 h.

SCHEME 10.9 The first glycosidation of polymer-supported glycosyl donors **15**(*4–6*).

was complete within 96 h. Acid cleavage provided linear precursor **38** with 53% purity (UV at 214 nm). The macrolactamization, using the preceding method, followed by deprotection of the methylthiomethyl group afforded aurilide (**37**) in 15% overall yield. The preceding two methods will be useful for the synthesis of a variety of cyclic depsipeptide analogs.

10.6 SUMMARY

Solid-phase syntheses of natural product analogs of vitamin D$_3$, trisaccharides, macrosphelides, and cyclic depsipeptides was demonstrated. The synthetic strategy was highly dependent on the structure of the targeted natural product. Combinatorial short syntheses using solid phase were achieved using efficient coupling reactions of various building blocks.

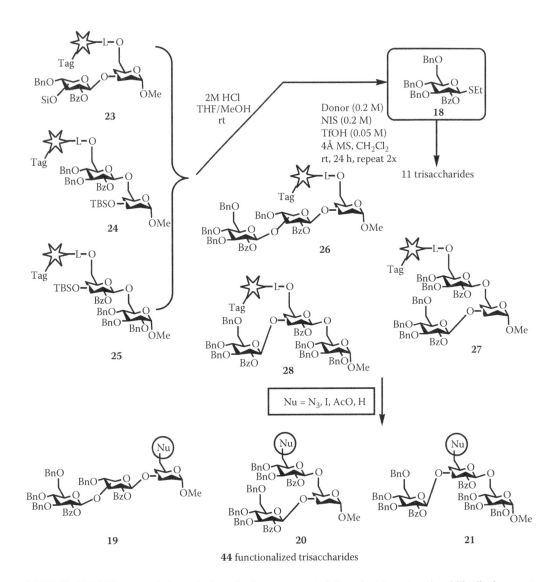

SCHEME 10.10 The second glycosylation of polymer-supported disaccharide and nucleophilic displacement of the phenylsulfonate traceless linker.

	Type	R^1	R^2	R^3	R^4
26a	A	β-OH	β-Me	β-OH	α-Me
26b	B	= O	β-Me	β-OH	α-Me
26c	C	β-OH	β-Me	H	α-Me
26d	D	α - or β-OH	β-Me	β-OH	α - or β-Me
26e	E	β-OH	α-Me	β-OH	α-Me
26f	F	β-OH	α-Me	β-OH	H
26g	H	β-OH	α-Ac	H	α-Me
26h	I	H	α-Me	β-OH	H

Macrosphelide (**26**)

FIGURE 10.5 Structures of macrophelide A–L.

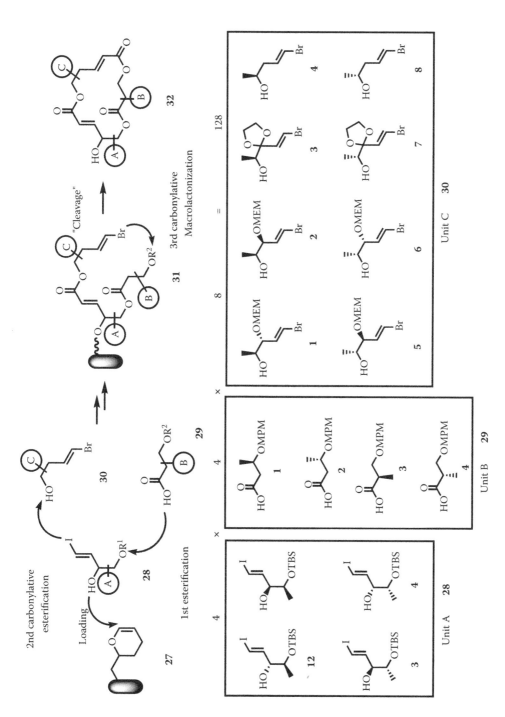

SCHEME 10.11 Molecular diversity design of macrosphelide analogs by a solid-phase combinatorial synthesis utilizing palladium-catalyzed carbonylation.

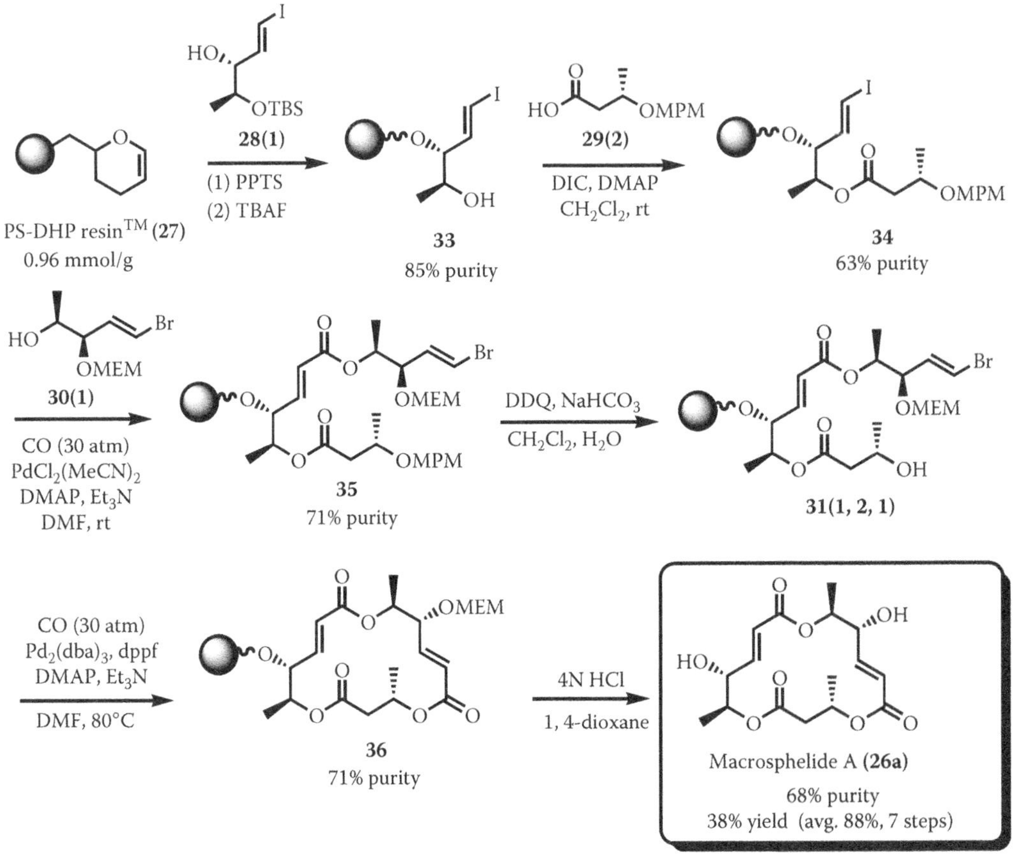

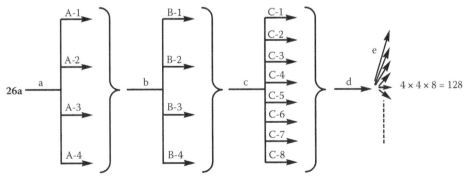

SCHEME 10.12 The total synthesis of macrosphelide A on solid phase. Purity was determined after cleavage by UV-HPLC (214 nm).

(a) Attachment to polymer-support. (b) (i) Deprotection of TBS; (ii) Esterification.
(c) Carbonylative esterification. (d) (i) Deprotection of MPM; (ii) Carbonylative macrolactonization.
(e) Acid hydrolysis.

SCHEME 10.13 A combinatorial synthesis of macrosphelide analogs using a split-and-mix method.

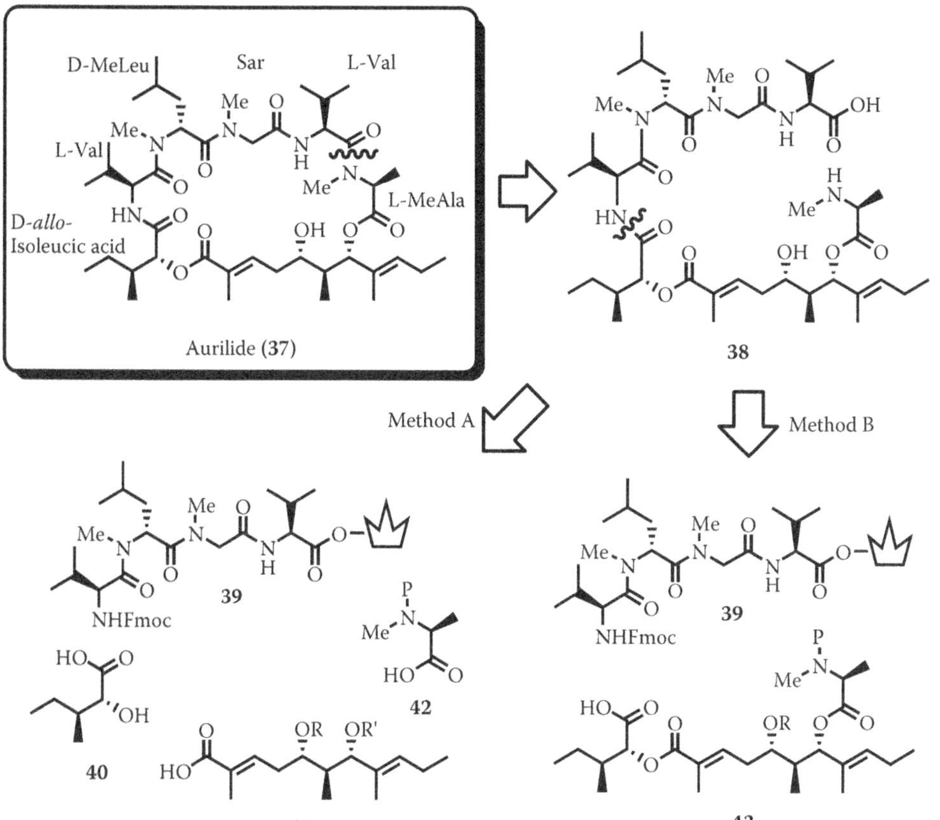

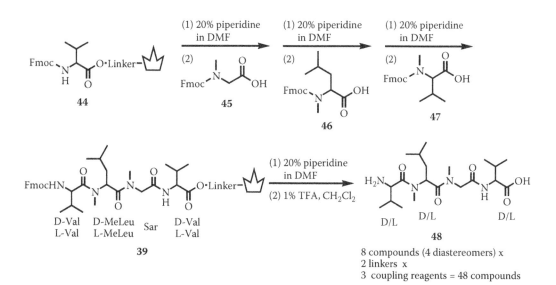

SCHEME 10.14 Strategy for a solid-phase synthesis of aurilide (**37**) and its analogs.

SCHEME 10.15 Combinatorial reaction optimization for a synthesis of polymer-supported tetrapeptides including two *N*-methylamino acids.

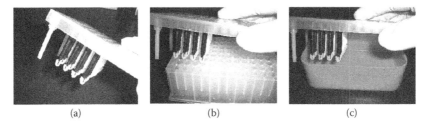

(a)   (b)   (c)

FIGURE 10.6 Parallel synthesis using a Multipin system. (a) Multipin setting, (b) parallel reaction, (c) washing all at once.

TABLE 10.1
Results of the Combinatorial Optimization of Tetrapeptide 48 (H-Val-MeLeu-Sar-Val-OH) Formation on a Trityl-Linked Polymer Support Using Various Coupling Reagents

Entry	Chirality of 48	HPLC Retention Time (min)	% Purity (UV at 214 nm)		
			DIC-HOBt	HBTU	TFFH
1	DL–D	4.9	87	79	71
2	DL–L	4.8	89	70	70
3	DD–D	4.3	81	70	70
4	DD–L	4.2	86	86	78
5	LL–D	4.2	81	79	80
6	LL–L	4.3	90	73	65
7	LD–D	4.8	89	66	68
8	LD–L	4.9	88	86	76

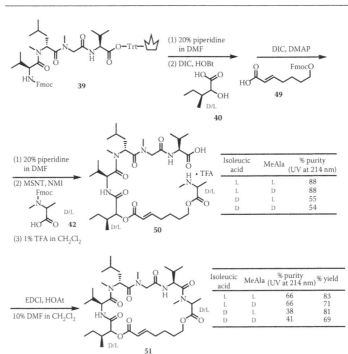

SCHEME 10.16 Solid-phase synthesis of linear precursors and solution-phase cyclization of aurilide analogs involving a simple aliphatic hydroxy acid.

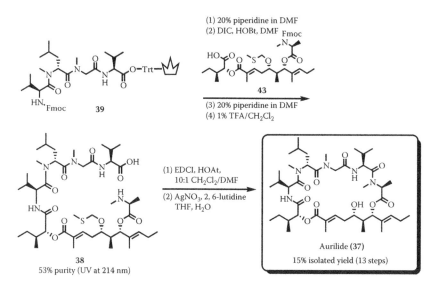

SCHEME 10.17 The total synthesis of aurilide using a polymer support.

REFERENCES

1. Ganesan, A., Natural products as a hunting group for combinatorial chemistry, *Curr. Opin. Biotechnol.*, 15, 584, 2004.
2. Ortholand J.-Y. and Ganesan, A., Natural products and combinatorial chemistry: back to the future, *Curr. Opin. Chem. Biol.*, 8, 271, 2004.
3. Boldi, A.M., Libraries from natural product-like scaffolds, *Curr. Opin. Chem. Biol.*, 8, 281, 2004.
4. Zhu, G.-D. and Okamura, W.H., Synthesis of vitamin D (Calciferol), *Chem. Rev.*, 95, 1877, 1995.
5. Doi, T., Hijikuro, I., and Takahashi, T., An efficient solid-phase synthesis of the vitamin D₃ system, *J. Am. Chem. Soc.*, 121, 6749, 1999.
6. Hijikuro, I., Doi, T., and Takahashi, T., Parallel synthesis of a vitamin D₃ library in the solid-phase, *J. Am. Chem. Soc.*, 123, 3716, 2001.
7. Nicolaou, K.C. et al., Solid and solution phase synthesis and biological evaluation of combinatorial sarcodictyin libraries, *J. Am. Chem. Soc.*, 120, 10814, 1998.
8. Takahashi, T. et al., Synthesis of a trisaccharide library by using a phenylsulfonate traceless linker on Synphase Crowns, *Angew. Chem., Int. Ed. Engl.*, 40, 3230, 2001.
9. Hunt, J.A. and Roush, W.R., Solid-phase synthesis of 6-deoxyoligosaccharides *J. Am. Chem. Soc.*, 118, 9998, 1996.
10. Takahashi, T. et al., Nucleophilic substitutions on Multipin systems linked with a traceless linker, *Synlett*, 1261, 1998.
11. Takahashi, T. et al., A combinatorial synthesis of a macrosphelide library utilizing a palladium-catalyzed carbonylation on a polymer support, *Angew. Chem., Int. Ed. Engl.*, 42, 5230, 2003.
12. Takahashi, T. et al., Palladium(0)-catalyzed carbonylation on the Multipin system, *Tetrahedron Lett.*, 40, 7843, 1999.
13. Takahashi, T. et al., Solid-phase library synthesis of cyclic depsipeptides: aurilide and aurilide analogs, *J. Comb. Chem.*, 5, 414, 2003.
14. Mutou, T. et al., Enantioselective synthesis of aurilide, a cytotoxic 26-membered cyclodepsipeptide of marine origin, *Synlett*, 1997, 199.
15. Bray, A.M. et al., Rapid optimization of organic reactions on solid-phase using the Multipin approach: synthesis of 4-aminoproline analogs by reductive amination, *Tetrahedron Lett.*, 36, 5081, 1995.
16. Nielsen, J. and Lyngsø, L.O., Combinatorial solid-phase synthesis of balanol analogs, *Tetrahedron Lett.*, 37, 8439, 1996.

11 Employing Natural Product-Like Combinatorial Libraries in the Discovery of Lead Libraries

Pedro M. Abreu, Paula S. Branco, and Susan Matthew

CONTENTS

11.1 INTRODUCTION

The structural diversity, wide range of biological properties, and pharmacological profiles displayed by natural products (secondary metabolites), in addition to their dominant role in the discovery of leads for the development of drugs, constitute reasons for their integration in combinatorial drug design approaches.[1–30] Moreover, from a combinatorial chemistry perspective, secondary metabolites are also an example of natural libraries, as illustrated by the pupal defensive secretion of *Epilachna borealis*, which is composed principally of several hundred polyazamacrolides generated by oligomerization of three building blocks,[31,32] and the aromadendrane-type sesquiterpenes from *Landolphia dulcis*

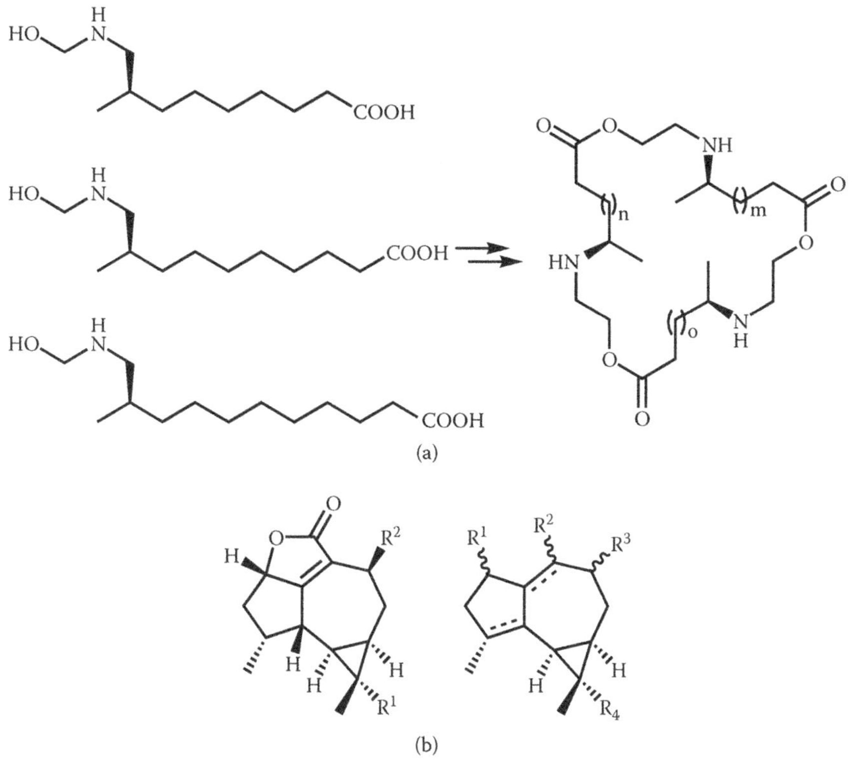

FIGURE 11.1 Natural combinatorial libraries from (a) *Epilachna borealis* (From Schröder, F.C. et al., A combinatorial library of macrocyclic polyamines produced by a ladybird beetle, *J. Am. Chem. Soc.*, 122, 3628, 2000); (b) *Landolphia dulcis*. (From Staerk, D. et al., Isolation of a library of aromadendranes from *Landolphia dulcis* and its characterization using the VolSurf approach, *J. Nat. Prod.*, 67, 799, 2004.)

(Figure 11.1).[33] The application of chemoinformatic techniques to this latter library revealed the presence of a common motif for possible interactions of the aromadendranes with a putative target receptor, and a considerable chemical diversity within the library. These results were interpreted in terms of evolutionary optimization of structures of secondary metabolites for interaction with macromolecular targets, and are of interest in terms of assessment of potential drug-likeness of natural products.

Although theories to explain the existence of the broad area of chemical space generated by the biosynthesis of secondary metabolites continue to be a matter of debate,[14,34,35] it is accepted that these compounds bind to specific receptors and find their way to the target final protein inside other organisms. Taking into account this biological relevance, combinatorial strategies have been developed in recent years to discover lead compounds from natural product and natural product-like libraries that match the elements of conservation and diversity simultaneously expressed by biological targets.[3,5,9,11–21,24–30,36–38]

At one end of the spectrum of library design approaches is combinatorial semisynthesis using individual natural products as core scaffolds for derivatization in either solid or solution phase.[11,17,39–42] In solid-phase derivatization, a natural product skeleton is immobilized onto a solid support to facilitate installation of diversity elements in either a parallel or split-and-pool format. The core structure must already be validated by the biological target, and the precursor molecule should be easily accessible either from natural sources, degradation chemistry, or solution-phase total synthesis. Other key factors for the success of this strategy include the identification of appropriate sites on the scaffold for attachment onto solid support. This facilitates reliable loading and release, and the application of an encoding technology for rapid structural identification of individual library members.

To make libraries from very complex natural product templates, hemi-synthetic approaches can combine solution- and solid-phase methods. In this case, an advanced natural product intermediate is synthesized in solution, attached to the support, and used as a scaffold for combinatorial diversification at several positions. Nevertheless, when mapped in chemical space, the resulting library constituents form clusters, because most of the structural diversity is due to appendage modification. In order to enhance the stereochemical and skeletal diversity obtained from combinatorial derivatization of natural templates, diversity-oriented synthesis is now being applied to generate hybrid libraries.[43-45] This approach involves the use of coupling reactions to attach different appendages to a common molecular skeleton possessing multiple reactive sites with potential for orthogonal functionalization. Stereochemical diversity can be generated by using stereospecific and stereoselective diversity-generating processes, whereas skeletal diversity can be achieved by both reagent-based (differentiating) and substrate-based (folding) strategies. The latter strategy has demonstrated potential for generating skeletal diversity combinatorially. In a similar approach designated as *diversity-modified natural scaffolds* (DYMONS),[27] natural scaffolds of microbial origin are chemically transformed to be orthogonally protected or to undergo chemoselective reactions. Substituents bearing pharmacophoric groups are then introduced combinatorially. A third approach for the construction of diversity-oriented natural product-based libraries uses highly structural diverse chiral building blocks obtained from selective chemical fragmentation of natural products. Hybrid structures are then generated from them combinatorially.[46]

Many biologically active natural products are also derived through mixed biosynthesis, either by integration of the different biosynthetic pathways to generate enmeshed structures, or in straightforward covalent linkage between components derived through different pathways. This biosynthetic feature has inspired another approach for the construction of hybrid libraries that incorporates diverse natural building blocks.[47,48]

In the course of pharmacological studies of benzodiazepines, it was recognized that derivatives within this compound class bind not only to benzodiazepine receptors of the central nervous system, but also to unrelated peripheral benzodiazepine receptors, and cholecystokinin receptors.[49] Although the natural substrates showed little resemblance to this compound class, this intrinsic binding affinity to several proteins that bind to similar regions of peptides or other proteins led to the search for common structural motifs in different families of natural products. These designed privileged structures can be optimized for their ligand affinity and selectivity. An example of such a privileged structure is the benzopyran moiety that can be found in more than 4000 compounds, including many bioactive natural products and pharmaceutically designed compounds. Based on this concept, strategies for the production of combinatorial libraries around privileged structures identified from natural products have been developed.[11,13-15,19,30]

Finally, at the other end of the spectrum of library design approaches lies the target-oriented total synthesis of natural product analogs.[13,15-17,28,39-42] In a typical case, the natural product itself is first targeted for synthesis on solid phase. Often requiring a multistep synthesis utilizing a large range of organic reactions and methods for loading, elaborating, and cleaving the final target, a pool of building blocks is then employed to construct the library.

The content of this chapter illustrates the preceding approaches for the construction of natural product-like combinatorial libraries. The emphasis is on the synthesis of relevant bioactive compounds, which are categorized according to the type of secondary metabolite.

11.2 CARBOHYDRATES

11.2.1 OLIGOSACCHARIDES

Carbohydrates are the most abundant group of natural compounds, and, as well as their glycoconjugates, are involved in such important functions as cell-to-cell recognition and communication, inflammation, immunological response, bacterial and viral infection, tumorigenesis, and metastasis.[50-58]

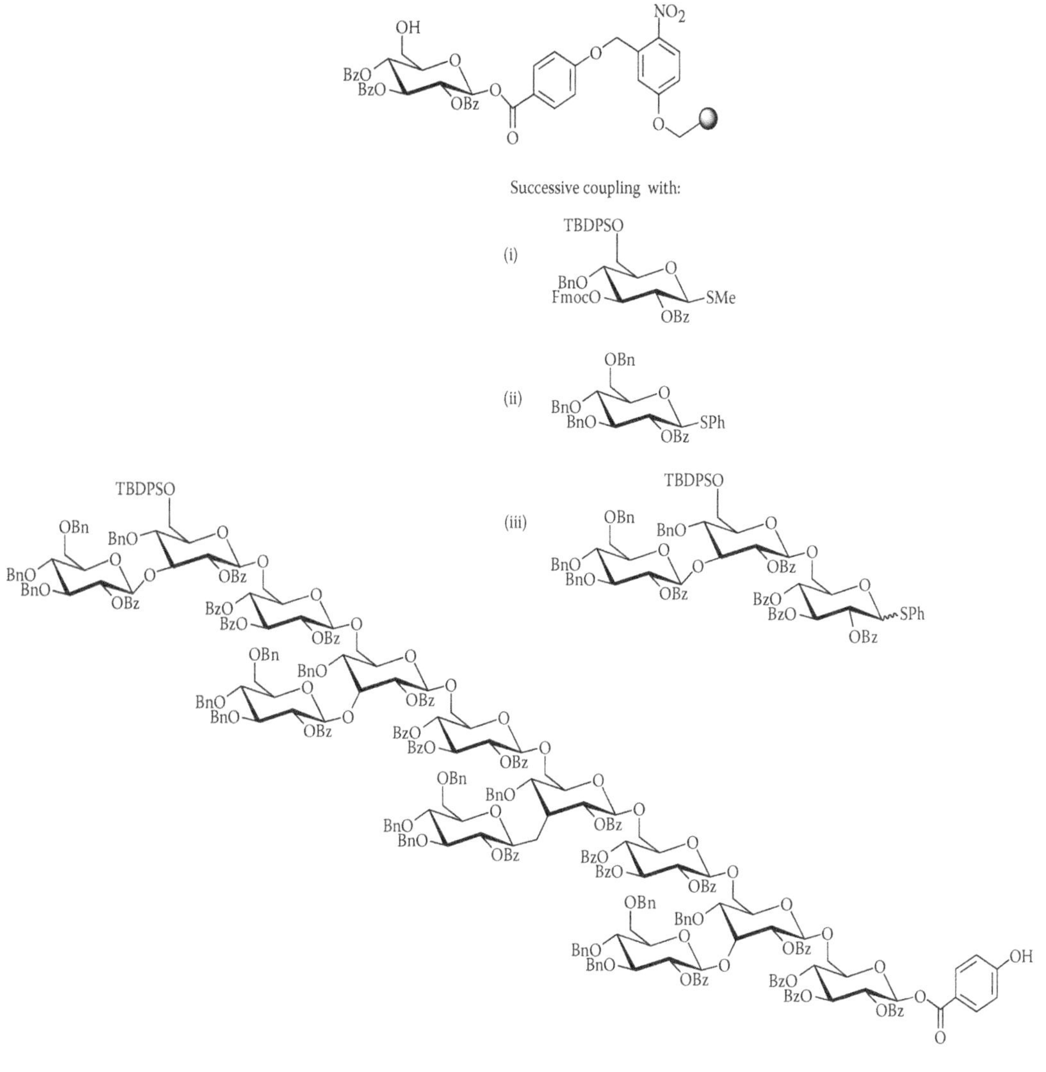

FIGURE 11.2 Solid-phase synthesis of a dodecasaccharide. (From Nicolaou, K.C. et al., Solid-phase synthesis of oligosaccharides: construction of a dodecasaccharide, *Angew. Chem. Int. Ed. Engl.*, 37, 1559, 1998.)

Also, the saccharide portions of various classes of natural products function as key molecular recognition elements important to the biological properties of the natural compound.[52,57,59,60] In addition, the conformation rigidity and functional richness of the carbohydrate system is unparalleled in value as a molecular scaffold for generating molecular diversity through combinatorial strategies (see Chapter 7).[52] So far, combinatorial chemistry of carbohydrates has primarily focused on the solid- and solution-phase synthesis of oligosaccharide libraries, as well as glycomimetics and peptidomimetics.[61–84] The largest oligosaccharide to be constructed on solid phase was achieved by Nicolaou et al.[85] Starting from a stereochemically homogenous conjugate consisting of the first monosaccharide unit, a benzoate spacer, and a photolabile *o*-nitrobenzene linked to a Merrifield resin, a dodecasaccharide was assembled in a building-block-type fashion and finally cleaved from the solid support by photolysis (Figure 11.2). The advantages of this method, highly enabling for the construction of large and diverse libraries of oligosaccharides, include convergence for block-type constructions, high-yielding glycosidation steps, maintenance of stereochemical integrity during loading and unloading, and flexibility.

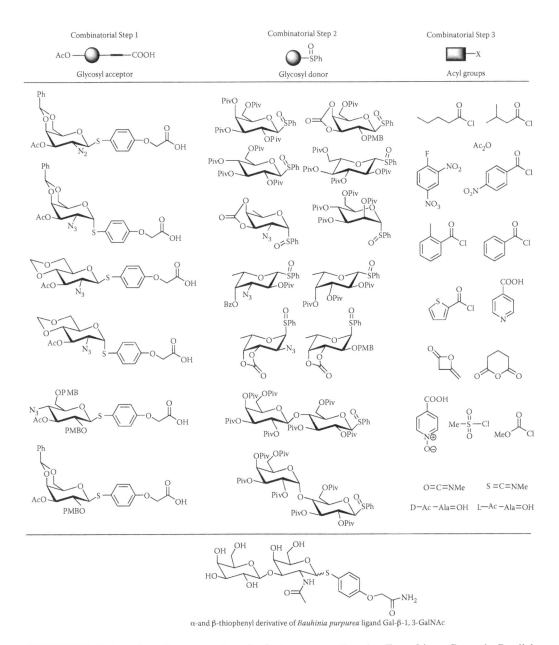

α-and β-thiophenyl derivative of *Bauhinia purpurea* ligand Gal-β-1, 3-GalNAc

FIGURE 11.3 Carbohydrate library based on *Bauhinia purpurea* ligands. (From Liang, R. et al., Parallel synthesis and screening of a solid phase carbohydrate library, *Science*, 274, 1520, 1996.)

The search for antibacterial agents led to the synthesis of disaccharides structurally related to lipid A, the bioactive principle of lipopolysaccharide (LPS) in the cell surface of Gram-X negative bacteria.[86–91] A combinatorial approach to studying cell surface carbohydrate binding to their protein targets involved the preparation of a library of approximately 1300 di- and trisaccharides, including both the α- and β-thiophenyl derivatives of *Bauhinia purpurea* ligand Gal-β-1,3-GalNAc (Figure 11.3).[91] The library was synthesized by a split-and-mix strategy from six glycosyl acceptors attached separately to TentaGel resin (step 1). Twelve different glycoside sulfoxide donors were coupled separately to mixtures of beads containing all the six monomers, recombined, and split again after reduction of sugar azides to amines (step 2). The separate pools of beads were then *N*-acylated (step 3), recombined, and finally deprotected. Chemical tags were introduced after every reaction for identification purposes. The outlined strategy, complemented with the library screening against *B. purpurea* lectin, can be used to identify carbohydrate-based ligands for any receptor and for discovering new compounds that bind to proteins participating in cell adhesion.

11.2.2 Dynamic Combinatorial Chemistry

A new and effective approach to combinatorial carbohydrate library generation for rapid ligand or receptor identification is dynamic combinatorial chemistry.[92,93] This concept is based on reversible connections between suitable building blocks, leading to spontaneous assembly of all their possible combinations and allowing for the simple one-step generation of extended libraries. To demonstrate the feasibility of this approach, and its potential in glycobiology, a dynamic combinatorial library was generated against concanavalin A (Con A), a plant lectin that binds a branched trimannoside unit located in certain glycoproteins.[94] The hydrazidecarbonyl–acylhydrazone interconversion was used as reversible chemistry. Aldehyde functionality can be reserved for the interactional components, and hydrazide functionality can be for the structural components, or *vice versa* (Figure 11.4). A complete acylhydrazone library containing up to 474 constituents was generated from the dynamic assembly of 9 oligohydrazide core building blocks and used to arrange the interactional components in a given geometry, with a set of 6 aldehyde counterparts potentially capable of interacting with the binding site of the target species. The interactional components were based on naturally occurring carbohydrates. The core components were chosen to contain one, two, or three

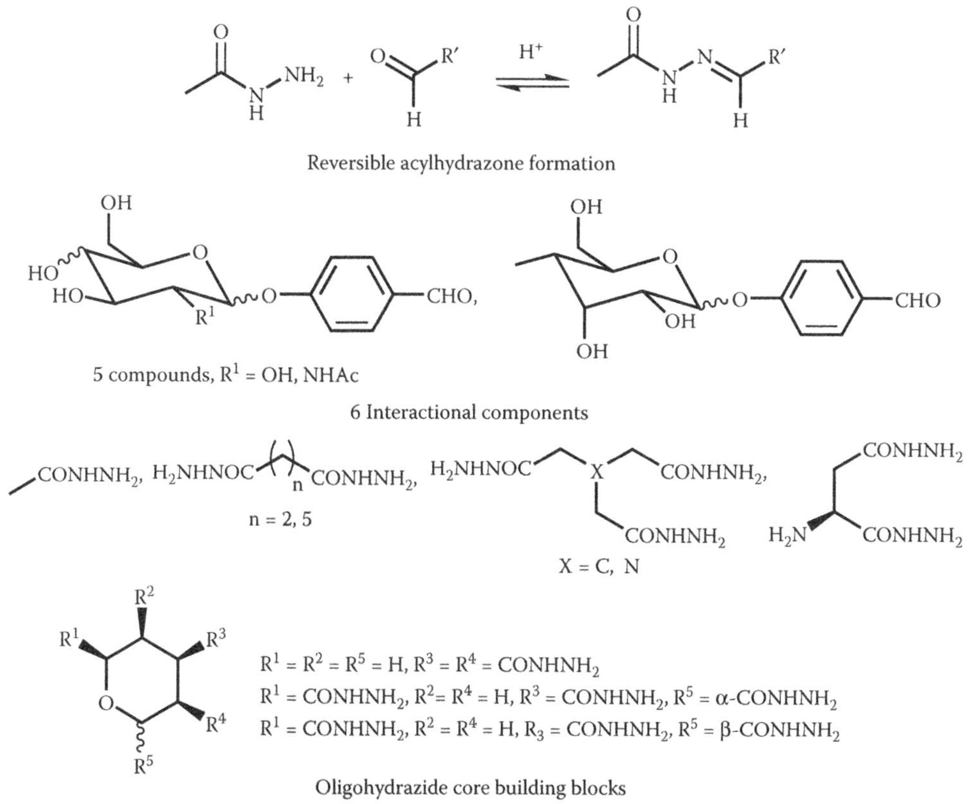

Reversible acylhydrazone formation

5 compounds, R^1 = OH, NHAc

6 Interactional components

n = 2, 5

X = C, N

$R^1 = R^2 = R^5 = H, R^3 = R^4 = CONHNH_2$
$R^1 = CONHNH_2, R^2 = R^4 = H, R^3 = CONHNH_2, R^5 = \alpha\text{-}CONHNH_2$
$R^1 = CONHNH_2, R^2 = R^4 = H, R_3 = CONHNH_2, R^5 = \beta\text{-}CONHNH_2$

Oligohydrazide core building blocks

FIGURE 11.4 Design of an acylhydrazone dynamic library. (From Ramstrom, O. et al., Dynamic combinatorial carbohydrate libraries: probing the binding site of the concanavalin A lectin, *Chem. Eur. J.*, 10, 1711, 2004.)

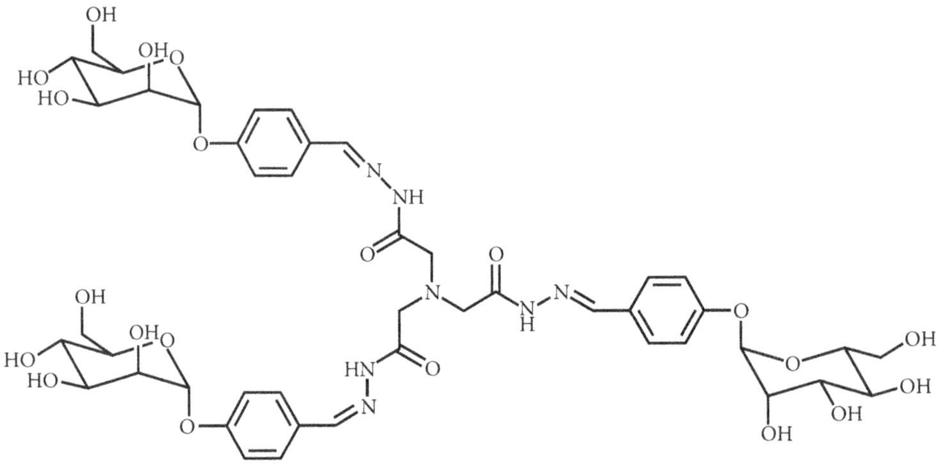

FIGURE 11.5 Tritopic mannoside with potent binding to concanavalin A. (From Ramstrom, O. et al., Dynamic combinatorial carbohydrate libraries: probing the binding site of the concanavalin A lectin, *Chem. Eur. J.*, 10, 1711, 2004.)

attachment groups in order to probe the effect of multivalency through the presence of several interacting groups in the same library constituent. Following the screening of the library against Con A, dynamic deconvolution allowed the identification of the structural features required for binding, and the selection of a strong binder (IC_{50} = 22 μM), the tritopic mannoside **1** (Figure 11.5).

The capacity of dynamic combinatorial chemistry to probe carbohydrate–protein interactions was also demonstrated with the glycosidase hen egg-white lysozyme (HEWL). HEWL is an enzyme involved in peptidoglycan degradation that cleaves *N*-acetyl-glucosamine oligomers into smaller units, which, up to the tetramer, behave as competitive inhibitors.[95] A small dynamic combinatorial library of potential HEWL binders was generated from the equilibration of two amino-derived carbohydrate compounds and six aromatic aldehydes. After reduction of the resulting iminium species, it was shown that the amine **2** was selected by HEWL as an optimal binder, and whose aryl group mimics a carbohydrate unit (Figure 11.6).

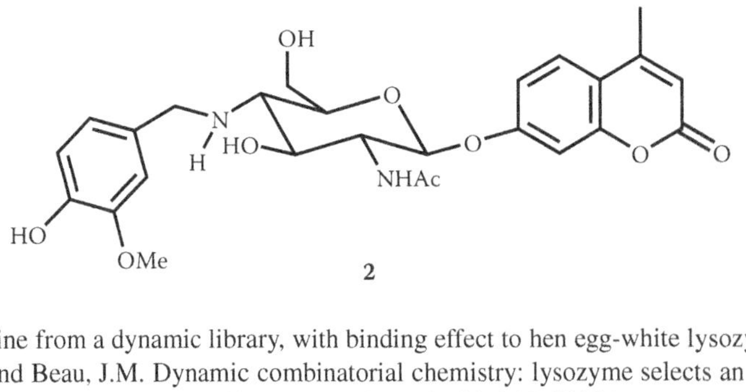

FIGURE 11.6 Amine from a dynamic library, with binding effect to hen egg-white lysozyme. (From Zameo, S., Vauzeilles, B., and Beau, J.M. Dynamic combinatorial chemistry: lysozyme selects an aromatic motif that mimics a carbohydrate residue, *Angew. Chem. Int. Ed. Engl.*, 44, 965, 2005.)

11.2.3 AMINOGLYCOSIDES AND GLYCOPEPTIDES

Examples of carbohydrate-based antibiotic libraries include aminoglycosides and glycopeptides. Moenomycins are a family of natural product antibiotics that are known to inhibit the synthesis of bacterial cell wall peptidoglycan through inhibition of transglycosylase (see Chapter 8).[96] Degradation studies of moenomycin A, a pentasaccharide containing a long lipid attached to the reducing sugar through a phosphoglycerate unit, showed that cell wall inhibitory activity was retained in a disaccharide core structure and that certain structural elements were important for retaining transglycosylase inhibitory activity. These facts led Sofia and coworkers[96] to use moenomycin A disaccharide analogs **3a–3c** as templates to build a combinatorial library of 1300 disaccharides employing the IRORI radio-frequency tagging method for directed-sorting mix and split (Figure 11.7).[97] Some of the library compounds (i.e., **4**) showed potent antibiotic activity against a panel of Gram-positive and Gram-negative bacteria with IC_{50} values below 15 µg/ml.

The naturally occurring pseudodisaccharide neamine has been used as a core structure for the generation of new libraries of aminoglycoside mimetics. In a representative synthesis of a neomycine B mimetic library, the protected substructure neamine **5**, bearing a free aldehyde, *tert*-butyl isocyanide **6** (or isocyanoacetic acid ethyl ester **7**), and a glycine derivative linked to a soluble polyethyleneglycol polymer **8**, were reacted with various *N*-protected amino acids **9** through Ugi-type multiple component condensation (Figure 11.8).[98] When screened for binding to the viral transactivator protein Rev of HIV mRNA, several library constituents were found to be more active than neamine. Following these results, a library of analogs with various polyamino, amino alcohol, or aromatic substitutions appended at C-5 of the 2-deoxystreptamine (2-DOS) unit of neamine, was prepared and analyzed for binding to models of the *Escherichia coli* 16S A-site ribosomal RNA, and for antibacterial activity.[99]

In pursuing their efforts to design small molecules that recognize RNA,[100] Wong et al. prepared a library of neamine-based antitumor compounds that target oncogenic RNA sequences (Figure 11.9).[101] Once again, the 5-position of 2-DOS of a protected neamine core **10** was selected for the introduction of library diversity. A parallel synthesis of a neamine library was carried out by reaction of the corresponding azide precursor with a variety of amines (RNH_2) after conversion of the 5-*O*-allyl group to a reactive chemical handle, cyclohexylcarbodiimide bound to a macroporous polystyrene resin. Several other aromatic and heteroaromatic analogs were prepared by reductive amination of an intermediate aldehyde derived from **10**. Final steps of azide reduction and hydrogenation yielded a library of compounds (**11** and **12**) modified at the C-5 position of neamine.

Through glycodiversification at the O-6 position of the neamine unit of kanamycin, a library of kanamycin B analogs was generated (Figure 11.10).[102] The library was prepared by reaction of phenylthioglucopyranose-based sugar donors having benzyl or azido groups at C-2, with the neamine acceptor whose amino and hydroxyl groups were previously converted into azide and acetate groups, respectively. Running the reaction in a 3:1 mixture of Et_2O/CH_2Cl_2, neamine underwent optimal regiospecific glycosylation at the O-6 position, and resulted in the desired 4,6-disubstituted 2-deoxystreptamine motif of kanamycin antibiotics. Evaluation of the antibacterial activity of these kanamycin analogs against *Escherichia coli* and *Staphylococcus aureus*, as well as molecular modeling for evaluating the binding toward the target site of 16SrRNA, led to the identification of several leads.

The role of natural carbohydrates in cell adhesion processes, specifically, between the selectins and the Lewis sugars and their derivatives,[103] led to the design and synthesis of mimics of sialyl Lewis X (sL^x) and related oligosaccharides via combinatorial strategies.[68,69] Armstrong et al.[104] and Wong et al.[105] used Ugi four-component condensation with a variety of functionalized C-glycosides to construct libraries of sL^x mimetics. The constituents were further tested as inhibitors of E- and P-selectin (Figure 11.11).

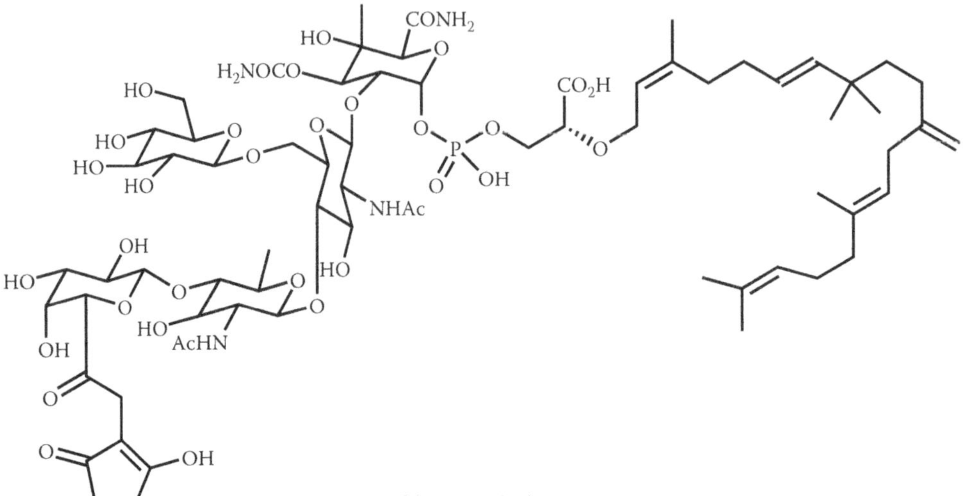

Moenomycin A

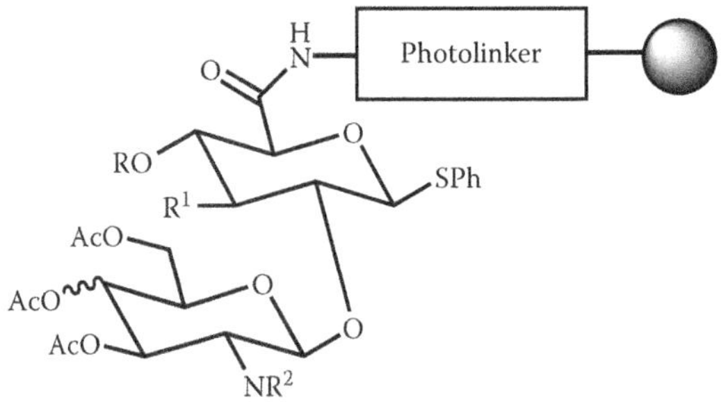

Moenomycin analogs
3a R = H, CH$_3$, Bz
3b R^1 = OLev, N$_3$
3c R^2 = Pht, HCOCF$_3$

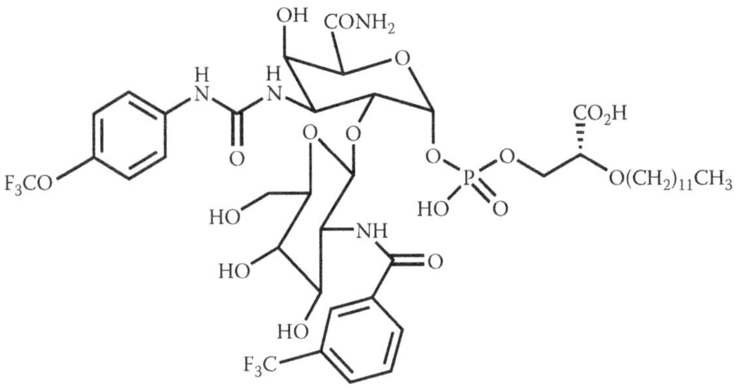

4

FIGURE 11.7 Moenomycin A disaccharide analogs. (From Nicolaou, K.C. et al., Radiofrequency encoded combinatorial chemistry, *Angew. Chem., Int. Ed. Engl.*, 34, 2289, 1995.)

11.3 FATTY ACIDS DERIVATIVES

11.3.1 PROSTAGLANDINS

The prostaglandin family constitutes one of the most pharmacologically active low molecular-weight natural products. Combinatorial access to prostaglandins is based on two main strategies of synthesis of its E- and F-series developed by Janda and Chen[106,107] and Ellman et al.[108] In the

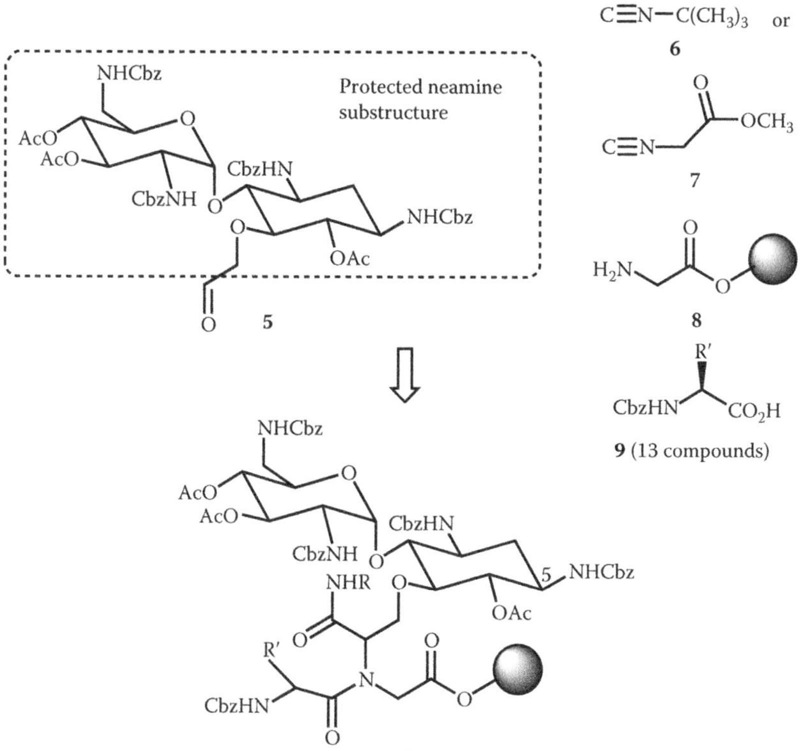

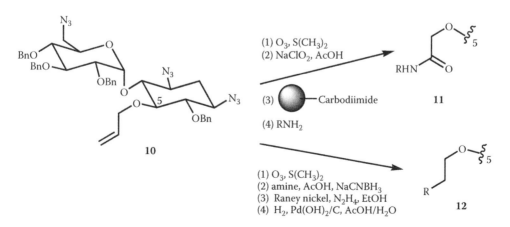

FIGURE 11.8 Synthesis of neomycin B mimetics using four-component condensation. (From Park, W.K.C. et al., Rapid combinatorial synthesis of aminoglycoside antibiotic mimetics: use of a polyethylene glycol-linked amine and a neamine-derived aldehyde in multiple component condensation as a strategy for the discovery of new inhibitors of the HIV RNA Rev responsive element, *J. Am. Chem. Soc.*, 118, 10150, 1996.)

FIGURE 11.9 Parallel synthesis of a 5-*O*-alkylated neamine library. (From Sucheck, S.J. et al., Design of small molecules that recognize RNA: development of aminoglycosides as potential antitumor agents that target oncogenic RNA sequences, *Angew. Chem. Int. Ed. Engl.*, 39, 1080, 2000.)

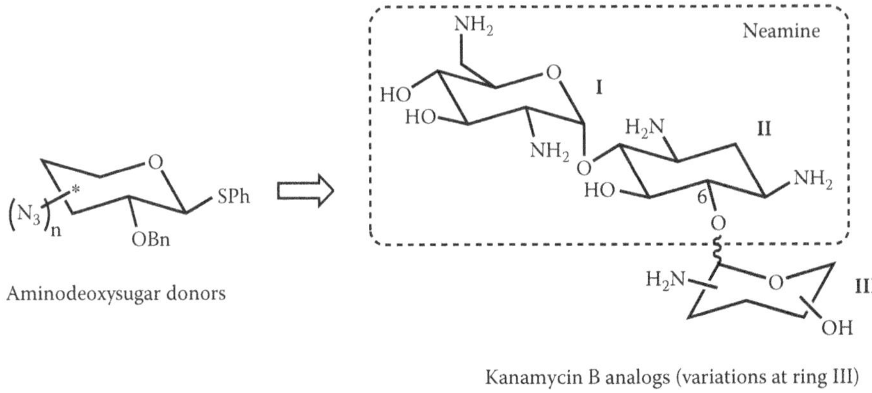

Kanamycin B analogs (variations at ring III)

FIGURE 11.10 Library of kanamycin analogs. (From Li, J. et al., Application of glycodiversification: expedient synthesis and antibacterial evaluation of a library of kanamycin B analogs, *Org. Lett.*, 6, 1381, 2004.)

approach of Janda and coworkers illustrated in Figure 11.12 for prostaglandin E_2 methyl ester and prostaglandin F_2, Noyori convergent three-component coupling[109] was carried out with a cyclopentene core linked to functionalized polystyrene resin **13**, long-chain cuprate **14** (ω side chain), and triflate **15** (side chain). Using a parallel-pool library strategy, the same research team prepared a library of sixteen prostaglandin derivatives distributed in four pools, each containing a constant side chain with four different ω side chains resulting from the Noyori reaction.[110] Some of these prostaglandin analogs proved to inhibit the growth of the herpes-family virus CMV on infected murines.

The same strategy was followed in the asymmetric synthesis of six-membered ring prostanoids, both in solution and using non-cross-linked polystyrene (NCPS) as soluble support (Figure 11.13).[111] Target molecule **19** was generated in a convergent fashion wherein chiral enone **16** was the precursor of the central ring, and cuprate **17** and triflate **18** were used to introduce the side chains at C-3 and C-2, respectively. The (*R*) configuration chiral center of **16** directed the facial selectivity of the conjugate addition reaction, which then dictated the stereochemical outcome of the subsequent alkylation. In order to obtain PGE-type prostanoids, the C-2 side chain was further modified by reduction of the triple bond to the *Z*-alkene via hydrogenation in the presence of 5% Pd/BaSO$_4$. This solution-phase methodology was further adapted to NCPS soluble support, which was used as a protecting group for the C-4 alcohol moiety.

Small prostanoid libraries were prepared by Ellman et al.,[108] using the bromocyclopentene core **20** linked to a chlorodibutylsilyl polystyrene resin (Figure 11.14). The α- and ω-side chains were successively inserted by Suzuki cross-coupling with alkylboranes **21a–21d** and conjugate addition with two types of high-order cuprates (R = A and B). A library of 20 prostanoids with amide α-side chain **24** was generated from mixture **23a**, which was activated for displacement by *N*-cyanomethylation and then treated independently with 10 diverse amines. Cleavage of mixture **23b–23d** from the triethylsilane linker afforded six α-alkylated side-chain compounds **25a–25c**.

11.3.2 Acetogenins

Acetogenins, one of the most rapidly growing classes of new natural products, exhibit remarkable cytotoxic, antitumor, antimalarial, immunosuppressive, pesticidal, and antifeedant activities.[112] A solution-phase synthesis with potential application in library generation of acetogenins, particularly those belonging to the *bis*-THF subgroup, was developed by Keinan et al. (Figure 11.15).[113] Based on retrosynthetic analysis of trilobacin (**31**), alkene **26** was selected as the "naked carbon skeleton" for successive introduction of stereogenic carbinol centers of the target molecules. Lactonization of **26** via the Sharpless asymmetric dihydroxylation reaction yielded the (4*R*, 5*R*) and (4*S*, 5*S*)

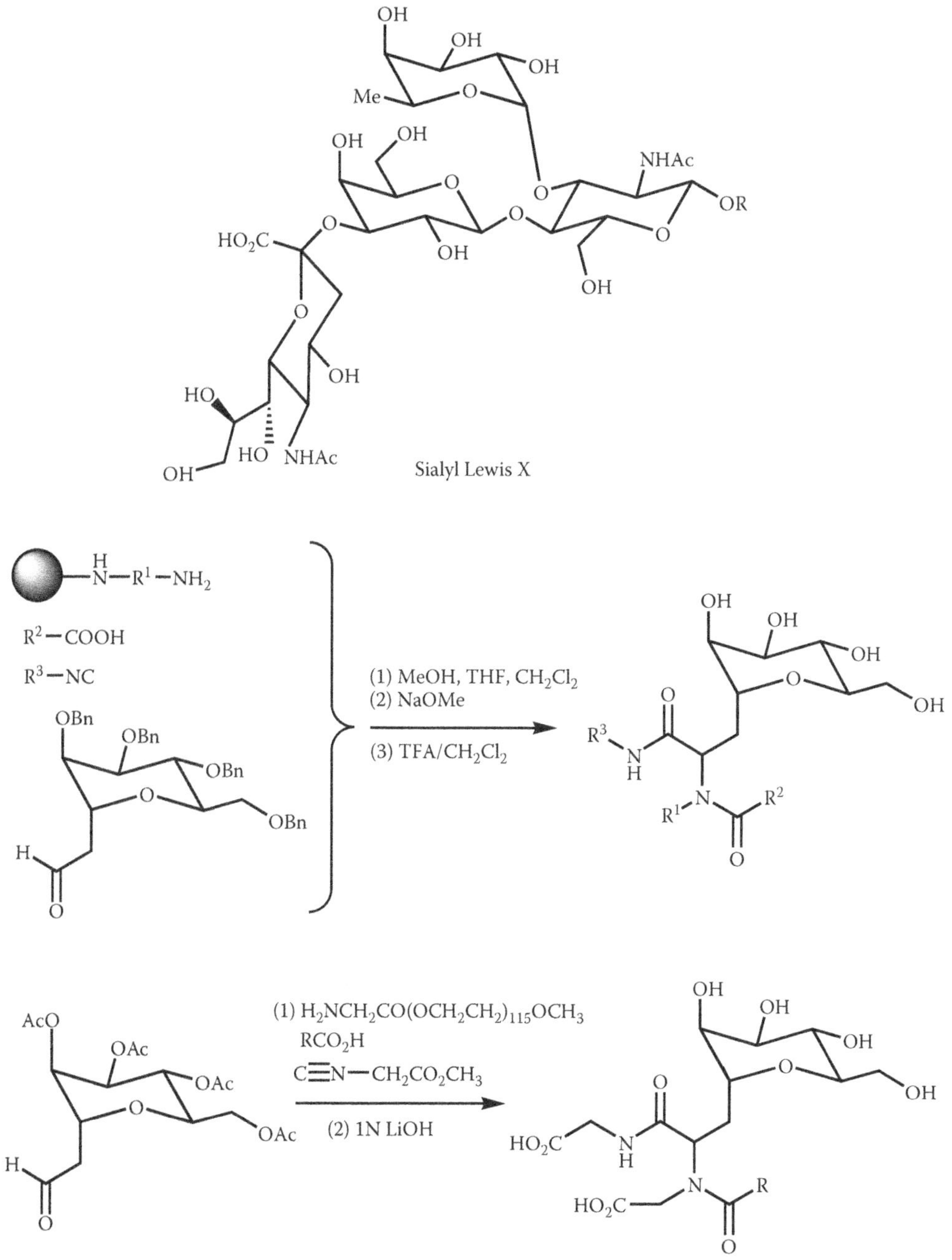

FIGURE 11.11 Construction of libraries of sL^x mimetics using Ugi four-component condensation. (From Sutherlin, D.P. et al., Generation of C-glycoside peptide ligands for cell surface carbohydrate receptors using a four-component condensation on solid-support, *J. Org. Chem.*, 61, 8350, 1996; Tsai, C-Y. et al., Synthesis of sialyl Lewis X mimetics using Ugi four-component reaction, *Bioorg. Med. Chem.*, 8, 2333, 1998.)

stereoisomers of **27**. These were respectively converted into the pair of isomers (4R, 5R)/(4R, 5S)-**27a** and (4S, 5S)/(4S, 5S)-**27a** by Mitsunobu inversion of the free alcohol's configuration. The *cis*-alkenes **28**, obtained by Wittig reaction of aldehydes **27a** with pentadec-4-ynyltriphenylphosphorane, were treated separately with vanadium and rhenium organometallics VO(acac)$_2$ and Re$_2$O$_7$,

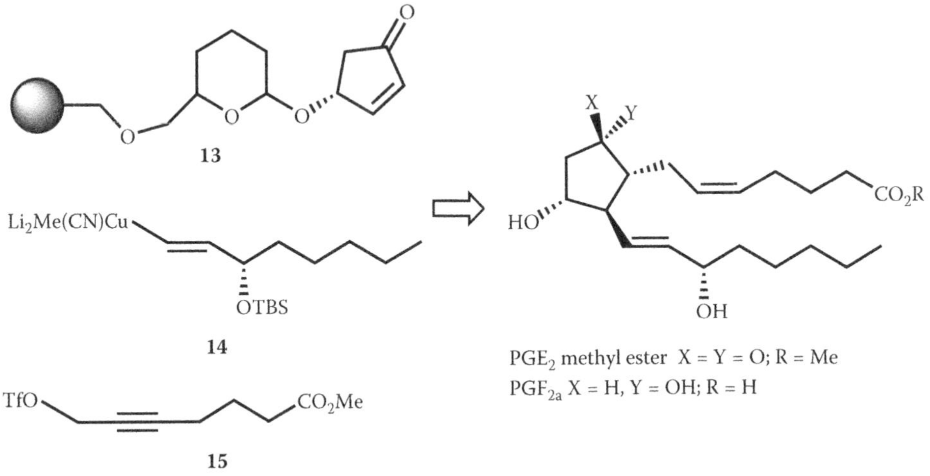

FIGURE 11.12 Prostaglandins E_2 and F_2 combinatorial libraries. (From Chen, S. and Janda, K.D., Synthesis of prostaglandin E_2 methyl ester on a soluble-polymer support for the construction of prostanoid libraries, *J. Am. Chem. Soc.*, 119, 8724, 1995; Chen, S. and Janda, K.D., Total synthesis of naturally occurring prostaglandin F_2 on a non-cross-linked polystyrene support, *Tetrahedron Lett.*, 39, 3943, 1998.)

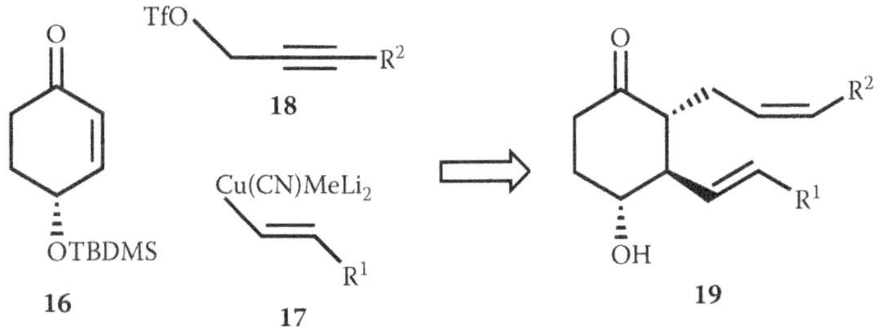

FIGURE 11.13 Solution asymmetric synthesis of six-membered ring prostanoids. (From Pelegrín, J.A.L. and Janda, K.D., Solution- and soluble-polymer supported asymmetric synthesis of six-membered ring prostanoids, *Chem. Eur. J.*, 6, 1917, 2000.)

to yield four (8*S*) and four (8*R*) isomers **29**, respectively. Each of these compounds was then converted to four *bis*-THF stereoisomers by using four different reaction conditions: Mitsunobu reaction followed by Re_2O_7, oxidative cyclization with Re_2O_7, oxidative cyclization with VO(acac)$_2$/TBHP, and Mitsunobu reaction followed by reaction with VO(acac)$_2$/TBHP. Overall, this approach led to the generation of a 64-member library of *bis*-THF acetogenin analogs **30**. Trilobacin (**31**) was further synthesized from the (4*R*, 5*R*, 8*R*, 9*S*, 12*R*) isomer of **30**.

11.3.3 CERAMIDES

In the search for fatty acids derivatives of biological importance, a 528-member library of ceramides **34** was generated using 33 acyl groups **32** and 16 sphingosine-like core structures **33**, and tested for induction of apoptosis and NF-κB signaling (Figure 11.16).[114] The most active ceramide analogs showed apoptotic activity in U937 leukemia cells with IC_{50} values ranging from 4 to $50\,\mu M$, whereas NF-κB activation was just induced by compounds ($10\,\mu M$) having a β-galactose head group (a more than eightfold increase of NF-κB activity).

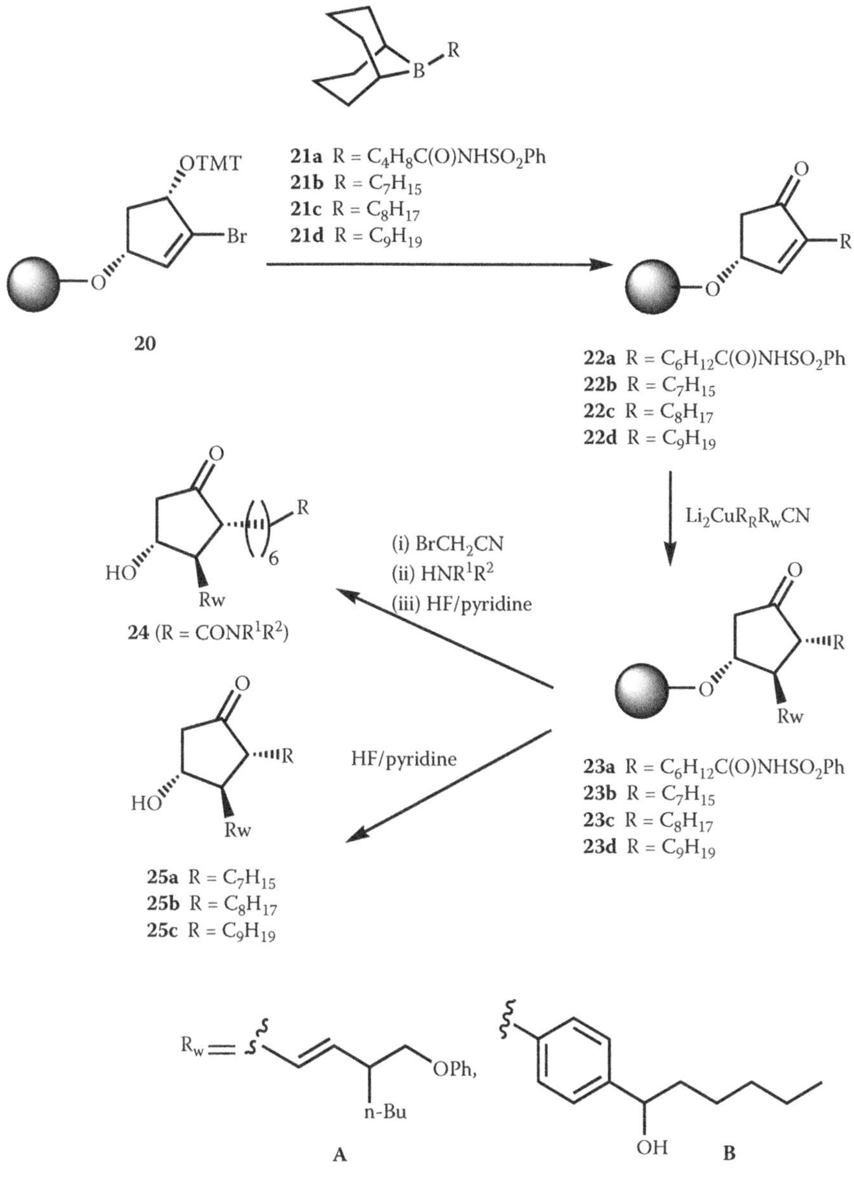

FIGURE 11.14 Synthesis of small prostanoid libraries. (From Thompson, L.A. et al., Solid-phase synthesis of diverse E- and F-series prostaglandins, *J. Org. Chem.*, 63, 2066, 1998.)

11.3.4 MUSCONE ANALOGS

Muscone (**40**) is a sex pheromone of the musk deer and a chemical component of cosmetics. A 12-member library of racemic muscone analogs was synthesized by Nicolaou et al.,[115] who employed a cyclorelease method on solid support to form the macrocycle scaffold (Figure 11.17). A phosphonate-functionalized resin loaded on encoded SMART microreactors **36** was coupled to olefinic esters **35** to form the β-ketophosphonates **37**. Sorting and cross olefin metathesis of **37** with two alkenols followed by oxidation with Dess–Martin reagent gave aldehydes **38**. An intramolecular ketophosphonate-aldehyde condensation (Horner–Emmons–Wadsworth reaction) of **38** caused smooth cyclorelease of macrocyclic enones **39**. Parallel solution-phase chemistry completed the sequence.

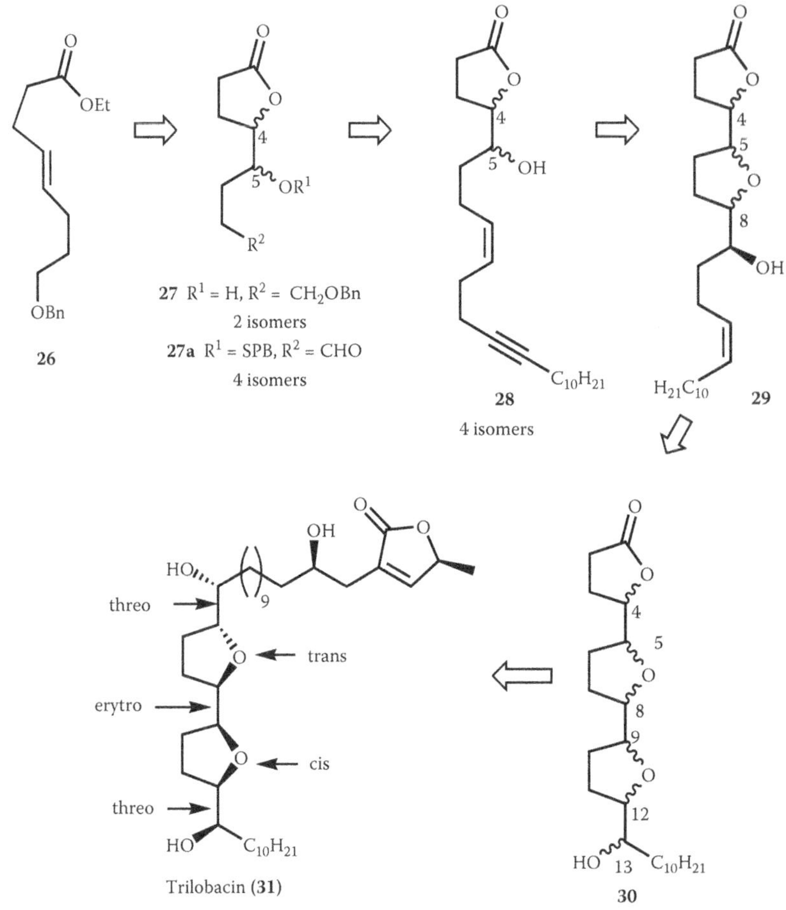

FIGURE 11.15 Chemical library of acetogenins. (From Keinan, E. et al., Towards chemical libraries of annonaceous acetogenins, *Pure Appl. Chem.*, 69, 423, 1997.)

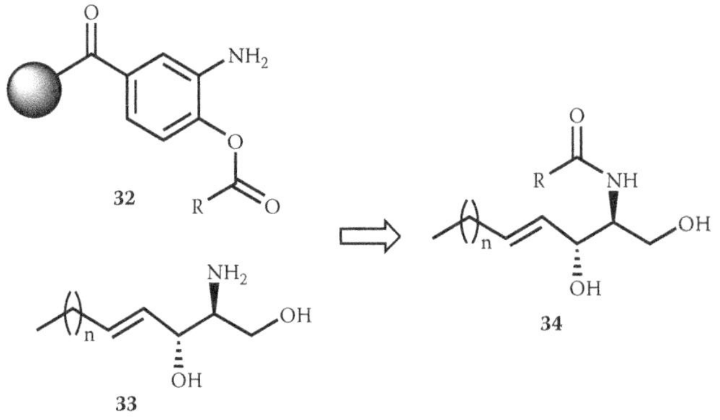

FIGURE 11.16 Ceramide library. (From Chang, Y.T. et al., The synthesis and biological characterization of a ceramide library, *J. Am. Chem. Soc.*, 124, 1856, 2002.)

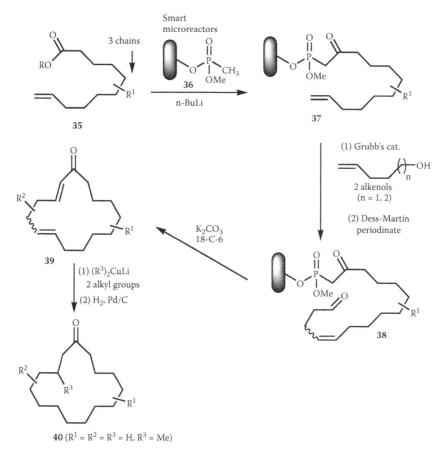

FIGURE 11.17 Muscone library. (From Nicolaou, K.C. et al., Solid phase synthesis of macrocycles by an intramolecular ketophosphonate reaction: synthesis of a (*dl*)-muscone library, *J. Am. Chem. Soc.*, 120, 10814, 1998.)

11.3.5 BONGKREKIC ACID ANALOGS

The natural toxin bongkrekic acid (BKA) is a potent inhibitor of adenine nucleotide translocase (ANT),[116] which is supposed to block the opening of mitochondrial permeability transition (MPT) pore through the electrostatic interactions between the terminal carboxylate moieties of BKA and positively charged amino acid residues of ANT, and hydrophobic interactions between the partially unsaturated middle segment of BKA and ANT. To verify these hypotheses, a library of *N*-acyl iminodiacetic acids, designed as BKA analogs, was synthesized and evaluated for their affinity to ANT. Starting from bromoacetate **41** coupled to Wang resin, three building blocks containing amino acid derivatives (R[1]), nitrophenyl-carboxylic acids or nitro-phenacyl chlorides (R[2]), and acyl-chlorides or anhydrides (R[3]) were successively assembled following a split-and-mix strategy (Figure 11.18).[117] Several lead compounds showing good binding affinity with the isoform ANT-1 ligand were identified from the library.

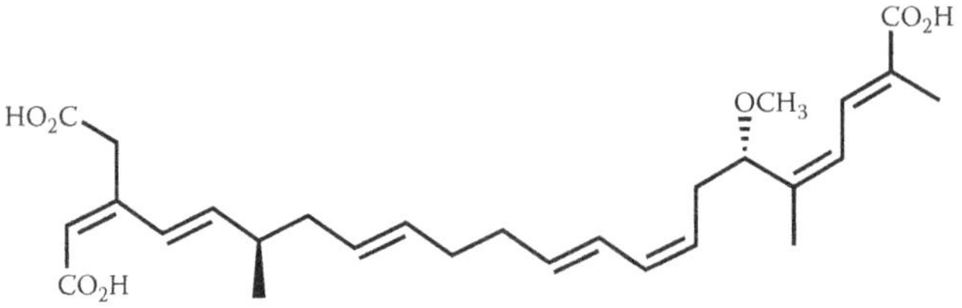

Bongkrekic acid

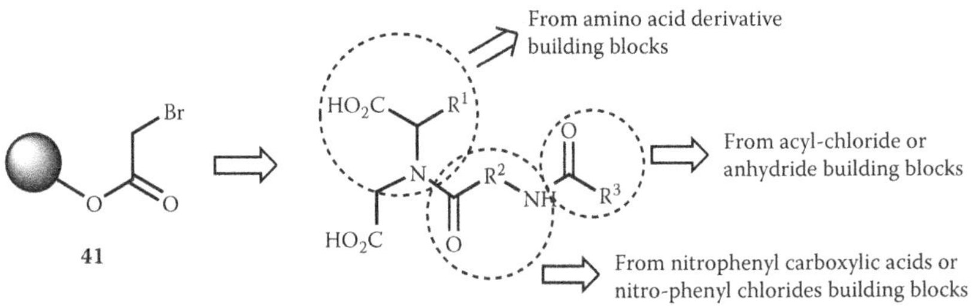

FIGURE 11.18 *N*-Acyl iminodiacetic acids library. (From Pei, Y. et al., Design and combinatorial synthesis of N-acyl iminodiacetic acids as bongkrekic acid analogs for the inhibition of adenine nucleotide translocase, *Synthesis*, 11, 1717, 2003.)

11.4 POLYKETIDES

11.4.1 EPOTHILONES

Polyketides, a large family of natural products with a wide range of pharmacological activity, occupy a relevant place in the pharmaceutical industry.[118–121] Although combinatorial biosynthesis is the method of choice for the generation of polyketide libraries,[122–130] synthetic protocols have been reported for the generation of macrolide libraries. Of these examples, epothilones are among the most interesting metabolites isolated from bacteria (*Sorangium sps.*) over the last few years.[131] The potential of the epothilones as novel anticancer drug candidates justified an intense design effort of classical and combinatorial synthetic strategies (see Chapter 6, Subsection 6.2.7). Presently, third- and fourth-generation compounds are heading toward clinical development.[132–135] Based on chemistry utilized in the synthesis of epothilone A,[136] Nicolaou et al. synthesized a split-pool library of 180 12,13-desoxyepothilones A from three key building blocks R[1], R[2], and R[3], using radio-frequency-encoded SMART microreactors (Figure 11.19).[137] Two types of macrocycles (**42** and **43**) were obtained, and each mixture contained 45 compounds with (*R*) and (*S*) configurations at the C-7 position. The *cis*-olefin was epoxidized in solution to deliver the final epothilone library. Further screening of the library in both tubulin as well as cytotoxicity assays against human ovarian and breast cancer cells led to the identification of nine epothilone analogs with IC_{50} values below 10 n*M*. Following these biological results, structure–activity relationships were deduced for epothilones.

11.4.2 MACROSPHELIDES

Macrosphelides A and B are other macrolides of microbial origin that have received much attention as lead compounds for the development of new anticancer drugs (see Chapter 10, Section 10.4).[138] Following a split-and-pool method, a radio-frequency-encoded combinatorial synthesis of a 128-member macrosphelide library employed a unique three-component coupling strategy

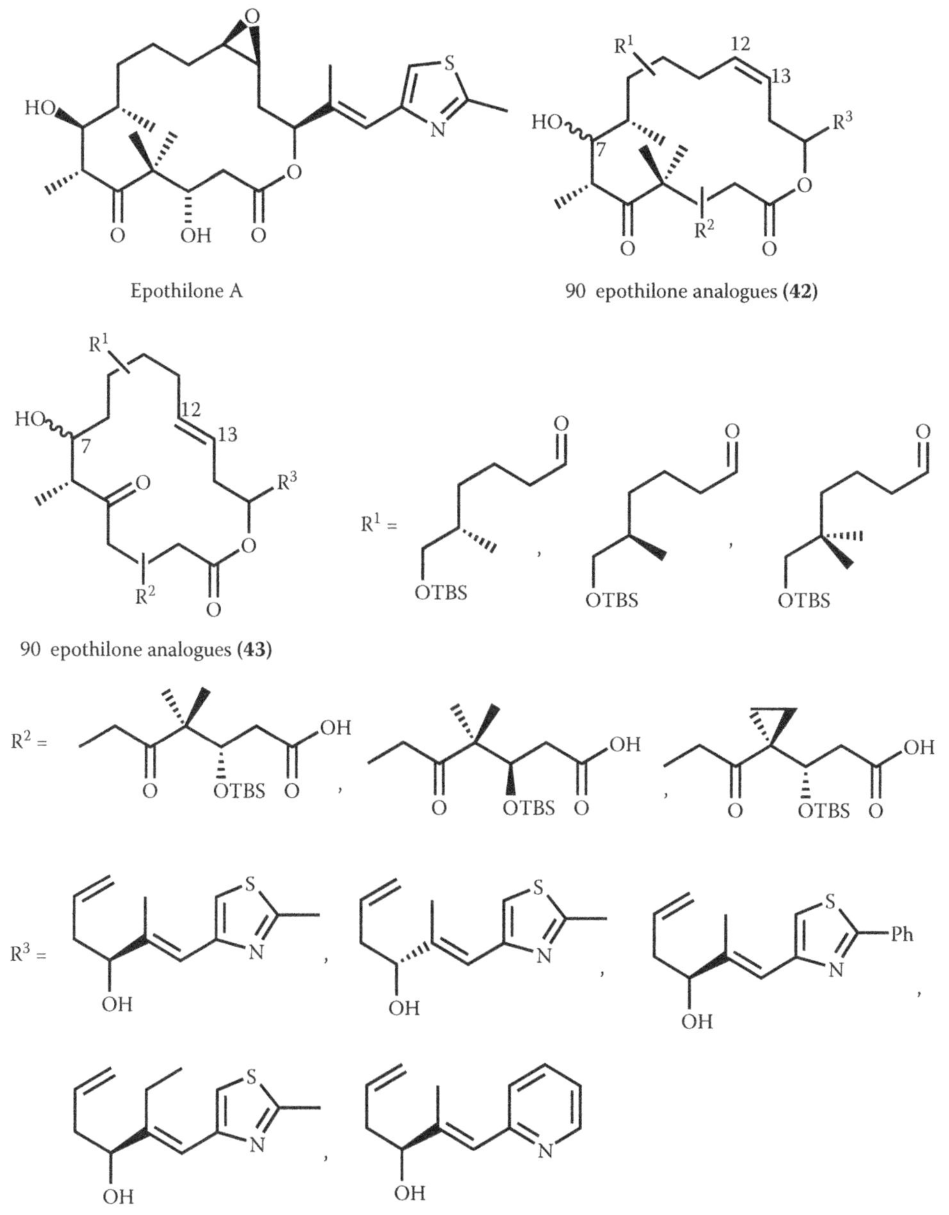

FIGURE 11.19 Epothilone-based library. (From Nicolaou, K.C. et al., Designed epothilones: combinatorial synthesis, tubulin assembly properties, and cytotoxic action against Taxol-resistant tumor cells, *Angew. Chem. Int. Ed. Engl.*, 36, 2097, 1997.)

(Figure 11.20). Both a palladium-catalyzed chemoselective carbonylation and a macrolactonization on a polymer support were featured in the synthesis.[139] Four **A** building blocks were attached to the PS-DHP resin, and then esterified with four **B** building blocks. Subsequent chemoselective carbonylation of the vinyl iodide of **A** with eight alcohols **C** utilized $PdCl_2(MeCN)_2$ as catalyst. A $Pd_2(dba)_3$/dppf-catalyzed macrolactonization of the resulting vinyl bromides **44** afforded the macrosphelide analogs.

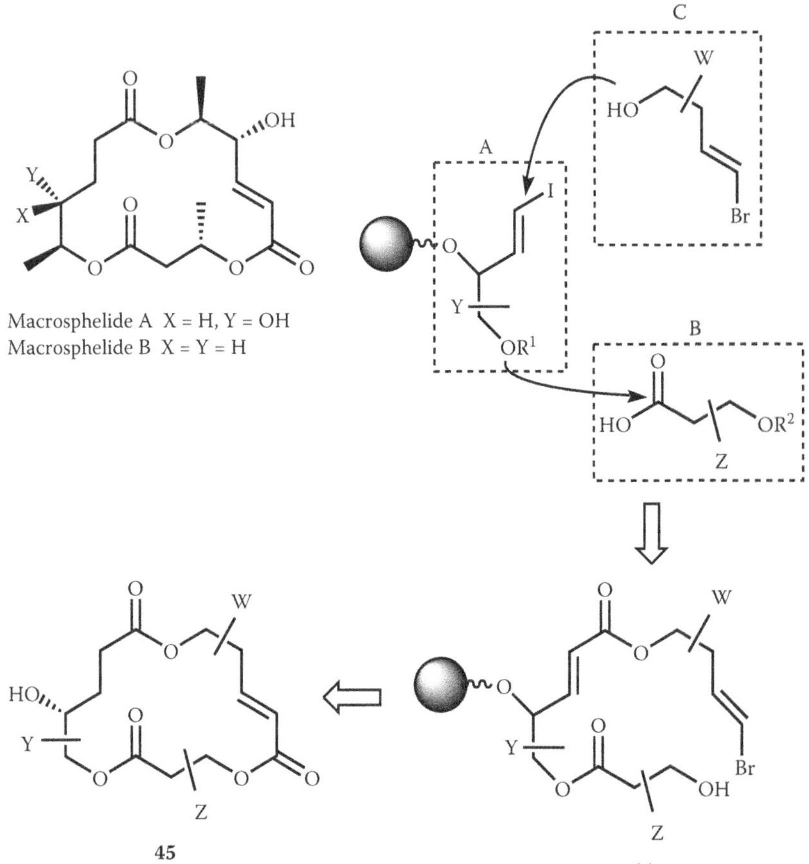

FIGURE 11.20 Combinatorial synthesis of macrosphelide analogs. (From Takahashi, T. et al., Combinatorial synthesis of a macrosphelide library utilizing a palladium-catalyzed carbonylation on a polymer support, *Angew. Chem. Int. Ed. Engl.*, 42, 5230, 2003.)

11.4.3 ERYTHROMYCIN ANALOGS

An example of the use of macrolides as scaffolds for combinatorial derivatization is illustrated by the solid-phase synthesis of a library of erythromycin analogs.[140] Starting from core **46**, obtained from 6-*O*-allyl-erythromycin A, diversity was introduced at remote sites (Figure 11.21). The first site of diversity was introduced by the reaction of the aldehyde group of **46** with a series of amino acids **47** attached to an aminomethylpolystyrene resin containing a fluorine tag and an extended modified Wang-type linker. The resulting secondary amines were reacted with aliphatic aldehydes under reductive amination conditions. The third diversity site was introduced through a reductive alkylation of the tethered primary amine with aromatic aldehydes to furnish **48**. The preceding methodology was developed for the production of a large mix-and-split combinatorial library of about 70,000 members.

11.4.4 METHODOLOGY FOR POLYKETIDE-TYPE LIBRARIES

In a novel strategy for the construction of natural product-based libraries by combinatorial chemical synthesis of hybrid structures, chiral building blocks obtained by selective chemical fragmentation and further modification of epothilone A and other myxobacterial natural products (soraphen A, sorangicin A, myxothiazol A and Z, apicularen A), were recombined on solid-phase support via a

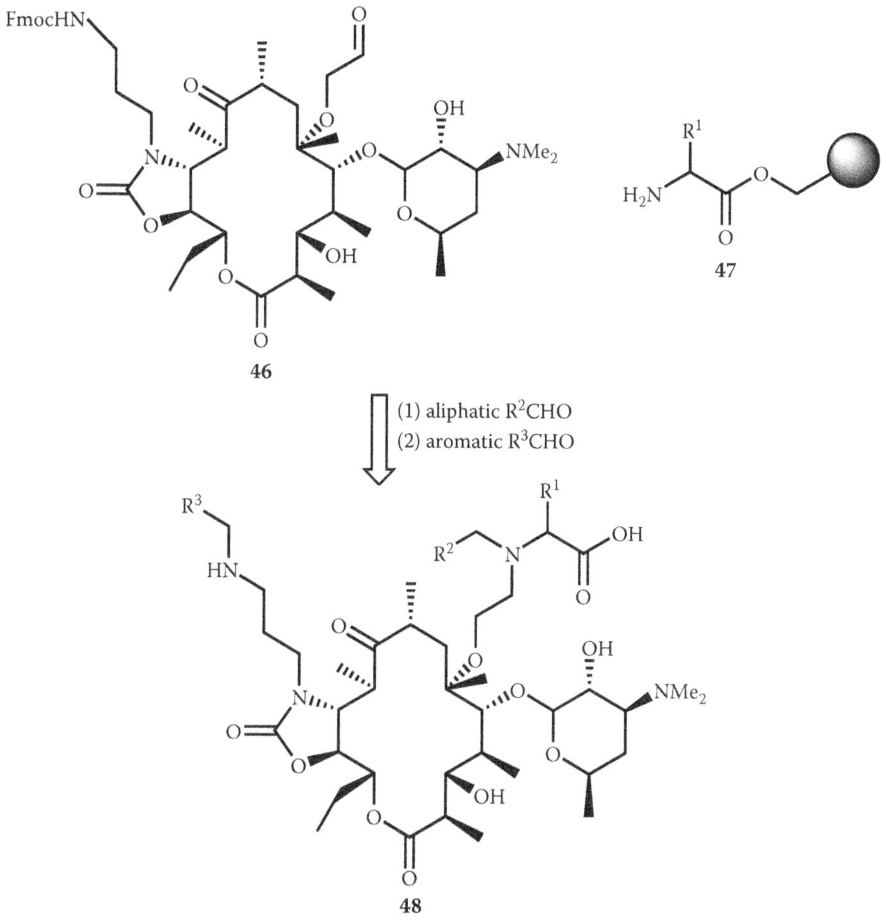

FIGURE 11.21 Solid-phase synthesis of erythromycin A analogs. (From Zanze, I.A. and Sowin, T.J., Solid-phase synthesis of macrolide analogs, *J. Comb. Chem.*, 3, 301, 2001.)

single SPOT synthesis.[46,141] The coupling of different natural product building blocks with a variety of membrane-linked spacers allowed the generation of a broad range of molecular structures.

One major challenge in polyketide synthesis is the elaboration of characteristic sequences of contiguous stereocenters that occur in structurally complex natural polyketides. Approaches based on the aldol reaction have been developed in order to construct chiral building blocks that can be used as precursors for segments' natural products such as the marine metabolites phorboxazole A and B, (+)-discodermolide, and spongistatin 1 (Figure 11.22). Misske and Hoffman[142] used one simple racemic substance **49** as the starting product for library generation of stereoisomeric bicyclic, monocyclic, and acyclic building blocks comprising eight stereochemical pentades of anomeric [3,3,1] lactone acetals, eight stereochemical tetrades of anomeric carbohydrate mimics, and eight stereotetrades of acyclic polypropionate units (Figure 11.23). Each type of these blocks can be easily transformed into segments of natural polyketides and analogs.

Diverse polyketide-type libraries were synthesized using α-chiral aldehydes attached by a silyl linker to a hydromethylpolysyrene resin (Figure 11.24).[143] Aldol chain extension with chiral ketone modules, and subsequent ketone reduction and manipulation on the solid support led to elaborated stereopentad sequences found in natural products such as 6-deoxyerythronolide B and discodermolide. Based on the biosynthesis of erythromycin, the same methodology was used in the combinatorial synthesis of polyketide sequence mimetics with a great variety of chain-extending units.[144]

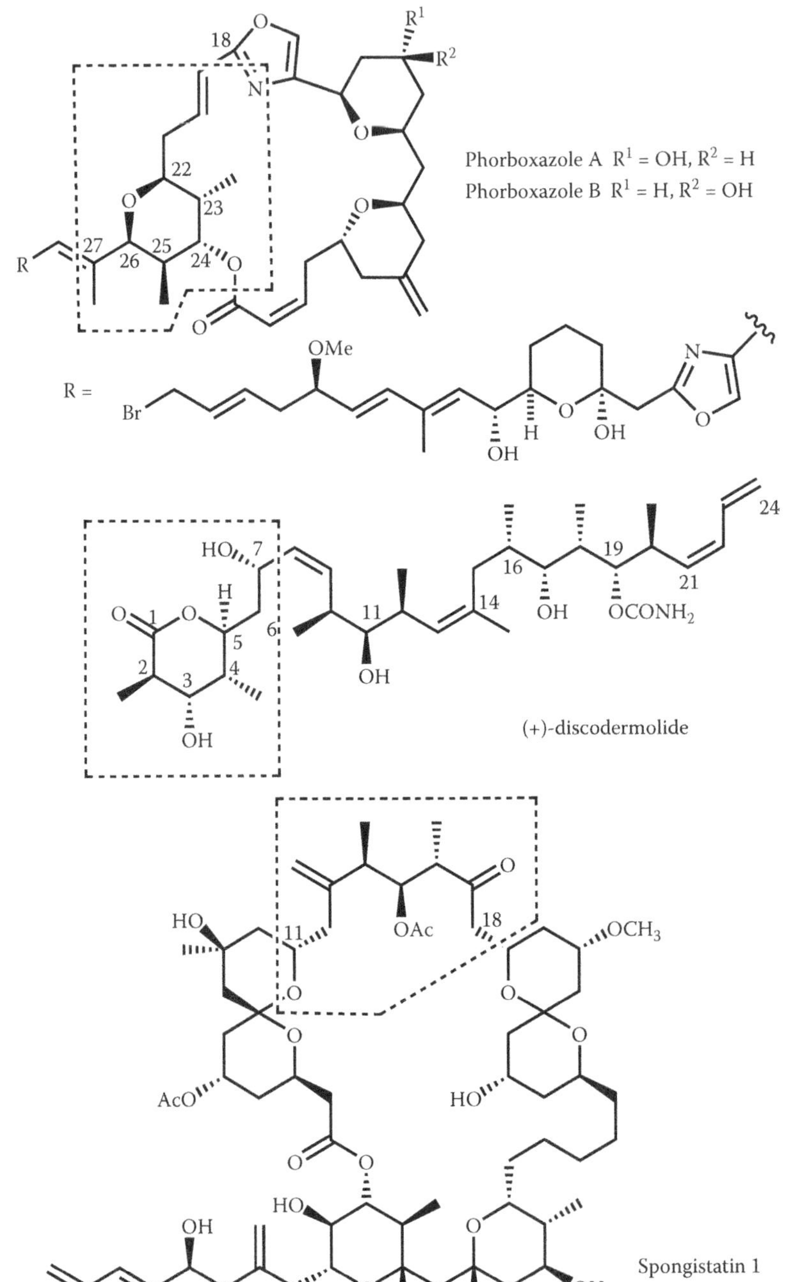

Phorboxazole A $R^1 = OH, R^2 = H$
Phorboxazole B $R^1 = H, R^2 = OH$

(+)-discodermolide

Spongistatin 1

FIGURE 11.22 Synthetic polyketide and carbohydrate segments of phorboxazole A and B, (+)-discodermolide, and spongistatin. (From Misske, A.M. and Hoffmann, H.M.R., High stereochemical diversity and applications for the synthesis of marine natural products: a library of carbohydrate mimics and polyketide segments, *Chem. Eur. J.*, 6, 3313, 2000.)

In this case, the aldehyde functionality was regenerated after the aldol reaction, and the cycle was repeated to furnish sequences of increasing length and complexity.

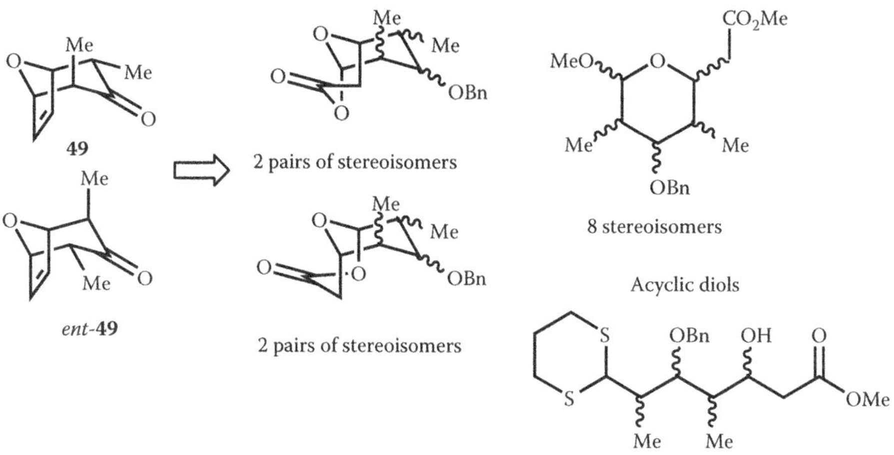

FIGURE 11.23 Library of polyketide fragments and carbohydrate mimics. (From Misske, A.M. and Hoffmann, H.M.R., High stereochemical diversity and applications for the synthesis of marine natural products: a library of carbohydrate mimics and polyketide segments, *Chem. Eur. J.*, 6, 3313, 2000.)

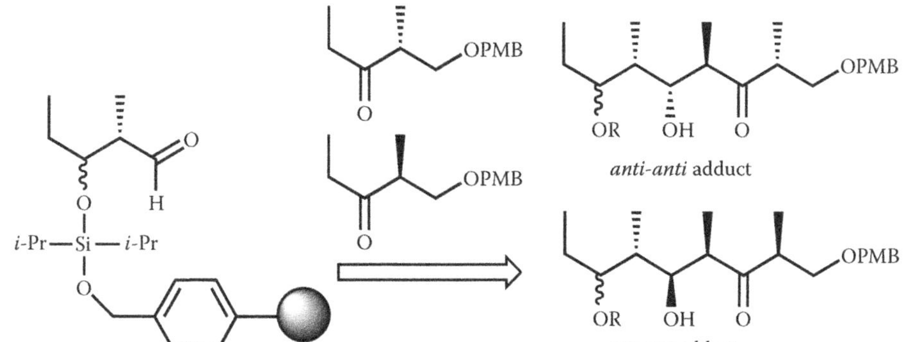

FIGURE 11.24 Solid-phase synthesis of polyketide fragments. (From Paterson, I. and Temal-Laïb, T., Toward the combinatorial synthesis of polyketide libraries: asymmetric aldol reactions with α-chiral aldehydes on solid-support, *Org. Lett.*, 4, 2473, 2002.)

11.4.5 DISCODERMOLIDE ANALOGS

The remarkable biological profile of discodermolide, in particular as a microtubule stabilizer, and the property of being synergistic with paclitaxel, has stimulated intensive synthetic effort, including the preparation of libraries of analogs with actions related to microtubule targeting.[145] In order to study how some areas of the molecule affect the microtubule assembly-inducing properties, the ability to displace paclitaxel from its binding site on tubulin, and antiproliferative activity, Curran et al. described small libraries of 28 analogs that differ from the natural product in which the methyl groups at C-14 and C-16 as well as C-7 hydroxyl group are omitted and the left-side lactone moiety is replaced by simple esters (domain A) (Figure 11.25).[145] In addition, three other domains of variation were introduced: replacement of the carbamate on C-19 with an acetoxy group (domain B); inversion of the configuration of the C-17 stereocenter and derivatization of its hydroxyl group, and keeping the hydroxyl group at C-19 free (domain C); and inversion of the C-11 stereocenter

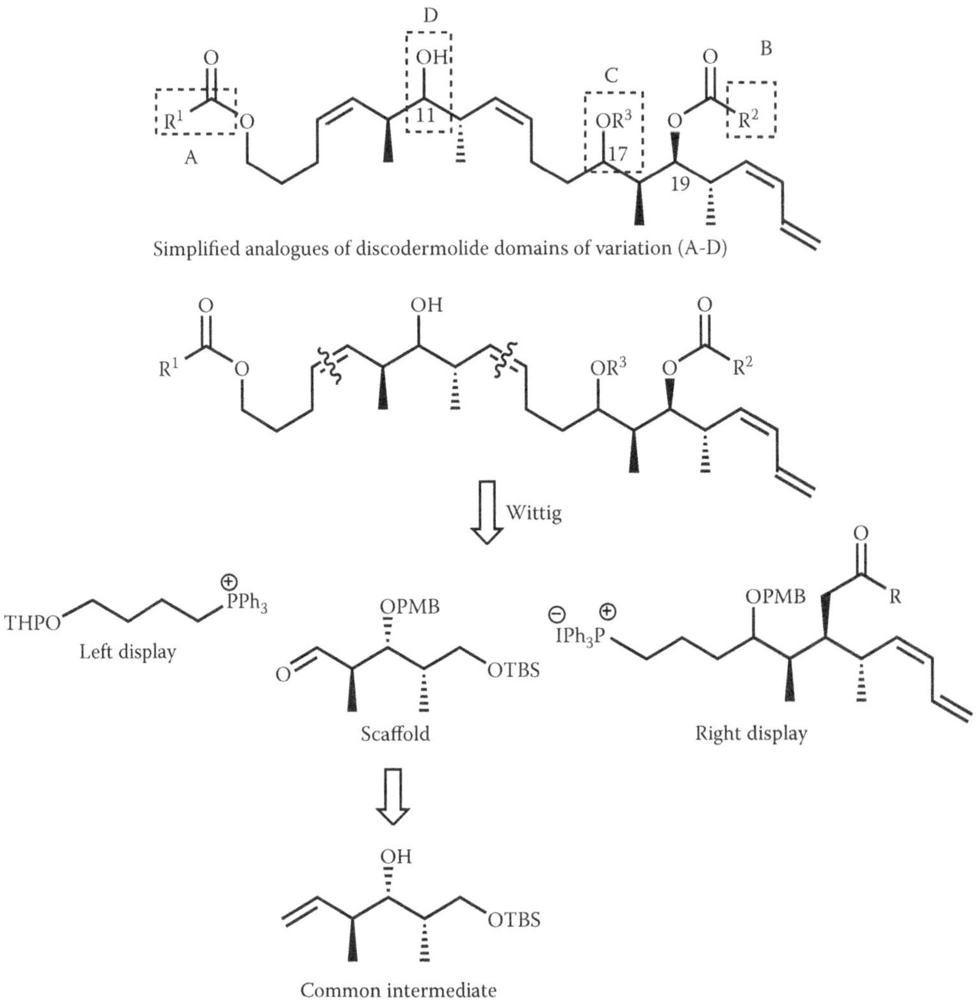

Simplified analogues of discodermolide domains of variation (A-D)

FIGURE 11.25 Synthetic design of discodermolide analogs. (From Minguez, J.M. et al., Synthesis and biological assessment of simplified analogs of the potent microtubule stabilizer (+)-discodermolide, *Bioorg. Med. Chem.*, 11, 3335, 2003.)

(domain D). The synthetic design of the simplified analogs was based on a highly convergent approach. Retrosynthetically, the polypropionate backbone was divided into three segments by disconnecting at the C-8-C-9 and C-13-C-14 double bonds: a THP-protected phosphonium salt as the "left display," a diene phosphonium salt as the "right display," and an aldehyde as the central fragment or "scaffold." A common stereochemical triad appears in the central and right fragments that can be retrosynthetically reduced to a common intermediate. Differently substituted subunits of the analogs were elaborated from this intermediate. Fragment couplings were achieved using Wittig reactions to install the (Z)-olefins. Many of the library constituents retained biological activity and exhibited discodermolide-like behavior in tubulin assays.

Curran's group also applied the concept of mixture synthesis with separation tags to synthesize four truncated analogs of discodermolide (Figure 11.26).[146] Precursors bearing four different groups at C-22, each with an unique fluorous *p*-methoxybenzyl substituent on the C-17 hydroxy group, were mixed and taken through a nine-step sequence. Demixing by fluorous chromatography followed by deprotection and purification provided the individual discodermolide analogs.

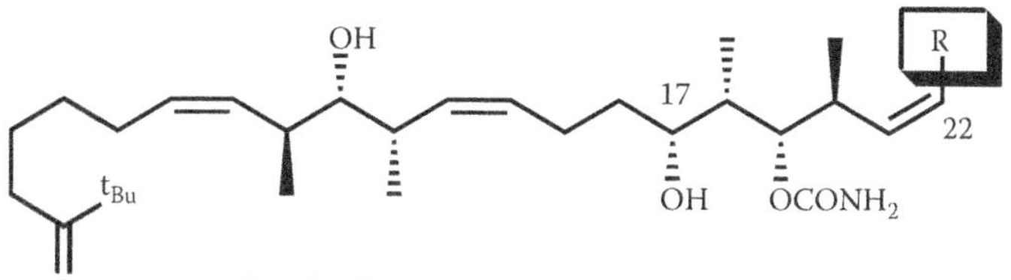

FIGURE 11.26 Truncate analogs of (+)-discodermolide. (From Curran, D.P. and Furukawa, T., Simultaneous preparation of four truncated analogs of discodermolide by fluorous mixture synthesis, *Org. Lett.*, 4, 2233, 2002.)

11.5 PEPTIDES

11.5.1 VANCOMYCIN ANALOGS

The search for new clinical antimicrobial agents has led to intense efforts to exploit the chemistry and biology of glycopeptide antibiotics,[147] a group of highly complex molecules whose structural features include cyclopeptide units with recognized potential as scaffolds for the generation of combinatorial libraries. Vancomycin (**50**), a popular glycopeptide antibiotic against Gram-positive pathogens, has been the target of intense synthetic effort in the last few years to find analogs effective against vancomycin-resistant bacteria.[148] The molecular basis for vancomycin resistance is the replacement of the D-Ala-D-Ala terminus of the bacterial cell wall peptidoglycan precursor with a D-Ala-D-Lac residue that results in an approximately 1000-fold reduction in binding affinity to the antibiotic (Figure 11.27).

A combinatorial approach to synthetic receptor molecules targeting vancomycin-resistant bacteria was therefore designed to bind to the mutated sequence D-Ala-D-Lac (see Chapter 1, Section 1.3 and Chapter 4, Section 4.3).[11,149] Using the biaryl ether core macrocycle **51** as an invariant building block, a library of 39,304 theoretical members was prepared by split synthesis on solid support with 34 amino acids inputs used to introduce diversity (Figure 11.28). The amino acids chosen for the tripeptide unit were selected on the basis of the side-chain display found in the proteogenic amino acids and on their ability to enhance the rigidity of the receptors. Screening of the resin bound library against tripeptides N-Ac$_2$-L-Lys-D-Ala-D-Lac (**52**) and N-Ac$_2$-L-Lys-D-Ala-D-Ala (**53**) labeled with the fluorophore nitrobenzodioxazole led to the identification of **54** and **55**, both exhibiting slightly lower binding affinity toward **52** than vancomycin. Yet, both **54** and **55** showed greater binding affinity toward **53**, and compound **55** had approximately five times greater affinity.

The significance of these results is that receptors less structurally complex than vancomycin can exhibit comparable and even enhanced binding affinity toward the same target. Further studies using other amino acid inputs led to the identification of a compound with the tripeptide segment L-Tpi-L-Phe-D-Dapa that had an IC$_{50}$ value four times more potent than vancomycin.[150] A similar approach to preparing peptide-bearing derivatives of glycopeptide antibiotics was described for the construction of carboxamide derivatives of chloroorienticin B.[151] Chloroorienticin B is structurally closely related to vancomycin. In this case, tripeptide elongation at the C-terminus was achieved using a standard Fmoc peptide synthesis protocol on Sasrin resin, and then coupled to chloroorienticin B. Among the tripeptides prepared, the compounds having both a tryptophan and a tyrosine residue were found to be potent against methicillin-resistant *Staphylococcus aureus* (MRSA) and vancomycin-resistant *Enterococci* (VRE).

In an effort to improve potency and to help elucidate the mechanism of action of synthetic vancomycin analogs, a series of covalent dimers were prepared through attachment of the macrocycle to a DHP resin bearing symmetrical N-Fmoc-protected amino acids (Figure 11.28).[150] Most

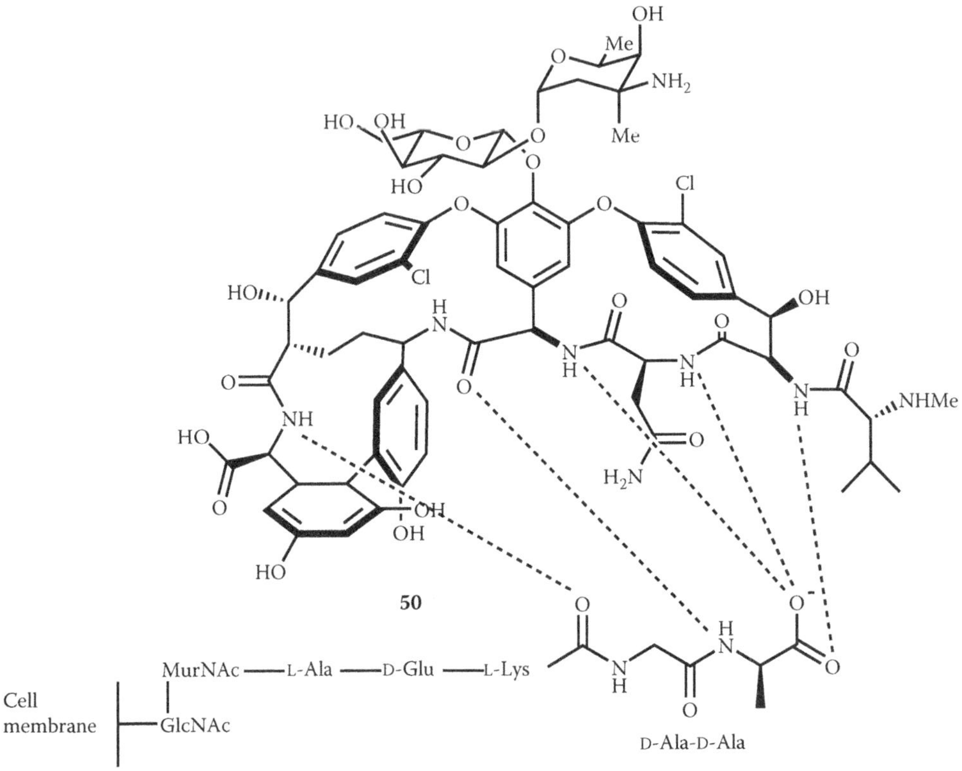

FIGURE 11.27 Binding of vancomycin to the D-Ala-D-Ala terminus of a peptidoglycan precursor. (From Xu, R. et al., Combinatorial library approach for the identification of synthetic receptors targeting vancomycin-resistant bacteria, *J. Am. Chem. Soc.*, 121, 4898, 1999.)

of the dimers were found to be up to 60-fold more active than vancomycin against vancomycin-resistant *Enterococcus faecium*. Two other vancomycin dimerization methods, disulfide formation [–(CH$_2$)n–S–S–(CH$_2$)n–] and olefin metathesis [(CH$_2$)n–CH=CH–S–(CH$_2$)n–)] for derivatives, were implemented by Nicolaou et al. (Figure 11.29).[152] Application to the target-accelerated combinatorial synthesis (TACS) of vancomycin dimer libraries would give heterodimers in which the vancomycin unit should be capable of binding the natural D-Ala-D-Ala segment of the cell-wall and the other vancomycin moiety would be poised to bind its mutant counterpart, D-Ala-D-Lac. To this end, three vancomycin analogs, 7-(LeuNMe)C$_2$, 7-(LeuNMe)C$_3$, and 7-(LeuNMe)C$_4$, bearing tethers of different length were mixed and subjected to target-accelerated covalent dimerization (Figure 11.30).[152] Following antibiotic assays, it was shown that dimers from both the olefinic and disulfide classes with optimal tether length of 16 to 18 atoms between the two nitrogen atoms of the vancosamine moiety exhibited potent antibacterial activities, particularly against VRE.

Another approach for the preparation of vancomycin analogs was described by Nicolaou et al. (Figure 11.31).[153] Following the development of a synthetic technology based on the design of a novel selenium safety catch linker, and its application to the solid-phase semisynthesis of vancomycin, vancomycin libraries with molecular diversity at the sugar moieties, the amino acid 1 site, and the N-terminus were prepared. First, vancomycin's disaccharide segment was replaced by a new monosaccharide library in a deglycosidation/reglycosidation sequence. The hydroxyl groups and the basic nitrogens of vancomycin were first protected, TBSOTf and alloc-Cl, respectively, before loading onto the selenium resin via the carboxylic acid group. The glycosyl moiety was then removed with TFA/Me$_2$S in CH$_2$Cl$_2$. Reglycosidation onto the phenol scaffold was carried out with trichloroacetimidate derivatives of a series of monosaccharides. A second series of vancomycin

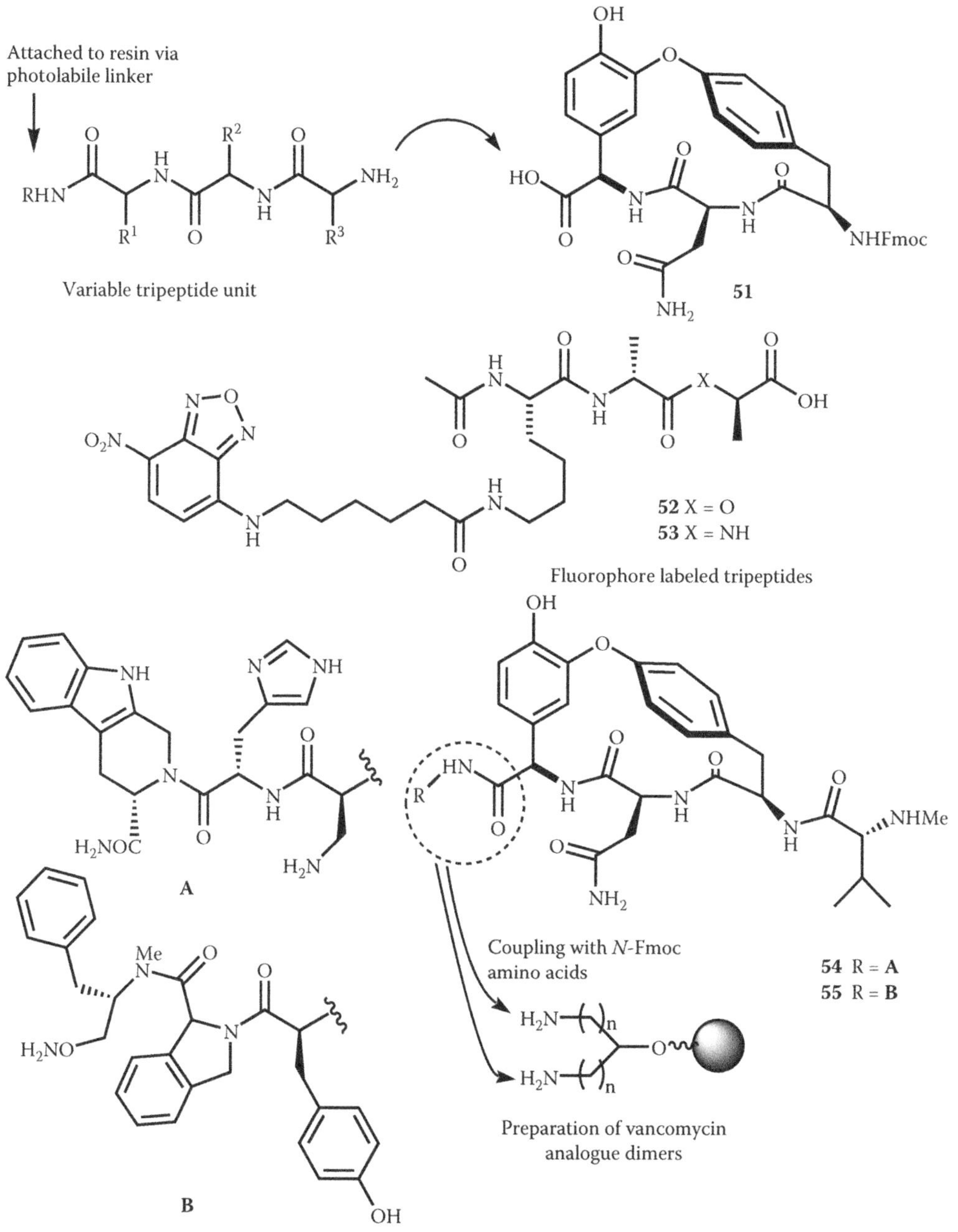

FIGURE 11.28 Vancomycin-based receptor libraries. (From Xu, R. et al., Combinatorial library approach for the identification of synthetic receptors targeting vancomycin-resistant bacteria, *J. Am. Chem. Soc.*, 121, 4898, 1999; Kateri, A.A. et al., Identification of potent and broad-spectrum antibiotics from SAR studies of a synthetic vancomycin analog, *Bioorg. Med. Chem. Lett.*, 13, 1683, 2003.)

analogs was synthesized in solution phase using the amino group of the vancosamine moiety as the site for diversification via reductive amination and leaving the rest of the skeleton intact. Solid-phase chemistry was applied to construct an expanded library of vancomycin analogs with various amino acids attached at the *N*-terminus and the carboxyl group of the molecule. Biological

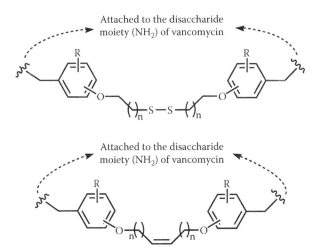

FIGURE 11.29 Dimerization of vancomycin analogs through disulfide bond formation and olefin metathesis. (From Nicolaou, K.C. et al., Synthesis and biological evaluation of vancomycin dimmers with potent activity against vancomycin-resistant bacteria: target-accelerated combinatorial synthesis, *Chem. Eur. J.*, 7, 3824, 2001.)

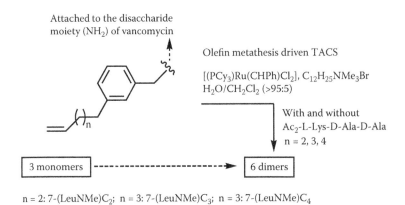

$$n = 2: 7\text{-}(\text{LeuNMe})C_2; \quad n = 3: 7\text{-}(\text{LeuNMe})C_3; \quad n = 3: 7\text{-}(\text{LeuNMe})C_4$$

FIGURE 11.30 Target-accelerated combinatorial synthesis (TACS) of vancomycin dimers. (From Nicolaou, K.C. et al., Synthesis and biological evaluation of vancomycin dimmers with potent activity against vancomycin-resistant bacteria: target-accelerated combinatorial synthesis, *Chem. Eur. J.*, 7, 3824, 2001.)

evaluation of these analogs against vancomycin-resistant strains revealed several highly potent compounds, particularly among vancosamine derivatives.

11.5.2 BIARYL-ETHER-CONTAINING MEDIUM RINGS

Besides vancomycin, several other macrocyclic glycopeptide antibiotics, such as teicoplanins, K-13, piperazinomycin, RA VII, and pterocaryanin C, contain the nonsymmetrical biaryl ether moiety bore.[149,154,155] The development of a diversity-oriented synthesis of biaryl-containing medium rings (9-, 10-, and 11-membered rings) led to the construction of a library of small molecules related to pterocaryanin C on polystyrene macrobeads functionalized with an alkyl aldehyde **56** (Figure 11.32).[154] A set of eight pairs of enantiomeric amino alcohols' building blocks **57** was first attached to the solid support via reductive amination to afford the secondary amines **58**. Upon reductive

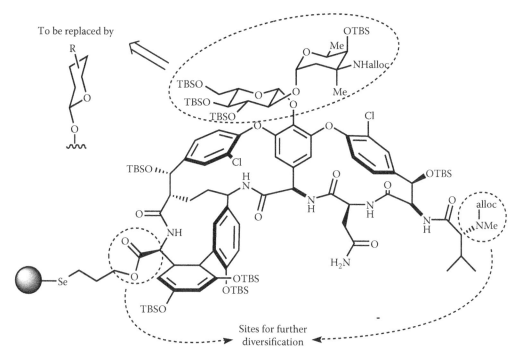

FIGURE 11.31 Solid-phase semisynthesis of a library of vancomycin monosaccharides. (From Nicolaou, K.C. et al., Solid- and solution-phase synthesis of vancomycin and vancomycin analogs with activity against vancomycin-resistant bacteria, *Chem Eur. J.*, 7, 3798, 2001.)

alkylation of amines **58** with bromobenzaldheydes **59** (R^2 diversity) and borane•pyridine complex as the reductant, amino alcohols **60** were obtained. The second aryl group was introduced by *O*-alkylation of **60** with benzyl bromides **61** (or **62** or **63**) as the R^3 diversity. Metalation of the resulting aryl bromides **64** followed by transmetalation with CuCN•2LiBr formed atropisomer biaryl compounds **65** with kinetic and thermodynamic opposite P:M ratios for both the (*S*) and (*R*) series. The chemical genetic approach was demonstrated in a series of phenotypic and protein-binding assays performed on members of the library.

An alternative strategy for the construction of the biaryl ether moiety was developed by Kiselyov et al. (Figure 11.33).[156] In this approach, an assembly of twenty 14-membered macrocycles **69** was achieved by initial amidation of solid-supported 3-hydroxytyrosine derivative **66** with bromoacetic acid followed by nucleophilic addition of primary amines ($R^1CH_2NH_2$). Subsequent coupling of 3-fluoro-4-nitrobenzoic acid (**67**) to the solid-supported secondary amine intermediate gave the cyclization precursor **68**. Intramolecular S_NAr between the fluorine and the phenolic moieties of **68** followed by *N*-alkyation with R^2CH_2Br resulted in cleavage off of solid support and furnished **69**.

In an alternative synthesis, a two-step sequence involved an Ugi four-component reaction and an intramolecular nucleophilic aromatic substitution to provide rapid access to biaryl-ether-containing macrocycles in solution phase and solid phase (Figure 11.34).[155] Using isonitriles, aldehydes, amines, and carboxylic acids as inputs for the Ugi reaction, dipeptides **70** were formed. Upon cyclization of **70** with K_2CO_3 in DMF, macrocycles **71** with four points of diversity were obtained. Moreover, the presence of the nitro group allowed the introduction of further structural variation. In the solid-phase synthesis, the Ugi condensation used isonitriles attached to a Wang resin.

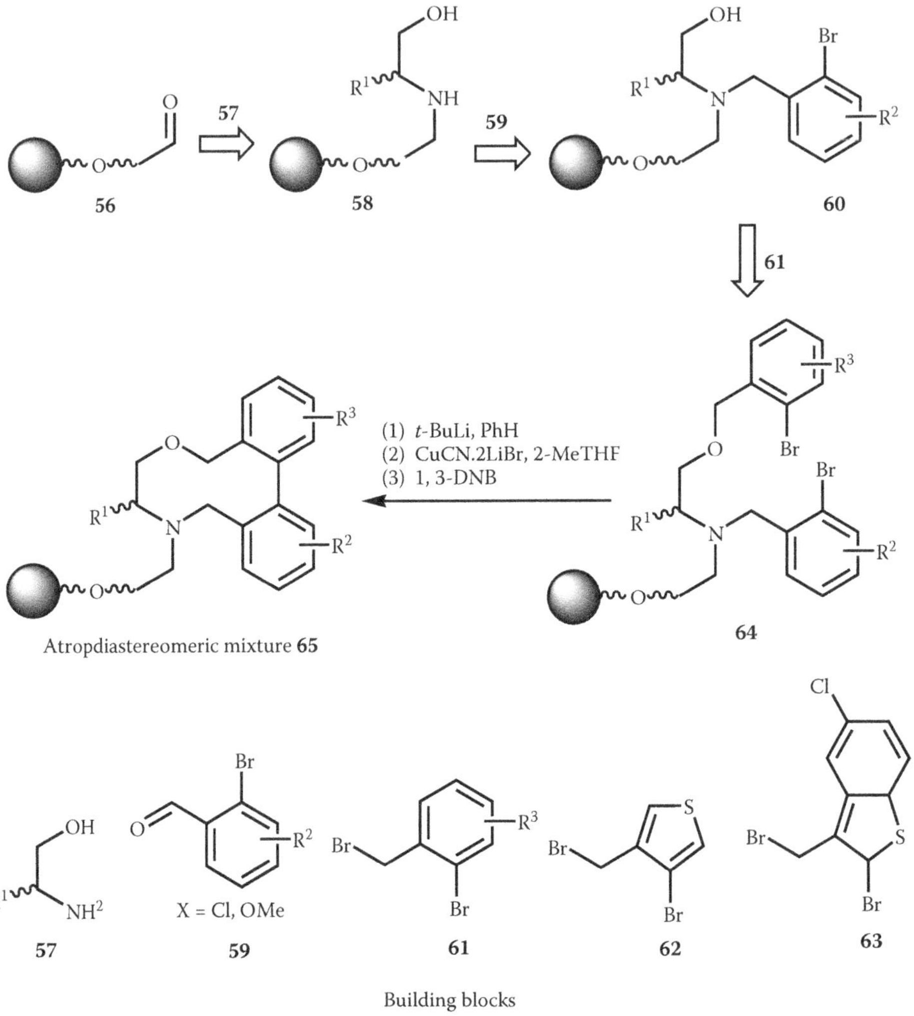

FIGURE 11.32 Diversity-oriented synthesis of biaryl-containing medium rings. (From Spring, D.R., Krishnan, S., and Schreiber, S.L., Towards diversity-oriented, stereoselective syntheses of biaryl- or bis(aryl)metalcontaining medium rings, *J. Am. Chem. Soc.*, 122, 5656, 2000.)

11.5.3 BLEOMYCINS

The bleomycins are a family of glycopeptide-derived antitumor antibiotics isolated from *Streptomyces verticillus* that mediate the sequence-selective oxidative damage of DNA and RNA. To discover analogs with improved biochemical properties through the solid-phase synthesis of deglycobleomycins, a 108-member deglycobleomycin library was prepared using three bithiazole (**A**), three threonine (**B**), three methylvalerate (**C**), and four β-hydroxyhistidine (**D**) derivatives as building blocks (Figure 11.35).[157] The bithiazole moiety **A** of deglycobleomycin was attached to a spermine-functionalized resin via HBTU-mediated coupling in DMF. Through subsequent coupling with the threonine unit **B** by use of HBTU, HOBt, and Hünig's base, the tripeptide unit was obtained. The methylvalerate moiety **C** was then added by the same sequence, followed by the coupling with histidine derivatives **D** to yield a resin-bound pentapeptide. The final coupling of Boc-pyrimidoblamic acid was performed at 0°C in the presence of BOP and Hünig's base. The library included two analogs that mediated plasmid relaxation to a greater extent than the parent deglycobleomycin molecule.

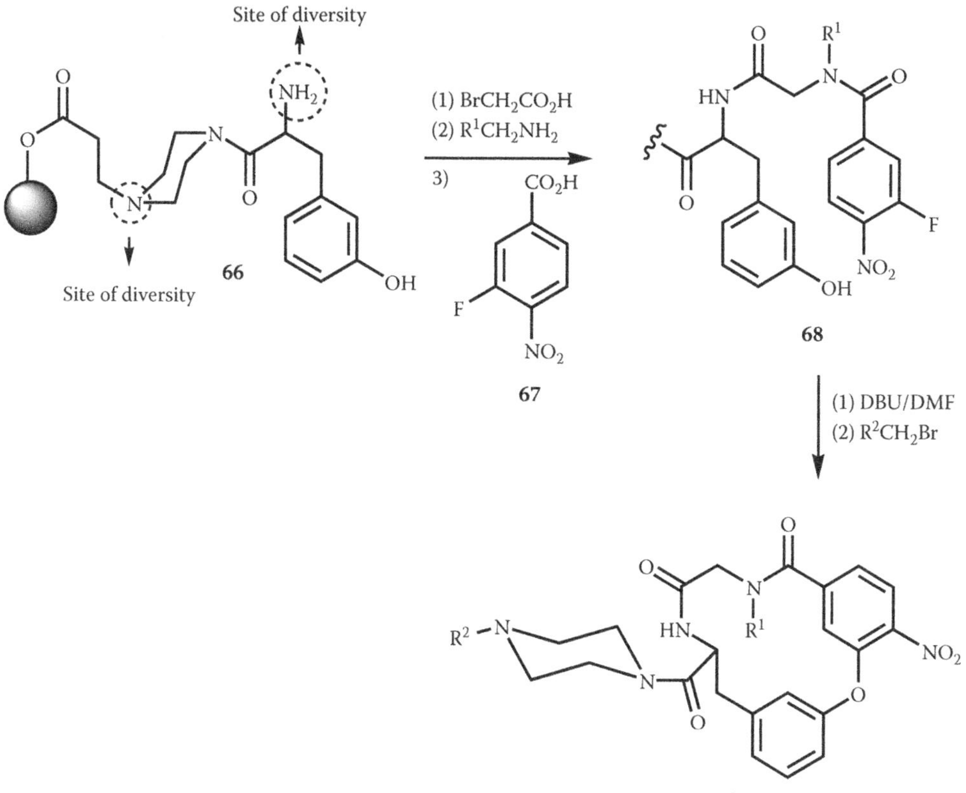

FIGURE 11.33 Library generation of 14-membered biaryl macrocycles. (From Kiselyov, A.S., Eisenberg, S., and Luo, Y., Solid-support synthesis of 14-membered macrocycles *via* S$_N$Ar methodology on acrylate resin, *Tetrahedron Lett.*, 40, 2465, 1999.)

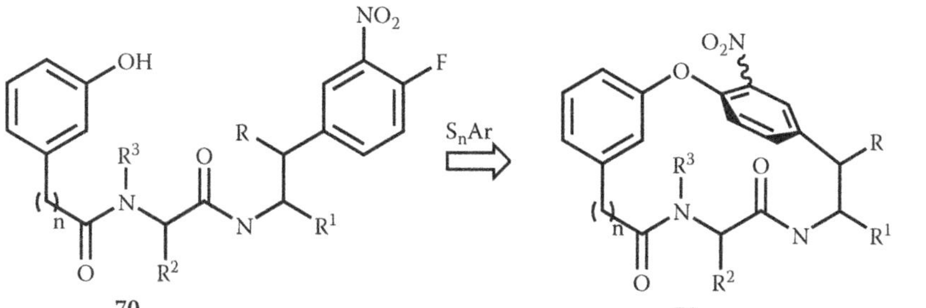

FIGURE 11.34 Synthesis of natural product-like biaryl-ether-containing macrocycles. (From Cristau, P., Vors, J.P., and Zhu, J., Rapid and diverse route to natural product-like biaryl ether containing macrocycles, *Tetrahedron,* 59, 7859, 2003.)

11.5.4 Tyrocidine Analogs

To increase the ability of broad-spectrum peptide antibiotics to differentiate bacteria from mammalian cells, a focused library of tyrocidine analogs was synthesized based on known structure–activity relationship information (Figure 11.33).[158] Significant side-chain changes of most constituent residues of the tyrocidine template result in the loss of antibiotic potency or the

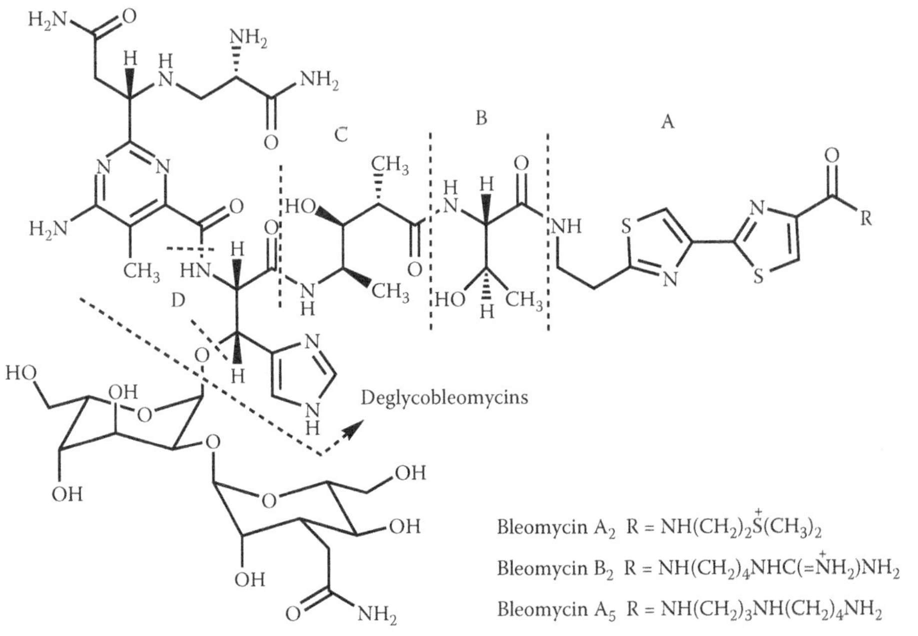

FIGURE 11.35 Structures of bleomycins. (From Leitheiser, C.J. et al., Solid phase synthesis of bleomycin group antibiotics: construction of a 108-member deglycobleomycin library, *J. Am. Chem. Soc.*, 12, 8218, 2003.)

inability to form the antiparallel β-sheeted structure necessary for biological activity. Moreover, upon comparison of tyrocidine A structure with other related cyclic peptides, small structural alterations occurred at the Tyr-4, Gln-5, DPhe-7, and Phe-8 positions of the cyclic scaffold. Other amino acid residues were highly conserved. Based on these observations, a library of 192 members was synthesized and screened against *Bacillus subtilis*. The library was constructed using IRORI's Accuatag-100 Combinatorial Chemistry System[97] and the conformation-dependent self-cyclization method previously developed for the solid-phase synthesis of tyrocidine A.[159] Typically, the first amino acid was attached to 4-sulfamylbutyryl resin, and peptide synthesis was carried out through repetitive cycling between Fmoc deprotection and coupling of an appropriate Fmoc amino acid residue according to the sequence of the target molecule. The head-to-tail cyclization of the linear precursor using DIEA/THF was preceded by cyanomethylation activation (ICH$_2$CN, NMP, DIEA) of the sulfonamide linker, and removal of the protecting groups. Positions Tyr-4, Gln-5, DPhe-7, and Phe-8 of the cyclic scaffold were chosen as sites to introduce different amino acids, AA$_1$, AA$_2$, AA$_3$, and AA$_4$ (Figure 11.36). Library screening identified nine analogs whose therapeutic indices increased by up to 90-fold in comparison to the natural products. Three of these analogs showed significant increase in antibacterial potency and concurrent drastic decrease in hemolytic activity.

11.5.5 SANGLIFEHRIN A ANALOGS

Sanglifehrin A (**72**), an immunosuppressive natural product isolated from *Streptomyces sp.*, has strong affinity for cyclophylin (CyP) by a mechanism whose details are currently under investigation.[160] As in the case of tyrocidine, the synthesis of a library of sanglifehrin A analogs provided valuable details regarding the structure–activity relationship of this macrocyclic peptide. Based upon retrosynthetic analysis of the 22-membered macrocycle, three building blocks were selected: three tripeptide segments (**A1–A3**) of one corresponds to the structure of **72,** and all form hydrogen bonds with CyP; three compounds (**B1–B3**) corresponding to the C-19 to C-26 fragment; and four

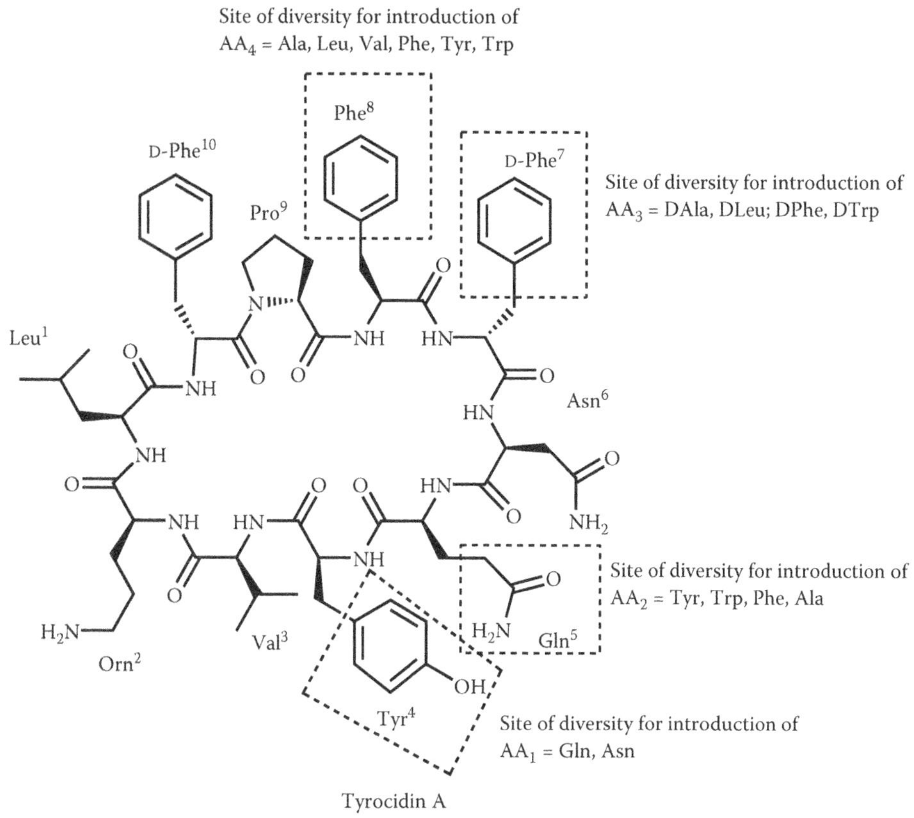

FIGURE 11.36 Design of an analog library of tyrocidine A. (From Qin, C. et al., A chemical approach to generate molecular diversity based on the scaffold of cyclic decapeptide antibiotic tyrocidine A, *J. Comb. Chem.*, 5, 353, 2003.)

fragments (**C1–C4**) as analogs of the polypropionate portion extending from C-13 to C-18 (Figure 11.37). Assembly of the macrolide library started with the removal of the Boc group of tripeptides **A1–A3** followed by coupling of the resulting amines to the carboxyl group of building blocks **C1–C4**. Methyl ester hydrolysis or selective deprotection of the trichloroethyl ester group with zinc provided the corresponding acids, which were esterified with alcohols **B1–B3** under Mitsunobu conditions. The critical macrocyclization step was achieved using a ring-closing metathesis (RCM) reaction with the ruthenium-based catalyst $(PCy_3)_2Cl_2Ru=CHPh$. Screening of the synthetic analogs of sanglifehrin A for their affinity for cyclophylin A revealed that binding is fully mediated by the macrocyclic portion. Simultaneous modification of several positions, including removal of substituents, configurational changes, and functional group modifications, resulted in a significant loss of affinity for CyP.

11.5.6 CYCLIC DEPSIPEPTIDES

Some cyclic depsipeptides, such as hapalosin, HUN-7293, and aurilide, were targeted for parallel and combinatorial synthesis strategies (see Chapter 10, Section 10.5). Analogs of hapalosin, a natural compound that is able to reverse multidrug resistance (MDR) in tumor cells, were described in a parallel synthesis report (Figure 11.38).[161] Three building blocks, β-hydroxy acids **A**, γ-amino-β-hydroxy acid **B**, and α-amino acids **C**, were assembled on Wang resin. Using Fmoc and THP protecting groups for the amino and hydroxyl functionality, respectively, tridepsipeptides with one ester and two amide bonds were formed. Following cleavage from the resin, macrocyclization

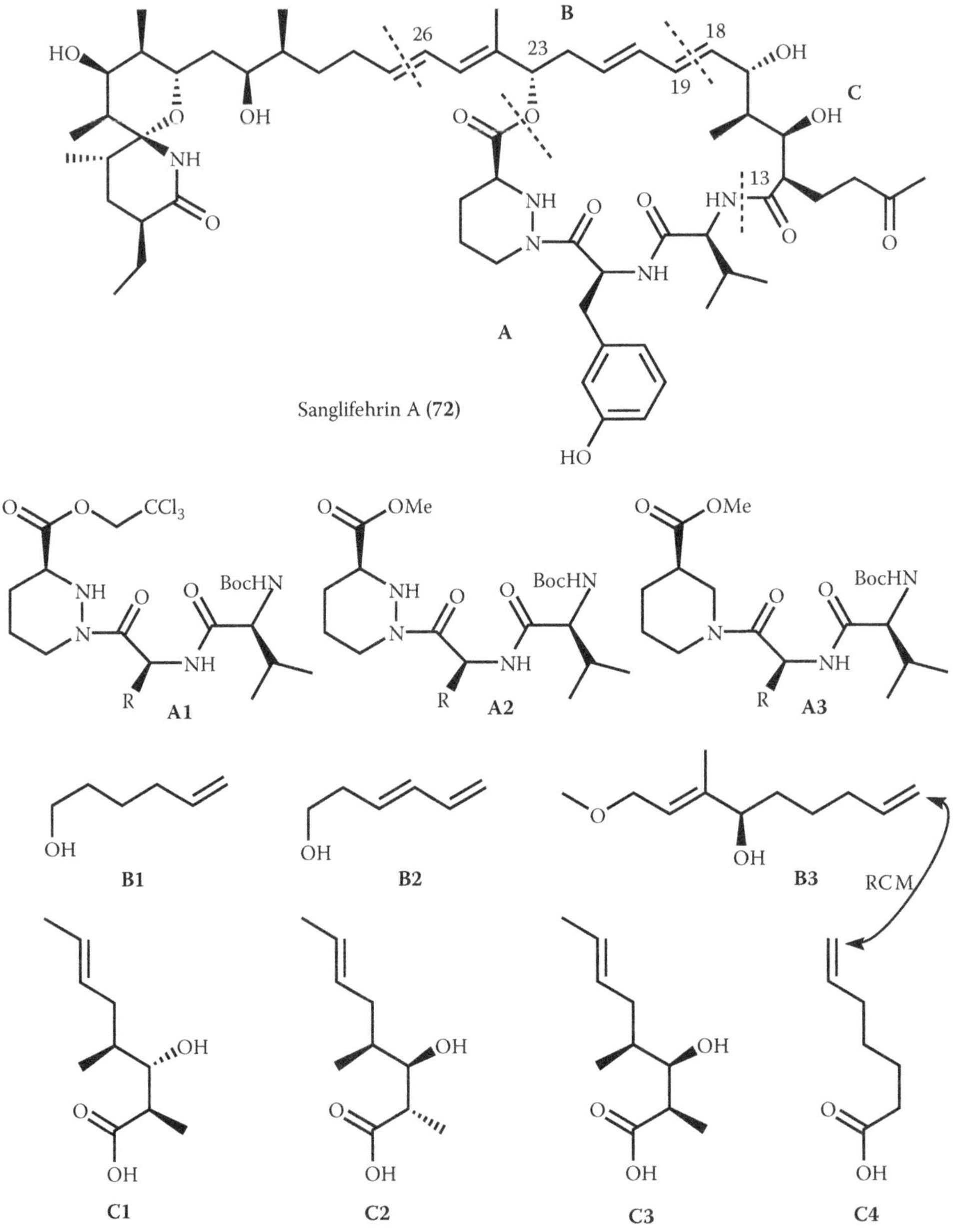

FIGURE 11.37 Design of a library of sanglifehrin A macrolide analogs. (From Sedrani, R. et al., Sanglifehrin-cyclophilin interaction: degradation work, synthetic macrocyclic analogs, X-ray crystal structure, and binding data, *J. Am. Chem. Soc.*, 125, 3849, 2003.)

reactions were run in DMF using 2-(1*H*-benzotriazole-1-yl)-1,1,3,3-tetramethyluronium tetrafluroborate (TBTU) as the condensing agent.

HUN-7293 is a fungal heptadepsipeptide with recognized inhibition of cell adhesion molecule expression, including intercellular cell adhesion molecule 1 (ICAM-1), vascular cell adhesion molecule 1 (VCAM-1), and E-selectin (Figure 11.39).[162] Representative of a general approach to defining structure–function relationships of such cyclic depsipeptides, a parallel synthesis and evaluation of a complete library of nearly 40 analogs of HUN-7293 was described. Based on a

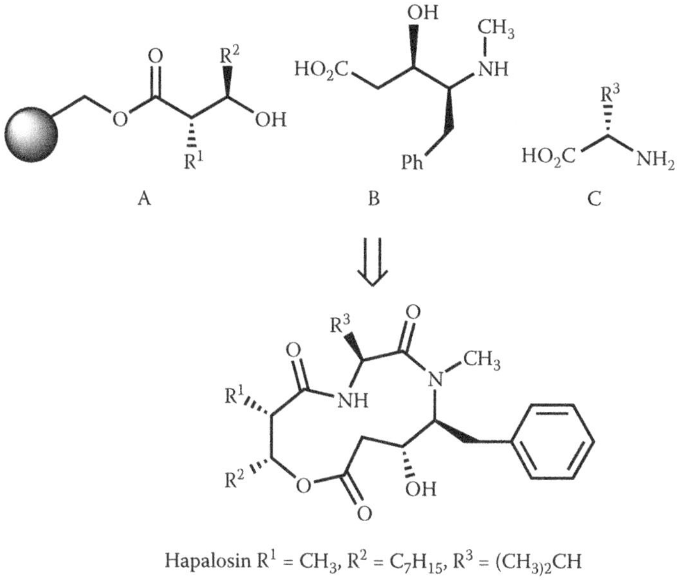

Hapalosin $R^1 = CH_3$, $R^2 = C_7H_{15}$, $R^3 = (CH_3)_2CH$

FIGURE 11.38 Design of hapalosin analogs. (From Herman, C. et al., Synthesis of hapalosin analogs by solid-phase assembly of acyclic precursors, *Tetrahedron*, 57, 8999, 2001.)

previous total synthesis of HUN-7293 using a convergent approach,[163] the backbone ester formation between the tetrapeptide **A** and the tripeptide **B** was achieved with a Mitsunobu reaction. Furthermore, the amide bond between the MLEU and LEU residues was identified as an effective macrocyclization site. For the preparation of the library, several analogs of **A** and **B** with diverse amino acid side chains (R^1-R^7) were synthesized and assembled. Structure–function relationships were assessed by alanine-scanning mutagenesis in which each amino acid side chain was replaced with a methyl group while retaining the stereochemistry of the peptide backbone. One member of the library displayed potent ICAM-1/VCAM-1 expression inhibitory activity.

Aurilide, a marine natural product that has potent cytotoxicity against HeLa S3 cells, is the third example of a cyclic depsipeptide that was targeted for combinatorial library synthesis of analogs (Figure 11.40).[164] Diversification of the tetrapeptide segment Val-MeLeu-Sar-Val of aurilide was explored for the synthesis of three analog libraries **75** in which AA_4-AA_3-AA_2-AA_1 was varied as follows:

- The first library contained all possible diastereomers based on the D and L configurations of the chiral amino acid residues.
- In the second library, the two *N*-methyl residues were replaced by the corresponding nonmethyl residues (Sar → Gly; MeLeu → Leu).
- In the third library, the amino acids were permutated.

Tetrapeptides **73** were first assembled on trityl linker functionalized SynPhase Crowns using an Fmoc strategy, and then coupled to the aliphatic moiety (7*R*)-**74** using DIC/HOBt. The 7*S* epimer of **74** was used in the solid-phase synthesis of aurilide. Following deprotection and cleavage from the solid support, solution-phase macrocyclization was achieved in the presence of EDCI and HOAt in 10% DMF under high-dilution conditions. Finally, deprotection of the methythiomethyl group with silver nitrate provided the library of aurilide analogs **75**.

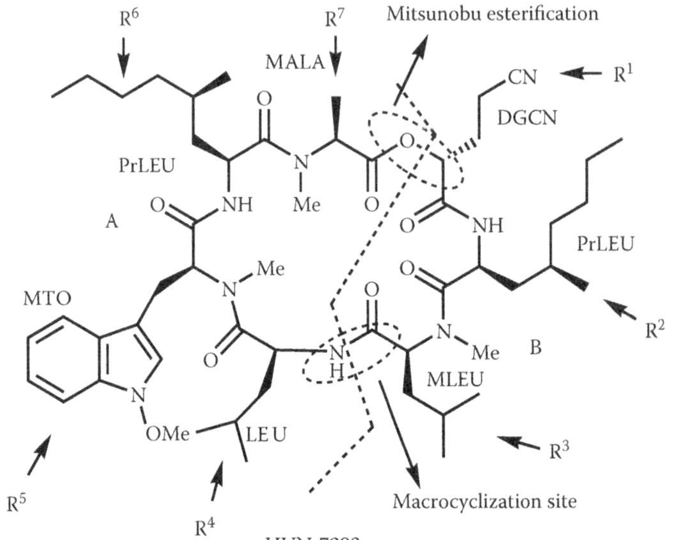

FIGURE 11.39 Design of a pharmacophore library of HUN-7293 analogs. (From Chen, Y. et al., Solution-phase parallel synthesis of a pharmacophore library of HUN-7293 analogs: a general chemical mutagenesis approach to defining structure-function properties of naturally occurring cyclic (depsi)peptides, *J. Am. Chem. Soc.*, 124, 5431, 2002.)

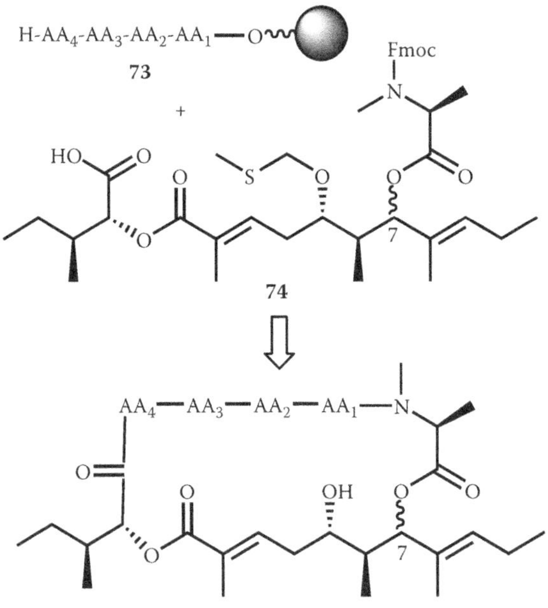

Aurilide (7R) AA$_1$ = AA$_4$ = Val, AA$_2$ = Sar, AA$_3$ = D-MeLeu

FIGURE 11.40 Library of aurilide analogs. (From Takahashi, T. et al., Solid phase library synthesis of cyclic depsipeptides: aurilide and aurilide analogs, *J. Comb. Chem.*, 5, 414, 2003.)

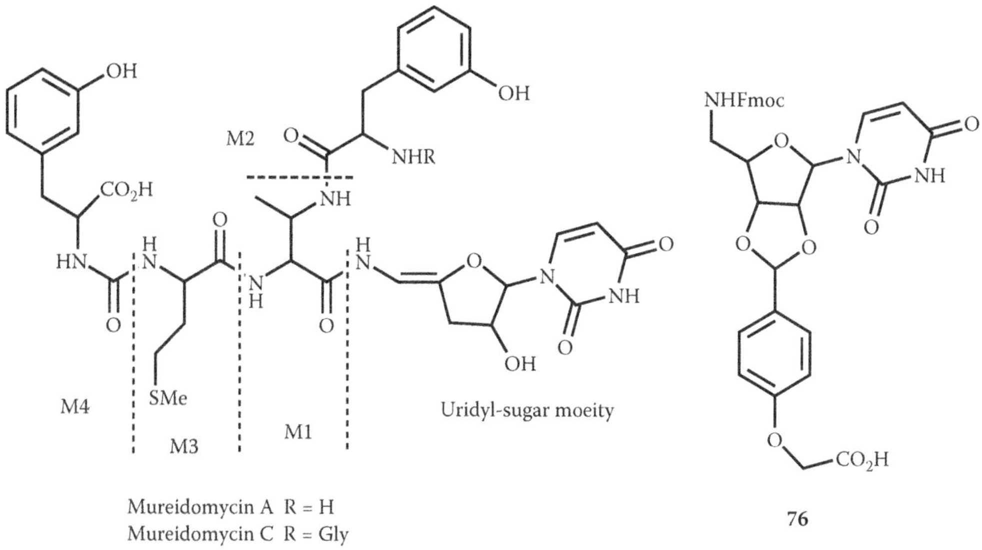

M2

M4

SMe

M3   M1

Uridyl-sugar moeity

NHFmoc

CO₂H

76

Mureidomycin A R = H
Mureidomycin C R = Gly

FIGURE 11.41 Design of a library of mureidomycin analogs. (From Bozzoli, A., et al., A solid-phase approach to analogs of the antibiotic mureidomycin, *Bioorg. Med. Chem. Lett.*, 10, 2759, 2000.)

11.5.7 MUREIDOMYCINS

In the search for novel antibacterial compounds, a solid-phase library of 80 analogs of the antibiotic mureidomycin was prepared (Figure 11.41).[165] The strategy employed the linkage of a suitably modified uridyl building block **76** to aminomethyl polystyrene resin. Assembly of the peptide chain was done in such a way as to allow deprotection and release of compounds in intermediate steps of the synthesis. The building blocks were chosen to most closely resemble the M1, M2, M3, and M4 residues present in the natural product and, because the absolute configuration of each residue of mureidomycin was unknown, both enantiomers of each building block were included in the design.

11.5.8 PEPTICINNAMIN E ANALOGS

Pepticinnamin E is a naturally occurring bi-substrate inhibitor of the enzyme protein farnesyl-transferase (PFT), and is of particular interest as an antitumor therapeutic agent. To elucidate the structure–activity relationships of PFT inhibitors, a library of pepticinnamin E analogs was prepared on solid phase, and investigated as potential inhibitors of PFT and inducers of apoptosis (Figure 11.42).[166–168] The peptide is composed of five building blocks $R^1 \ldots R^5$. The central tripeptide has the aromatic amino acids D-tyrosine (R^2), the N-methylated DOPA-derivative (R^3), and N-methyl-phenylalanaine (R^4). A lipophilic cinnamic acid derivative (R^1) was attached to the N-terminus of the tripeptide, and the C-terminus was esterified with a diketopiperazine R^5 via the side-chain alcohol of a serine residue incorporated into this heterocycle. Based on the strategy previously followed in the total synthesis of pepticinnamin E, a library of 51 analogs was developed by variation of all five building blocks. In particular, the lipophilic N-terminal group, the amino acid side chains, the polar C-terminus, and the degree of N-methylation were varied. To this end, the first amino acid was attached to a Wang resin, and the tripeptide was assembled through chain elongation in the C-terminal, N-terminal, or alternating C- and N-terminal directions. Different immobilized tripeptide intermediates **77–79** were deprotected at their N-termini. Subsequent attachment of lipophilic carboxylic acids and liberation of C-terminus (step A) gave **80**. This tripeptide could be released from the solid support (step B) to form a chain-shortened compound **81** or coupled

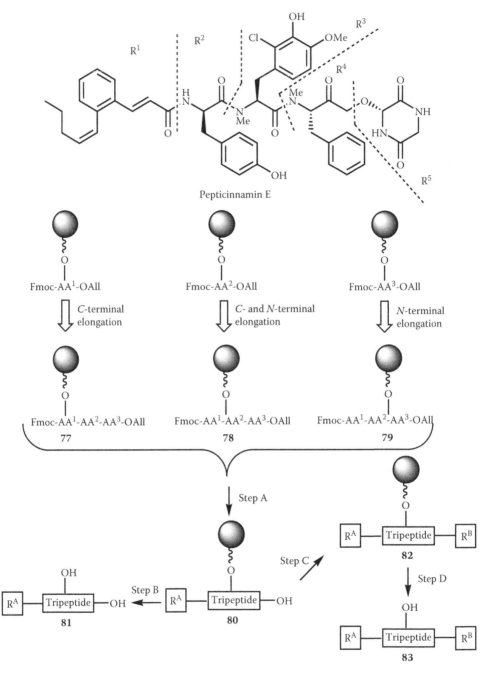

FIGURE 11.42 Library of pepticinnamin E analogs. (From Thutewohl, M. and Waldmann, H., Solid-phase synthesis of a pepticinnamin E library, *Bioorg. Med. Chem.*, 11, 2591, 2003.)

to polar building blocks (step C) to yield immobilized esters and amides **82**. After cleavage from the solid support, pepticinnamin E analogs **83** resulted. Several compounds showed pronounced PFT inhibitory activity, with the lowest IC$_{50}$ value reaching 1 μM, and induction of apoptosis in a Ras-transformed tumor cell line.[168]

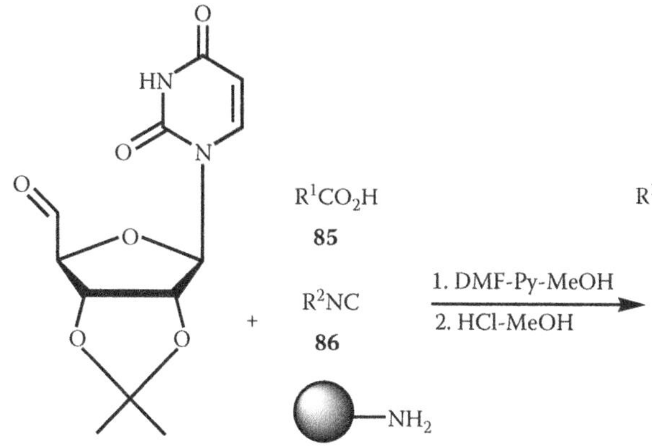

FIGURE 11.43 Combinatorial synthesis of nikkomycin derivatives using Ugi reaction. (From Suda, A. et al., Combinatorial synthesis of nikkomycin analogs on solid-support, *Heterocycles,* 55 1023, 2001.)

11.5.9 Nikkomycin Analogs

In the discovery of new antifungal compounds, Tsukuda and coworkers synthesized a nikkomycin analog library by Ugi condensation of uridine derivative **84**, 59 carboxylic acids **85**, 15 isocyanides **86**, and an amino component attached to a Rink amide resin **87** (Figure 11.43).[169] After cleavage from solid support, compounds **88** were tested for the enzyme inhibitory activity against *Candida albicans* chitin synthase 1. Among these derivatives, 246 samples exhibited more than 50% inhibition against this enzyme at a concentration of 10 µg/ml.

11.5.10 Distamycin A Analogs

Small molecules that are able to selectively bind DNA and activate (block a repressor) or inhibit (block an activator) gene expression hold great promise as therapeutics. The natural antibiotic distamycin A was chosen as a lead structure for the development of new bioactive DNA-binding agents.[170] Two prototypical solution-phase libraries containing 2640 distamycin A analogs were prepared in a small mixture format (Figure 11.44).[171] Using 11 *N*-BOC heterocyclic amino acids and 12 amino esters, the individual subunits were coupled to provide all possible 132 individual dipeptides in parallel. These dimers were deprotected and coupled to a mixture of 10 *N*-BOC carboxylic acids to give 132 mixtures of 10 *N*-BOC trimers in which only the last position (subunit A) was not defined (1320 compounds). This first-generation library was further coupled to the basic side chain *N,N*-dimethylaminobutyric acid (DMABA) to afford a second-generation DMABA-trimer library. For screening of DNA-binding affinity, a rapid high-throughput assay was developed based on the loss of fluorescence from a target oligonucleotide presaturated with ethidium bromide. Deconvolution of the most potent mixtures by resynthesis led to the identification of compounds that were 1000 times more potent than distamycin A in cytotoxic assays ($IC_{50} = 29$ nM in the L1210 assay) that bind to poly[dA]- poly[dT] with comparable affinity, and exhibit an altered DNA-binding sequence selectivity.

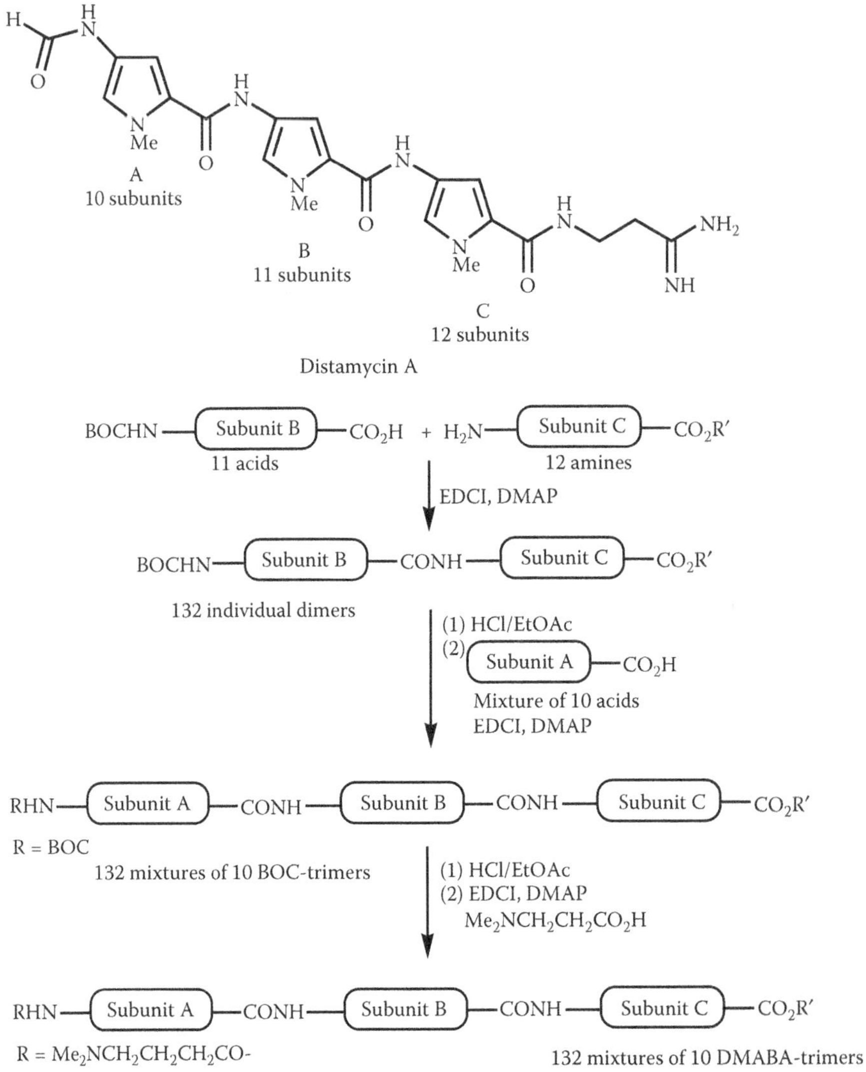

FIGURE 11.44 Reaction sequence for preparation of distamycin A analogs. (From Dale, D.L., Fink, B.E., and Hedrick. J.L., Total synthesis of distamycin A and 2640 analogs: a solution-phase combinatorial approach to the discovery of new, bioactive DNA binding agents and development of a rapid, high-throughput screen for determining relative DNA binding affinity or DNA binding sequence selectivity, *J. Am. Chem. Soc.*, 122, 6382, 2000.)

11.6 TERPENOIDS

11.6.1 DITERPENES AND TRITERPENES

11.6.1.1 Paclitaxel (Taxol®) Derivatives

Several examples of combinatorial diversification of terpenoid scaffolds have been reported. The diterpenoid paclitaxel (Taxol®), first isolated from the bark of the Pacific Yew (*Taxus brevifolia*), has proved to be the most important drug introduced in the last 10 yr with sales of over $1.5 billion in 1999. The chemistry of Taxol® has been thoroughly investigated to determine its structure–activity relationships and thus to define the pharmacophore in chemical terms.[172,173] However, the poor water solubility of Taxol® and its susceptibility to drug resistance led to the search for analogs with improved pharmacological properties and, hence, to the design and synthesis of taxoid libraries. Using radio-frequency-encoded combinatorial chemistry, a 400-membered taxoid library from baccatin III (**89**) was synthesized and included design criteria to potentially enhance water solubility, modulate biological activity, and utilize novel solid-phase synthesis methodology (Figure 11.45).[174] The solid-phase synthesis of targeted library **92** required the preparation of core structure **90** from baccatin III (**89**), which, after removal of both the Troc and trichloroethyl groups, reacted with 2-chlorotrityl resin in the presence of N,N-diisopropylethylamine to afford resin-bound ester **91**. The loaded resin was distributed into 400 microreactors, Fmoc-deprotected, and subjected to successive couplings with a set of 20 carboxylic acids (R^1CO_2H and R^2CO_2H) at the side-chain amino group and another set of 20 carboxylic acids at C-7 and C-2 to yield **92**.

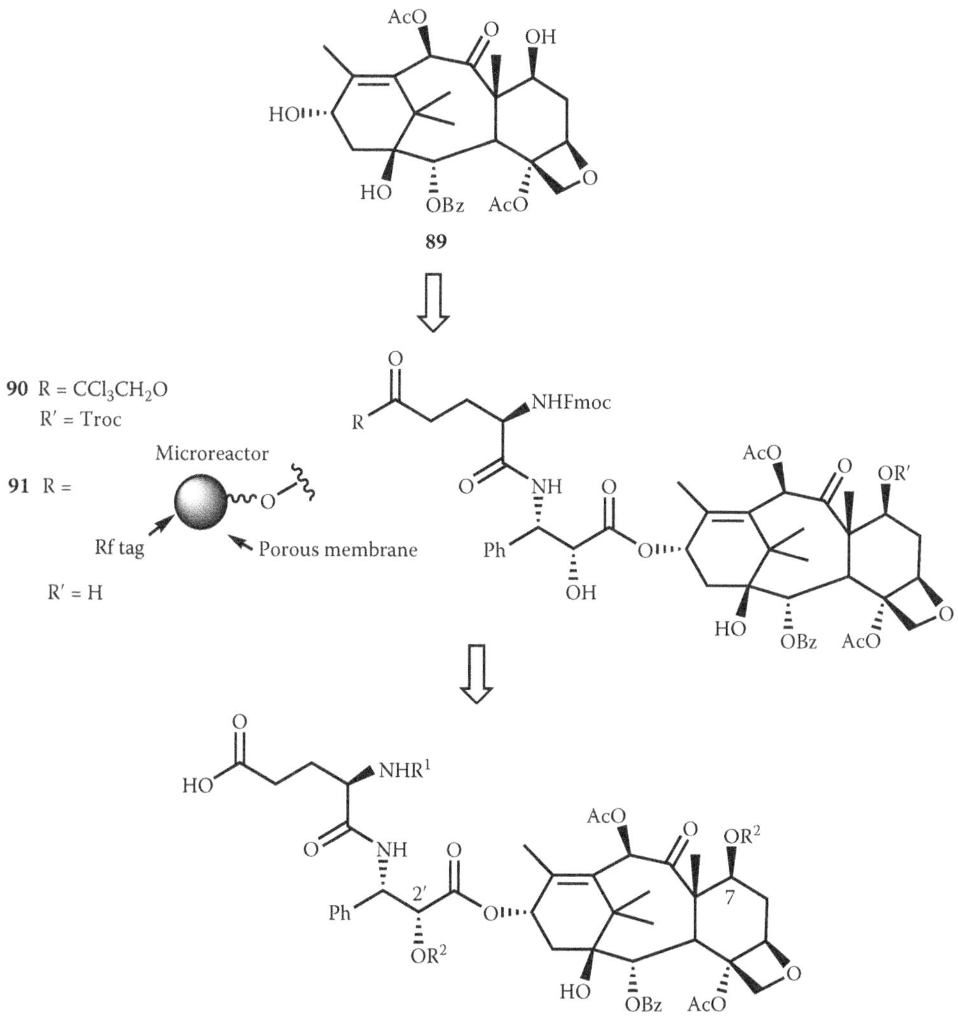

FIGURE 11.45 Taxoid library. (From Xiao, X.-Y., Parandoosh, Z., and Nova, M.P., Design and synthesis of a taxoid library using radiofrequency encoded combinatorial chemistry, *J. Org. Chem.*, 62, 6029, 1997.)

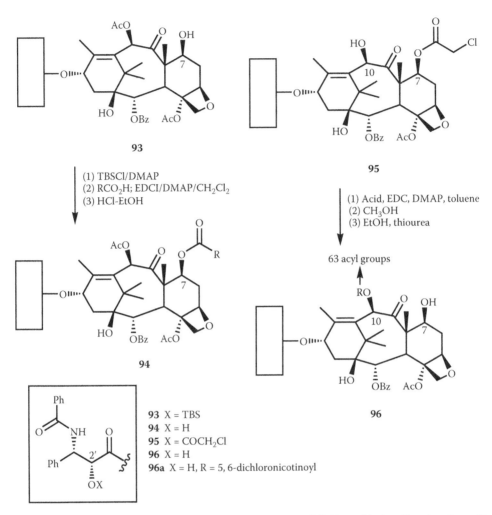

FIGURE 11.46 Solution-phase synthesis of paclitaxel C-7 esters and C-10 modified paclitaxel analogs. (From Bhat, L. et al., Synthesis and evaluation of paclitaxel C7 derivatives: solution phase synthesis of combinatorial libraries, *Bioorg. Med. Chem. Lett.*, 8, 3181, 1998; Liu, Y. et al., A systematic SAR study of C10 modified paclitaxel analogs using a combinatorial approach, *Comb. Chem. High Throughput Screen.*, 5, 39, 2002.)

An automated solution-phase synthesis of a 26-memberd library of paclitaxel C-7 esters was accomplished by Georg and coworkers (Figure 11.46).[175] Condensation of 2′-O-(*tert*-butyldimethylsilyl) paclitaxel **93** with a set of 26 carboxylic acids in the presence of the dehydrating agent 1-(3-dimethylaminopropyl)-3-ethylcarbodiimide (EDCI) and DMAP afforded the corresponding C-7 esters. Upon deprotection of the 2′-*O*-TBS group, the desired paclitaxel library **94** was obtained. The paclitaxel C-7 esters were purified by HPLC, evaluated for their ability to initiate the polymerization of tubulin in the tubulin assembly assay, and screened against the human breast cancer cell line MCF7 and the resistant human breast cancer cell line MCF7-R. Higher activity than paclitaxel was observed for compounds derivatized with acetic, *p*-toluic, valeric, and methylthioacetic acids.

To further understand the structure–activity relationship at C-10, especially the cytotoxicity against drug-resistant cancer cell lines, the same research group prepared a library with 63 paclitaxel analogs modified at this position using parallel solution-phase synthesis (Figure 11.46).[176] Library **96** was synthesized in three steps from 2′,7-*bis*-protected 10-deacetylpaclitaxel **95** in a single-flask procedure. The new C-10-modified taxanes were evaluated for their ability to promote tubulin

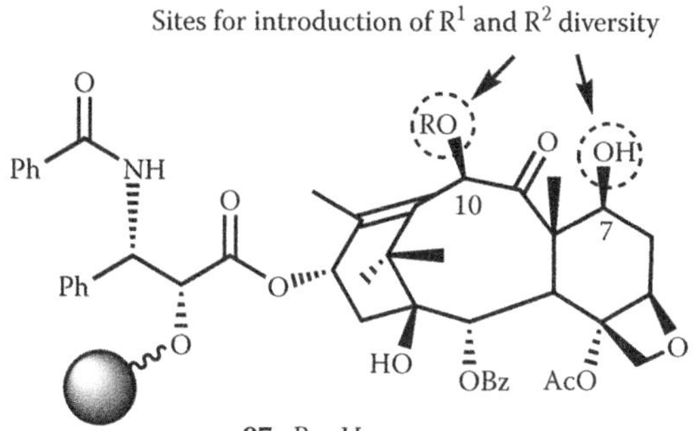

FIGURE 11.47 Solid-phase synthesis of 7,10-diacyl analogs of paclitaxel. (From Jagtap, P.G. et al., Design and synthesis of a combinatorial chemistry library of 7-acyl, 10-acyl, and 7,10-diacyl analogs of paclitaxel (Taxol) using solid phase synthesis, *J. Nat. Prod.*, 65, 1136, 2002.)

polymerization, and for their cytotoxicities against the B16 melanoma cell line, the drug-resistant human breast cancer cell line MCF7-ADR, and the drug-sensitive human breast cancer cell line MCF7. About 50% of the analogs demonstrated better activity against MCF7-ADR than paclitaxel, and suggested that their P-glycoprotein affinities change with structural modification at C-10. Derivative **96a**, compared to paclitaxel, was found to be the most active compound against MCF7-ADR cells with a tenfold improved potency.

On the basis of the known structure–activity relationships of paclitaxel and previous resin-based approaches to the synthesis of taxoid analogs, an attachment to the 2-hydroxyl group appeared to be most desirable because modifications at this position are usually deleterious to activity. Nevertheless, the steric hindrance caused by THP ether linker or related alkoxy linkers precluded the synthesis of analogs with a normal *N*-benzoyl function. To overcome this limitation, Kingston and coworkers[177] used an alternate solid support, a polystyrene-divinyl benzene resin functionalized with a butyldiethylchlorosilane linker (Figure 11.47). Treatment of the resin with 10-deacetylpa-clitaxel provided resin-bound compound **97**. Using Holton's method ($R^1COOCOR^1$, $CeCl_3$, THF), **97** was converted to the corresponding 10-acyl analogs **97a**. Acylation at the C-7 position was achieved using 1,3-diisopropylcarbodiimide (DIPC) and acid to generate a 21-member library. The authors developed a methodology to determine the tubulin assembly of compounds synthesized, and discovered three 10-acyl derivatives ($R^1 = COC_6H_5$, $COCH=CHCH_3$, and $COOCH_2C_6H_5$) and one 7,10-diacyl derivative ($R^1 = COCH_2CH_3$, $R^2 = COCH_2Cl$) with improved tubulin assembly activity compared to paclitaxel.

11.6.1.2 Sarcodictyins

Sarcodictyins, isolated from certain species of soft corals,[178,179] constitute a group of diterpenoids that display potent antitumor activity and Taxol®-like mode of action, and are important synthetic targets.[180] In order to discover analogs possessing activities superior to those of the natural products, Nicolaou and coworkers generated sarcodictyin-based libraries for further evaluation of their biological activity (Figure 11.48).[181,182] Starting from solid-supported scaffold **98**, a combinatorial library of about 100 sarcodictyin analogs of the general structures **100–102** was constructed by modifying the C-8 ester, C-15 ester, and C-4 ketal functionalities.[182] Functionalization of the hydroxyl group at C-8 with acetic acid anhydrides, acid chlorides, carboxylic acids, or isocyanates, yielded **99**. After this first step for introducing diversity in the library, this intermediate was further modified at the C-15 and C-4 positions. In some cases, low yields were observed for DCC-coupled products on solid phase, and these compounds were prepared in solution. Based on the screening results for induction of tubulin polymerization and cytotoxicity studies with ovarian cancer cell

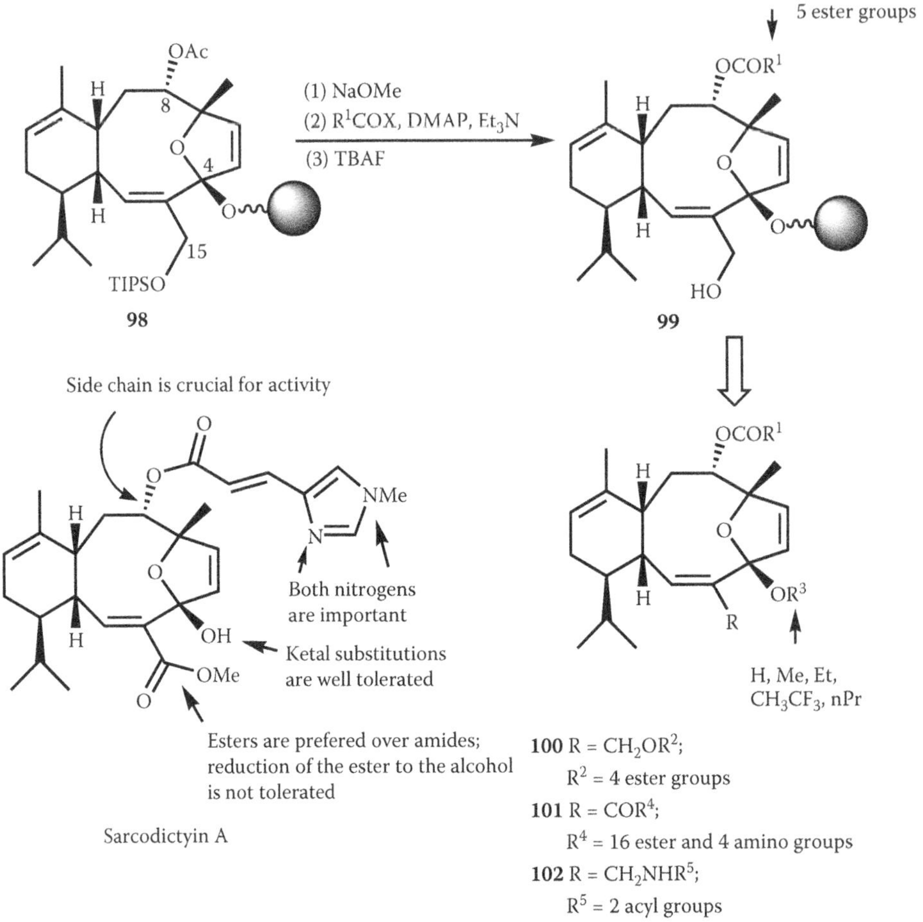

FIGURE 11.48 Solid- and solution-phase sarcodictyin libraries, and SAR of sarcodictyin A. (From Nicolaou, K.C. et al., Solid and solution phase synthesis and biological evaluation of combinatorial sarcodictyin libraries, *J. Am. Chem. Soc.*, 120, 10814, 1998.)

lines, structure–activity relationships of sarcodictyins, illustrated in Figure 11.48 for sarcodictyin A, were proposed.

11.6.1.3 Various Diterpenes and Triterpenes

The labdane diterpenoid 14-deoxyandrographolide (**103**)[183] and pentacyclic triterpenes ursolic acid, betulinic acid (see Chapter 5, Subsection 5.3.4.1), and lupeol were also used as scaffolds for the generation of solid-phase combinatorial libraries (Figure 11.49). The derivatization of **103** involved coupling of the C-19 hydroxyl group on the 2-chlorotrityl resin, oxidation of the secondary alcohol to form a ketone precursor, and introduction of diversity through the corresponding oxime esters **104**.[184] A small library of 20 derivatives was generated from **104** using 5 alkyl and 5 aryl carboxylic acids.

An identical approach was followed for the preparation of combinatorial libraries from ursolic acid, betulinic acid, and lupeol (Figure 11.50). In the case of betulinic acid (**105**) and ursolic acid, the triterpene scaffold was immobilized onto a prederivatized amino acid on 2-chlorotrityl or Sieber amide resin, and treated with a variety of aliphatic, aromatic, and amino acids, to generate a library of C-3 and C-28 derivatives **106**.[185] In another series, C-3 oxime esters **107** were prepared from

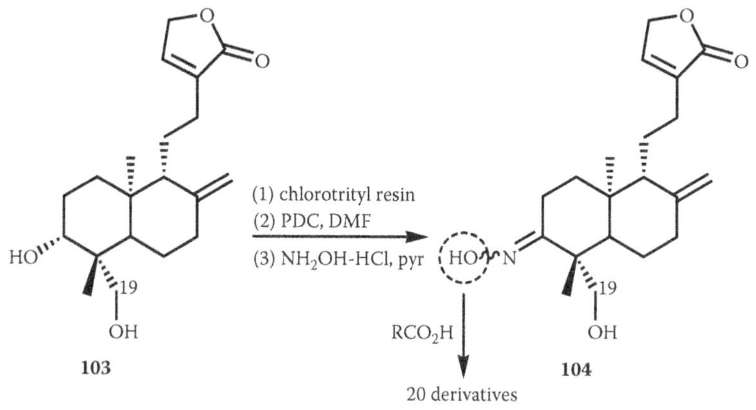

FIGURE 11.49 Library generation from the labdane diterpenoid 14-deoxyandrographolide. (From Biabani, M.A.F. et al., A novel diterpenoid lactone-based scaffold for the generation of combinatorial libraries B, *Tetrahedron Lett.*, 42, 7119, 2001.)

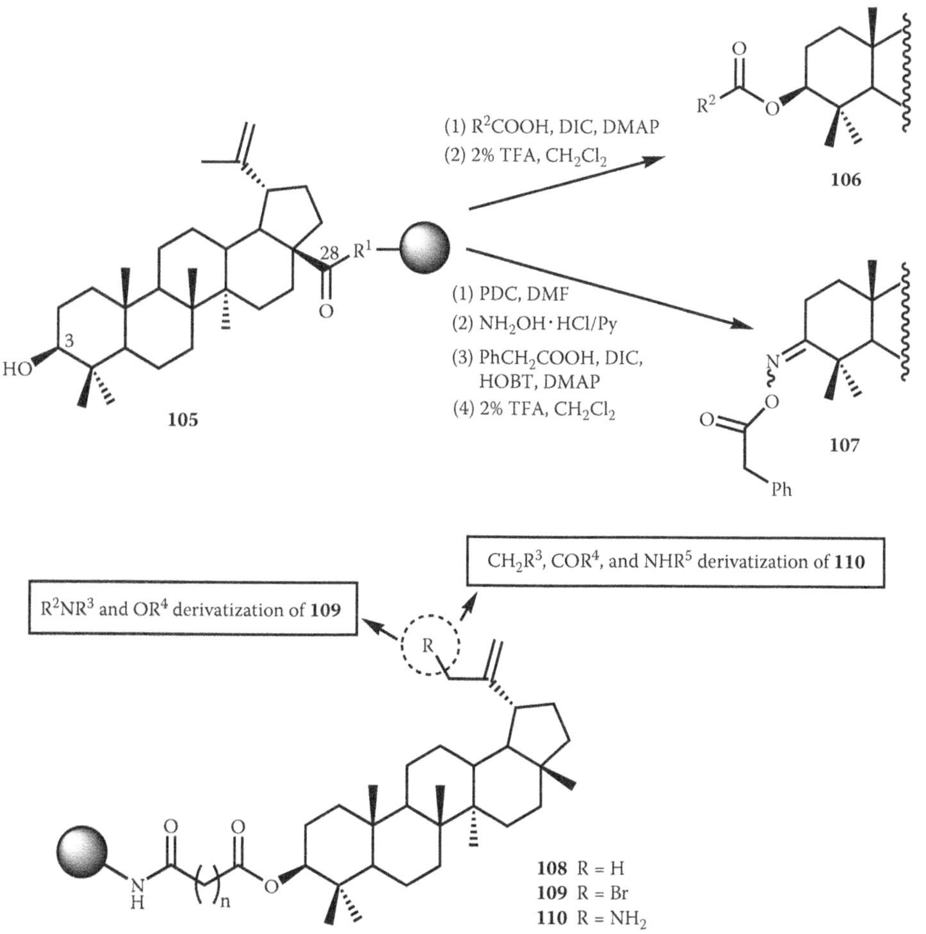

FIGURE 11.50 Betulinic acid and lupeol libraries. (From Pathak, A. et al., Synthesis of combinatorial libraries based on terpenoid scaffolds, *Comb. Chem. High Throughput Screen*, 5, 241, 2002; Srinivasan, T et al., Solid-phase synthesis and bioevaluation of lupeol-based libraries as antimalarial agents, *Bioorg. Med. Chem. Lett.*, 12, 2803, 2002.)

the immobilized 3-keto derivatives of **105**. Two betulinic acid and two ursolic acid derivatives exhibited fivefold increase in the antimalarial activity (MIC = 10 µg/ml) in comparison to the parent molecules. For lupeol, diversity was introduced at C-3 and C-30.[186] The triterpenoid was anchored to a Rink amide Sieber amide linker through aliphatic dicarboxylic acids moieties **108**, which also served as a site for introducing diversity. The aliphatic side chain was then halogenated and aminated to afford intermediates **109** and **110**. Further derivatization yielded 2 libraries of 48 compounds each. *In vitro* screening of the libraries for antimalarial activity against *Plasmodium falciparum* led to the identification of compounds with seven- to ninefold increase in biological activity compared to lupeol.

11.6.2 SESQUITERPENES

11.6.2.1 Dysidiolides

The marine sesquiterpene dysidiolide, the first naturally occurring inhibitor of the dual-specificity cdc25 protein phosphatase family, plays a crucial role in the regulation of the cell cycle.[187] To determine if analogs are biologically active, 6-*epi*-dysidiolide (**116a**) and a small solid-phase library of dysidiolide analogs were synthesized on Merrifield resin using a Diels–Alder cycloaddition route (Figure 11.51).[188,189] For the solid-phase synthesis of 6-*epi*-dysidiolide (**116a**), the polymer-bound diene **111** was subjected to Diels–Alder reaction conditions with the quasi-C_2-symmetric acetal **112** to yield the aldehyde **113** after hydrolytic removal of the chiral auxiliary. The carbon chain of **113** was elongated by means of a two-step sequence including a Wittig reaction and hydrolysis of the enol ether formed therein to yield aldehyde **114**. Nucleophilic addition of 3-lithiofuran resulted in a 2:1 mixture of epimeric alcohols **115**. Finally, the furan was oxidized with singlet oxygen in the presence of Hünig's base to give γ-hydroxy-butenolides **116a** and **116b** after release from the polymeric carrier by an olefin metathesis reaction with Grubbs catalyst. Based on the preceding strategy, seven analogs were synthesized from intermediate aldehydes **114** and **115**, and further tested for inhibition of protein phosphatase cdc25C and cytotoxicity.

11.6.2.2 Nakijiquinones

The marine sesquiterpenes nakijiquinones are the only natural products known to be inhibitors of the Her-2/Neu receptor tyrosine kinase, one of the proteins that play a crucial role in the control of cell growth and differentiation.[190] Based on the modular structure of nakijiquinones, a small solution-phase library of 74 analogs was made and evaluated for inhibition of kinases with highly similar ATP-binding domains (Figure 11.52).[191] In the first series of compounds, the decalin core was only modified, and analogs with altered configurations at C-2 and an exocyclic double bond instead of an endocyclic one were synthesized. Also, different amino acids were introduced, or omitted. Finally, the *p*-quinoid system was replaced by an *o*-quinoid structure or an aromatic group. In a second series of analogs, changes from the natural product were more pronounced. In particular, variations were made on the hydrophobic sesquiterpene unit which, according to the accepted binding mode of kinase inhibitors to the ATP-binding domain, may occupy a hydrophobic pocket close to the ATP binding site. Also, the quinone-type building block was varied; it should contribute essential hydrogen bonds to the hinge region of receptor tyrosine kinases, and, therefore, serve to orient the inhibitor within the binding site. Finally, size and stereochemistry of the amino acids were changed, which might both enhance binding and improve selectivity via formation of hydrogen bonds and side-chain interactions with the protein.

In general, these compounds were synthesized from appropriately substituted and selectively protected phenolic esters. A reaction sequence that consisted of selective deprotection, oxidation to the quinoid system, and addition of an amino acid gave vinylogous esters. Linkage to the *trans*-decalin building blocks with an exocyclic double bond was achieved with a Suzuki coupling of substituted arylboronic acids to the corresponding decalin-derived vinyl bromides. Reduction of

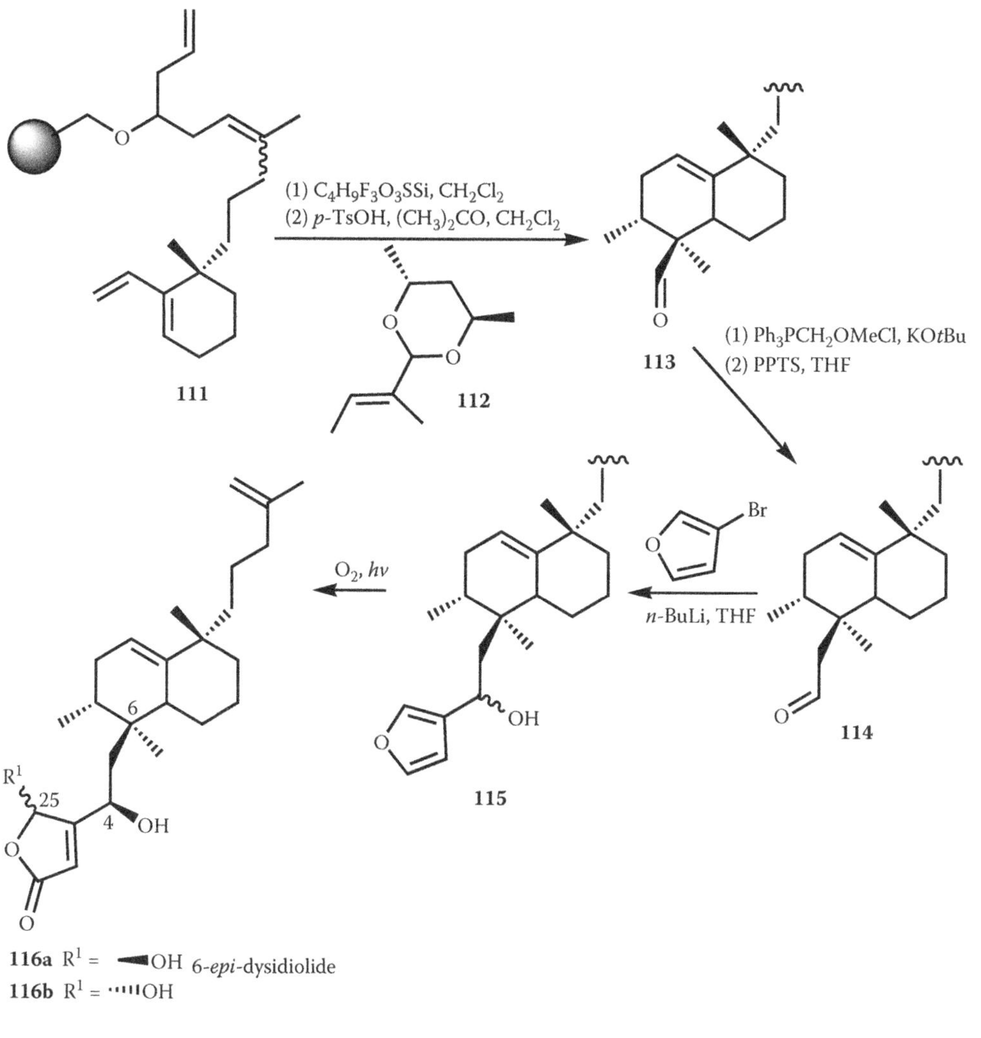

FIGURE 11.51 Solid-phase synthesis of dysidiolide analogs. (From Brohm, D. et al., Natural products are biologically validated starting points in structural space for compound library development: solid-phase synthesis of dysidiolide-derived phosphatase inhibitors, *Angew. Chem., Int. Ed. Engl.*, 41, 307, 2002; Brohm, D. et al., Solid-phase synthesis of dysidiolide-derived protein phosphatase inhibitors, *J. Am. Chem. Soc.*, 124, 13171, 2002.)

the double bond in the Suzuki products yielded analogs with an alkyl side chain attached to the decalin core.

The most potent inhibitor identified in the library was compound **117**, which inhibited the Tie-2 receptor with an IC_{50} value of 5 μM as well as the VEGFR-3 and IGF1R receptor tyrosine kinases. The investigation illustrated the concept that natural products are biologically validated starting points in structural space for compound library development.[13] Within this "domain concept," they are regarded by Waldmann as molecular entities selected by evolution for binding to structurally conserved yet genetically mobile protein domains. Owing to the high homology of the proteins selected for this study with respect to their ATP binding kinase domains, the results obtained with nakijiquinone C analogs suggest that a library synthesized around a natural product binding to such domains should yield, with a high hit rate, ligands for similar domains in other proteins.

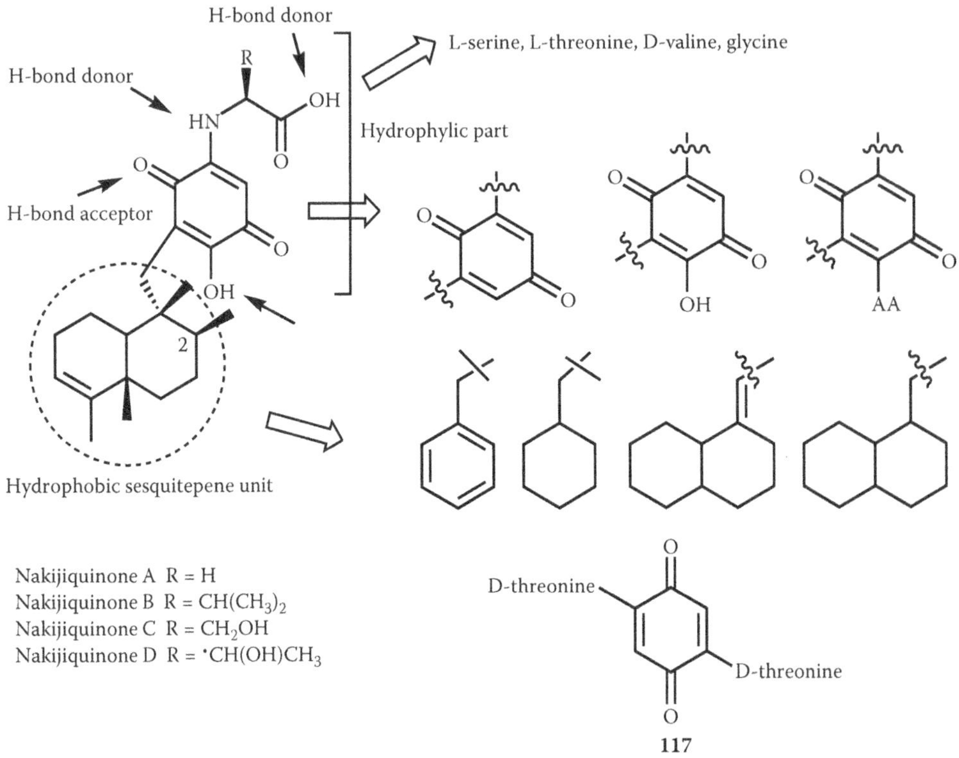

FIGURE 11.52 Modular composition of a nakijiquinone library. (From Kissau L. et al., Development of natural product-derived receptor tyrosine kinase inhibitors based on conservation of protein domain fold, *J. Med. Chem.*, 2003, 46, 2917.)

11.7 STEROIDS

Mainly due to their rigid nucleus, which is versatile in terms of functionality and side-chain substitution, steroids are of particular interest for the design and synthesis of scaffolds for combinatorial chemistry. They generally possess at least a ketone, hydroxy, or carboxylic acid group at various carbon positions (typically C-3, C-11, C-17, C-20, C-21, or C-27) that can be used to simultaneously protect a chemically sensitive group and link the steroid to solid support. Most studies published so far report the diversification of the steroid nuclei at a single site. Few combinatorial approaches thus far allow the generation of highly diverse natural or nonnatural steroid nuclei starting from a single building block.[192,193]

11.7.1 ESTRADIOLS

In the course of studies on therapeutic agents for the treatment of breast cancer, Poirier and coworkers prepared hydroxysteroid libraries with two levels of molecular diversity (Figure 11.53). The libraries were designed as potential inhibitors of steroid sulfatase and 17β-hydroxysteroid dehydrogenase, two key enzymes involved in the biosynthesis of estradiol.[194,195] Estradiol libraries bearing functionalized side chains at C-16 (compounds **118** and **119**), C-17 (compound **120**), and C-7 (compound **121**) were synthesized from estrone and 6-dehydro-19-nor-testosterone acetate attached to a polystyrene resin through the phenolic function at C-3. The functionality at 16β and 7α was introduced from precursors bearing an azidoalkyl side chain in those two positions. An oxirane group was chosen as the key precursor function to provide the 17 amide side chain of

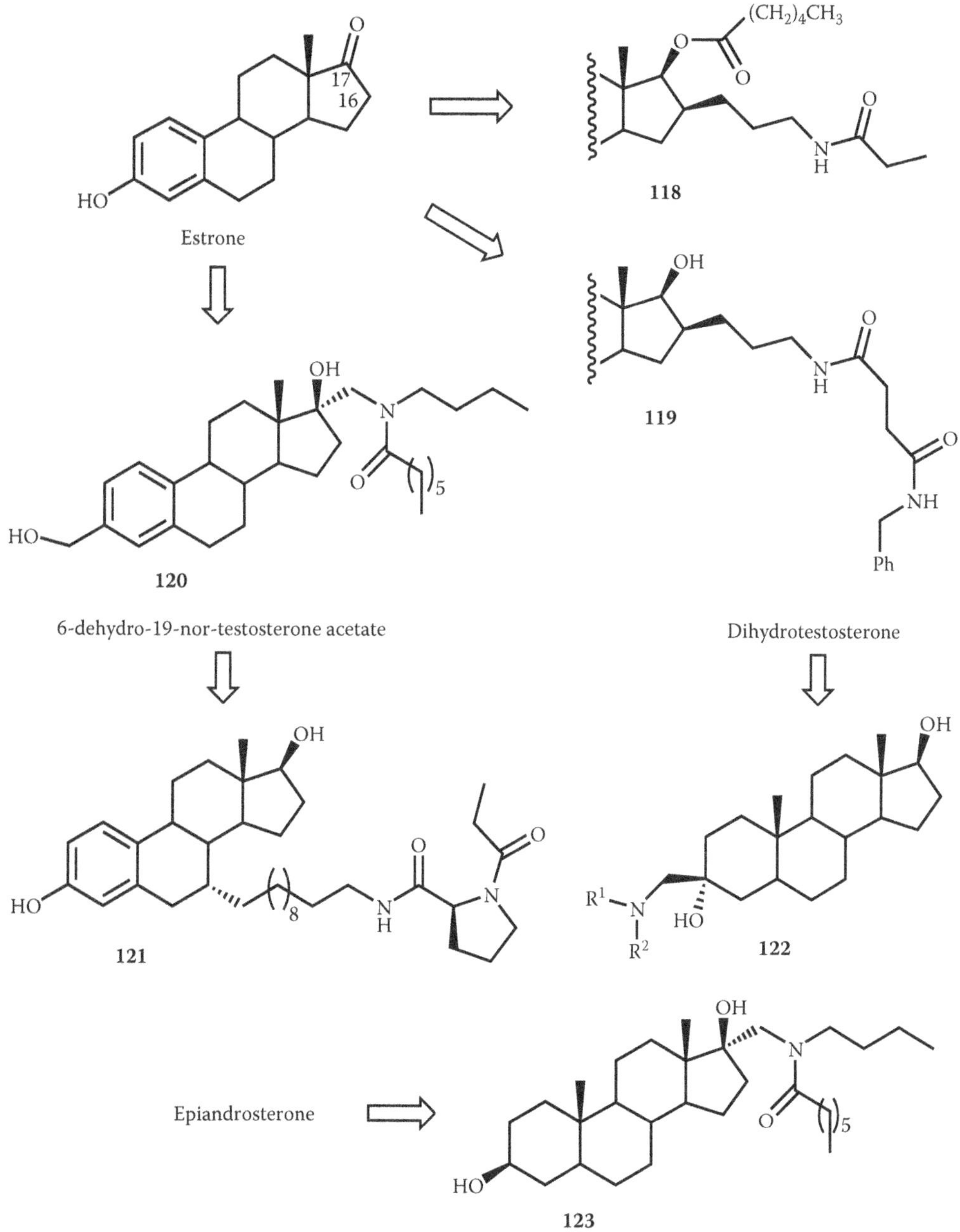

FIGURE 11.53 Hydroxysteroid derivatives with two levels of molecular diversity. (From Tremblay, M.R. and Poirier, D., Solid-phase synthesis of phenolic steroids: from optimization studies to a convenient procedure for combinatorial synthesis of biologically relevant estradiol derivatives, *J. Comb. Chem.*, 2, 48, 2000; Maltais, R., Tremblay, M.R., and Poirier, D., Solid-phase synthesis of hydroxysteroid derivatives using the diethylsilyloxy linker, *J. Comb. Chem.*, 2, 604, 2000.)

compound **120**. This latter methodology was also employed in the construction of libraries from nonphenolic steroids bearing 3β and 17α side chains (compounds **122** and **123**).

This investigation was further extended to the synthesis of inhibitors of type 3 17β-hydroxysteroid dehydrogenase (17β-HSD), a key steroidogenic enzyme transforming the inactive androgen 4-androstene-3,17-dione (Δ_4-dione) into testosterone (Figure 11.54). This enzyme constitutes an interesting target for blocking the biosynthesis of testosterone and, thus, its effects in androgen-sensitive diseases such as prostate cancer.[196,197] Once androsterone (ADT) had been identified as an inhibitor of type 3 17β-HSD and a lead compound for further development, several model libraries of 3 β-peptido ADT derivatives were developed by parallel solid- and solution-phase synthesis. In the solid-phase synthesis,[197] azide precursor **124** derived from dihydrotestosterone was coupled through its C-17 ketonic group onto polymer-bound glycerol using acetal exchange conditions, trimethylorthoformate, and Se(OTf)$_3$ in toluene, and reduced to amine **125**. Amine **125** was, in turn, subsequently transformed into amines **126** and **127** by Fmoc amino acid coupling. The three libraries **A** (R^1 diversity), **B** (R^1, R^2 diversity), and **C** (R^1, R^2, R^3 diversity) were obtained from amines **125**, **126**, and **127**, respectively, upon treatment with mixed anhydrides resulting from a carboxylic acid, PyBOP, HOBt, and DIPEA in DMF and cleavage from the solid support. Potent inhibitors were identified from these model libraries, especially six members from library **B** having at least one phenyl group.

Using parallel solution-phase approach, the same research group prepared four libraries of 3β-substituted-ADT.[198] The first library of 3β-amidomethyl-ADT derivatives (**D**) consisted of 168 members. This library had two sites of molecular diversity on the amide (R^1 and R^2), which were introduced by regioselective opening of the oxirane precursor **128** with a series of primary amines and by a regioselective reaction of amines **129** with acyl chlorides, respectively. The screening of library **D** revealed that relatively small hydrophobic chains at R^1 (5-8 carbons) and small hydrophobic substituents at R^2 (1-4 carbons) were characteristics of the most potent inhibitors. According to these inhibition results, a second library of 3β-amidomethyl-ADT derivatives (**E**) consisted of 56 members, and was generated from the dioxolane protected epiandrosterone **130** using an improved method that used scavenger resins and solution-phase parallel chemistry. In this method, R^2 diversity was added using acid chlorides with an acid scavenger resin, piperidonomethyl polystyrene, and a nucleophilic resin, aminomethyl polystyrene. Library **E** generated more potent inhibitors than library **D**, and provided useful information on structure–activity relationships for the design of a new library **F** consisting of 49 members. A fourth library of 3-carbamate-ADT-derivatives (**G**) was finally designed using a method that allowed the preparation of more rigid molecules with two sites of molecular diversity (R^1/R^2 and R^3) in the area occupied by the adamantane group in library **F**. The parallel synthesis of library **G** started with precursor **131**, prepared from **130**, and whose aminolysis with a secondary amine (R^{1}R^{2}NH) introduced the first site of diversity. From the resulting β-amino-alcohol **132**, the second site of diversity was obtained using the optimized acylation conditions previously used for libraries **D** and **E**. Libraries **F** and **G** contained the two most potent inhibitors with activity approximately 18-fold higher than the natural substrate.

During the development of steroid sulfatase inhibitors as therapeutic agents for estrogen-sensitive cancers, the sulfamate group was identified as a good pharmacophore, and later it was demonstrated that the combination of this group at C-3 and a hydrophobic substituent at C-17 within the same steroidal molecule resulted in significant improvement of steroidal sulfatase inhibition compared to compounds using only one of these two substituents. Furthermore, the sulfamate group can be used as an anchoring group or a linker in the solid-phase synthesis of sulfamates and sterols. The synthesis of model libraries of *N*-derivatized 17 α-piperazinomethyl estradiols by solid-phase parallel chemistry illustrated the utility of the sulfamate linker strategy (Figure 11.55).[198] A 3-sulfamoyl precursor, derived from estrone and linked to trityl chloride resin **133**, was acylated with Fmoc-protected amino acids after cleavage of the trifluoroacetyl protecting group. After removal of the Fmoc protecting group, the R^2 carboxylic acids were introduced to resins **134** through an amidation step. At the end of this second acylation step, and following a split methodology, a library of 25 sulfamates **135** and a library of 25 phenols **136** were obtained from acidic and nucleophilic cleavage, respectively. Screening of the sulfamate analogs in homogenates of HEK-293 cells transfected with steroid sulfatase identified some potent inhibitors.

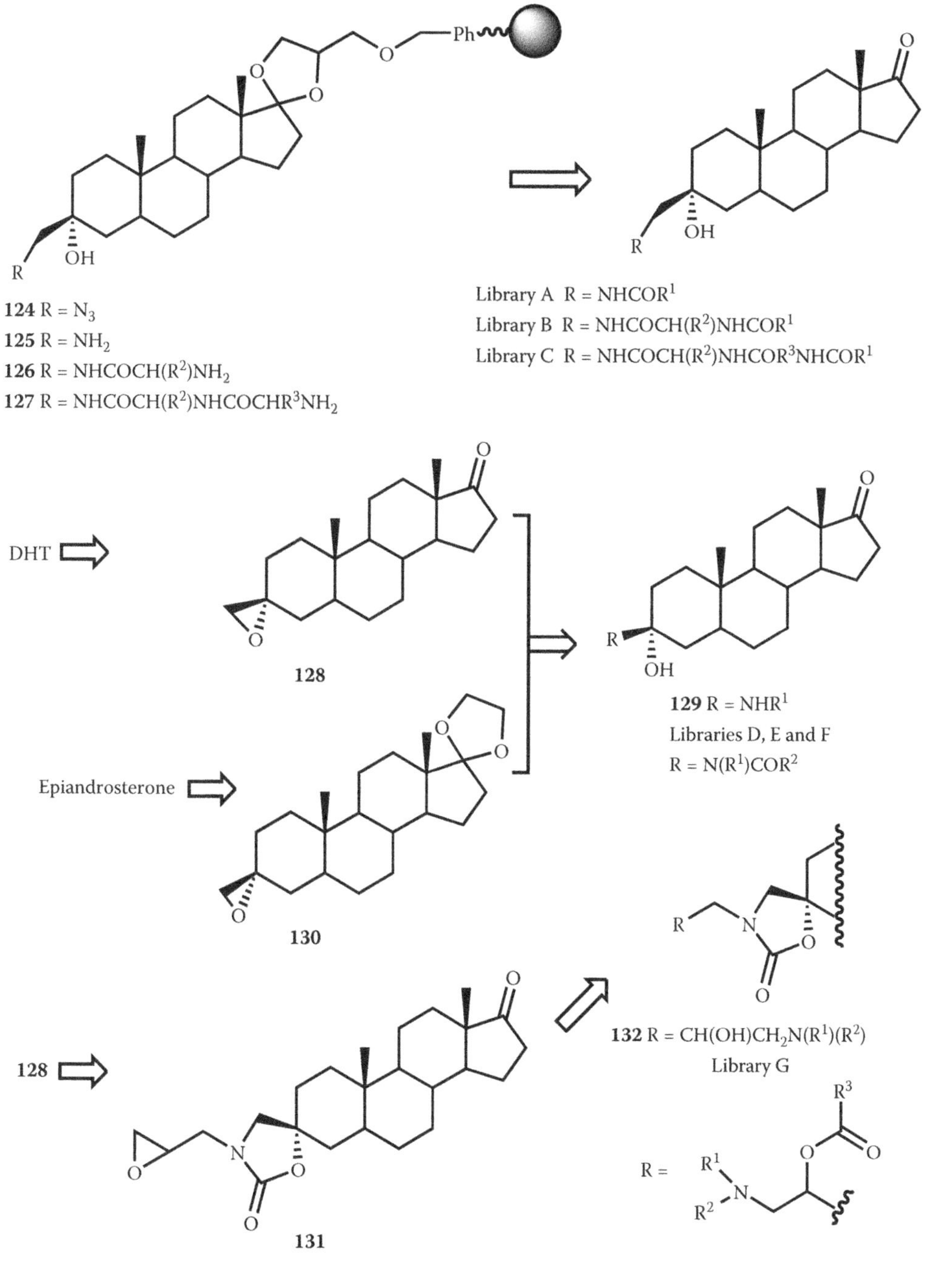

124 R = N$_3$
125 R = NH$_2$
126 R = NHCOCH(R^2)NH$_2$
127 R = NHCOCH(R^2)NHCOCHR^{3}NH$_2$

Library A R = NHCOR^1
Library B R = NHCOCH(R^2)NHCOR^1
Library C R = NHCOCH(R^2)NHCOR^{3}NHCOR^1

DHT

128

Epiandrosterone

130

129 R = NHR^1
Libraries D, E and F
R = N(R^1)COR^2

128

131

132 R = CH(OH)CH$_2$N(R^1)(R^2)
Library G

R =

FIGURE 11.54 Libraries of type 3 17β-hydroxysteroid dehydrogenase inhibitors. (From Maltais, R., Luu-The, V., and Poirier, D., Parallel solid-phase synthesis of 3β-peptido-3-hydroxy-5-androstan-17-one derivatives for inhibition of type 3 17β-hydroxysteroid dehydrogenase, *Bioorg. Med. Chem.,* 9, 3101, 2001; Maltais, R., Luu-The, V., and Poirier, D., Synthesis and optimization of a new family of type 17β-hydroxysteroid dehydrogenase inhibitors by parallel liquid phase chemistry, *J. Med. Chem.,* 45, 640, 2002.)

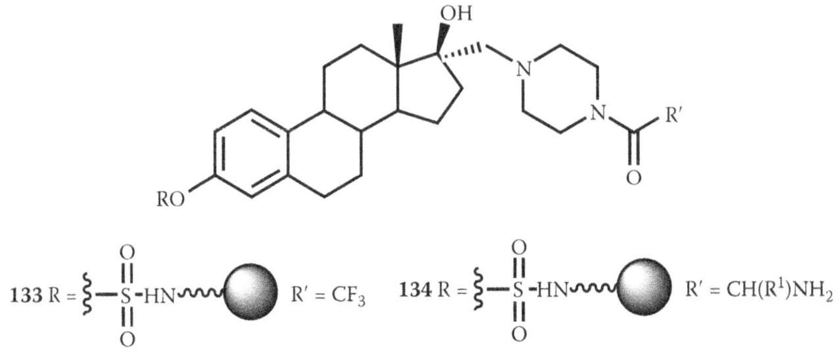

135 R = SO_2NH_2 R′ = $CH(R^1)NHCOR^2$ **136** R = H R′ = $CH(R^1)NHCOR^2$

FIGURE 11.55 Libraries of 17α-substituted estradiol sulfamates and phenols. (From Ciobanu, L.C. and Poirier, D., Solid-phase parallel synthesis of 17α-substituted estradiol sulfamate and phenol libraries using the multidetachable sulfamate linker, *J. Comb. Chem.*, 5, 429, 2003.)

As part of a program to develop probes for the hormone-binding domain of the estrogen receptor (ER-HBD), a series of 4-*para*-substituted phenylvinyl estradiol derivatives were prepared in order to provide information regarding the interactions between ligands and the estrogen receptor isoforms (Figure 11.56).[199] Three approaches were followed for the preparation of these derivatives. The solution-phase synthesis used the Stille coupling to introduce side-chain diversity in a stannated estradiol intermediate **137** prepared from ethynyl estradiol. In a solid-phase approach, ethynyl estradiol was hydrostannated to give predominantly the *E*-stannylvinyl derivative **138**. This intermediate was coupled to a carboxylated polystyrene resin to give intermediate **139**. Stille coupling with aryl halides followed by cleavage from the resin gave targeted estradiol derivatives **140**. A third approach utilized the Suzuki coupling reaction. This involved first performing iododestannylation of **137** to give an iodovinyl estradiol derivative. This intermediate then underwent facile Suzuki coupling with 4-fluorophenyl-boronic acid to afford, after hydrolysis, 17α-(4-fluorophenyl) vinyl estradiol. Most of these 4-*para*-substituted phenylvinyl estradiol derivatives displayed high relative binding affinity for Er-HBD with values in the range of 25 to 60% and exceeded the binding affinity of the unsubstituted parent.

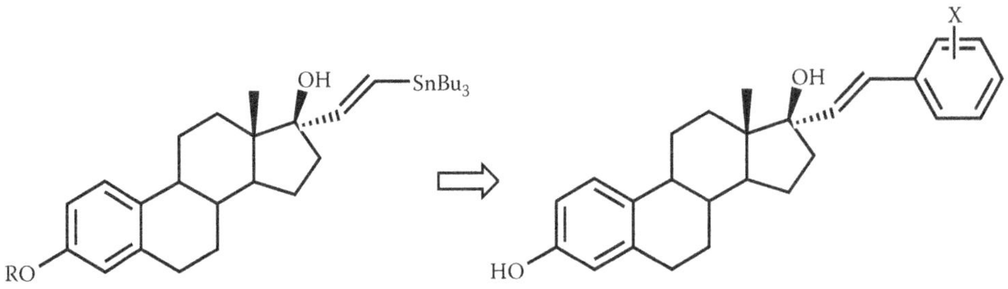

137 R = Ac
138 R = H
139 R = PS resin

140

FIGURE 11.56 Library of 4-*para*-substituted phenylvinyl estradiol derivatives. (From Hanson, R.N. et al., Synthesis and evaluation of 17-20*E*-21-(4-substituted phenyl)-19-norpregna-1,3,5(10), 20-tetraene-3,17β-diols as probes for the estrogen receptor a hormone binding domain, *J. Med. Chem.*, 46, 2865, 2003.)

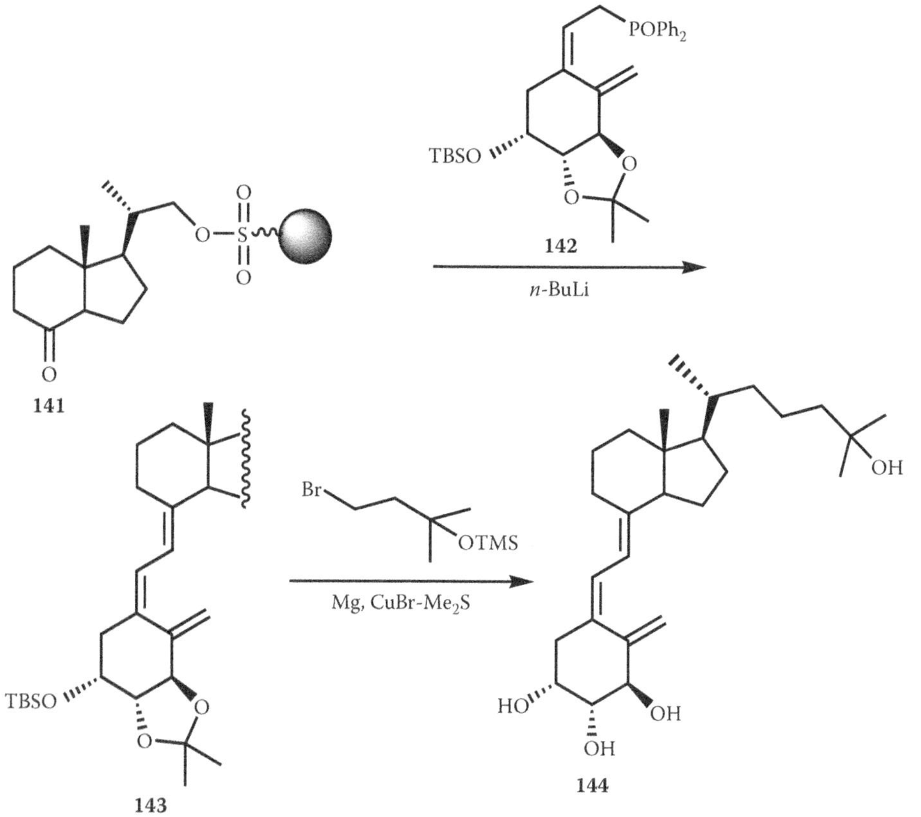

FIGURE 11.57 Solid-phase synthesis of a vitamin D$_3$ system. (From Hijikuro, I., Doi, T., and Takahashi, T., Parallel synthesis of a vitamin D$_3$ library in the solid-phase, *J. Am. Chem. Soc.*, 123, 3716, 2001.)

11.7.2 Vitamin D$_3$

The cholesterol derivative 1α,25-dihydroxyvitamin D$_3$ (Vitamin D$_3$) is a natural hormone that exhibits a variety of physiological activities such as regulation of calcium and phosphorous metabolism, cell differentiation and proliferation, and the immune system. In recent years, a number of analogs of vitamin D$_3$ were synthesized by many laboratories in order to investigate their structure–activity relationships and enhance specific activities. Takahashi and coworkers (see Chapter 10, Section 10.2) constructed a vitamin D$_3$ combinatorial library by means of a split-and-pool methodology utilizing radio-frequency-encoded combinatorial (REC) chemistry, and a manual parallel synthesizer for side-chain diversification and deprotection (Figure 11.57).[200] Horner–Wadsworth–Emmons olefination of the solid-supported CD-ring building block **141** with lithiated A-ring phosphine oxide **142** followed by simultaneous introduction of the side chain and cleavage from resin with a Cu (I)-catalyzed Grignard reaction afforded the vitamin D$_3$ analog **144**. Three and four CD- and A-ring building blocks, respectively, and six different side chains were used in a parallel synthesis to generate a 72-member vitamin D$_3$ combinatorial library.

11.8 ALKALOIDS

11.8.1 Indoles

One of the earliest reports on the combinatorial modification of natural products described the derivatization of the Rauwolfia indole alkaloids yohimbine and rauwolscine (Figure 11.58).[201] The

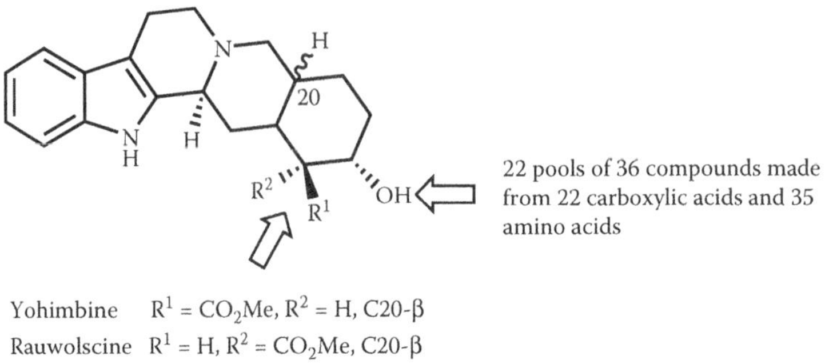

22 pools of 36 compounds made from 22 carboxylic acids and 35 amino acids

Yohimbine $R^1 = CO_2Me$, $R^2 = H$, C20-β
Rauwolscine $R^1 = H$, $R^2 = CO_2Me$, C20-β

FIGURE 11.58 *Rauwolfia* alkaloids library — sites of oriented diversity. (From Atuegbu, A. et al., Combinatorial modification of natural products: preparation of unencoded and encoded libraries of *Rauwolfia* alkaloids, *Bioorg. Med. Chem. Lett.*, 4, 1097, 1996.)

alkaloid templates were attached to Tentagel resin-bound α-amino acid via the E-ring carboxylic group and subsequently derivatized with a range of carboxylic acids and amino acids to afford a 792-member library.

The Pictet–Spengler reaction has remained a valuable synthetic tool for the preparation of indole-based compounds from resin-bound tryptophan.[202] Koomen and coworkers[203,204] reported a solid-phase synthesis toward fumitremorgin analogs based on a cyclization/cleavage strategy.[205] Extensively used to generate compounds without leaving any trace of the linker used for tethering to solid support, the synthesis started with resin-bound L-tryptophan (Figure 11.59). A library of 42 diastereoisomeric mixtures of fumitremorgin-type indolyl diketopiperazines was prepared by solid-phase synthesis (Figure 11.59), and screened for breast cancer resistance protein (BCRP) inhibitory activity. Interesting, potent, and selective leads were identified. A similar solid-phase approach was described for the synthesis of demethoxy fumitremorgin, the most active of this series of alkaloids, allowing the use of a larger variety of aldehydes for the Pictet–Spengler condensation with *N*-acyliminium species.[206]

In a subsequent study, Bonnet and Ganesan[207] employed a modified *N*-acyliminium Pictet–Spengler reaction to synthesize 1,2,3,4-tetrahydro-β-carboline hydantoins, small-molecule inhibitors of cell division (see Chapter 2). A highly diastereoselective solid-phase Pictet–Spengler reaction for the synthesis of 1,2,3,4-tetrahydro-β-carboline hydantoins was reported by Kapoor et al.[208] They investigated the effects of varying the N-protecting group of the solid support on the diastereoselectivity of the Pictet–Spengler reaction. They found that the allyl group, easily removed with $Pd(PPh_3)_4$, furnished the tetrahydro-β-carboline group with high diastereoselectivity. The 1,2,3,4-tetrahydro-β-carboline hydantoins were formed by reaction with isocyanates and simultaneous cyclorelease of the products from solid support.

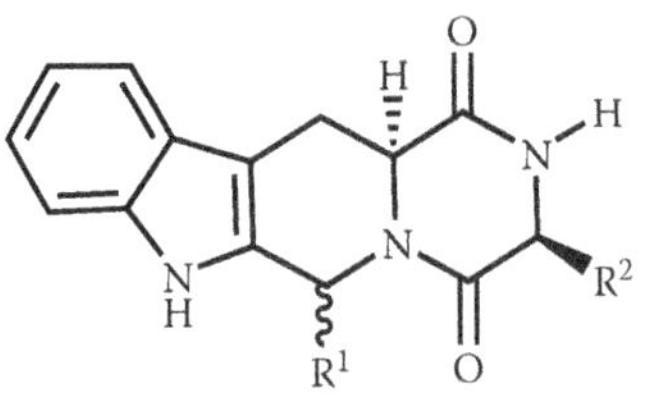

FIGURE 11.59 1,2,3,4-Tetrahydro-β-carboline-derived diketopiperazines library. (From Kundu, B., Solid-phase strategies for the design and synthesis of heterocyclic molecules of medicinal interest, *Curr. Opin. Drug Discovery Dev.*, 6, 815, 2003.)

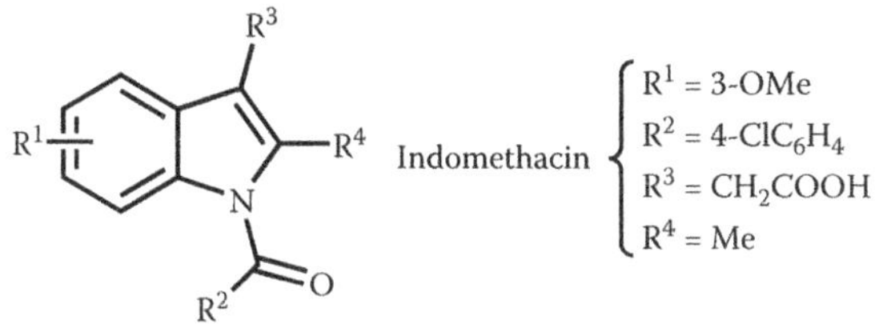

Indomethacin
$$R^1 = 3\text{-OMe}$$
$$R^2 = 4\text{-ClC}_6\text{H}_4$$
$$R^3 = \text{CH}_2\text{COOH}$$
$$R^4 = \text{Me}$$

FIGURE 11.60 Indomethacin library. (From Rosenbaum, C. et al., Synthesis and biological evaluation of an indomethacin library reveals a new class of angiogenesis-related kinase inhibitors, *Angew. Chem. Int. Ed. Engl.*, 43, 224, 2004.)

The extensive number of indole-based biologically active natural products and indole-derived drugs are of special interest in the arenas of chemical biology and medicinal chemistry research. Waldmann and coworkers developed a synthesis of an indomethacin library,[209] active indole derivatives for the treatment of, for example, pain, arthritis, cardiovascular diseases, and Alzheimer's disease, and for the treatment and prevention of cancer (Figure 11.60).[210–214] A library of 197 indomethacin analogs was synthesized on solid support starting from a polystyrene aldehyde resin and using a resin-capture-release Fisher indole synthesis reaction sequence. Of the 134 compounds investigated, 6 indomethacin indole derivative were inhibitors of angiogenesis receptor tyrosine kinases.

Willoughby and coworkers reported a solid-phase strategy for the preparation of 2-arylindole combinatorial libraries in a split-and-pool format (Figure 11.61).[215] Twenty different arylalkyl ketoacids were immobilized onto a sulfonamide resin, mixed, and separated into twenty equal portions. The Fischer indole cyclization was carried out with 20 arylhydrazines in the presence of $ZnCl_2$ and acetic acid. The resin was mixed, separated into 80 equal portions, and finally alkylated under Mitsunobu conditions. A library of 12,800 compounds was generated, and led to the discovery of potent ligands for a variety of G-protein coupled receptors.

In order to generate libraries of potential activators and inhibitors of protein kinase C (PKC),[216] Waldmann and coworkers developed a solid-phase synthesis of teleocidins.[217,218] In particular, (–)-indolactam V, a metabolite that possesses the core structure of the tumor-promoting teleocidins and which has been recognized as a PKC activator, and its N-13 des(methyl) analogs were targeted (Figure 11.62). Based on known structural requirements of indolactams to bind to PKC, a 31-member library of analogs with substituents at C-12, C-7, and N-13 was prepared. Some analogs were as effective as natural PKC activators such as phorbol dibutyrate and (–)-indolactam V.

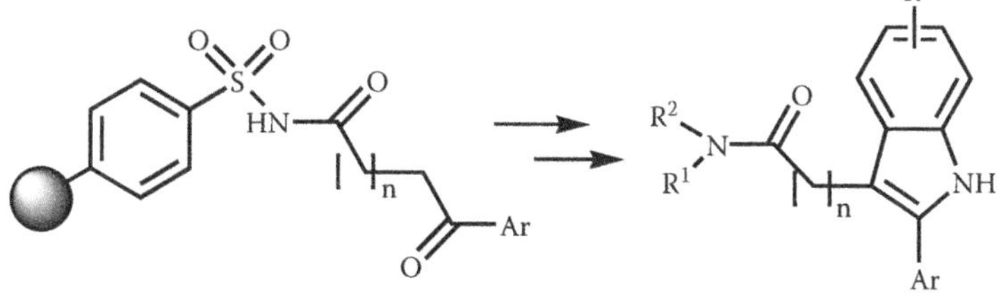

FIGURE 11.61 2-Arylindole combinatorial library. (From Willoughby, C.A. et al., Combinatorial synthesis of 3-(amidoalkyl) and (aminoalkyl)-2-arylindole derivatives: discovery of potent ligands for a variety of G-protein coupled receptors, *Bioorg. Med. Chem. Lett.*, 12, 93, 2002.)

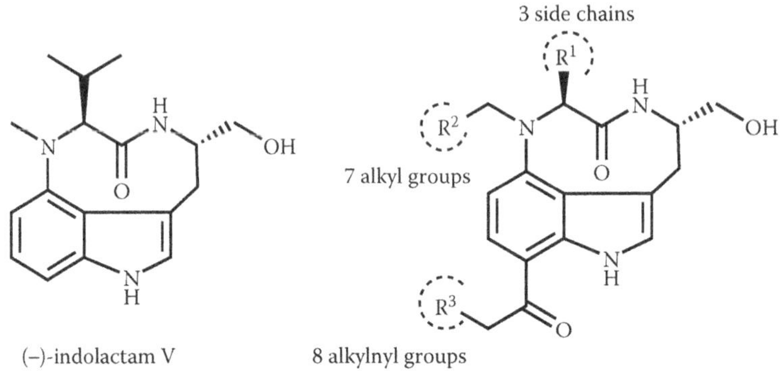

FIGURE 11.62 Indolactam derivatives. (From Meseguer, B. et al., Natural product synthesis on polymeric supports-synthesis and biological evaluation of an indolactam library, *Angew. Chem. Int. Ed. Engl.,* 38, 2902, 1999.)

11.8.2 INDOLINES

In addition to indoles, there are several natural products known to possess the indoline structure. A number of derivatives exhibit a wide range of biological activities.[219] Arya et al. reported the use of a hydroxyindoline-derived scaffold to generate a library of indoline-based natural product-like tricyclic derivatives (Figure 11.63).[220] The synthesis was achieved by first immobilizing the hydroxyindoline scaffold onto solid support, and performing a series of reactions, including a Mitsunobu reaction. Using the IRORI split-and-mix approach, two sites of diversity were introduced. The library is being screened for eukaryotic protein translation synthesis inhibition.

11.8.3 SPIROXINDOLES

Schreiber and coworkers[221] reported the split-pool synthesis of more than 3000 spirooxindoles (Figure 11.64). This interesting scaffold exhibiting antimitotic activity is found in some tryptophan-derived natural products isolated from the fermentation broth of *Aspergillus fumigatus.*[222] The key reaction in assembling the spirooxindole core is a stereoselective Lewis acid variant of the Williams' three-component coupling reaction.[223] After formation of the spirocyclic skeleton, Sonogashira couplings, amide formation, and *N*-acylations of the lactam moiety enabled further derivatization. To demonstrate the utility of the overall discovery process, a cell-based screen was performed with library members in order to identify enhancers of the growth arrest induced by latrunculin B, a natural product that sequesters monomeric actin and prevents the formation of actin microfilaments. Through resynthesis, one of the active compounds was confirmed and found to have an EC_{50} value in the submicromolar range.

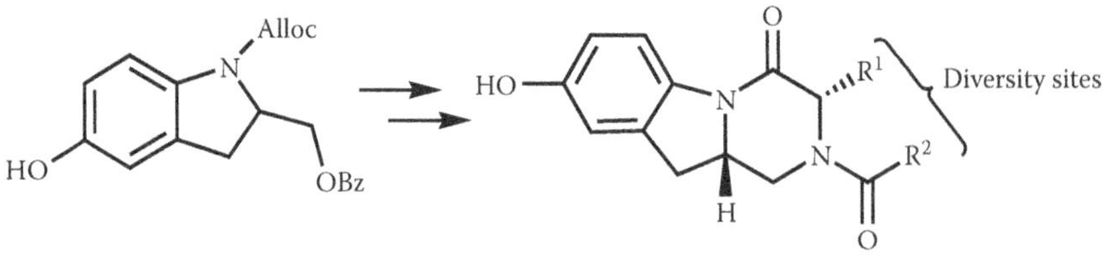

FIGURE 11.63 Library synthesis of indoline-derived tricyclic derivatives using a Mitsunobu approach. (From Arya, P. et al., A solid-phase library synthesis of hydroxyindoline-derived tricyclic derivatives by Mitsunobu approach, *J. Am. Chem. Soc.,* 6, 65, 2004.)

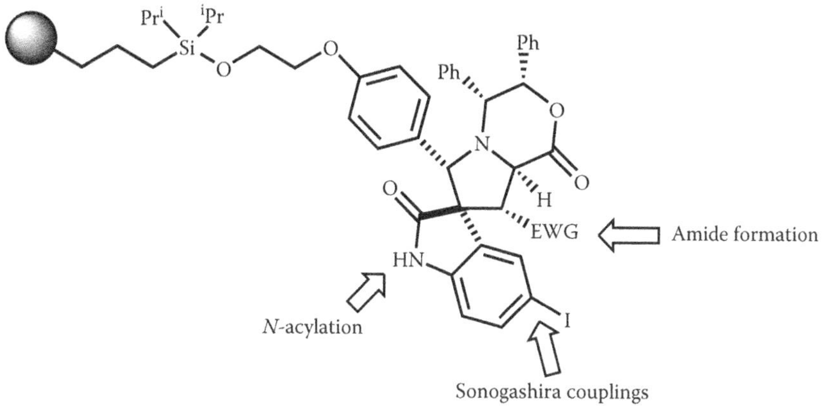

FIGURE 11.64 A spirooxindole library. (From Lo, M.M.-C. et al., A library of spirooxindoles based on a stereoselective three-component coupling reaction, *J. Am. Chem. Soc.*, 126, 16077, 2004.)

11.8.4 Mappicine and Nothapodytine B Analogs

(*S*)-Mappicine[224] and its ketone analog nothapodytine B[225] are two tryptophan-derived alkaloids with antiviral activity against herpes viruses (HSV) and human cytomegalovirus (HCMV). Curran and coworkers reported the generation of a 128-member solution-phase library of mappicine analogs (64 racemates) and a 48-member library of nothapodytine B analogs based on a radical cascade annulation.[226] Recently, Curran reported a new solution-phase synthesis of 560 library mappicine analogs using fluorous mixture chemistry (Figure 11.65).[227]

Fluorous mixture synthesis is the first solution-phase mixture technique that allows the isolation of individual pure components at the end of the mixture exercise. Briefly, a series of *n* initial substrates (S^1–S^n) are tagged with different fluorous tags (F^1–F^n). The tagged substrates (S^1F^1–S^nF^n) are mixed to give a single mixture (M1) that is taken through a series of mixture steps. The mixture steps can be either one-pot or split-parallel reactions with a set of *x* building blocks to generate *x*

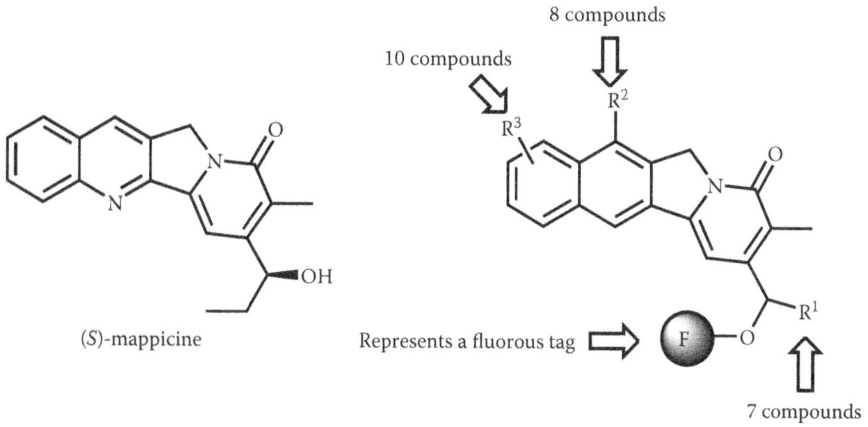

FIGURE 11.65 Library of mappicine analogs. (From Zhang, W. et al., Solution phase preparation of a 560-compound library of individual pure mappicine analogs by fluorous mixture synthesis, *J. Am. Chem. Soc.*, 124, 10443, 2002.)

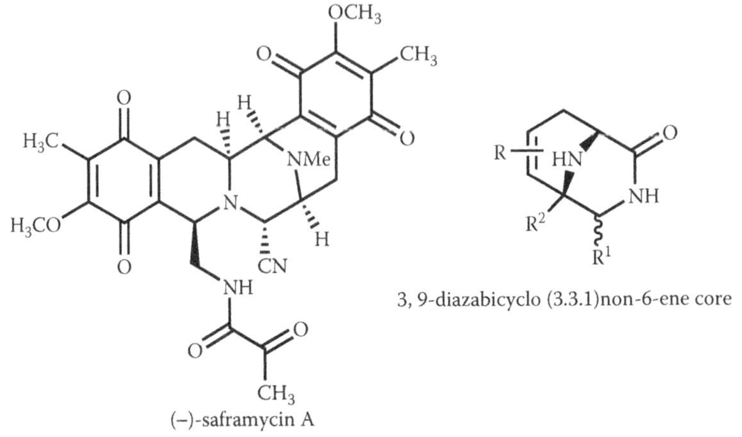

(−)-saframycin A

3, 9-diazabicyclo (3.3.1)non-6-ene core

FIGURE 11.66 Saframycin A structure and 3,9-diazabicyclo[3.3.1]non-6-ene core. (From Myers, A.G. and Lanman, B.A., A solid-supported, enantioselective synthesis suitable for the rapid preparation of large numbers of diverse structural analogs of (−)-saframycin A, *J. Am. Chem. Soc.*, 124, 12969, 2002; Orain, D., Koch, G., and Giger, R., From solution phase studies to solid-phase synthesis: a new indole based scaffold for combinatorial chemistry, *Chimia,* 57, 255, 2003.)

numbers of new mixtures. At the end of the synthesis, all of the mixtures are separated chromatographically over fluorous silica gel. For the preparation of the mappicine analogs library, a seven-component mixture was carried through a four-step mixture synthesis (two one-pot and two parallel steps) to incorporate two additional points of diversity onto the tetracyclic core.

11.8.5 SAFRAMYCIN ANALOGS

Saframycin A, a structurally complex alkaloid with a polycyclic framework, is a member of a series of natural antiproliferative agents that shows promising clinical efficacy in the treatment of solid tumors (Figure 11.66).[228] Saframycin, safracin, renieramycin, and ecteinascidin families, which show powerful antiproliferative and antitumor properties, have a 3,9-diazabicyclo [3.3.1]non-6-ene core structural element in common. Recently Myers and Lanman reported a very elegant solid-phase synthesis of a small series of 23 (−)-saframycin A analogs that included 16 which were prepared by simultaneous parallel synthesis.[229] The ten-step solid-phase sequence involved two highly diastereo-selective Pictet–Spengler reactions and a final cyclorelease of *bis*-tetrahydroisoquinoline. Furthermore, Giger et al. reported solution- and solid-phase syntheses of polycyclic compounds bearing the common 3,9-diazabicyclo[3.3.1]non-6-ene core (Figure 11.66).[230] The key step of the synthesis was a novel sequential Dakin–West/Pictet–Spengler reaction. The feasibility of further functionalization of the bridgehead nitrogen by solid-phase acylation and alkylation was also demonstrated.

11.8.6 TETRAHYDROQUINOLINES

Several naturally occurring tetrahydroquinolines possess interesting biological activities, such as the antitumor antibiotic dynemycin and a bradykinin antagonist isolated from the tropical plant *Martinella iquitosensis* (Figure 11.67).[219] Tetrahydroquinolines offer most of the required properties for a combinatorial scaffold — rigid and compact, hydrogen bond donor and acceptor sites, and aromatic rings. Baudelle et al. developed a parallel solution-phase synthesis of polysubstituted tetrahydroquinolines utilizing a multiple-component condensation (MCC) with aldehydes, amines, and alkenes (Figure 11.67).[231,232] Following the preceding strategy, these authors prepared a lead discovery library containing 1920 diastereoisomeric pairs of tetrahydroquinolines prepared in a 96-well plate format, using a high-throughput robot starting with 80 aldehydes, 24 anilines, and

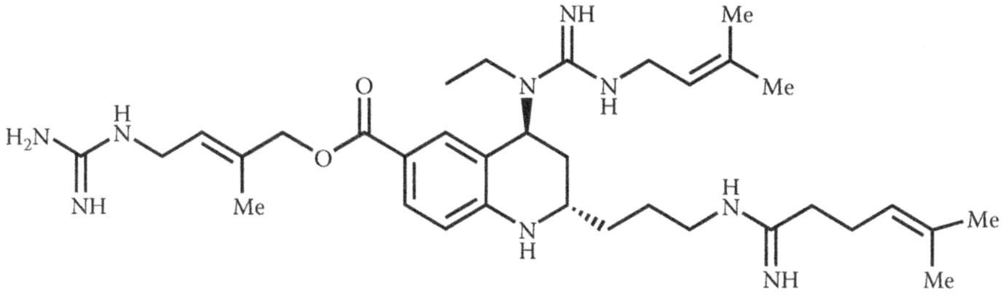

Bradykinin antagonist isolated from *Martinella iquitosensis*

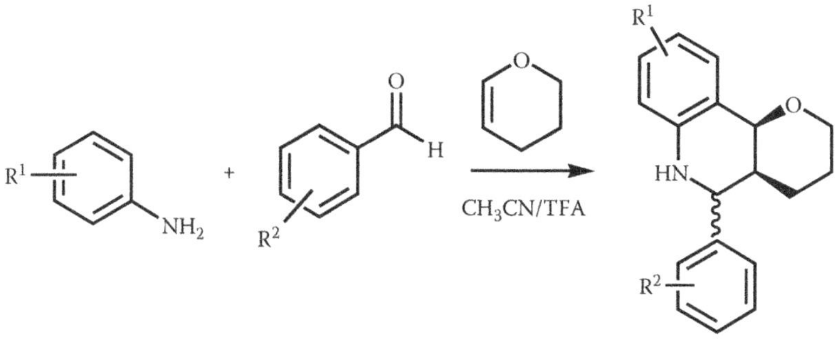

FIGURE 11.67 Polysubstituted tetrahydroquinolines library. (From Baudelle, R. et al., Parallel synthesis of polysubstituted tetrahydroquinolines, *Tetrahedron*, 54, 4125, 1998.)

2,3-dihydrofuran.[232] A similar methodology was developed by Kobayashi et al.[233] and Armstrong et al.[234] using Ln(OTf)₃ and Yb(OTf)₃ in solution phase and solid phase, respectively, as the Lewis acid catalyst for three-component reactions between aldehydes, amines and alkenes. To develop a library synthesis of tetrahydroquinoline-derived natural product-like small molecules, Arya et al. reported the synthesis of enantiomerically pure tetrahydroquinoline scaffold (Figure 11.68).[235] The

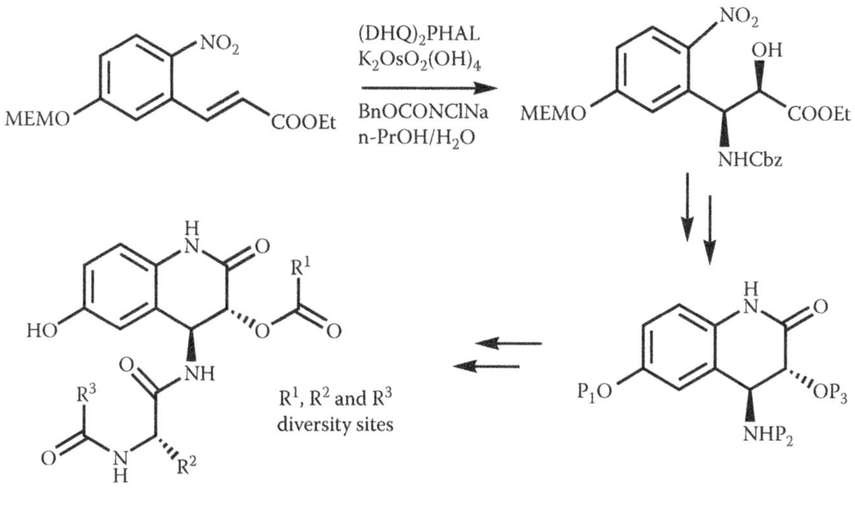

FIGURE 11.68 Library of tetrahydroquinoline-derived natural product-like derivatives. (From Couve-Bonnaire, S. et al., A solid-phase library synthesis of natural product-like derivatives from an enantiomerically pure tetrahydroquinoline scaffold, *J. Am. Chem. Soc.*, 6, 73, 2004.)

key step involved an asymmetric aminohydroxylation reaction. This scaffold was immobilized onto solid support to diversify the amino alcohol functionality following an IRORI split-and-mix-approach for the library generation. Further development led to complex, polycyclic, natural product-like derivatives that utilized a tetrahydroquinoline-based scaffold as the starting material.[236,237]

11.8.7 ISATINS

Ivachtchenko et al. used functionalized isatins as versatile building blocks for polycyclic fused heteroaromatics with potential biological activity (Figure 11.69).[238] Two combinatorial libraries of 2-arylimidazotriazinoindoles (12-member) and 3-(alkylthio)triazinoindoles (11-member) were generated from reaction of isatins derivatives with 1,2-diaminoimidazole and aminoguanidine hydrochloride. These herteroaromatics are interesting because of their potential biological activity in analogy to the antiviral drug VP32947 and the DNA intercalating drug/DNA topoisomerase II inhibitor NCA0424 (Figure 11.69). The study seems promising for the generation of a wide variety of small-molecule combinatorial libraries.

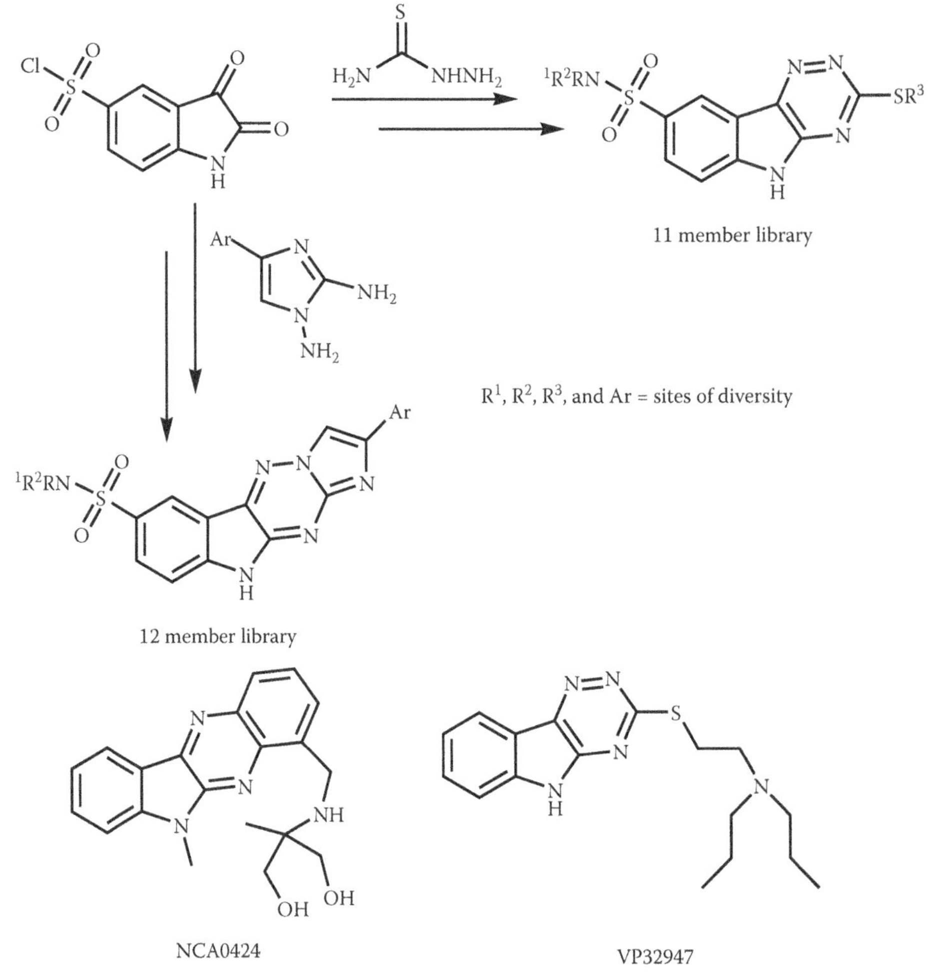

FIGURE 11.69 5-Sulfamoylisatins as promising scaffolds for the syntheses of new combinatorial libraries of various heterocycles. (From Ivatchenko, A.V. et al., New scaffolds for combinatorial synthesis. 1. 5-Sulfamoylisatins and their reactions with 1,2-diamines, *J. Comb. Chem.*, 4, 419, 2002.)

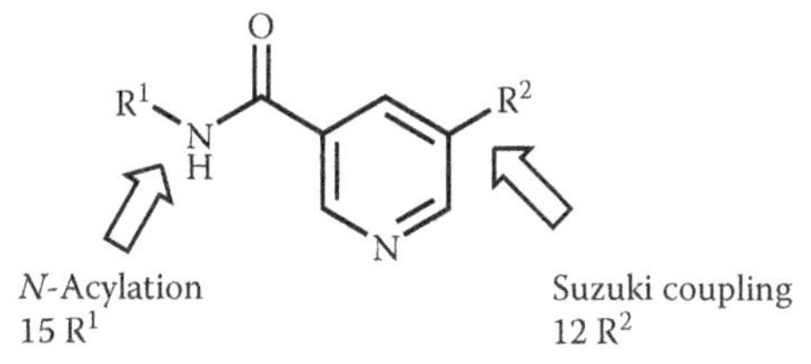

FIGURE 11.70 5-Substituted nicotinic acid and nicotinamide derivatives library. (From Fernàndez, J.-C. et al., Suzuki coupling reaction for the solid-phase preparation of 5-substituted nicotinic acid derivatives, *Tetrahedron Lett.*, 46, 581, 2005.)

11.8.8 NICOTINIC ACIDS

Nicotinic acid is a B vitamin that has useful clinical applications for treating hyperlipidemia, and several of its derivatives (e.g., nicotinamide) have been evaluated as drug carriers in chemical delivery systems.[239] A very simple method was reported by Fernàndez et al. for the preparation of 5-substituted nicotinic acid derivatives in extremely good yield and high purity on three different solid supports via the Suzuki cross-coupling reaction (Figure 11.70).[240] Briefly, a primary amine was attached to a functionalized polystyrene resin via reductive amination and used to make alkylamino resins. Subsequent acylation with 5-bromonicotinic acid and Suzuki coupling gave a small library of 180 secondary nicotinamides.

11.8.9 BISPIDINES

The bispidine skeleton constitutes the B and C rings of a variety of lupanine alkaloids such as sparteine (Figure 11.71). The parallel solution-phase synthesis of libraries containing the bispidine (3,7-diazabicyclo [3.3.1]nonane) moiety[241] as the core structure with three points of diversity were synthesized starting from (–)-cytisine, readily available from natural sources[242] or by synthesis.[243] (–)-Cytisine was subjected to a catalytic hydrogenation followed by alkali-mediated ring cleavage. Applying the "library from library" concept, a series of novel combinatorial libraries containing the bispidine moiety, including a 25-member library of 3-*N*-acylated derivatives and a 23-member library of 3-*N*-sulfonylated derivatives, were prepared. These libraries represent a very useful source of bispidine derivatives for biological testing and further derivatization to novel polycyclic structures.

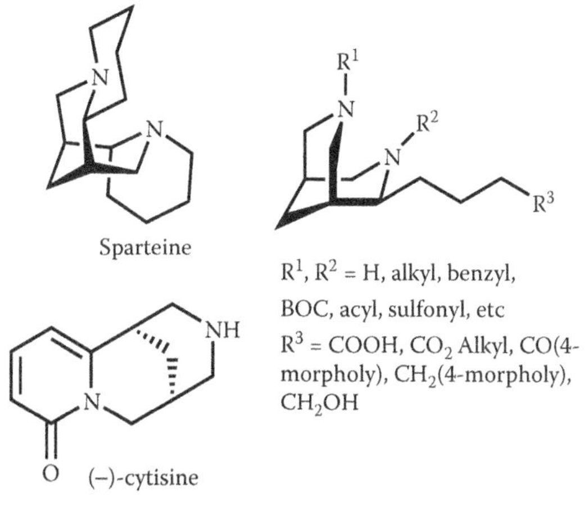

Sparteine

R^1, R^2 = H, alkyl, benzyl, BOC, acyl, sulfonyl, etc

R^3 = COOH, CO_2 Alkyl, CO(4-morpholy), CH_2(4-morpholy), CH_2OH

(–)-cytisine

FIGURE 11.71 Bispidine derivatives library. (From Ivachtchenko, A.V. et al., A parallel solution phase synthesis of substituted 3,7-diazabicyclo[3.3.1]nonanes, *J. Comb. Chem.*, 6, 828, 2004.)

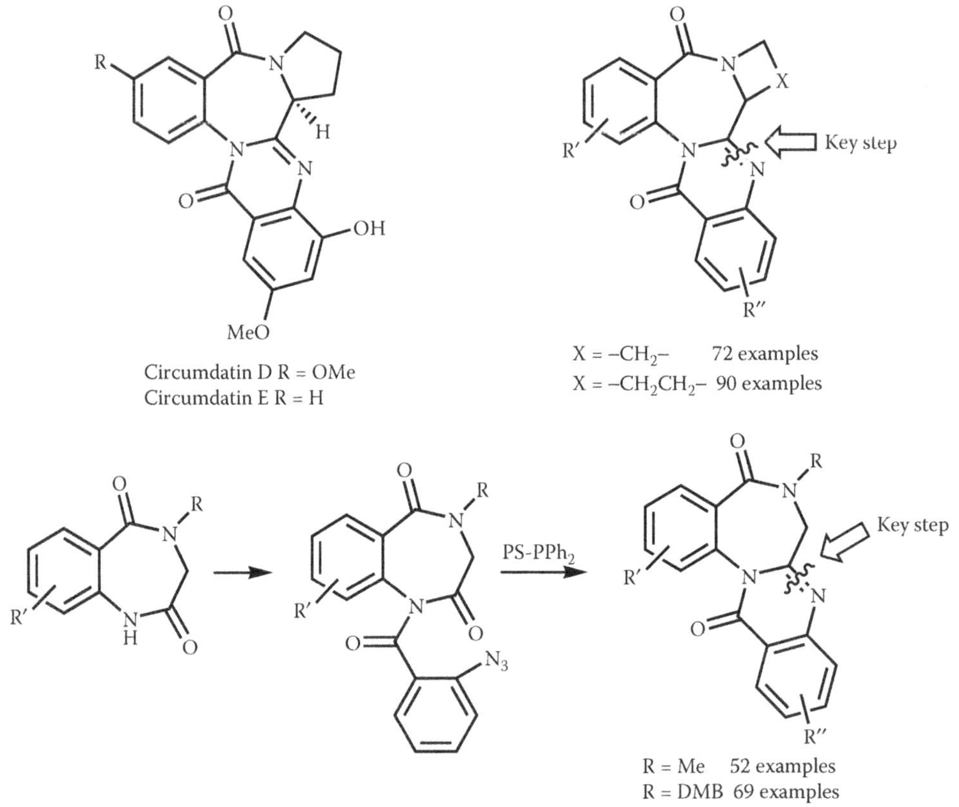

Circumdatin D R = OMe
Circumdatin E R = H

X = –CH$_2$–   72 examples
X = –CH$_2$CH$_2$– 90 examples

PS-PPh$_2$

Key step

R = Me   52 examples
R = DMB 69 examples

FIGURE 11.72 Sublibraries of the circumdatin family of natural products. (From Grieder, A. and Thomas, A.W., A concise building block approach to a diverse multi-arrayed library of the circumdatin family of natural products, *Synthesis,* 11, 1707, 2003.)

11.8.10 CIRCUMDATINS

Grieder and Thomas prepared a diverse library of natural product-like benzodiazepine-quinazolinone alkaloids similar to the family of circumdatins (Figure 11.72).[244] Starting from benzodiazepinedione derivatives, the synthesis relied on a modified Eguchi protocol using a polymer-supported phosphine-mediated intramolecular aza-Wittig reaction in a key step of the reaction sequence. Following this route two sublibraries of tricyclic derivatives similar to the circumdatins D and E were prepared.

11.8.11 GALANTHAMINES

Based on the Amaryllidaceae alkaloid galanthamine, a biomimetic solid-phase synthesis of 2527 compounds was reported by Shair and coworkers (Figure 11.73).[245] The core scaffold, initially prepared in several steps, was diversified by means of four successive reactions: Mitsunobu reaction of the phenolic moiety with five primary alcohols, Michael addition of the α, β-unsaturated cyclohexenone with thiols, N-acylation or N-alkylation of the cyclic secondary amine, and treatment of the ketone with hydrazines and hydroxylamines. Further evaluation of library constituents for their ability to block protein trafficking in the secretory pathway of mammalian cells led to the discovery of sercramine as a potent inhibitor of the VSVG-GFP protein movement from the Golgi apparatus to the plasma membrane.

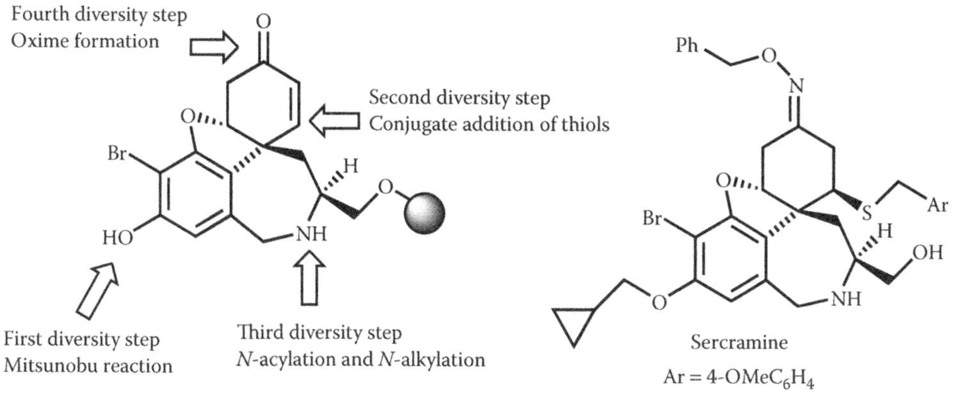

FIGURE 11.73 Diversity-oriented synthesis of a galanthamine-based library. (From Pelish, H.E. et al., Use of biomimetic diversity-oriented synthesis to discover galanthamine-like molecules with biological properties beyond those of the natural product, *J. Am. Chem. Soc.*, 123, 6740, 2001.)

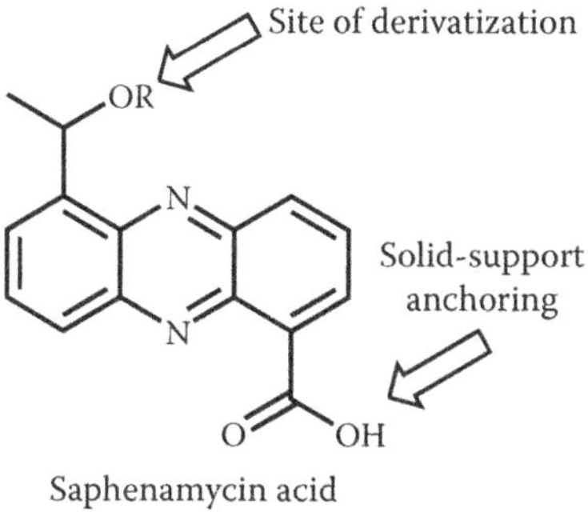

FIGURE 11.74 Saphenamycin-based library. (From Laursen, J.B. et al., Solid-phase synthesis of new saphenamycin analogs with antimicrobial activity, *Bioorg. Med. Chem. Lett.*, 12, 171, 2002.)

11.8.12 SAPHENAMYCINS

Saphenamycin A, a tricyclic heteroaromatic phenazine that exhibits potent antimicrobial activity toward a broad range of bacteria, was isolated from the marine microorganism *Streptomyces antibioticus* strain Tü 2706 (Figure 11.74).[246,247] Nielsen and coworkers synthesized on solid support an array of 12 saphenamycin analogs modified at the benzoate moiety.[248] The secondary alcohol was acylated with a series of substituted benzoic acids after chemoselective anchoring of saphenamycin acid to solid-support through its carboxylic group. Eight analogs exhibited MIC values against *Bacillus subtilis* from 0.07 to 3.93 µg/ml.

11.9 FLAVONOIDS

11.9.1 BENZOPYRANS

The benzopyran moiety is a structural motif of flavonoids, one of the largest group of naturally occurring phenols. Found in many medicinally relevant compounds, the benzopyran is a privileged structure,[249–253] a term originally introduced to describe select structural types that bind to multiple, unrelated classes of proteins receptors as high-affinity ligands.[249] So far, most benzopyran combinatorial libraries (see Chapter 1, Section 1.4) have focused on the synthesis of coumarin and chromone analogs.[250–256]

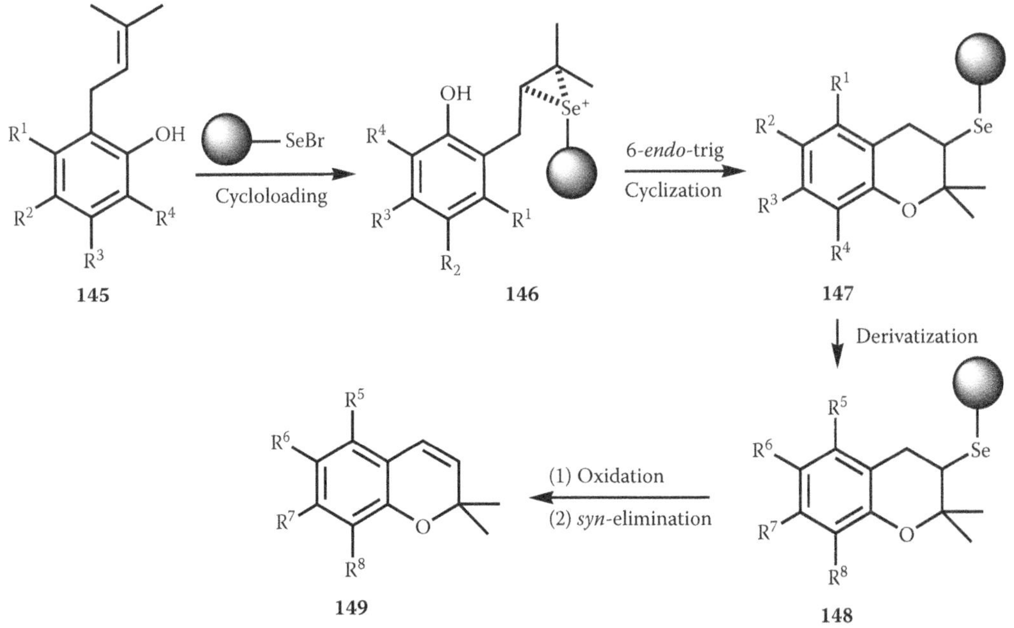

FIGURE 11.75 Synthesis of 2,2-dimethylbenzopyran scaffolds. (From Nicolaou, K.C., Cao, G.-Q., and Pfefferkorn, J.A., Selenium-based solid-phase synthesis of benzopyrans II: applications to combinatorial synthesis of medicinally relevant small organic molecules, *Angew. Chem. Int. Ed. Engl.*, 39, 739, 2000; Nicolaou, K.C. et al., Natural product-like combinatorial libraries based on privileged structures. 2. Construction of a 10,000-membered benzopyran library by directed split-and-pool chemistry using NanoKans and optical encoding, *J. Am. Chem. Soc.*, 122, 9954, 2000.)

Employing a cycloloading strategy that relied on the use of a polystyrene-based selenenyl bromide resin, Nicolaou and coworkers synthesized the 2,2-dimethylbenzopyran motif **149** as a template for the construction of libraries of chalcones, pyranocoumarins, chromene glycosides, stilbenoids, polycyclic steroid biosynthesis inhibitors, *N*-heterocycles, and pyranoflavones (Figure 11.75).[250–253] The *ortho*-prenylated phenol **145** was attached to the solid support, and the resultant compound **146** then underwent a 6-*endo*-trig cyclization to furnish resin-bound benzopyrans **147**. After derivatization via condensation, annulation, glycosidation, aryl/vinyl couplings, or organometallic addition reactions to provide **148**, oxidation of the selenium to selenoxide and spontaneous *syn*-elimination occurred to provide the dihydrobenzopyrans **149**. The linkage through the pyran ring allowed diversification of all four positions of the benzene ring in contrast to more traditional linking strategies.

Based on the 2,2-dimethylbenzopyran template, a family of chalcones **152**, which includes the Cubé resin components lonchocarpin, 4-hydroxylonchocarpin, 4-hydroxy-3-methoxy-lonchocarpin, and paratocarpin,[257–260] with important biological activities was synthesized. Condensation of a variety of 11 aldehydes **151** with a set of resin-bound benzopyrans bearing a methyl ketone substituent **150** followed by cleavage from the resin by hydrogen-peroxide-promoted selenoxide elimination gave chalcones **152** (Figure 11.76).[252]

A second library demonstrated the construction of linear and angular pyranocoumarins similar to the Sri Lankan bioactive metabolites seselin, xanthyletin, and xanthoxyletin (Figure 11.77).[97,252,260] The library was prepared by a split-pool strategy with the aid of radio frequency-encoded IRORI tags and MacroKans.[97,261] A set of five resin-bound benzopyrans possessing an *o*-hydroxy aldehyde functionality **153** were treated with aryl, alkyl, or alkoxy β-ketoesters **154** to provide the lactones **157** or **158** through a Knoevanagel condensation and concomitant transesterification. Alternatively,

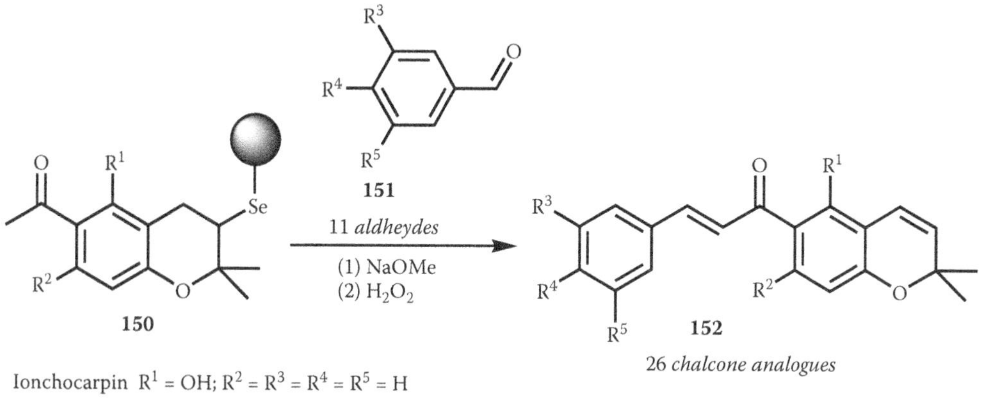

Ionchocarpin R^1 = OH; R^2 = R^3 = R^4 = R^5 = H

4-hydroxylonchocarpin R^1 = R^4 = OH; R^2 = R^3 = R^5 = H

4-hydroxy-3-methoxy-lonchocarpin R^1 = R^4 = OH; R^2 = R^5 = H; R^3 = OMe;

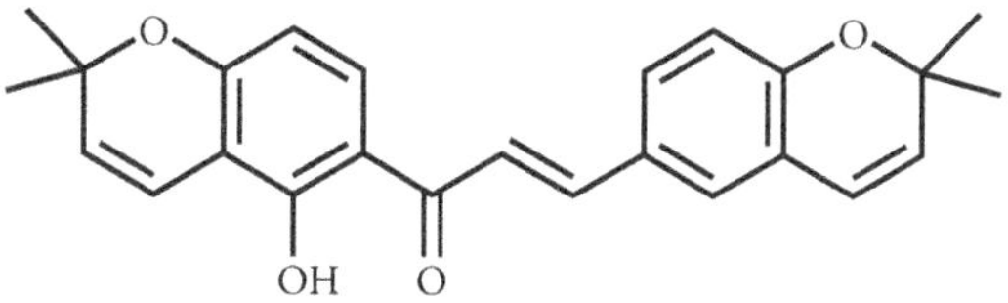

Paratocarpin A

FIGURE 11.76 Chalcone libraries. (From Nicolaou, K.C. et al., Natural product-like combinatorial libraries based on privileged structures. 1. General principles and solid-phase synthesis of benzopyrans, *J. Am. Chem. Soc.*, 122, 9939, 2000.)

these lactones were formed by coupling **153** to four different phenylacetic acids **155** or stabilized Wittig reagents **156**. At this stage, structures with no additional functionality were cleaved under standard conditions. Phenol substituents on the scaffold were further derivatized with bromide **159** or by Mitsunobu reaction with alcohol **160**.

A third library was constructed based on a chromene glycoside isolated from *Ageratum conyzoides* (Figure 11.78).[262] Using IRORI radio-frequency tags and MacroKans, D-glucose, D-xylose, and L-rhamnose trichloroacetimidates **164** were coupled onto three types of phenol-containing scaffolds **163** to afford selectively the β-glycosides **165**. After acetylation and cleavage from the resin, the chromene glycoside and eight analogs **166** were obtained.[252] As an alternative to the radio-frequency tags, the authors introduced the use of the IRORI NanoKan optical encoding system for high-throughput nonchemical tagging and sorting of library members during split-and-pool synthesis. The integration of NanoKan microreactors in the solid-phase method described in the preceding text allowed the rapid construction of a 10,000-membered benzopyran library.[253]

Initial screening of this library for farnesoid X receptor (FXR) activation utilizing a cell-based reporter assay led to the identification of several lead compounds possessing low micromolar activity (EC_{50} values = 5 to 10 μM). Guided by the preliminary structure–activity relationships gained from the evaluation of this initial library, a follow-up focused library of about 200 benzopyran-based compounds was synthesized.[256] The most active compounds (**167** and **168**) discovered from this second round of screening contained three regions that were systematically optimized by parallel solution- and solid-phase syntheses in order to evaluate the structural requirements for potent FXR agonism (Figure 11.79). These studies indicated that in region I, methyl acrylate or allylic methyl ether is necessary for optimum activity. In some instances, when other areas are optimized, the olefin could be removed with some potency being retained. In region II, an amide or urea was essential for maximum activity. Region III must have a functional group in the *para*-position for activity.

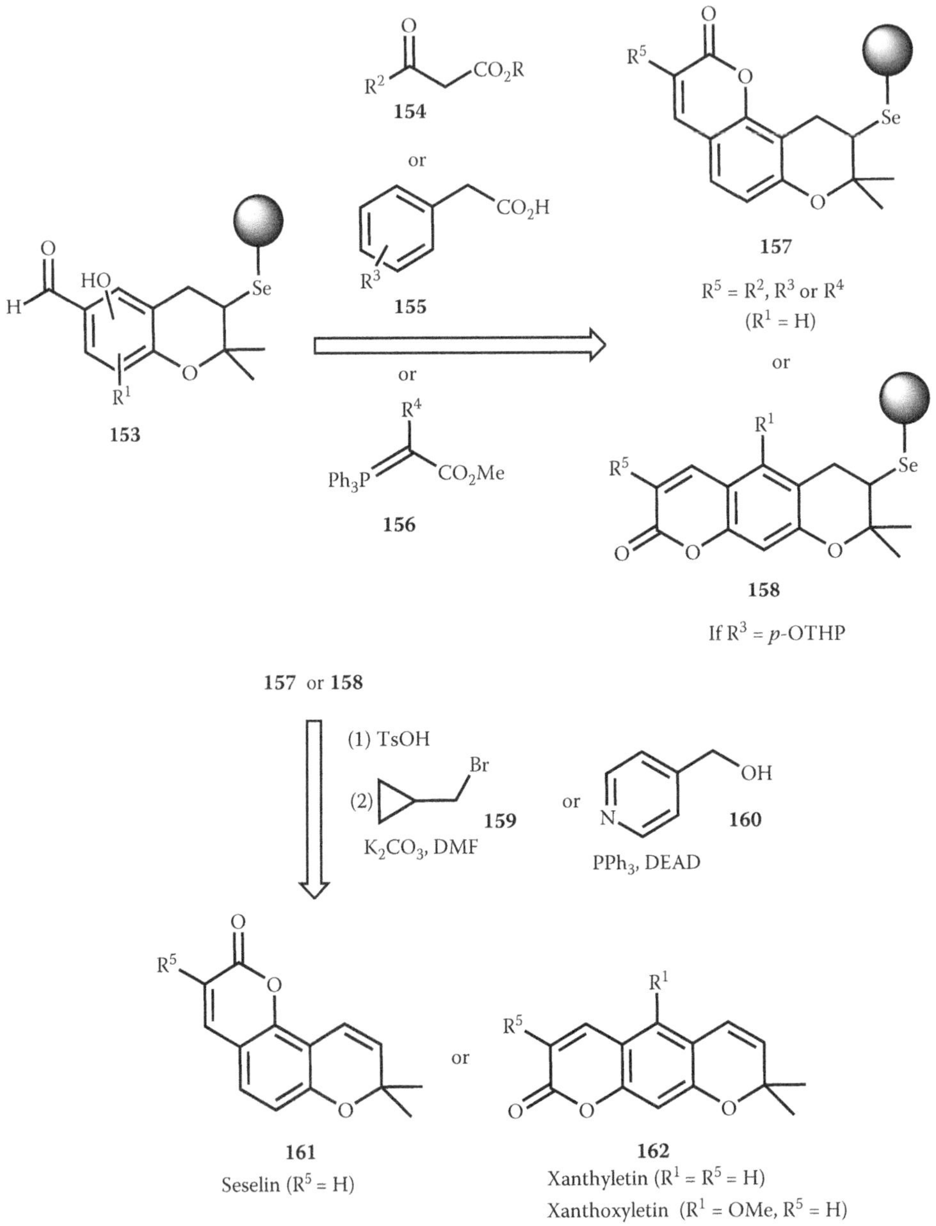

FIGURE 11.77 Chalcone and pyranocoumarin libraries. (From Nicolaou, K.C. et al., Natural product-like combinatorial libraries based on privileged structures. 1. General principles and solid-phase synthesis of benzopyrans, *J. Am. Chem. Soc.*, 122, 9939, 2000.)

Because many natural and synthetic benzopyrans with modifications of the pyran olefin exhibit higher activity than (or different activity from) that of the parent molecule (e.g., β-lapachone, (+)-kellactone), the Nicolaou group developed a solution-phase "library-from-library" strategy. Previous libraries were employed for further derivatization at the pyran olefin, thereby increasing the structural diversity (Figure 11.80).[254,255,263–266] Starting from benzopyrans **169** released from the selenenyl resin into 96-well plates, a new library of functionalized benzopyrans **170** was prepared by epoxidation of **169**, nucleophilic cleavage of the epoxide, and derivatization of the resulting secondary alcohol with electrophiles to furnish **170**.

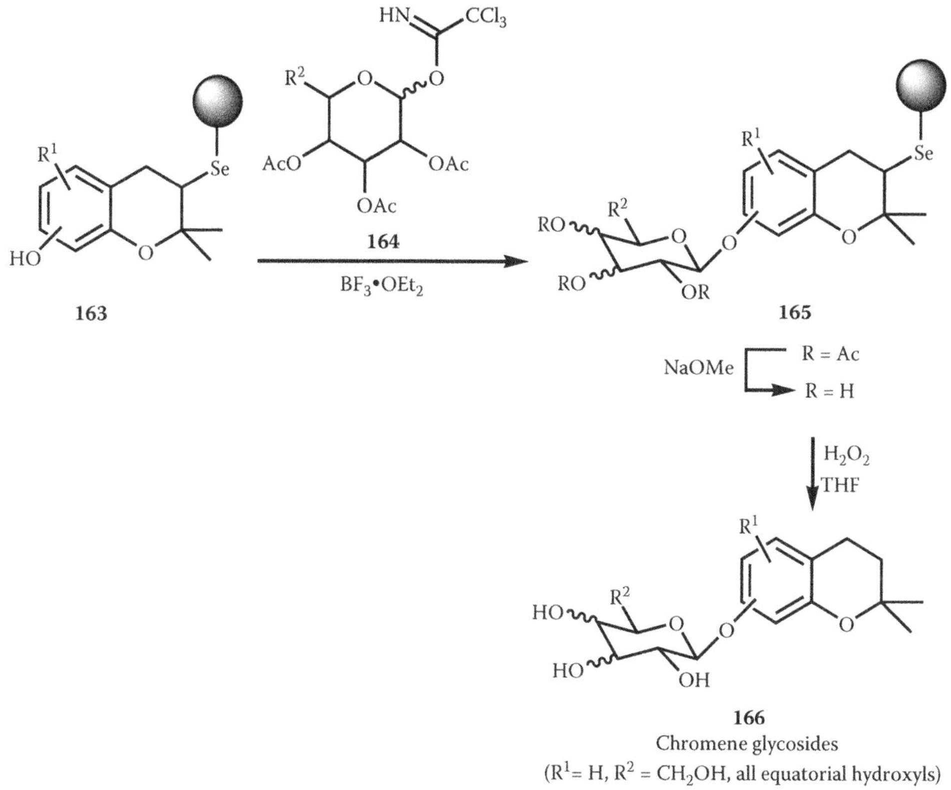

FIGURE 11.78 Library of chromene glycosides. (From Nicolaou, K.C. et al., Natural product-like combinatorial libraries based on privileged structures. 1. General principles and solid-phase synthesis of benzopyrans, *J. Am. Chem. Soc.*, 122, 9939, 2000.)

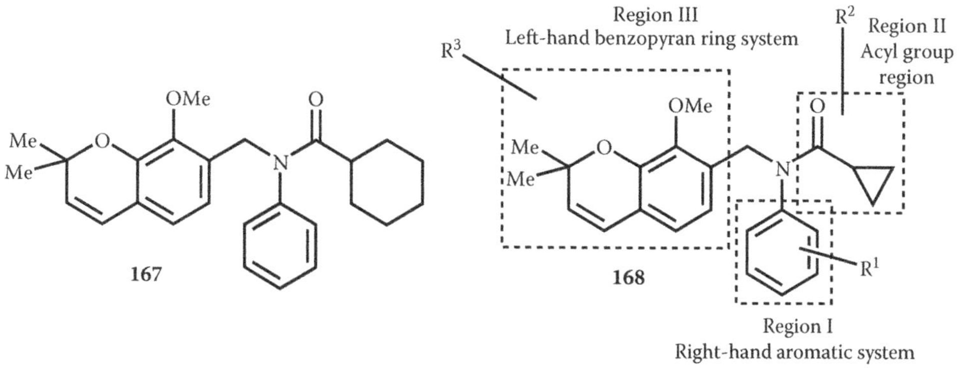

FIGURE 11.79 Farnesoid X receptor agonists from benzopyran libraries. (From Nicolaou, K.C. et al., Discovery and optimization of non-steroidal FXR agonists from natural product-like libraries, *Org. Biomol. Chem.*, 1, 908, 2003.)

In an alternative approach to the synthesis of dihydrobenzopyran libraries, the dihydrobenzopyranone core was synthesized by Breitenbucher and Hui in solution via acylation of *p*-hydroxybenzoic acid (**171**) and subsequent condensation with a ketone or aldehyde (Figure 11.81).[267] After coupling of carboxylic acid **172** to the Marshall linker, the carbonyl of the dihydrobenzopyranone

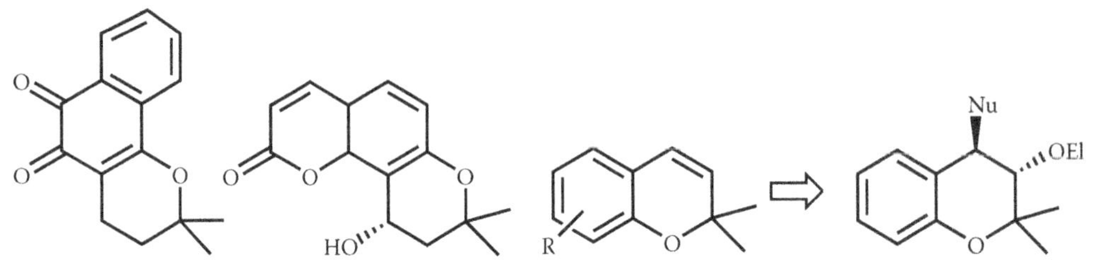

FIGURE 11.80 Combinatorial derivatization of benzopyran libraries. (From Nicolaou, K.C. et al., Natural product-like combinatorial libraries based on privileged structures. 3. The "libraries from libraries" principle for diversity enhancement of benzopyran libraries, *J. Am. Chem. Soc.*, 122, 9968, 2000.)

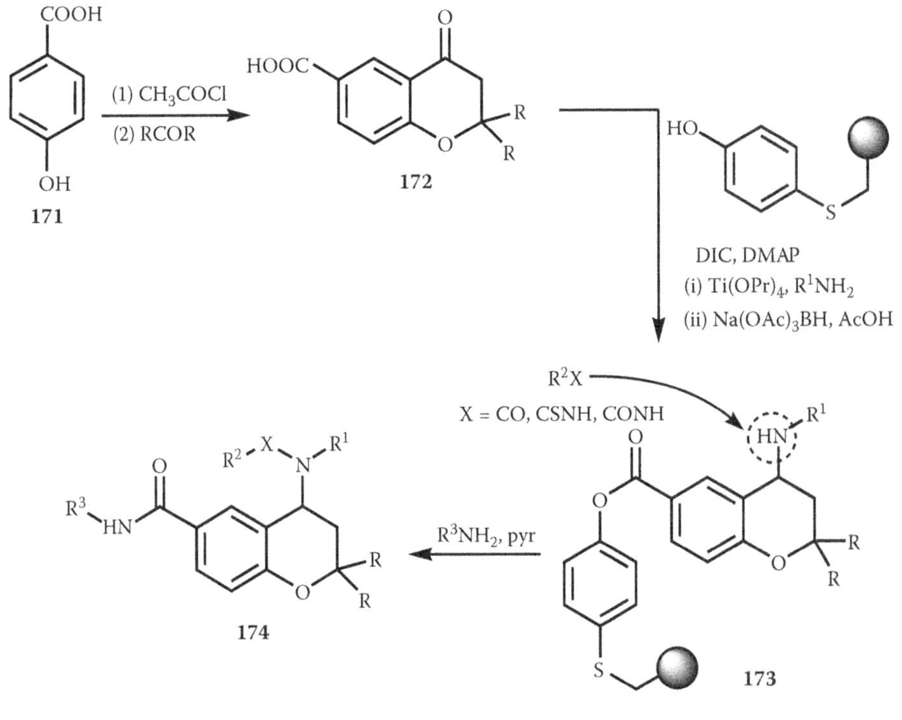

FIGURE 11.81 Solid-phase synthesis of dihydrobenzopyrans. (From Breitenbucher, J.G. and Hui, H.C., Titanium mediated reductive amination on solid-support: extending the utility of the 4-hydroxy-thiophenol linker, *Tetrahedron Lett.*, 39, 8207, 1998.)

ring was converted to a secondary amine by Ti(OiPr)$_4$-mediated reductive amination to yield **173**. Further derivatization of the secondary amine and nucleophilic cleavage from the resin by treatment with amines gave a library of 8448 substituted dihydrobenzopyrans **174**.

11.9.2 FLAVONES

Some naturally occurring flavonoids, as well as the flavone nucleus itself, were found to be ligands to central benzodiazepine receptors (BDZ-Rs). In the search for new compounds with affinity for the BDZ-Rs, Marder and coworkers described a solution-phase synthesis of flavone derivatives (Figure 11.82).[268] In a three-step sequence, mixtures containing equimolar amounts of four

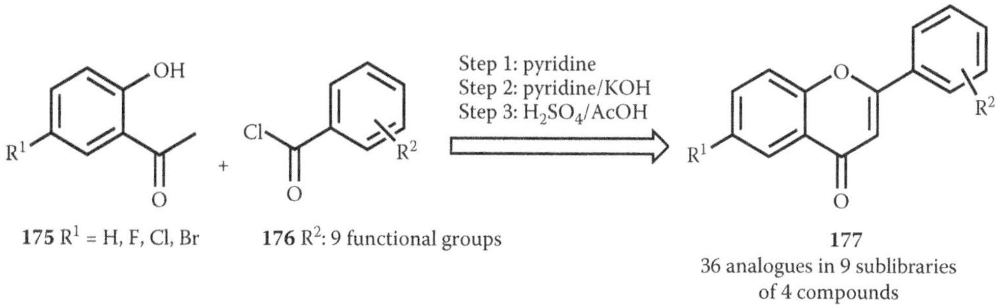

FIGURE 11.82 Library of flavonoids with affinity for benzodiazepine receptors. (From Marder, M. et al., Detection of benzodiazepine receptor ligands in small libraries of flavone derivatives synthesized by solution phase combinatorial chemistry, *Biochem. Biophys. Res. Commun.*, 249, 481, 1998.)

5′-substituted-2′-hydroxyacetophenones **175** were independently reacted in pyridine with one of nine different benzoyl chlorides **176** to yield ester intermediates (step 1). These intermediates were cyclized into flavone derivatives **177** by successive addition of hydroxide base (step 2) and heating with acid (step 3). Several library members showed high binding affinity to rat cerebral cortex BDZ-Rs with K_i values in the range of 17 to 23 n*M*. Furthermore, pharmacological experiments in mice revealed that the 6,3′-dibromoflavone had anxiolytic effect.

11.9.3 PSORALENS

In search of drugs useful for photodynamic therapy, Lam and coworkers developed an efficient solid-phase synthesis approach to a series of psoralen analogs (Figure 11.83).[269–271] This scaffold has a 3-carboxylic acid that can readily be coupled to solid support and an *ortho*-nitro arylfluoride that can undergo facile aromatic nucleophilic substitution for the introduction of additional diversity.

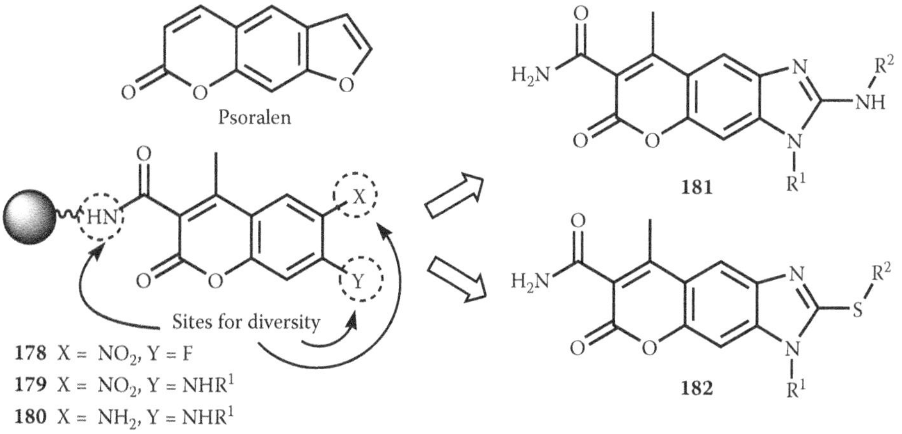

FIGURE 11.83 Solid-phase synthesis of psoralen analogs. (From Song, A., Zhang, J., and Lam, K.S., Synthesis and reactions of 7-fluoro-4-methyl-6-nitro-2-oxo-2*H*-1-benzopyran-3-carboxylic acid: a novel scaffold for combinatorial synthesis of coumarins, *J. Comb. Chem.*, 6, 112, 2004; Song, A. et al., Solid-phase synthesis and spectral properties of 2-alkylthio-6*H*-pyrano[2,3-*f*]benzimidazole-6-ones: a combinatorial approach for 2-alkylthioimidazo coumarins, *J. Comb. Chem.*, 6, 604, 2004; Song, A. and Lam, K.S., Parallel solid-phase synthesis of 2-arylamino-6H-pyrano [2,3-f] benzimidazole-6-ones, *Tetrahedron*, 60, 8605, 2004.)

Starting from scaffold **178** linked to Rink amide resin, two libraries of psolaren analogs (**181** and **182**) were prepared from a common *o*-dianilino intermediate **180**. The benzoimidazole ring of psolaren analogs **181** was formed upon treatment of **180** with aryl isothiocyanates in the presence of DIC, whereas for analogs **182** cyclization was carried out with 1,1′-thiocarbonyldiimidazole (TCD) followed by S-alkylation with alkyl halides in the presence of *N*,*N*-diisopropylethylamine. The unique spectral properties of these coumarin derivatives indicated that they may be useful in photochemotherapy.

11.9.4 KHELACTONES

Substituted khelactones constitute a group of coumarins with a broad range of biological activities, and their (*3′R*, *4′R*)-di-*O*-*cis* derivatives are crucial drug leads.[265,272] An asymmetric solid-phase synthetic route to (*3′R*, *4′R*)-di-*O*-*cis*-acyl-3-carboxyl khelactones was reported (Figure 11.84). [272] The synthesis highlighted a Knoevenagel condensation, asymmetric dihydroxylation, acylation, and product cleavage from solid support. The strategy consisted of attaching the khelactone ring skeleton to the Wang resin through the 3-carboxylate group. The asymmetric dihydroxylation of khelactone scaffold **183** using (DHQ)$_2$-PHAL as ligand and catalytic OsO$_4$ yielded **184**. Subsequent acylation with anhydrides and cleavage off solid support gave **185**.

11.9.5 ISOFLAVONES

In the search to discover a novel and effective lead for the treatment of giardiasis, a chronic gastrointestinal disease caused by the protozoan *Giardia intestinalis*, and in view of the antigiardial activity shown by some naturally occurring isoflavones (i.e., formononetin and pseudobaptigenin), a solution-phase synthesis isoflavone derivatives was accomplished (Figure 11.85).[273] Based on the structure–activity relationships of formononetin, the library was designed to introduce fluoro, chloro, and bromo substituents into the C-6 and C-8 positions and modify the C ring. Moreover, compounds with improved water solubility in order to enhance oral bioavailability were included in the library. In the first step of the solution-phase parallel synthesis, Friedel–Crafts acylation of halogen-substituted resorcinols **186** with substituted phenylacetic acids **187** was carried out in the presence of BF$_3$.Et$_2$O. The resulting intermediates **188** were subjected to Vilsmeier–Haack cyclization conditions (DMF, MeSO$_2$Cl) to furnish the final products **189**. Some of these compounds showed potent antigiardial activity.

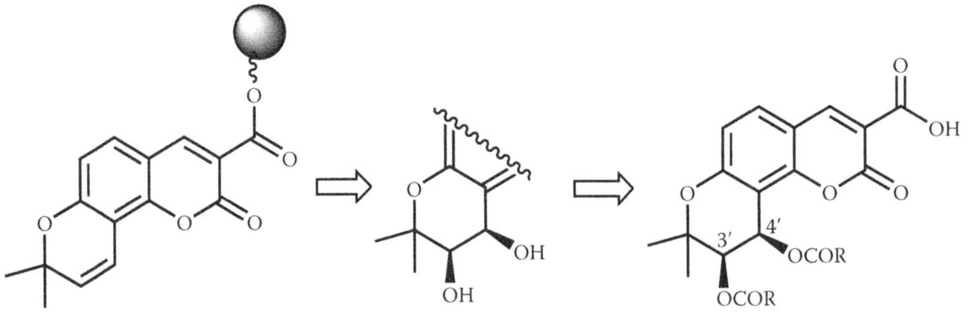

FIGURE 11.84 Solid-phase synthesis of (*3′R*, *4′R*)-di-*O*-*cis*-acyl-3-carboxyl khelactones. (From Xia Y. et al., Asymmetric solid-phase synthesis of (*3′R*, *4′R*)-di-*O*-*cis*-acyl 3-carboxyl khellactone, *Org. Lett.*, 1, 2113, 1999.)

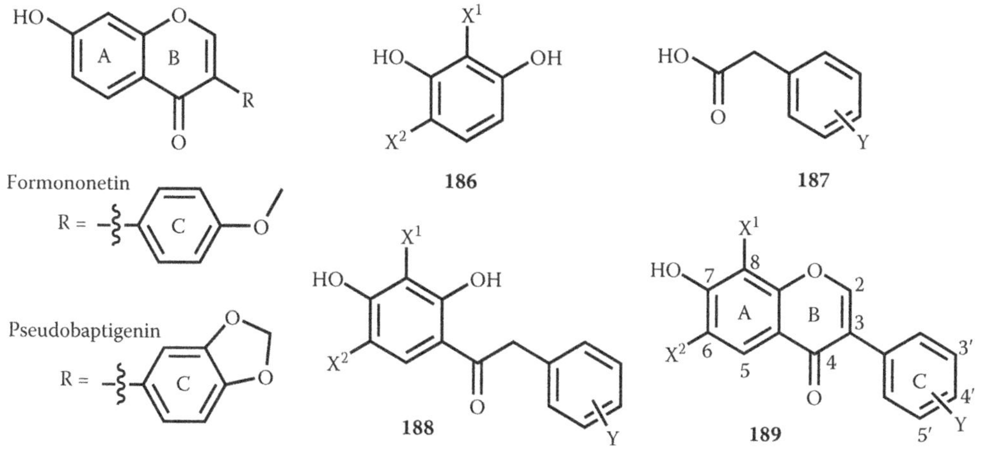

FIGURE 11.85 Isoflavone library with antigiardial activity. (From Mineno, T. et al., Solution-phase parallel synthesis of an isoflavone library for the discovery of novel antigiardial agents, *Comb. Chem. High Throughput Screen.*, 5, 481, 2002.)

11.10 OTHER NATURAL PRODUCTS

11.10.1 Heterocyclic Kinase Inhibitors

Small heterocyclic molecules are extremely important to elucidate the function of novel proteins being identified from genomics, proteomics, and traditional biochemical approaches. The vast majority of kinase inhibitor scaffolds consist of planar heterocycles with the desired hydrogen bond donating and accepting functionality and proper hydrophobicity.[274,275] Schultz and Gray reported a simple synthetic approach whereby the heterocyclic scaffold was the diversity element. Thus, the synthesis of diverse heterocycle libraries with many members exhibiting a wide variety of kinase inhibition was enabled.[276] Their strategy involved the capture of dichloroheterocyclic scaffolds, purines, pyrimidines, quinazolines, pyrazines, phthalazines, pyridazines, and quinoxalines onto solid support. Subsequent elaboration at the chloro positions gave a library consisting of 45,140 discrete and highly diverse heterocyclic small molecules. These compounds evaluated in a variety of cell- and protein-based assays.

11.10.2 Tetracyclic Template for Chemical Genetics

Libraries of small molecules with structural features reminiscent of natural products were designed and synthesized for use in chemical genetic assays in order to activate or inactivate proteins by direct interactions. This chemical genetic approach involves the search for small molecules that modulate specific biological pathways or processes ("forward chemical genetic" approach). The active small molecules and their protein partners are then identified, thereby leading to an understanding of the protein's role in the pathway. On the other hand, a small molecule having a known and specific protein target can be used to alter the function of its target. By determining the pathways and processes altered by the small molecules, the functions of its target can be inferred ("reverse chemical genetic" approach).[277,278]

These approaches are illustrated by the work of Schreiber and coworkers. These authors reported a six-step reaction sequence using the split-pool techniques that gave a binary-encoded library **190** calculated to contain 2.18 million polycyclic compounds (Figure 11.86).[279,280] These compounds are compatible with miniaturized cell-based "forward" chemical genetic assays designed to explore biological pathways and "reverse" chemical genetic assays designed to explore protein function.

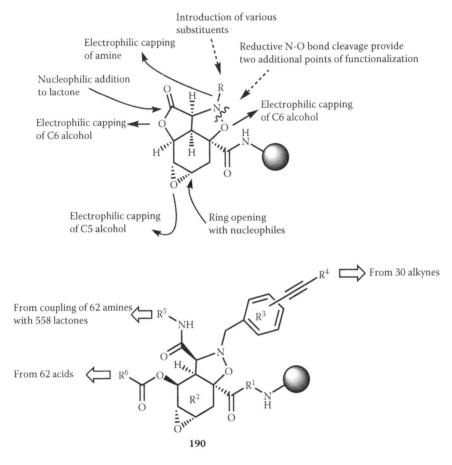

FIGURE 11.86 Octahydrobenzisoxazole scaffold for the construction of library **190** of over 2 million compounds. (From Tan, D.S. et al., Stereoselective synthesis of over two million compounds having structural features both reminiscent of natural products and compatible with miniaturized cell-based assays, *J. Am. Chem. Soc.*, 120, 8565, 1998; Tan, D.S., Synthesis and preliminary evaluation of a library of polycyclic small molecules for the use in chemical genetic assays, *J. Am. Chem. Soc.*, 121, 9073, 1999.)

As a simple illustration of the potential of these compounds, several compounds were shown to activate a TGF-β-responsive reporter gene in a stably transfected mink lung cell line.[281] The synthesis relied in part on simple acylation chemistry. Shikimic acid was converted in epoxycyclohexenol carboxylic acid, which was then coupled to a photocleavable linker on solid support (Tentagel S NH$_2$). Treatment of the resin-bound epoxycyclohexenol with various nitrone carboxylic acids under esterification conditions yielded tetracyclic compounds with complete regio- and stereoselectivity via tandem acylation/1,3-dipolar cycloaddition. The tetracyclic template is rigid and densely functionalized. A variety of building blocks were coupled to the central octahydrobenzisoxazole structure, potentially without the use of protecting groups. The isoxazoline nitrogen atom was substituted with various functional groups, installed at either the nitrone or the tetracyclic stage. The electrophilic lactone and epoxide were treated with nucleophiles while unmasking the C-6 and C-5 alcohols for subsequent reactions. Furthermore, reductive N-O bond cleavage provided two additional points for functionalization.

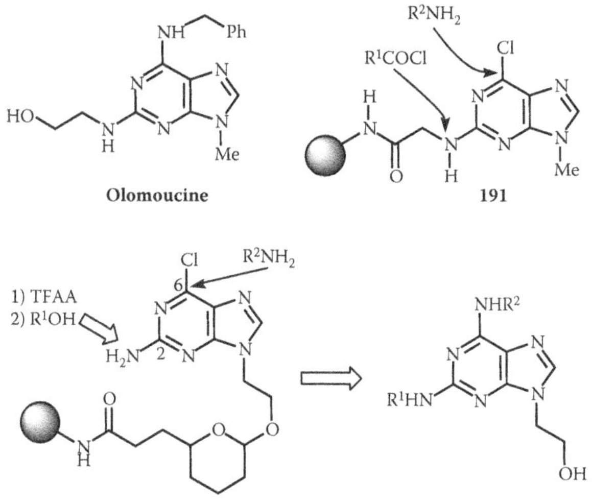

FIGURE 11.87 Solid-phase olomucine-related libraries. (From Norman, T.C. et al., A structure-based library approach to kinase inhibitors, *J. Am. Chem. Soc.*, 118, 7430, 1996.)

11.10.3 PURINES AND PYRIMIDINES

A number of other libraries were reported for natural product analogs of diverse biosynthetic origin. The adenine derivative olomoucine was found to inhibit several cyclin-dependent kinases, such CDK2-cyclin A,[282] a group of enzymes suspected to be implicated in several diseases including cancer. From the crystal structure of the olomoucine–CDK2 complex, it was reasoned that diversification at C-6, C-2, and N-9 of the purine scaffold may improve binding affinity and selectivity with the enzyme. With this objective in mind, Schultz and coworkers[283] described a solid-phase approach for the generation of a library of olomoucine analogs from purine scaffold **191** coupled to the Rink amide linker (Figure 11.87). After acylation of the amine with 5 acid chlorides, the second element of diversity was introduced by aromatic substitution on C-6 with 58 amines. After cleavage from the resin, a 348-member library of purine analogs resulted. In a different approach, purine scaffold **192** was attached to solid support via the N-9 position. Mitsunobu alkylation of the C-2 trifluoroacetanilide and aromatic amination at C-6, followed by aminolysis of the trifluoroacetamide and cleavage from the solid support, provided 2-hydroxyethyl purines **193**. Other approaches employing 2-fluoro-6-chloropurine as the template led to the discovery of purvalanol B, a highly potent inhibitor of the human CDK2-cyclin A (IC$_{50}$ = 6 nM).

A synthetic route for the high-throughput derivatization of purine and pyrimidine scaffolds was described.[284] Diversity was introduced at the C-5′ and C-6 positions of the purine and C-5′ and C-4 position of the pyrimidine scaffolds (Figure 11.88). Chemical transformations were carried out on low-cross-linked polystyrene-based macroporous support with the ribose subunit of the nucleoside attached via a 2′,3′-acetal linkage. Nanokan technology developed by IRORI allowed encapsulation of up to 10 mg of resin in two-dimensional bar-coded microreactors.[97] Derivatization on the solid support gave a library of 25,000 compounds that was screened in high-throughput cell-based assays.

A solid-phase combinatorial approach was used to synthesize nucleosides containing triazine- and pyrimidine-based exocyclic analogs of clitocine (**194**, R^4 = R^5 = H), a fungal metabolite that exhibits potent cytostatic effects against leukemia cell lines through the inhibition of adenosine kinase as well as insecticidal activity (Figure 11.89).[288,289] The initial library **195** of 1234 compounds was prepared by successive reaction of a 4,6-dichloro-substituted precursor attached to

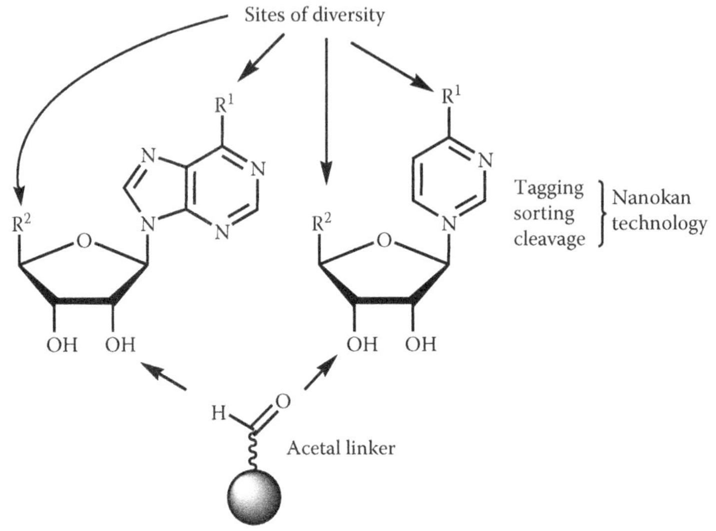

FIGURE 11.88 General strategy for the solid-supported synthesis of nucleoside analogs. (From Epple, R., Kudirka, R., and Greenberg, W.A., Solid-phase synthesis of nucleoside analogs, *J. Comb. Chem.*, 5, 292, 2003.)

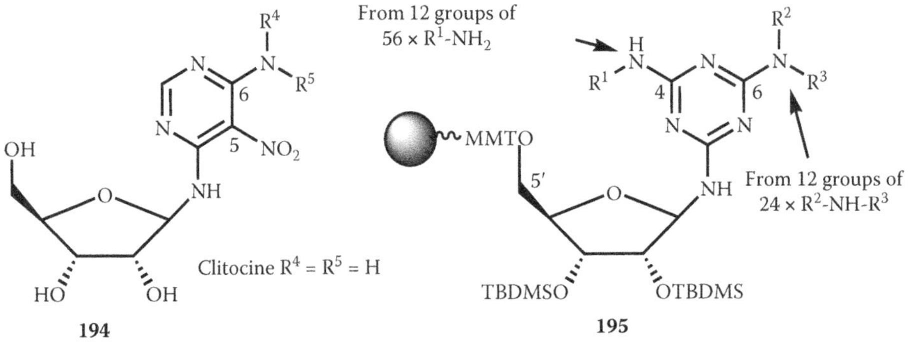

FIGURE 11.89 Novel exocyclic amino nucleoside libraries. (From Varaprasad, C.V. et al., Synthesis of novel exocyclic amino nucleosides by parallel solid-phase combinatorial strategy, *Tetrahedron*, 59, 2297, 2003.)

monomethoxytrityl resin with 12 groups of 56 primary amines and 24 secondary amines. A second library **194**, structurally related to clitocine, was prepared by reaction of the 5-nitro-6-chloro substituted precursor with 82 amines HNR^4R^5.

11.10.4 HYDANTOINS

Based on the "libraries from libraries" concept, Houghten and coworkers developed a synthetic route for the solid-phase synthesis of hydantoin and thiohydantoin libraries **196** from resin-bound dipeptides **195** (Figure 11.90).[287–289] Using 54 different amino acids for the first site of diversity (R^1), 60 different amino acids for the second site of diversity (R^2) and 4 different alkyl groups for the third site of diversity (R^3), a library of 38,880 compounds was prepared and examined in a sigma opioid radioreceptor-binding assay. The screening results revealed the importance of basic or N-benzylated hydrophobic amino acids in the R^1 position (IC_{50} values in the 1 to 10 μM range) and N-benzylated basic amino acids in the R^2 position (IC_{50} values in the 100 to 500 nM range).

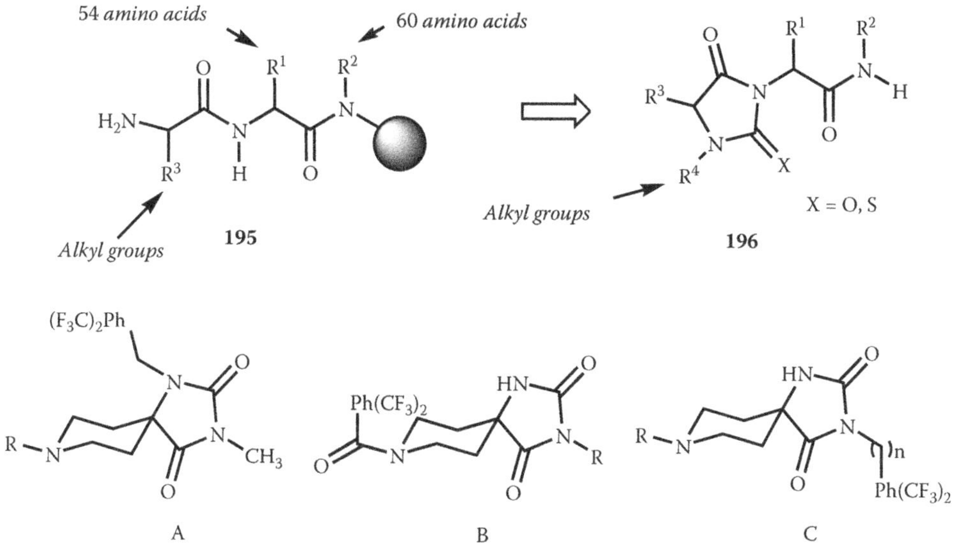

FIGURE 11.90 Libraries of hydantoin, thiohydantoin, and spirohydantoin derivatives. (From Boehm, H.-J. et al., Novel inhibitors of DNA gyrase: 3D structure based biased needle screening, hit validation by biophysical methods, and 3D guided optimization: a promising alternative to random screening, *J. Med. Chem.*, 43, 2664, 2000; Bleicher, K.H. et al., Parallel solution- and solid-phase synthesis of spirohydantoin derivatives as neurokinin-1 receptor ligands, *Bioorg. Med. Chem. Lett.*, 12, 2519, 2002.)

11.10.5 SPIROPIPERIDINES

A well-known class of pharmacologically relevant molecules showing biological activity for various target families ranging from enzyme inhibitors to ion channel blockers are spiropiperidines, which include spiropiperidine hydantoins that are neurokinin receptor (NK-1) antagonist. In the search of novel small-molecule ligands targeting the NK-1 receptor, a focused library synthesis of spiro-hydantoins was designed that followed a strategy which combined the concept of "privileged structure," here illustrated by the spiropiperidine motive, with the "needle" approach (Figure 11.90).[290,291] A needle is a fragment of an active molecule showing very specific interactions with one particular biological target. Here, the 3,5-*bis*(trifluoromethyl)phenyl fragment was coupled to all available positions on the core structure, and the remaining nitrogen was decorated with various R groups. Three libraries of spiropiperidine hydantoin derivatives (**A**, **B**, and **C**) were prepared using solution- and solid-phase methodologies. Several compounds showed moderate-to-high NK-1 binding affinities.

11.10.6 CARPANONE AND KRAMERIXIN

Carpanone and kramerixin are two lignan compounds isolated from *Cinnamomum sp.* and *Krameria sp.*, respectively.[292,293] Based on the reactions implicated in the biosynthesis of carpanone, Shair and coworkers exploited the generality of the solid-phase biomimetic reaction to contract a 100,000-member library of carpanone-like molecules (Figure 11.91).[294] The strategy involved a solid-phase heterocoupling between electron-deficient phenols **197** and an electron-rich phenol immobilized on a silicon-linked resin **198** using PhI(OAc)$_2$ as oxidant. The tetracyclic adducts **199** were obtained as single isomers through complete electronic control inverse electron-demand Diels–Alder cycloadditions.

A library of 120 analogs of kramerixin was prepared following a solution-phase strategy in which the phenolic moiety **200** and the nature and number of substituents X in **201** were varied (Figure 11.92).[293] The key step, which produces an intermediate that reacts by an intramolecular

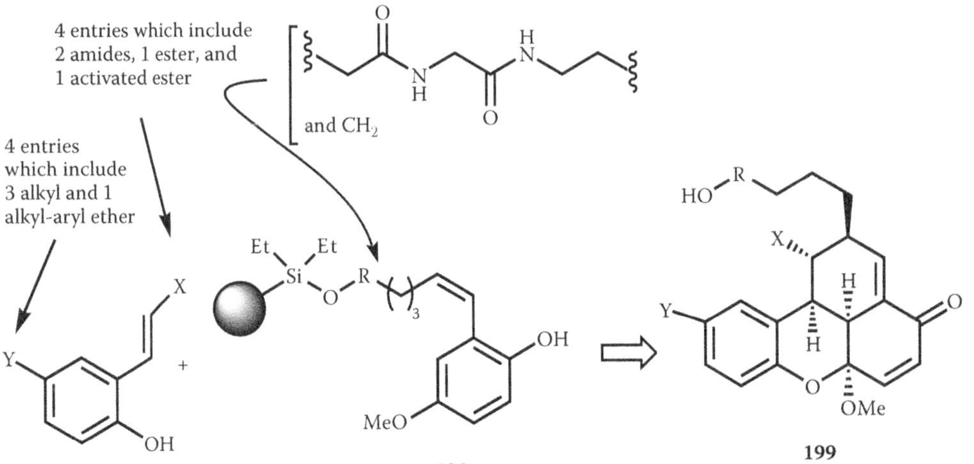

FIGURE 11.91 Solid-phase biomimetic synthesis of carpanone-like molecules. (From Lindslay, C.W. et al., Solid-phase biomimetic synthesis of carpanone-like molecules, *J. Am. Chem. Soc.*, 122, 422, 2000.)

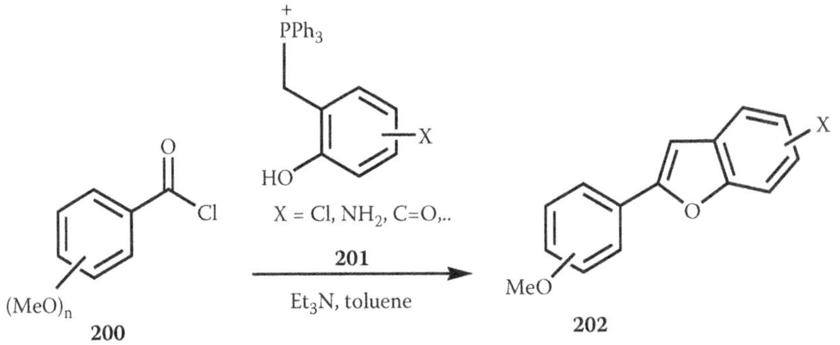

FIGURE 11.92 Multiple parallel synthesis of kramerixin and analogs. (From Fecik, R.A. et al., Use of combinatorial and multiple parallel synthesis methodologies for the development of anti-infective natural products, *Pure Appl. Chem.*, 71, 559, 1999.)

Wittig reaction to form the furan ring, was the tandem base-promoted reaction of acid chlorides **200** with the benzylidenephosphonium-containing phenols **201**. Demethylation of the methoxy groups with PyBr-HBr yielded the final phenolic compounds **202**. Further antifungal screening identified several active compounds with one derivative, the 5,7-dichloro analog with the conserved 2,4-dihydroxy-phenyl furan substituent, that provided improved activity compared to kramerixin (MIC 1.25 μg/ml vs. 3.12 μg/ml of respiration of *C. albicans*).

11.10.7 Curacin A Analogs

The marine natural product curacin A is an antimitotic agent that promotes arrest of the cell cycle at the G2/M checkpoint and competitively inhibits the binding of [³H]-colchicine to tubulin.[295] Therefore, it can be considered a colchicine site agent. Because curacin A binds on the colchicine site on tubulin, and the alkenyl thiazoline moiety of the molecule is largely responsible for the chemical instability of the natural product, Wipf and coworkers planned to replace the heterocycle with electron-rich arenes, reminiscent of the trimethoxyphenyl ring A of colchicine.[296] Furthermore, the homoallylic methyl ether terminus of curacin A was substituted with a broad range of more

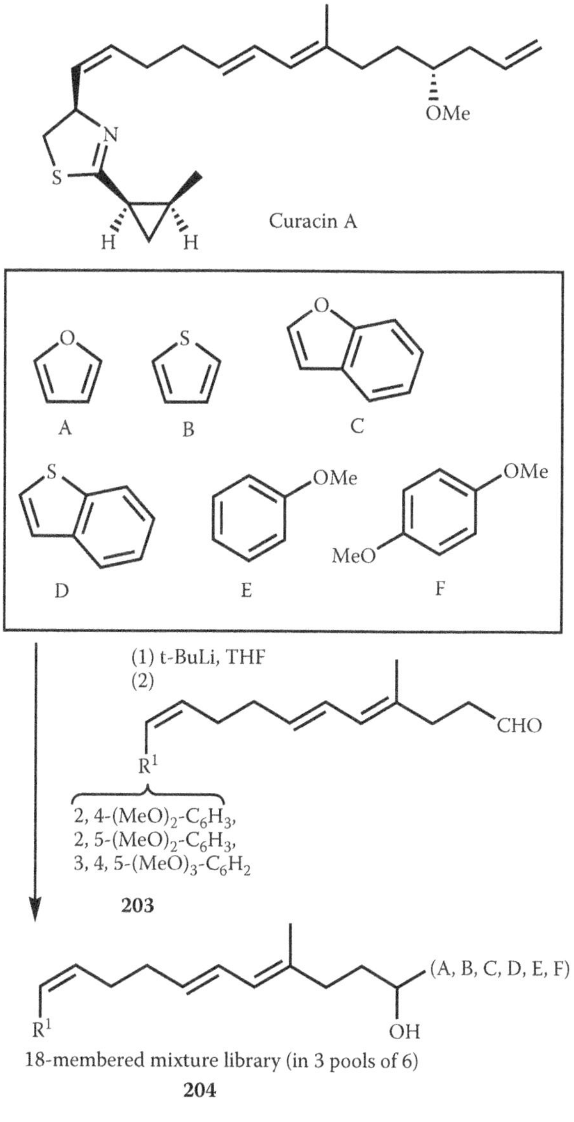

FIGURE 11.93 Curacin-A-based library. (From Wipf, P. et al., Synthesis and biological evaluation of a focused mixture library of analogs of the antimitotic marine natural product curacin A, *J. Am. Chem. Soc.*, 122, 9391, 2000.)

hydrophilic benzyl alcohols, whereas the diene unit was kept intact, proven to be essential in SAR studies.[297] In accordance with these structural requirements, a solution-phase library based on curacin A was prepared in three mixtures of six compounds by reacting the three key building blocks **203** separately with the lithiated arenes (**A-F**) (Figure 11.93). A streamlined purification of the benzylic alcohol products **204** was achieved using a fluorous trapping protocol. To this end, mixture **204** was attached to a vinyl ether tag and extracted with the perfluorinated solvent FC-72. Hydrolysis and elimination of the fluorinated acetal afforded pure final mixtures. The most active library components [**204**, $R^1 = 3,4,5\text{-}(MeO)_3C_6H_2$), **B**; $R^1 = 3,4,5\text{-}(MeO)_3C_6H_2$), **D**] inhibited tubulin polymerization with an IC_{50} value of about 1 μM and showed an average growth inhibition activity GI_{50} of about 250 nM. The molecules inhibited colchicine binding to tubulin, and blocked mitotic progression at nanomolar concentrations.

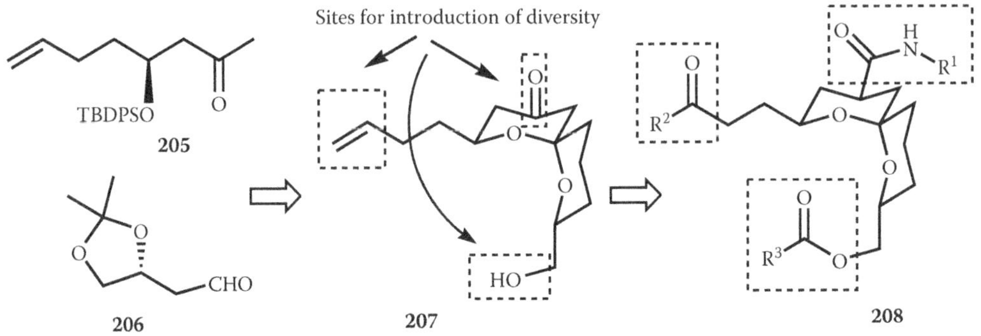

FIGURE 11.94 Spiroketal library. (From Kulkarni, B.A. et al., Combinatorial synthesis of natural product-like molecules using a first-generation spiroketal scaffold, *J. Comb. Chem.*, 4, 56, 2002.)

11.10.8 SPIROKETALS

6,6-Spiroketals are the prevalent underlying structural element in a wide range of important natural products with differing biological activity, such as okadaic acid, and the macrolide antibiotics cytovaricin, spongistatins (altohyrtins), and rutamycins/oligomycins (see Chapter 7, Subsection 7.2.4). Importantly, structurally simplified spiroketals derived from natural products retain their biological activity and, therefore, the underlying 6,6-spiroketal structure is a suitable starting point for the development of natural product compound libraries.[298,299] Starting from scaffold **207**, prepared from condensation of chiral ketone **205** with aldehyde **206**, Porco and coworkers described the synthesis of 90 highly functionalized spiroketals with the general structure **208** (Figure 11.94).[298] The strategy consisted of the introduction of diversity at three sites of spriroketal **207**.

11.10.9 LAVENDUSTIN A AND BALANOL ANALOGS

Examples of combinatorial solid-phase synthesis of small molecules based on the retrosynthetic analysis of natural product structures are the fungal metabolites lavendustin A and balanol analogs, both displaying high protein-kinase-inhibiting activity,[299–301] and the cytotoxic tripeptide analogs of aspergillamide A (Figure 11.95).[302,303] The framework of lavendustin A is synthetically accessible from three subunits (X, Y, and Z). Using the carboxyl group of subunit X for resin attachment, subunits Y and Z were incorporated in **209** by successive reaction of the amine group with five substituted benzaldehydes (subunit Y) and five substituted benzyl bromides (subunit Z). A library of 60 lavendustin A analogs **210** were obtained.[299] The solid-phase synthesis of a 36-member library of balanol analogs was achieved from the assembly of three key balanol building blocks X, Y, Z (mono-protected diacids, protected amino alcohols, and benzoic acid derivatives) through routine transformations.[301] A solution-phase Ugi multicomponent reaction (MCR) was designed for the construction of a 224-member library of aspergillamide analogs using a pool of 8 Schiff-bases, 4 carboxylic acids, and 7 isocyanides, as MCR building blocks (Figure 11.95).[303] Some of these compounds, further synthesized on larger scale, exhibited antibiotic activity and cytotoxic activity against MCF-7 cells.

11.10.10 PSAMMAPLIN A ANALOGS

Psammaplin A is a symmetrical bromotyrosine-derived disulfide dimer isolated from a marine sponge *Psammalysilla*, which exhibits *in vitro* antibacterial activity against methicillin-resistant *Staphylococcus aureus* (Figure 11.96).[304] In order to find potent and structurally simplified analogs of psammaplin A that could be used as tools for SAR studies, Nicolaou et al. undertook the combinatorial synthesis of a library of disulfide-containing molecules using a pool of 88 synthetic

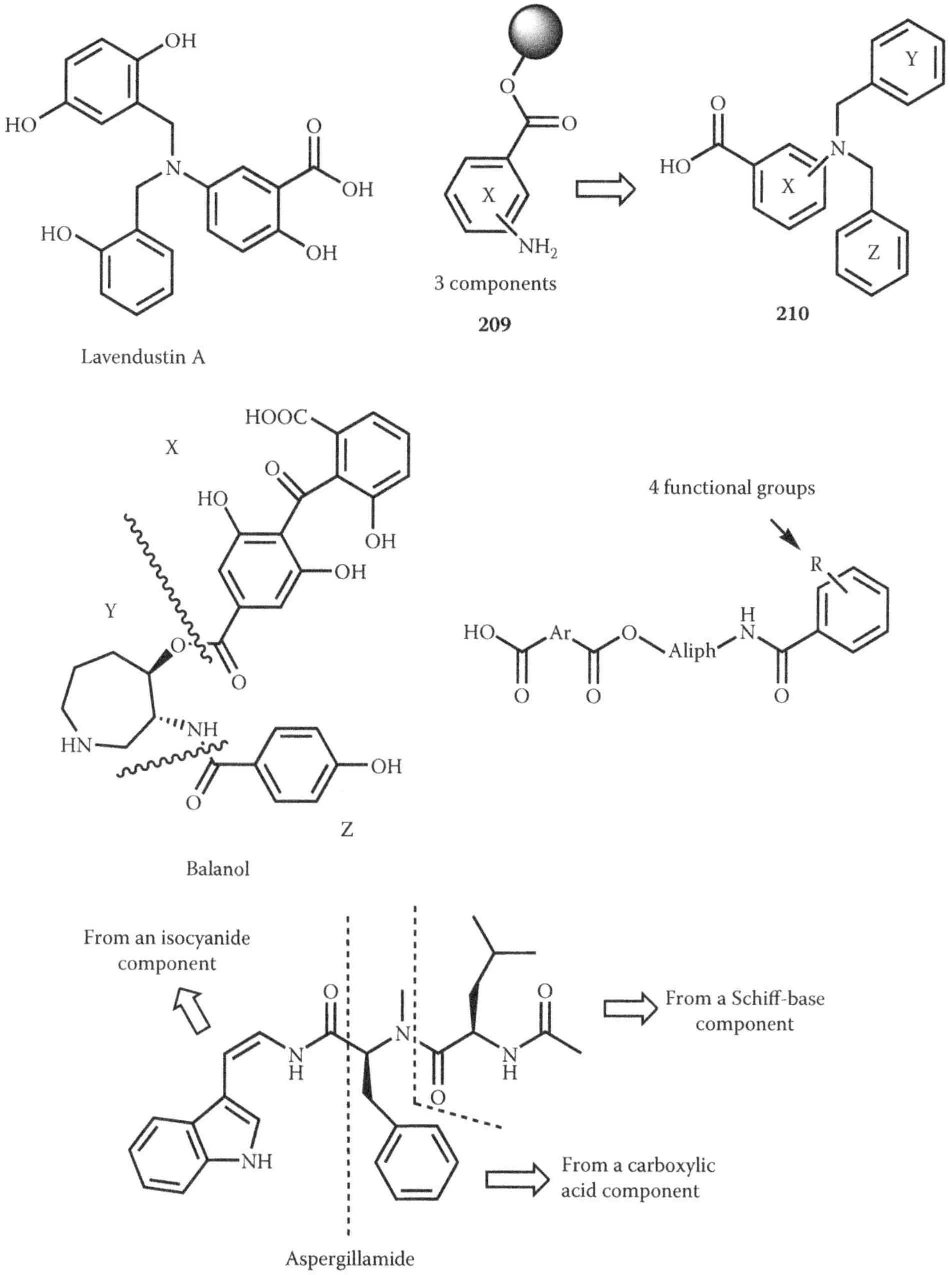

FIGURE 11.95 Library design of lavendustin A, balanol, and aspergillamide analogs. (From Green, J., Solid phase synthesis of lavendustin A and analogs, *J. Org. Chem.*, 60, 4287, 1995; Nielsen, J. and Lyngsø, L.O., Combinatorial solid-phase synthesis of balanol analogs, *Tetrahedron Lett.*, 37, 8439, 1996 and references cited therein; Beck, B., Hess, S., and Dömling, A., One-pot synthesis and biological evaluation of aspergillamides and analogs, *Bioorg. Med. Chem. Lett.*, 10, 1701, 2000.)

homodimeric disulfides. These disulfides were systematically scrambled to create a large (> 3800-membered) library of heterodimeric analogs.[305] The combinatorial scrambling strategy employed here was based on the exchange chemistry of the disulfide motif, which undergoes facile exchange reactions with other disulfides in high yield under mild conditions. The 88 homodimeric disulfide

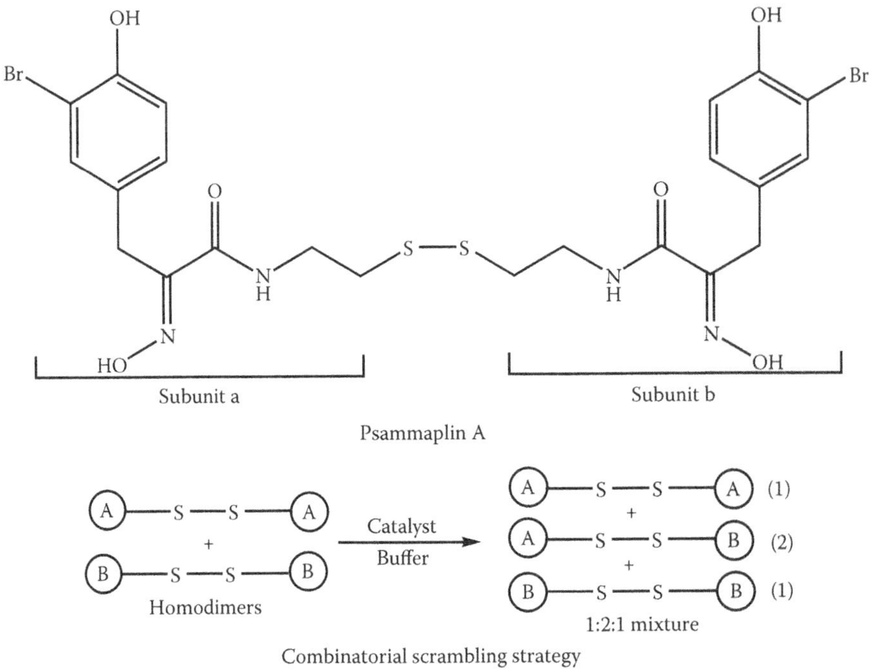

FIGURE 11.96 Design of a library of psammaplin A analogs through disulfide exchange. (From Nicolaou, K.C. et al., Combinatorial synthesis through disulfide exchange: discovery of potent psammaplin A type antibacterial agents active against methicillin-resistant *Staphylococcus aureus* (MRSA), *Chem. Eur. J.*, 7, 4280, 2001.)

analogs (building blocks) were dispensed into 96-well plates filled with DMSO/buffer solution (pH 8.3). Catalytic amount of dithiothreitol was added, and the compounds were scrambled. Direct antibacterial screening of the library identified new leads whose structures were optimized for issues such as potency, toxicity and nonspecific protein binding.[306]

11.10.11 CINNAMAMIDES

In search of molecules that selectively induce apoptosis in cancer cells, Hergenrother and coworkers synthesized a combinatorial library based on the *N*-acylated aromatic amine **211**, isolated from *Isodon excisus*, which displayed activity in either pro- or antiapoptotic assays (Figure 11.97).[307,308] The library was created from the parallel coupling of 8 acids and 11 amine building blocks, using

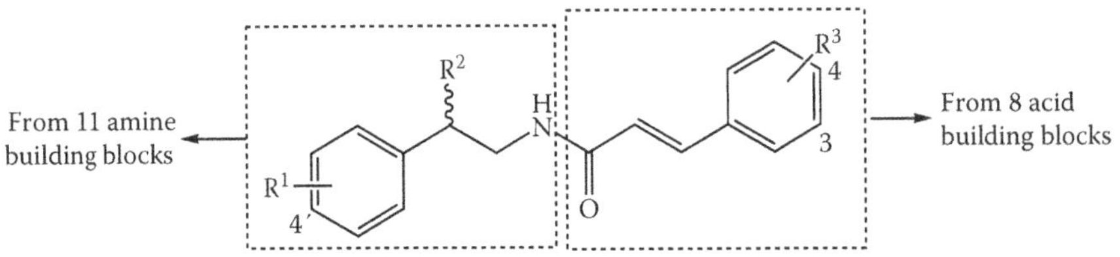

211 R^1 = 4′-OH, R^2 = ·····ΙΙΙΙOMe, R^3 = 3-OMe, 4-OH

FIGURE 11.97 Library of active compounds that induces apoptosis in cancer cells. (From Nesterenko, V., Putt, K.S., and Hergenrother, P.J., Identification from a combinatorial library of a small molecule that selectively induces apoptosis in cancer cells, *J. Am. Chem. Soc.*, 125, 14672, 2003.)

a carbodiimide polymeric support. Following high-throughput screening against HL-60 (leukemia) and U-937 (lymphoma) cell lines, a compound that selectively induced apoptosis in cancerous white blood cells but was nontoxic toward noncancerous white blood cells, was identified (R^1 = 4-OH, R^2 = R-OH, R^3 = 3-OMe-4-OH; Figure 11.97).

The modular structure of philanthotoxins was exploited for the construction of the first combinatorial library of these compounds. The prototype structure is philanthotoxin-433 (PhTX-433), a venom constituent of the Egyptian digger wasp *Philanthus triangulum* (Figure 11.98).[309] PhTX-433, as well as some synthetic analogs, were found to antagonize various types of nicotinic acetylcholine receptors (nAChRs) and ionotropic glutamate receptors (iGluRs). This led to SAR studies on these compounds. For this purpose, a combinatorial library of philanthotoxin analogs was constructed using three sets of building blocks: (*S*)-tyrosine and (*S*)-3-hydroxylphenylalanine as amino acid components; spermine, 1,12-dodecanediamine, and 4,9-dioxa-1,12-dodecanediamine as amine components; and butanoyl, phenylacetyl, and cyclohexylacetyl as *N*-acyl groups (Figure 11.98).[310] Amines **212** were attached to trityl chloride resin and coupled with Fmoc-protected amino acids to give the intermediates **213**. Deprotection of the primary amino group and subsequent *N*-acylation and cleavage from solid support yielded the 18-member library **214**.

An example of formation of "libraries from sublibraries" is illustrated by the solution-phase synthesis of antibiotic polyenes based on the structure of (–)-stipiamide (Figure 11.99).[311] (–)-Stipiamide has moderate multidrug reversal activity with colchicine-resistant cells. Its truncated 6,7-dehydro derivative **215** is potent with a variety of drugs with resistant human breast cancer MCF7-adrR cells, and was found to bind to the P-glycoprotein expressed at the surface of these cells. Based on this evidence, the two end sites of **215** were chosen as diversification points by the assembly of the two sublibraries **216** and **217** via the Sonagashira cross-coupling reaction.[312] Each individual acetylene was reacted with a mixture of 6 vinyl iodides to create 13 pools indexed in 2 dimensions for a total of 42 compounds. These compounds were isolated by radial chromatography and tested against MCF7-adrR cells. Library members with R = Ph or naphthyl and R = alaninol were among the most active compounds.

11.11 CONCLUSIONS

In the last few years, the emergence of new combinatorial approaches for the synthesis of natural product analogs has begun to address the problem of efficiently finding hits and leads from large combinatorial libraries. Today, it is recognized that the design of a library should be biologically relevant, diverse, and drug-like. This implies that a significant investment needs to be made in order to identify suitable synthetic strategies to construct specific libraries. New methods and reagents for solid-phase synthesis, resins and linkers for convenient loading and cleavage, and diversity analysis tools used to optimize molecular properties help accelerate the process of lead discovery and optimization.[13,15,17,24,37,44,313, 314]

In pursuing these objectives, diversity-oriented synthesis and chemical genetic approaches have made significant contributions to expanding the chemical structure space occupied by natural product-like libraries. A wide range of bioactive small-molecule chemical probes whose potential as specific binding agents and selective modulators of macromolecular interactions should be further explored.[45,315–317] Because the same structural domain is often found in many proteins in, more or less, a modified form, Waldmann and coworkers (see Chapter 1, Section 1.6) recently proposed that protein domains or cores with similar three-dimensional structures can be clustered into "protein structure similarity clusters" (PSSCs). Knowledge about known ligands for members of such a cluster can be employed to guide compound library development for other members of the cluster. From this perspective, PSSC can be exploited to select biologically prevalidated starting points for compound library synthesis. Natural products binding to one PSSC member protein provide such a starting point.[318]

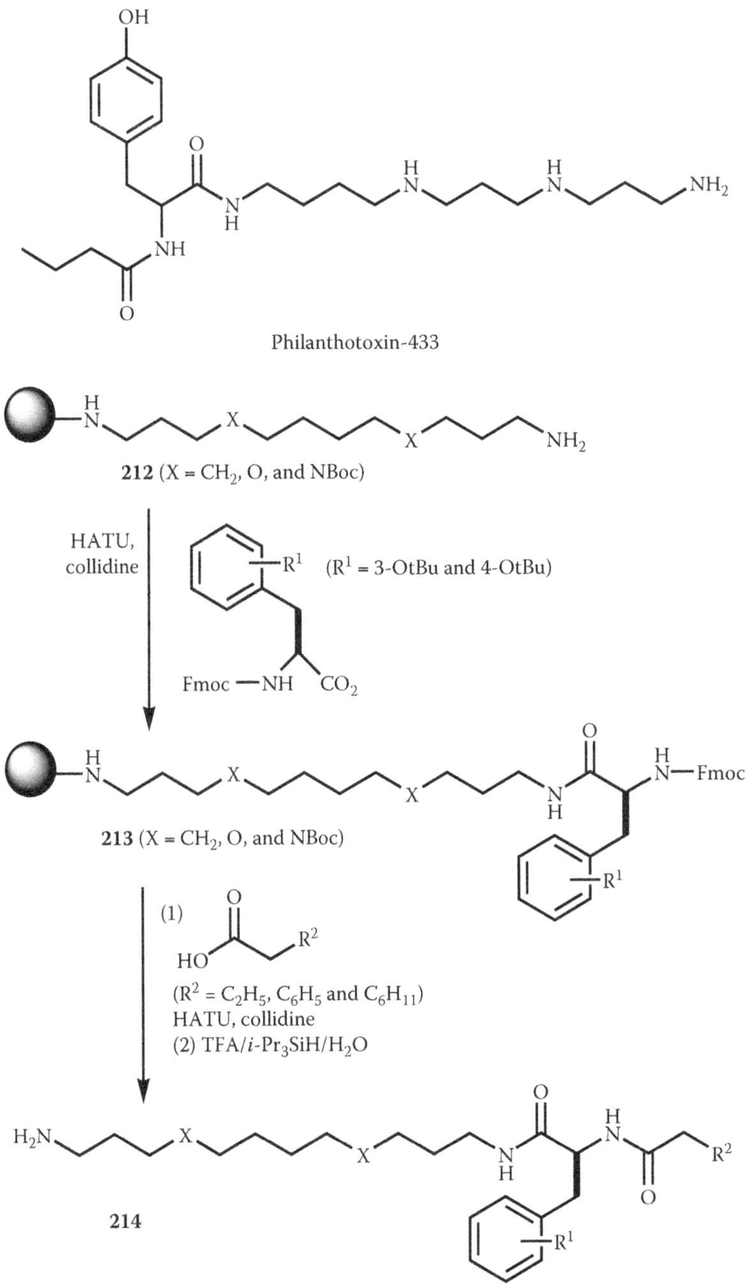

FIGURE 11.98 Solid-phase synthesis of philanthotoxin analogs. (From Strømgaard, K. et al., Solid-phase synthesis and biological evaluation of a combinatorial library of philanthotoxin analogs, *J. Med. Chem.*, 43, 4526, 2000.)

The introduction of semiautomatic combinatorial synthesis systems using microreactors, radio-frequency tagging, directed sorting,[319] as well as the development of analytical and screening techniques,[320–325] have helped overcome the immense task of identifying and isolating bioactive agents from large libraries. With the continued need for novel drug-like lead structures, and elucidation of structure–activity relationships, in particular those related with biomacromolecular

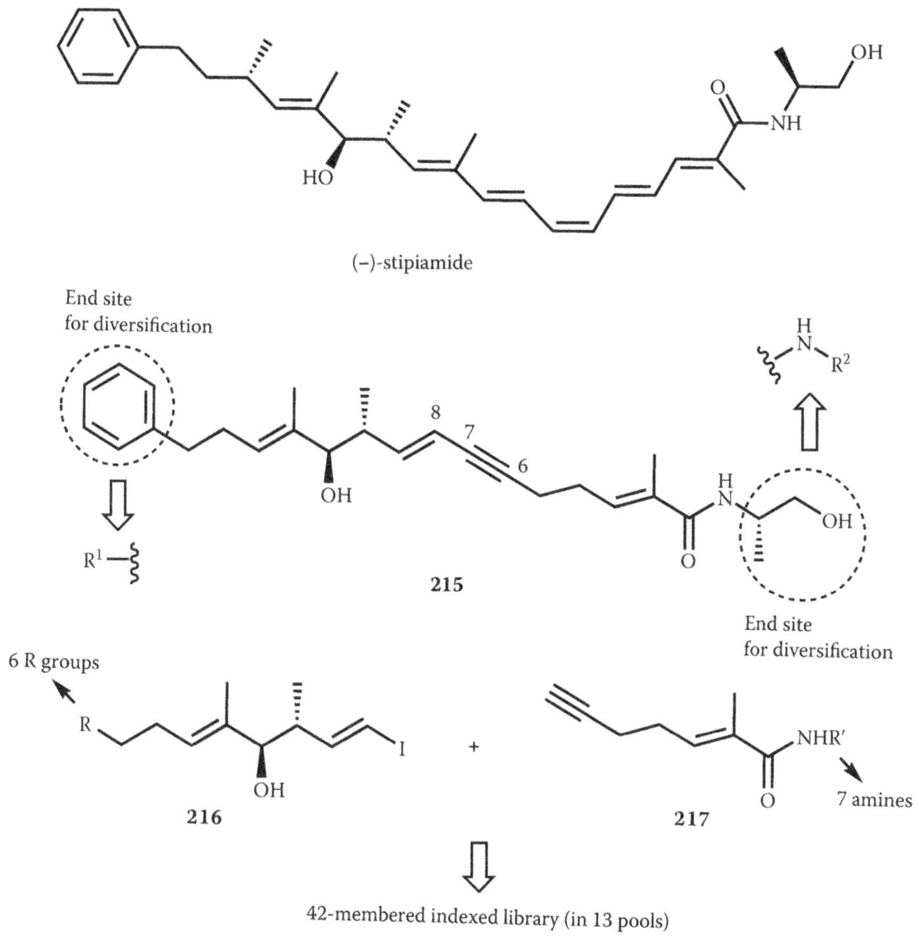

FIGURE 11.99 Stipiamide-based library. (From Andrus, M.B. et al., The synthesis and evaluation of a solution phase indexed combinatorial library of non-natural polyenes for reversal of P-glycoprotein-mediated multidrug resistance, *J. Org. Chem.*, 65, 4973, 2000.)

interactions, the synthesis of natural product-like combinatorial libraries as useful small-molecule chemical probes continues to be a promising area for future development.

ACKNOWLEDGMENT

Fundação para a Ciência e Tecnologia for a postdoctoral grant (S. Matthew, SFRH/BPD/8570/2002).

REFERENCES

1. Cragg, G.M., Newman, D., and Snader, K.M., Natural products in drug discovery and development, *J. Nat. Prod.*, 60, 52, 1997.
2. Waterman, P.G., Natural products for drug discovery, in *Advances in Drug Discovery Techniques*, Harvey, A.L., Ed., John Wiley & Sons, Chichester, U.K., 1998, chap. 2.
3. Bertels, S. et al., Synergistic use of combinatorial and natural product chemistry, in *Drug Discovery from Nature*, Grabley, S. and Thiericke, R., Eds., Springer-Verlag, Heidelberg, 1999, chap. 5.

4. Grabley, S. and Thiericke, R., Bioactive agents from natural sources: trends in discovery and application, *Adv. Biochem. Eng. Biotechnol.*, 64, 102, 1999.

5. Paululat, T. et al., Combinatorial chemistry: the impact of natural products, *Chimica Oggi*, 17, 52, 1999.

6. Newman, D.J., Cragg, G.M., and Snader, K.M., Natural products as sources of new drugs over the period 1981–2002, *J. Nat. Prod.*, 66, 1022, 2003.

7. Butler, M.S., Natural products to drugs: natural product derived compounds in clinical trials, *Nat. Prod. Rep.*, 22, 135, 2005.

8. Harvey, A., Strategies for discovering drugs from previously unexplored natural products, *Drug Discovery Today*, 5, 294, 2000.

9. Wessjohann, L.A., Synthesis of natural product-based compound libraries, *Curr. Opin. Chem. Biol.*, 4, 303, 2000.

10. Bindseil, K.U. et al., Pure compound libraries: a new perspective for natural product based drug discovery, *Drug Discovery Today*, 6, 840, 2001.

11. Hall, D.G., Manku, S., and Wang, F., Solution- and solid-phase strategies for the design, synthesis, and screening of libraries based on natural product templates: a comprehensive survey, *J. Comb. Chem.*, 3, 125, 2001.

12. Coffen, D.L. and Xiao, X.-Y., A natural approach to combinatorial chemistry, *Chimica Oggi*, 19, 9, 2001.

13. Breinbauer, R, Vetter, I.R., and Waldmann, H., From protein domains to drug candidates-natural products as guiding principles in the design and synthesis of compound libraries, *Angew. Chem. Int. Ed. Engl.*, 41, 2878, 2002.

14. Kingston, D.G.I. and Newman, D.J., Mother nature's libraries; their influence on the synthesis of drugs, *Curr. Opin. Drug. Discovery Dev.*, 5, 304, 2002.

15. Breinbauer, R. et al., Natural product guided compound library development, *Curr. Med. Chem.*, 9, 2129, 2002.

16. Nielsen, J., Combinatorial synthesis of natural products, *Curr. Opin. Chem. Biol.*, 6, 297, 2002.

17. Nicolaou, K.C. and Pfefferkorn, J.A., Solid-phase synthesis of natural products and natural product-like libraries, in *Handbook of Combinatorial Chemistry*, Nicolaou, K.C., Hanko, R., and Hartwig, W., Eds., Wiley-VCH, Weinheim, 2002, chap. 21.

18. Knepper, K., Gil, C., and Brase, S., Natural product-like and other biologically active heterocyclic libraries using solid-phase techniques in the post-genomic era, *Comb. Chem. High Throughput Screen.*, 6, 673, 2003.

19. Abreu, P.M. and Branco, P.S., Natural product-like combinatorial libraries, *J. Braz. Chem. Soc.*, 14, 675, 2003.

20. Myles, D.C., Novel biologically active natural and unnatural products, *Curr. Opin. Biotechnol.*, 14, 627, 2003.

21. Grabley, S. and Sattler, I., Natural products for lead identification: nature is a valuable resource for providing tools, in *Modern Methods of Drug Discovery*, Hillisch, A. and Hilgenfeld, R., Eds., Birkhauser Verlag, Switzerland, 2003, chap. 5.

22. Rouhi, M., Rediscovering natural products, *Chem. Eng. News*, 81, 77, 2003.

23. Rouhi, M., Moving beyond natural products, *Chem. Eng. News*, 81, 104, 2003.

24. Ganesan, A., Natural products as a hunting ground for combinatorial chemistry, *Curr. Opin. Biotechnol.*, 15, 584, 2004.

25. Boldi, A.M., Libraries from natural product-like scaffolds, *Curr. Opin. Chem. Biol.*, 8, 281, 2004.

26. Vuorela, P. et al., Natural products in the process of finding new drug candidates, *Curr. Med. Chem.*, 11, 1375, 2004.

27. Abel, U. et al., Modern methods to produce natural product libraries, *Curr. Opin. Chem. Biol.*, 6, 453, 2002.

28. Ortholand, J.Y. and Ganesan, A., Natural products and combinatorial chemistry: back to the future, *Curr. Opin. Chem. Biol.,* 8, 271, 2004.

29. Piggott, A.M. and Karuso, P., Quality, not quantity: the role of natural products and chemical proteomics in modern drug discovery, *Comb. Chem. High Throughput Screen.*, 7, 607, 2004.

30. Tan, D.S., Current progress in natural product-like libraries for discovery screening, *Comb. Chem. High Throughput Screen.*, 7, 631, 2004.

31. Schröder, F.C. et al., Combinatorial chemistry in insects: a library of defensive macrocyclic polyamines, *Science*, 281, 428, 1998.
32. Schröder, F.C. et al., A combinatorial library of macrocyclic polyamines produced by a ladybird beetle, *J. Am. Chem. Soc.*, 122, 3628, 2000.
33. Staerk, D. et al., Isolation of a library of aromadendranes from *Landolphia dulcis* and its characterization using the VolSurf approach, *J. Nat. Prod.*, 67, 799, 2004.
34. Feher, M. and Schmidt, J.M., Property distributions: differences between drugs, natural products, and molecules from combinatorial chemistry, *J. Chem. Inf. Comput. Sci.*, 43, 218, 2003.
35. Williams, D.H. et al., Why are secondary metabolites (natural products) biosynthesized?, *J. Nat. Prod.*, 52, 1189, 1989.
36. Ganesan, A., Recent developments in combinatorial organic synthesis, *Drug Discovery Today*, 7, 47, 2002.
37. Goodnow Jr., R.A., Guba, W., and Haap, W., Library design practices for success in lead generation with small molecule libraries, *Comb. Chem. High Throughput Screen.*, 6, 649, 2003.
38. Arya, P., Joseph, R., and Chou, D.T., Toward high-throughput synthesis of complex natural product-like compounds in the genomics and proteomics age, *Chem. Biol.*, 9, 145, 2002.
39. Watson, C., Polymer-supported synthesis of non-oligomeric natural products, *Angew. Chem., Int. Ed. Engl.*, 38, 1903, 1999.
40. Thompson, L.A., Recent applications of polymer-supported reagents and scavengers in combinatorial, parallel, or multistep synthesis, *Curr. Opin. Biotechnol.*, 4, 324, 2000.
41. Wilson, L.J., Recent advances in solid-phase synthesis of natural products, in *Solid-Phase Organic Synthesis*, Burgess, K., Ed., John Wiley & Sons, New York, 2000, chap. 8.
42. Sun, C.-M., Recent advances in liquid-phase combinatorial chemistry, *Comb. Chem. High Throughput Screen.*, 2, 299, 1999.
43. Schreiber, S.L., Target-oriented and diversity-oriented organic synthesis in drug discovery, *Science*, 287, 1964, 2000.
44. Liao, Y. et al., Diversity oriented synthesis and branching reaction pathway to generate natural product-like compounds, *Curr. Med. Chem.*, 10, 2285, 2003.
45. Burke, M.D. and Schreiber, S.L., A planning strategy for diversity-oriented synthesis, *Angew. Chem. Int. Ed. Engl.*, 43, 46, 2004.
46. Niggemann, J. et al., Natural product-derived building blocks for combinatorial synthesis. Part 1. Fragmentation of natural products from myxobacteria, *J. Chem. Soc., Perkin Trans. 1*, 22, 2490, 2002.
47. Metha, G. and Singh, V., Hybrid systems through natural product leads: an approach towards new molecular entities, *Chem. Soc. Rev.*, 31, 324, 2002.
48. Tietze, L.F., Bell, H.P., and Chandrasekhar S., Natural product hybrids as new leads for drug discovery, *Angew. Chem., Int. Ed. Engl.*, 42, 3996, 2003.
49. Evans, B.E. et al., Methods for drug discovery: development of potent, selective, orally effective cholecystokinin antagonists, *J. Med. Chem.*, 31, 2235, 1998.
50. Varki, A., Biological roles of oligosaccharides: all of the theories are correct, *Glycobiology*, 3, 97, 1993.
51. Dwek, R.A., Glycobiology: toward understanding the function of sugars, *Chem. Rev.*, 96, 683, 1996.
52. Sofia, M.J., Oligosaccharide and glycoconjugate solid-phase synthesis technologies for drug discovery, in *Combinatorial Chemistry and Molecular Diversity in Drug Discovery*; Gordon, E.M. and Kerwin, J.F., Jr, Eds., Wiley-Liss, New York, 1998, chap. 13.
53. Zhang, Z. et al., Programmable one-pot oligosaccharide synthesis, *J. Am. Chem. Soc.*, 121, 734, 1999.
54. Karlsson, K.-A., Glycobiology: a growing field for drug design, *Trends Pharmacol. Sci.*, 12, 265, 1991.
55. Feizi, T.J., Oligosaccharides that mediates mammalian cell-cell adhesion, *Curr. Opin. Struct. Biol.*, 3, 701, 1993.
56. Sharon, N. and Lis, H., Carbohydrates in cell recognition, *Sci. Am.*, 268, 82, 1993.
57. Weymouth-Wilson, A.C., The role of carbohydrates in biologically active natural products, *Nat. Prod. Rep.*, 14, 99, 1997.
58. Bechthold, A. and Fernández, J.A.S., Combinatorial biosynthesis of microbial metabolites, in *Combinatorial Chemistry: Synthesis, Analysis, Screening*; Jung, G., Ed., Wiley-VCH, Weinheim, 1999; chap. 12.

59. Sofia, M.J., Chemical strategies for introducing carbohydrate molecular diversity into the drug discovery process, *Network Sci.* 1996, http://www.netsci.org/Science/Combichem/feature12.html, accessed on September 22, 2003.

60. Kahne, D., Combinatorial approaches to carbohydrates, *Curr. Opin. Chem. Biol.*, 1, 130, 1997.

61. Arya, P. and Ben, R.N., Combinatorial chemistry for the synthesis of carbohydrate libraries, *Angew. Chem. Int. Ed. Engl.*, 36, 1280, 1997.

62. Wang, Z.G. and Hindsgaul, O., Combinatorial carbohydrate chemistry, *Adv. Exp. Med. Biol.*, 435, 219, 1998.

63. Sofia, M.J., Carbohydrate-based combinatorial libraries, *Mol. Diversity,* 3, 75, 1997–1998.

64. Obrecht, D. and Villalgordo, J.M., Solid-supported combinatorial and parallel synthesis of small-molecular-weight compound libraries, in *Tetrahedron Organic Chemistry Series*, Vol. 17, Baldwin, J.E., Williams, F.R.S., and R. M., Eds., Pergamon, Elsevier Science, Oxford, 1998.

65. Bunin, B.A., *The Combinatorial Index*, Academic Press, San Diego, CA, 1998.

66. Ito, Y. and Manabe, S., Solid-phase oligosaccharide synthesis and related technologies, *Curr. Opin. Chem. Biol.*, 2, 701, 1998.

67. Osborn, H.M.I. and Khan, T.H., Recent developments in polymer supported syntheses of oligosaccharides and glycopeptides, *Tetrahedron*, 55, 1807, 1999.

68. Schweizer, F. and Hindsgaul, O., Combinatorial synthesis of carbohydrates, *Curr. Opin. Chem. Biol.*, 3, 291, 1999.

69. Takuya, K. and Osamu, K., Carbohydrate-related libraries, *Trends Glycosci. Glycotechnol.*, 11, 267, 1999.

70. Edwards, P.J. et al., Applications of combinatorial chemistry to drug design and development, *Curr. Opin. Drug Discovery Dev.*, 2, 321, 1999.

71. Haase, W.-C. and Seeberger, P.H., Recent progress in polymer-supported synthesis of oligosaccharides and carbohydrate libraries, *Curr. Org. Chem.*, 4, 481, 2000.

72. Hilaire, P.M.St. and Meldal, M., Gylcopeptide and oligosaccharide libraries, *Angew. Chem. Int. Ed. Engl.*, 39, 1162, 2000.

73. Hilaire, P.M.St. and Meldal, M., Glycopeptide and oligosaccharide libraries, in *Combinatorial Chemistry: Synthesis, Analysis, Screening*, Jung, G., Ed., Wiley-VCH, Weinheim, 1999, chap. 8.

74. Brown, F., Gund, P., and Maliski, E., The management of chemical and biological information, *Curr. Opin. Drug. Discovery Dev.*, 3, 268, 2000.

75. Schriemer, D.C. and Hindsgaul, O., Deconvolution approaches in screening compound mixtures, *Comb. Chem. High Throughput Screen.*, 1, 155, 1998.

76. Tseng, K., Hedrick, J.L., and Lebrilla, C.B., Catalog-library approach for the rapid and sensitive structural elucidation of oligosaccharides, *Anal. Chem.*, 71, 3747, 1999.

77. Barkley, A. and Arya, P., Combinatorial chemistry toward understanding the function(s) of carbohydrate conjugates, *Chem. Eur. J.*, 7, 555, 2001.

78. Kanemitsu, T. and Kanie, O., Recent developments in oligosaccharide synthesis: tactics, solid-phase synthesis and library synthesis, *Comb. Chem. High Throughput Screen.*, 5, 339, 2002.

79. Marcaurelle L.A. and Seeberger P.H., Combinatorial carbohydrate chemistry, *Curr. Opin. Chem. Biol.*, 6, 289, 2002.

80. Randell, K.D., Barkley, A., and Arya, P., High-throughput chemistry toward complex carbohydrates and carbohydrate-like compounds, *Comb. Chem. High Throughput Screen.*, 5, 179, 2002.

81. Chakraborty, T.K., Ghosh, S., and Jayaprakash, S., Sugar amino acids and their uses in designing bioactive molecules, *Curr. Med. Chem.*, 9, 421, 2002.

82. Chakraborty, T.K., Jayaprakash, S., and Ghosh, S., Sugar amino acid based scaffolds — novel peptidomimetics and their potential in combinatorial synthesis, *Comb. Chem. High Throughput Screening*, 5, 373, 2002.

83. Gruner, S.A. et al., Carbohydrate-based mimetics in drug design: sugar amino acids and carbohydrate scaffolds, *Chem. Rev.*, 102, 491, 2002.

84. Schweizer, F., Glycosamino acids: building blocks for combinatorial synthesis — implications for drug discovery, *Angew. Chem. Int. Ed. Engl.*, 41, 230, 2002.

85. Nicolaou, K.C. et al., Solid-phase synthesis of oligosaccharides: construction of a dodecasaccharide, *Angew. Chem. Int. Ed. Engl.*, 37, 1559, 1998.

86. Christ, W.J. et al., E5531, a pure endotoxin antagonist of high potency, *Science*, 268, 80, 1995.

87. Fukase, K. et al., Synthesis of an analog of biosynthetic precursor IA of lipid A by an improved method: a novel antagonist containing four (*S*)-3-hydroxy fatty acids, *Tetrahedron Lett.*, 36, 7455, 1995.

88. Oikawa, M. et al., New efficient synthesis of a biosynthetic precursor of lipid A, *Bull. Chem. Soc. Jpn.*, 70, 1435, 1997.

89. Fukase, K. et al., Divergent synthesis and biological activities of lipid A analogs of shorter acyl chains, *Tetrahedron*, 54, 4033, 1998.

90. Liu, W-C. et al., A divergent synthesis of lipid A and its chemically stable unnatural analogs, *Bull. Chem. Soc. Jpn.*, 72, 1377, 1999.

91. Liang, R. et al., Parallel synthesis and screening of a solid phase carbohydrate library, *Science*, 274, 1520, 1996.

92. Lehn, J.-M., Dynamic combinatorial chemistry and virtual combinatorial libraries, *Chem. Eur. J.*, 5, 2455, 1999.

93. Ramstrom, O. and Lehn, J.M., Drug discovery by dynamic combinatorial library, *Nat. Rev. Drug Discovery*, 1, 26, 2002.

94. Ramstrom, O. et al., Dynamic combinatorial carbohydrate libraries: probing the binding site of the concanavalin A lectin, *Chem. Eur. J.*, 10, 1711, 2004.

95. Zameo, S., Vauzeilles, B., and Beau, J.M., Dynamic combinatorial chemistry: lysozyme selects an aromatic motif that mimics a carbohydrate residue, *Angew. Chem. Int. Ed. Engl.*, 44, 965, 2005.

96. Sofia, M.J. et al., Discovery of novel disaccharide antibacterial agents using a combinatorial library approach, *J. Med. Chem.*, 42, 3193, 1999, and references cited therein.

97. Nicolaou, K.C. et al., Radiofrequency encoded combinatorial chemistry, *Angew. Chem., Int. Ed. Engl.*, 34, 2289, 1995.

98. Park, W.K.C. et al., Rapid combinatorial synthesis of aminoglycoside antibiotic mimetics: use of a polyethylene glycol-linked amine and a neamine-derived aldehyde in multiple component condensation as a strategy for the discovery of new inhibitors of the HIV RNA Rev responsive element, *J. Am. Chem. Soc.*, 118, 10150, 1996.

99. Greenberg, W.C. et al., Design and synthesis of new aminoglycoside antibiotics containing neamine as an optimal core structure: correlation of antibiotic activity with in vitro inhibition of translation, *J. Am. Chem. Soc.*, 121, 6527, 1999.

100. Wong, C.–H. et al., A library approach to the discovery of small molecules that recognize RNA: use of a 1,3-hydroxyamine motif as core, *J. Am. Chem. Soc.*, 120, 8319, 1998.

101. Sucheck, S.J. et al., Design of small molecules that recognize RNA: development of aminoglycosides as potential antitumor agents that target oncogenic RNA sequences, *Angew. Chem. Int. Ed. Engl.*, 39, 1080, 2000.

102. Li, J. et al., Application of glycodiversification: expedient synthesis and antibacterial evaluation of a library of kanamycin B analogs, *Org. Lett.*, 6, 1381, 2004.

103. Simanek, E.E. et al., Selectin-carbohydrate interactions: from natural ligands to designed mimics, *Chem. Rev.*, 98, 833, 1998.

104. Sutherlin, D.P. et al., Generation of C-glycoside peptide ligands for cell surface carbohydrate receptors using a four-component condensation on solid-support, *J. Org. Chem.*, 61, 8350, 1996.

105. Tsai, C-Y. et al., Synthesis of sialyl Lewis X memetics using Ugi four-component reaction, *Bioorg. Med. Chem.*, 8, 2333, 1998.

106. Chen, S. and Janda, K.D., Synthesis of prostaglandin E_2 methyl ester on a soluble-polymer support for the construction of prostanoid libraries, *J. Am. Chem. Soc.*, 119, 8724, 1995.

107. Chen, S. and Janda, K.D., Total synthesis of naturally occurring prostaglandin F_2 on a non-cross-linked polystyrene support, *Tetrahedron Lett.*, 39, 3943, 1998.

108. Thompson, L.A. et al., Solid-phase synthesis of diverse E- and F-series prostaglandins, *J. Org. Chem.*, 63, 2066, 1998.

109. Suzuki, M. et al., Three-component coupling synthesis of prostaglandins: a simplified, general procedure, *Tetrahedron*, 46, 4809, 1990.

110. Lee, K. J. et al., Soluble-polymer supported synthesis of a prostanoid library: identification of antiviral activity, *Org. Lett.*, 1, 1859, 1999.

111. Pelegrín, J.A.L. and Janda, K.D., Solution- and soluble-polymer supported asymmetric synthesis of six-membered ring prostanoids, *Chem. Eur. J.*, 6, 1917, 2000.

112. Alali, F.Q., Liu, X.X., and McLaughlin, J.L., Annonaceous acetogenins: recent progress, *J. Nat. Prod.*, 62, 504, 1999.
113. Keinan, E. et al., Towards chemical libraries of annonaceous acetogenins, *Pure Appl. Chem.*, 69, 423, 1997.
114. Chang, Y.T. et al., The synthesis and biological characterization of a ceramide library, *J. Am. Chem. Soc.*, 124, 1856, 2002.
115. Nicolaou, K.C. et al., Solid phase synthesis of macrocycles by an intramolecular ketophosphonate reaction: synthesis of a (*dl*)-muscone library, *J. Am. Chem. Soc.*, 120, 10814, 1998.
116. Lumbach, G.W.M., Berends, W., and Cox, H.C., Elucidation of the chemical structure of bongkrekic acid-I: isolation, purification and properties of bongkrekic acid, *Tetrahedron*, 26, 5993, 1969.
117. Pei, Y. et al., Design and combinatorial synthesis of N-acyl iminodiacetic acids as bongkrekic acid analogs for the inhibition of adenine nucleotide translocase, *Synthesis*, 11, 1717, 2003.
118. O'Hagan, D., Biosynthesis of fatty acid and polyketide metabolites, *Nat. Prod. Rep.*, 12, 1, 1995.
119. Rawlings, B.J., Biosynthesis of polyketides (other than actinomycete macrolides), *Nat. Prod. Rep.*, 16, 425, 1999.
120. Borchardt, J.K., Combinatorial biosynthesis: panning for pharmaceutical industry, *Mod. Drug Discovery*, 2, 22, 1999.
121. Bentley, R. and Bennett, J.W., Constructing polyketides: from collie to combinatorial biosynthesis, *Annu. Rev. Microbiol.*, 53, 411, 1999.
122. Tsoi, C.J. and Khosla, C., Combinatorial biosynthesis of "unnatural" natural products: the polyketide example, *Chem. Biol.*, 2, 355, 1995.
123. Fu, H. and Khosla, C., Antibiotic activity of polyketide products derived from combinatorial biosynthesis: implications for directed evolution, *Mol. Diversity*, 1, 121, 1996.
124. Staunton, J., Combinatorial biosynthesis of erythromycin and complex polyketides, *Curr. Opin. Chem. Biol.*, 2, 339, 1998.
125. Hutchinson, C.R., Combinatorial biosynthesis of antibiotics, in *Drug Discovery from Nature*, Grabley, S. and Thiericke, R., Eds., Springer-Verlag, Heidelberg, 1999, chap. 13.
126. Khosla, C., Combinatorial biosynthesis: new tools for the medicinal chemists, *Chemtracts Org. Chem.*, 11, 1, 1998.
127. Lowe, G., Oligomeric and biogenetic combinatorial libraries, *Nat. Prod. Rep.*, 16, 641, 1999.
128. Khosla, C., Combinatorial chemistry of "unnatural" natural products, in *Combinatorial Chemistry and Molecular Diversity in Drug Discovery*; Gordon, E.M and Kerwin Jr, J.F., Eds., Wiley-Liss, New York, 1998, chap. 21.
129. Reynolds, K.A., Combinatorial biosynthesis: lesson learned from nature, *Proc. Natl. Acad. Sci. U.S.A.*, 95, 12744, 1998.
130. Shen, B., Liu, W., and Nonaka, K., Enedyne natural products: biosynthesis and prospect towards engineering novel antitumor agents, *Curr. Med. Chem.*, 10, 2317, 2003.
131. Newman, D.J., Cragg, G.M., and Snader, K.M., The influence of natural products upon drug discovery, *Nat. Prod. Rep.*, 17, 215, 2000.
132. Nicolaou, K.C., Roschangar, F., and Vourloumis, D., Chemical biology of epothilones, *Angew. Chem. Int. Ed. Engl.*, 37, 2014, 1998.
133. Nicolaou, K.C., Sorensen, E.J., and Winssinger, N., The art and science of organic and natural products synthesis, *J. Chem. Educ.*, 75, 1225, 1998.
134. Storer R.I. et al., Multi-step application of immobilized reagents and scavengers: a total synthesis of epothilone C, *Chem. Eur. J.*, 10, 2529, 2004.
135. Storer R.I. et al., A total synthesis of epothilones using solid-supported reagents and scavengers, *Angew. Chem. Int. Ed. Engl.*, 42, 2521, 2003.
136. Nicolaou, K.C. et al., Synthesis of epothilones A and B in solid and solution phase, *Nature*, 387, 268, 1997.
137. Nicolaou, K.C. et al., Designed epothilones: combinatorial synthesis, tubulin assembly properties, and cytotoxic action against Taxol-resistant tumor cells, *Angew. Chem. Int. Ed. Engl.*, 36, 2097, 1997.
138. Fukami, A. et al., Macrosphelide B suppressed metastasis through inhibition of adhesion of sLe^x/E-selectin molecules, *Biochem. Biophys. Res. Commun.*, 291, 1065, 2002.
139. Takahashi, T. et al., Combinatorial synthesis of a macrosphelide library utilizing a palladium-catalyzed carbonylation on a polymer support, *Angew. Chem. Int. Ed. Engl.*, 42, 5230, 2003.

140. Zanze, I.A. and Sowin, T.J., Solid-phase synthesis of macrolide analogs, *J. Comb. Chem.*, 3, 301, 2001.
141. Frank, R., Spot-synthesis: an easy technique for the positionally addressable, parallel chemical synthesis on a membrane support, *Tetrahedron*, 48, 9217, 1992.
142. Misske, A.M. and Hoffmann, H.M.R., High stereochemical diversity and applications for the synthesis of marine natural products: a library of carbohydrate mimics and polyketide segments, *Chem. Eur. J.*, 6, 3313, 2000.
143. Paterson, I. and Temal-Laïb, T., Toward the combinatorial synthesis of polyketide libraries: asymmetric aldol reactions with α-chiral aldehydes on solid-support, *Org. Lett.*, 4, 2473, 2002.
144. Paterson, I., Donghi, M., and Gerlach, K., A combinatorial approach to polyketide-type libraries by iterative asymmetric aldol reactions performed on solid-support, *Angew. Chem. Int. Ed. Engl.*, 39, 3315, 2000.
145. Minguez, J.M. et al., Synthesis and biological assessment of simplified analogs of the potent microtubule stabilizer (+)-discodermolide, *Bioorg. Med. Chem.*, 11, 3335, 2003.
146. Curran, D.P. and Furukawa, T., Simultaneous preparation of four truncated analogs of discodermolide by fluorous mixture synthesis, *Org. Lett.*, 4, 2233, 2002.
147. Nicolaou, K.C. et al., Chemistry, biology, and medicine of the glycopeptide antibiotics, *Angew. Chem. Int. Ed. Engl.*, 38, 2096, 1999.
148. Gao, Y., Glycopeptide antibiotics and development of inhibitors to overcome vancomycin resistance, *Nat. Prod. Rep.*, 19, 100, 2002.
149. Xu, R. et al., Combinatorial library approach for the identification of synthetic receptors targeting vancomycin-resistant bacteria, *J. Am. Chem. Soc.*, 121, 4898, 1999.
150. Kateri, A.A. et al., Identification of potent and broad-spectrum antibiotics from SAR studies of a synthetic vancomycin analog, *Bioorg. Med. Chem. Lett.*, 13, 1683, 2003.
151. Yasukata, T. et al., An efficient and practical method for solid-phase synthesis of tripeptide-bearing glycopeptide antibiotics: combinatorial parallel synthesis of carboxamide derivatives of chloroorientacin B, *Bioorg. Med. Chem. Lett.*, 12, 3033, 2002.
152. Nicolaou, K.C. et al., Synthesis and biological evaluation of vancomycin dimmers with potent activity against vancomycin-resistant bacteria: target-accelerated combinatorial synthesis, *Chem. Eur. J.*, 7, 3824, 2001.
153. Nicolaou, K.C. et al., Solid- and solution-phase synthesis of vancomycin and vancomycin analogs with activity against vancomycin-resistant bacteria, *Chem Eur. J.*, 7, 3798, 2001.
154. Spring, D.R., Krishnan, S., and. Schreiber, S.L., Towards diversity-oriented, stereoselective syntheses of biaryl- or bis(aryl)metal-containing medium rings, *J. Am. Chem. Soc.*, 122, 5656, 2000.
155. Cristau, P., Vors, J.P., and Zhu, J., Rapid and diverse route to natural product-like biaryl ether containing macrocycles, *Tetrahedron,* 59, 7859, 2003.
156. Kiselyov, A.S., Eisenberg, S., and Luo, Y., Solid-support synthesis of 14-membered macrocycles *via* S_NAr methodology on acrylate resin, *Tetrahedron Lett.*, 40, 2465, 1999.
157. Leitheiser, C.J. et al., Solid phase synthesis of bleomycin group antibiotics: construction of a 108-member deglycobleomycin library, *J. Am. Chem. Soc.*, 12, 8218, 2003.
158. Qin, C. et al., Optimization of antibacterial cyclic decapeptides, *J. Comb. Chem.*, 6, 398, 2004.
159. Qin, C. et al., A chemical approach to generate molecular diversity based on the scaffold of cyclic decapeptide antibiotic tyrocidine A, *J. Comb. Chem.*, 5, 353, 2003.
160. Sedrani, R. et al., Sanglifehrin-cyclophilin interaction: degradation work, synthetic macrocyclic analogs, X-ray crystal structure, and binding data, *J. Am. Chem. Soc.*, 125, 3849, 2003.
161. Herman, C. et al., Synthesis of hapalosin analogs by solid-phase assembly of acyclic precursors, *Tetrahedron*, 57, 8999, 2001.
162. Chen, Y. et al., Solution-phase parallel synthesis of a pharmacophore library of HUN-7293 analogs: a general chemical mutagenesis approach to defining structure-function properties of naturally occurring cyclic (depsi)peptides, *J. Am. Chem. Soc.*, 124, 5431, 2002.
163. Boger, D.L. et al., Total synthesis of HUN-7293, *J. Am. Chem. Soc.*, 121, 6197, 1999.
164. Takahashi, T. et al., Solid phase library synthesis of cyclic depsipeptides: aurilide and aurilide analogs, *J. Comb. Chem.*, 5, 414, 2003.
165. Bozzoli, A., et al., A solid-phase approach to analogs of the antibiotic mureidomycin, *Bioorg. Med. Chem. Lett.*, 10, 2759, 2000.

166. Thutewohl, M. et al., Solid-phase synthesis and biological evaluation of a pepticinnamin E library, *Angew. Chem. Int. Ed. Engl.*, 41, 3616, 2002.

167. Thutewohl, M. and Waldmann, H., Solid-phase synthesis of a pepticinnamin E library, *Bioorg. Med. Chem.*, 11, 2591, 2003.

168. Thutewohl, M. et al., Identification of mono- and bisubstrate inhibitors of protein farnesyltransferase and inducers of apoptosis from a pepticinnamin E library, *Bioorg. Med. Chem.*, 11, 2617, 2003.

169. Suda, A. et al., Combinatorial synthesis of nikkomycin analogs on solid-support, *Heterocycles,* 55 1023, 2001.

170. Arcamone, F. et al., Structure and synthesis of distamycin, *Nature*, 203, 1064, 1964.

171. Dale, D.L., Fink, B.E., and Hedrick. J.L., Total synthesis of distamycin A and 2640 analogs: a solution-phase combinatorial approach to the discovery of new, bioactive DNA binding agents and development of a rapid, high-throughput screen for determining relative DNA binding affinity or DNA binding sequence selectivity, *J. Am. Chem. Soc.*, 122, 6382, 2000.

172. Kingston, D.G.I., Recent advances in the chemistry of Taxol, *J. Nat. Prod.*, 63, 726, 2000.

173. Kingston, D.G.I., Taxol, a molecule for all seasons, *Chem. Commun.*, 867, 2001.

174. Xiao, X.-Y., Parandoosh, Z., and Nova, M.P., Design and synthesis of a taxoid library using radio-frequency encoded combinatorial chemistry, *J. Org. Chem.*, 62, 6029, 1997.

175. Bhat, L. et al., Synthesis and evaluation of paclitaxel C7 derivatives: solution phase synthesis of combinatorial libraries, *Bioorg. Med. Chem. Lett.*, 8, 3181, 1998.

176. Liu, Y. et al., A systematic SAR study of C10 modified paclitaxel analogs using a combinatorial approach, *Comb. Chem. High Throughput Screen.*, 5, 39, 2002.

177. Jagtap, P.G. et al., Design and synthesis of a combinatorial chemistry library of 7-acyl, 10-acyl, and 7,10-diacyl analogs of paclitaxel (taxol) using solid phase synthesis, *J. Nat. Prod.*, 65, 1136, 2002.

178. D'Ambrosio, M., Guerriero, A., and Pietra, F., Sarcodictyin A and sarcodictyin B, novel diterpenoidic alcohols esterified by (*E*)-*N*(1)-methylurocanic acid: isolation from the Mediterranean stolonifer *Sarcodictyon roseum*, *Helv. Chim. Acta*, 70, 2019, 1987.

179. D'Ambrosio, M., Guerriero, A., and Pietra, F., Isolation from the Mediterranean stoloniferan coral *Sarcodictyon roseum,* of sarcodictyin C, D, E, and F, novel diterpenoidic alcohols esterified by (*E*)- or (*Z*)- *N*(1)-methylurocanic acid: failure of the carbon-skeleton type as a classification criterion, *Helv. Chim. Acta*, 71, 964, 1988.

180. Schiff, P.B., Fant, J., and Horwitz, S.B., Promotion of microtubule assembly in vitro by Taxol, *Nature*, 277, 665, 1979.

181. Nicolaou, K.C. et al., Synthesis of the tricyclic core of eleutherobin and sarcodictyins and total synthesis of sarcodictyin A, *J. Am. Chem. Soc.*, 119, 11353, 1997.

182. Nicolaou, K.C. et al., Solid and solution phase synthesis and biological evaluation of combinatorial sarcodictyin libraries, *J. Am. Chem. Soc.*, 120, 10814, 1998.

183. Amroyan, E. et al., Inhibitory effect of andrographolide from *Andrographis paniculata* on PAF-induced platelet aggregation, *Phytomedicine*, 6, 27, 1999.

184. Biabani, M.A.F. et al., A novel diterpenoid lactone-based scaffold for the generation of combinatorial libraries B, *Tetrahedron Lett.*, 42, 7119, 2001.

185. Pathak, A. et al., Synthesis of combinatorial libraries based on terpenoid scaffolds, *Comb. Chem. High Throughput Screen*, 5, 241, 2002.

186. Srinivasan, T et al., Solid-phase synthesis and bioevaluation of lupeol-based libraries as antimalarial agents, *Bioorg. Med. Chem. Lett.*, 12, 2803, 2002.

187. Gunasekera, S.P. et al., Dysidiolide: a novel protein phosphatase inhibitor from the Caribbean sponge *Dysidea etheria* de Laubenfels, *J. Am. Chem. Soc.*, 118, 8759, 1996.

188. Brohm, D. et al., Natural products are biologically validated starting points in structural space for compound library development: solid-phase synthesis of dysidiolide-derived phosphatase inhibitors, *Angew. Chem., Int. Ed. Engl.*, 41, 307, 2002.

189. Brohm, D. et al., Solid-phase synthesis of dysidiolide-derived protein phosphatase inhibitor*s, J. Am. Chem. Soc.*, 124, 13171, 2002.

190. Kobayashi, J., Madono T., and Shigemori H., Nakijiquinones C and D, new sesquiterpenoid quinones with a hydroxy amino acid residue from a marine sponge inhibiting c-erbB-2 kinase, *Tetrahedron*, 51, 10867, 1995.

191. Kissau L. et al., Development of natural product-derived receptor tyrosine kinase inhibitors based on conservation of protein domain fold, *J. Med. Chem.*, 46, 2917, 2003.
192. Hirschmann, R. et al., The versatile steroid nucleus: design and synthesis of a peptidomimetic employing this novel scaffold, *Tetrahedron*, 49, 3665, 1993.
193. Maltais, R. et al., Steroids and combinatorial chemistry, *J. Comb. Chem.*, 6, 443, 2004.
194. Tremblay, M.R. and Poirier, D., Solid-phase synthesis of phenolic steroids: from optimization studies to a convenient procedure for combinatorial synthesis of biologically relevant estradiol derivatives, *J. Comb. Chem.*, 2, 48, 2000.
195. Maltais, R., Tremblay, M.R., and Poirier, D., Solid-phase synthesis of hydroxysteroid derivatives using the diethylsilyloxy linker, *J. Comb. Chem.*, 2, 604, 2000.
196. Maltais, R., Luu-The, V., and Poirier, D., Parallel solid-phase synthesis of 3β-peptido-3α-hydroxy-5α-androstan-17-one derivatives for inhibition of type 3 17β-hydroxysteroid dehydrogenase, *Bioorg. Med. Chem.*, 9, 3101, 2001.
197. Maltais, R., Luu-The, V., and Poirier, D., Synthesis and optimization of a new family of type 17β-hydroxysteroid dehydrogenase inhibitors by parallel liquid phase chemistry, *J. Med. Chem.*, 45, 640, 2002.
198. Ciobanu, L.C. and Poirier, D., Solid-phase parallel synthesis of 17α-substituted estradiol sulfamate and phenol libraries using the multidetachable sulfamate linker, *J. Comb. Chem.*, 5, 429, 2003.
199. Hanson, R.N. et al., Synthesis and evaluation of 17α-20E-21-(4-substituted phenyl)-19-norpregna-1,3,5(10), 20-tetraene-3,17β-diols as probes for the estrogen receptor a hormone binding domain, *J. Med. Chem.*, 46, 2865, 2003.
200. Hijikuro, I., Doi, T., and Takahashi, T., Parallel synthesis of a vitamin D_3 library in the solid-phase, *J. Am. Chem. Soc.*, 123, 3716, 2001.
201. Atuegbu, A. et al., Combinatorial modification of natural products: preparation of unencoded and encoded libraries of *Rauwolfia* alkaloids, *Bioorg. Med. Chem. Lett.*, 4, 1097, 1996.
202. Nielsen, T.E., Diness, F., and Meldal, M., The Pictet-Spengler reaction in solid-phase combinatorial chemistry, *Curr. Opin. Drug Discovery Dev.*, 6, 801, 2003.
203. Loevezijn, A.V. et al., Solid phase synthesis of fumitremorgin, verruculogen and tryprostatin analogs based on a cyclization/cleavage strategy, *Tetrahedron Lett.*, 39, 4737, 1998.
204. Loevezijn, A.V. et al., Inhibition of BCRP-mediated drug efflux by fumitremorgin-type indolyl diketopiperazines, *Bioorg. Med. Chem. Lett.*, 11, 29, 2001.
205. Kundu, B., Solid-phase strategies for the design and synthesis of heterocyclic molecules of medicinal interest, *Curr. Opin. Drug Discovery Dev.*, 6, 815, 2003.
206. Wang, H. and Ganesan, A., The *N*-acyliminium Pictet-Spengler condensation as a multicomponent combinatorial reaction on solid phase and its application to the synthesis of demethoxyfumitremorgin C analogs, *Org. Lett.*, 1, 1647, 1999.
207. Bonnet, D. and Ganesan, A., Solid-phase synthesis of tetrahydro-beta-carbolinehydantoins via the N-acyliminium Pictet-Spengler reaction and cyclative cleavage, *J. Comb. Chem.*, 4, 546, 2002.
208. Hotha, S. et al., HR22C16: a potent small-molecule probe for the dynamics of cell division, *Angew. Chem. Int. Ed. Engl.*, 42, 2379, 2003.
209. Rosenbaum, C. et al., Synthesis and biological evaluation of an indomethacin library reveals a new class of angiogenesis-related kinase inhibitors, *Angew. Chem. Int. Ed. Engl.*, 43, 224, 2004.
210. Schuna, A.A., Update on treatment of rheumatoid arthritis, *J. Am. Pharm. Assoc.*, 38, 728, 1998.
211. Goodnight, S.H., Aspirin therapy for cardiovascular disease, *Curr. Opin. Hematol.*, 3, 355, 1996.
212. Flynn, B.L. and Theesen, K.A., Pharmacologic management of Alzheimer disease part III: nonsteroidal antiinflammatory drugs — Emerging protective evidence?, *Ann. Pharmacother.*, 33, 840, 1999.
213. Sloane, P.D., Advances in the treatment of Alzheimer's disease, *Am. Fam. Physician*, 58, 1577, 1998.
214. Jones, M.K. et al., Inhibition of angiogenesis by nonsteroidal anti-inflammatory drugs: insight into mechanisms and implications for cancer growth and ulcer healing, *Nat. Med.*, 5, 1418, 1999.
215. Willoughby, C.A. et al., Combinatorial synthesis of 3-(amidoalkyl) and (aminoalkyl)-2-arylindole derivatives: discovery of potent ligands for a variety of G-protein coupled receptors, *Bioorg. Med. Chem. Lett.*, 12, 93, 2002.
216. Nishizuka, Y., Intracellular signaling by hydrolysis of phospholipids and activation of protein-kinase-C, *Science*, 258, 607, 1992.

217. Meseguer, B. et al., Natural product synthesis on polymeric supports-synthesis and biological evaluation of an indolactam library, *Angew. Chem. Int. Ed. Engl.,* 38, 2902, 1999.
218. Meseguer, B. et al., Solid-phase synthesis and biological evaluation of a teleocidin library-discovery of a selective PKCδ down regulator, *Chem. Eur. J.,* 6, 3943, 2000.
219. Dewick, P.M., *Medicinal Natural Products: A Biosynthetic Approach,* Wiley-Interscience, New York, 2002.
220. Arya, P. et al., A solid-phase library synthesis of hydroxyindoline-derived tricyclic derivatives by Mitsunobu approach, *J. Am. Chem. Soc.,* 6, 65, 2004.
221. Lo, M.M.-C. et al., A library of spirooxindoles based on a stereoselective three-component coupling reaction, *J. Am. Chem. Soc.,* 126, 16077, 2004.
222. Cui, C.-B., Kakeya, H., and Osada, H., Novel mammalian cell cycle inhibitors, cyclotroprostatins A–D, produced by *Aspergillus fumigatus,* which inhibit mammalian cell cycle at G2/M phase, *Tetrahedron,* 53, 59, 1997.
223. Sebahar, P.R. et al., Asymmetric, stereocontrolled total synthesis of (+) and (–)-spirotryprostatin B via a diastereoselective azomethine ylide [1,3]-dipolar cycloaddition reaction, *Tetrahedron,* 58, 6311, 2002.
224. Govindachari, T.R., Ravindranath, K.R., and Viswanathan, N., Mappicine, a minor alkaloid from *Mappia foetida* miers, *J. Chem. Soc., Perkin Trans I,* 1215, 1974.
225. Wu, T.S. et al., Nothapodytines A and B from *Nothapodytes foetida, Phytochemistry,* 42, 907, 1996.
226. Frutos, O. and Curran, D.P., Solution-phase synthesis of libraries of polycyclic natural product analogs by cascade radical annulation: synthesis of a 64-member library of mappicine analogs and a 48-Member library of mappicine ketone analogs, *J. Comb. Chem.,* 2, 639, 2000.
227. Zhang, W. et al., Solution-phase preparation of a 560-compound library of individual pure mappicine analogs by fluorous mixture synthesis, *J. Am. Chem. Soc.,* 124, 10443, 2002.
228. Remers, A.W., *The Chemistry of Antitumor Antibiotics,* Wiley-Interscience, New York, 1988.
229. Myers, A.G. and Lanman, B.A., A solid-supported, enantioselective synthesis suitable for the rapid preparation of large numbers of diverse structural analogs of (-)-saframycin A, *J. Am. Chem. Soc.,* 124, 12969, 2002.
230. Orain, D., Koch, G., and Giger, R., From solution-phase studies to solid-phase synthesis: a new indole based scaffold for combinatorial chemistry, *Chimia,* 57, 255, 2003.
231. Armstrong, R.W. et al., Multiple-component condensation strategies for combinatorial library synthesis, *Acc. Chem. Res.,* 29, 123, 1996.
232. Baudelle, R. et al., Parallel synthesis of polysubstituted tetrahydroquinolines, *Tetrahedron,* 54, 4125, 1998.
233. Kobayashi, S., Komiyama, S., and Ishitani, H., A convenient method for library construction: parallel synthesis of β-amino ester and quinoline derivatives in liquid phase using Ln(OTf)₃-catalyzed three-component reactions, *Biotechnol. Bioeng. (Com. Chem.),* 61, 23, 1998.
234. Kiselyov, A.S., Smith L., II, and Armstrong, R.W., Solid-support synthesis of polysubstituted tetrahydroquinolines via three-component condensation catalyzed by Yb(OTf)₃, *Tetrahedron,* 54, 5089, 1998.
235. Couve-Bonnaire, S. et al., A solid-phase library synthesis of natural product-like derivatives from an enantiomerically pure tetrahydroquinoline scaffold, *J. Am. Chem. Soc.,* 6, 73, 2004.
236. Khadem, S. et al., A solution- and solid-phase approach to tetrahydroquinoline-derived polycyclics having a 10-membered ring, *J. Am. Chem. Soc.,* 6, 724, 2004.
237. Arya, P. et al., Solution- and solid phase synthesis of natural product-like tetrahydroquinoline-based polycyclic having a medium size ring, *J. Am. Chem. Soc.,* 6, 735, 2004.
238. Ivatchenko, A.V. et al., New scaffolds for combinatorial synthesis. 1. 5-Sulfamoylisatins and their reactions with 1,2-diamines, *J. Comb. Chem.,* 4, 419, 2002.
239. Bartzat, R., Applying pattern recognition methods and structure property correlations to determine drug carrier potential of nicotinic acid and analogize to dihydropyridine, *Eur. J. Pharm. Biopharm.,* 59, 63, 2005.
240. Fernàndez, J.-C. et al., Suzuki coupling reaction for the solid-phase preparation of 5-substituted nicotinic acid derivatives, *Tetrahedron Lett.,* 46, 581, 2005.
241. Ivachtchenko, A.V. et al., A parallel solution-phase synthesis of substituted 3,7-diazabicyclo[3.3.1]nonanes, *J. Comb. Chem.,* 6, 828, 2004.

242. Bojadschiewa, M. et al., New method for obtaining cytisine from seeds of *Cytisus laburnum, Pharmazie*, 26, 643, 1971.

243. O'Neill, B.T. et al., Total synthesis of (+/-)-cytisine, *Org. Lett.*, 2, 4201, 2000.

244. Grieder, A. and Thomas, A.W., A concise building block approach to a diverse multi-arrayed library of the circumdatin family of natural products, *Synthesis*, 11, 1707, 2003.

245. Pelish, H.E. et al., Use of biomimetic diversity-oriented synthesis to discover galanthamine-like molecules with biological properties beyond those of the natural product, *J. Am. Chem. Soc.*, 123, 6740, 2001.

246. Geiger, A. et al., Metabolites of microorganisms. 247, Phenazines from *Streptomyces-antibioticus*, Strain Tu 2706, *J. Antibiot.*, 41, 1542, 1988.

247. Bahnmuller, U. et al., Metabolites of microorganisms. 248. Synthetic analogs of saphenamycin, *J. Antibiot.*, 41, 1552, 1998.

248. Laursen, J.B. et al., Solid-phase synthesis of new saphenamycin analogs with antimicrobial activity, *Bioorg. Med. Chem. Lett.*, 12, 171, 2002.

249. Horton, D.A., Bourne, G.T., and Smythe, M.L., The combinatorial synthesis of bicyclic privileged structures or privileged substructures, *Chem. Rev.*, 103, 893, 2003.

250. Nicolaou, K.C., Pfefferkorn, J.A., and Cao, G.-Q., Selenium-based solid-phase synthesis of benzopyrans I: applications to combinatorial synthesis of natural products, *Angew. Chem. Int. Ed. Engl.*, 39, 734, 2000.

251. Nicolaou, K.C., Cao, G.-Q., and Pfefferkorn, J.A., Selenium-based solid-phase synthesis of benzopyrans II: applications to combinatorial synthesis of medicinally relevant small organic molecules, *Angew. Chem. Int. Ed. Engl.*, 39, 739, 2000.

252. Nicolaou, K.C. et al., Natural product-like combinatorial libraries based on privileged structures. 1. General principles and solid-phase synthesis of benzopyrans, *J. Am. Chem. Soc.*, 122, 9939, 2000.

253. Nicolaou, K.C. et al., Natural product-like combinatorial libraries based on privileged structures. 2. Construction of a 10,000-membered benzopyran library by directed split-and-pool chemistry using NanoKans and optical encoding, *J. Am. Chem. Soc.*, 122, 9954, 2000.

254. Nicolaou, K.C. et al., Natural product-like combinatorial libraries based on privileged structures. 3. The "libraries from libraries" principle for diversity enhancement of benzopyran libraries, *J. Am. Chem. Soc.*, 122, 9968, 2000.

255. Nicolaou, K.C. et al., Discovery of novel antibacterial agents active against methicillin-resistant *Staphylococcus aureus* from combinatorial benzopyran libraries, *ChemBioChem.*, 2, 460, 2001.

256. Nicolaou, K.C. et al., Discovery and optimization of non-steroidal FXR agonists from natural product-like libraries, *Org. Biomol. Chem.*, 1, 908, 2003.

257. Fang, N. and Casida, J.E., Anticancer action of cube insecticide: correlation for rotenoid constituents between inhibition of NADH:ubiquinone oxidoreductase and induced ornithine decarboxylase activities, *Proc. Natl. Acad. Sci. U.S.A.*, 95, 3380, 1998.

258. Fang, N. and Casida, J.E., New bioactive flavonoids and stilbenes in Cubé resin insecticide, *J. Nat. Prod.*, 62, 205, 1999.

259. Parmar, V.S. et al., Anti-invasive activity of alkaloids and polyphenolics in vitro, *Bioorg. Med. Chem.*, 5, 1609, 1997.

260. Gunatilaka, A.A.L. et al., Biological activity of some coumarins from Sri Lankan rutaceae, *J. Nat. Prod.*, 57, 518, 1994.

261. Moran, E.J. et al., Radiofrequency tag encoded combinatorial library method for the discovery of tripeptide-substituted cinnamic acid inhibitors of the protein tyrosine phosphatase PTP1B, *J. Am. Chem. Soc.*, 117, 10787, 1995.

262. Ahmed, A.A. et al., A new chromene glucoside from *Ageratum conyzoides, Planta Med.*, 65, 171, 1999.

263. Wuerzberger, S.M. et al., Induction of apoptosis in MCF-7:WS8 breast cancer cells by beta-lapachone, *Cancer Res.*, 58, 1876, 1998.

264. Li, C.J. et al., Potent inhibition of tumor survival in vivo by beta-lapachone plus taxol: combining drugs imposes different artificial checkpoints, *Proc. Natl. Acad. Sci. U.S.A.*, 96, 13369, 1999.

265. Xie, L. et al., Anti-AIDS agents. 37. Synthesis and structure-activity relationships of (3'R,4'R)-(+)-*cis*-khellactone derivatives as novel potent anti-HIV agents, *J. Med. Chem.*, 42, 2662, 1999.

266. Ostresh, J.M. et al., "Libraries from libraries": chemical transformation of combinatorial libraries to extend the range and repertoire of chemical diversity, *Proc. Natl. Acad. Sci. U.S.A.*, 91, 11138, 1994.

267. Breitenbucher, J.G. and Hui, H.C., Titanium mediated reductive amination on solid-support: extending the utility of the 4-hydroxy-thiophenol linker, *Tetrahedron Lett.*, 39, 8207, 1998.

268. Marder, M. et al., Detection of benzodiazepine receptor ligands in small libraries of flavone derivatives synthesized by solution phase combinatorial chemistry, *Biochem. Biophys. Res. Commun.*, 249, 481, 1998.

269. Song, A., Zhang, J., and Lam, K.S., Synthesis and reactions of 7-fluoro-4-methyl-6-nitro-2-oxo-2*H*-1-benzopyran-3-carboxylic acid: a novel scaffold for combinatorial synthesis of coumarins, *J. Comb. Chem.*, 6, 112, 2004.

270. Song, A. et al., Solid-phase synthesis and spectral properties of 2-alkylthio-6*H*-pyrano[2,3-*f*]benzimidazole-6-ones: a combinatorial approach for 2-alkylthioimidazo coumarins, *J. Comb. Chem.*, 6, 604, 2004.

271. Song, A. and Lam, K.S., Parallel solid-phase synthesis of 2-arylamino-6H-pyrano [2,3-f] benzimidazole-6-ones, *Tetrahedron*, 60, 8605, 2004.

272. Xia Y. et al., Asymmetric solid-phase synthesis of (3′R, 4′R)-di-*O*-*cis*-acyl 3-carboxyl khellactone, *Org. Lett.*, 1, 2113, 1999.

273. Mineno, T. et al., Solution-phase parallel synthesis of an isoflavone library for the discovery of novel antigiardial agents, *Comb. Chem. High Throughput Screen.*, 5, 481, 2002.

274. Adams, J.L. and Lee, D., Recent progress towards the identification of selective inhibitors of serine/threonine protein kinases, *Curr. Opin. Drug Discovery Dev.*, 2, 96, 1999.

275. Garcia-Echeverria, C., Taxler, P., and Evans, D.B., ATP site-directed competitive and irreversible inhibitors of protein kinases, *Med. Res. Rev.* 20, 28, 2000.

276. Ding, S. et al., A combinatorial scaffold approach toward kinase-directed heterocycle libraries, *J. Am. Chem. Soc.*, 124, 1594, 2002.

277. Schreiber, S.L., Chemical genetics resulting from a passion for synthetic organic chemistry, *Bioorg. Med. Chem.*, 6, 1127, 1998.

278. Khersonsky, S.M. and Chang, Y-T., Forward chemical genetics: library scaffold design, *Comb. Chem. High Throughput Screen.*, 7, 645, 2004.

279. Tan, D.S. et al., Stereoselective synthesis of over two million compounds having structural features both reminiscent of natural products and compatible with miniaturized cell-based assays, *J. Am. Chem. Soc.*, 120, 8565, 1998.

280. Tan, D.S., Synthesis and preliminary evaluation of a library of polycyclic small molecules for the use in chemical genetic assays, *J. Am. Chem. Soc.*, 121, 9073, 1999.

281. New, D.C., Miller-Martini, D.M., and Wong, Y.H., Reporter gene assays and their applications to bioassays of natural products, *Phytother. Res.*, 17, 439, 2003.

282. Vesely, J. et al., Inhibition of cyclin-dependent kinases by purine analogs, *Eur. J. Biochem.*, 224, 771, 1994.

283. Norman, T.C. et al., A structure-based library approach to kinase inhibitors, *J. Am. Chem. Soc.*, 118, 7430, 1996.

284. Epple, R., Kudirka, R., and Greenberg, W.A., Solid-phase synthesis of nucleoside analogs, *J. Comb. Chem.*, 5, 292, 2003.

285. Kubo, I. et al., Clitocine, a new insecticidal nucleoside from the mushroom *Clitocybe inversa*, *Tetrahedron Lett.*, 27, 4277, 1986.

286. Varaprasad, C.V. et al., Synthesis of novel exocyclic amino nucleosides by parallel solid-phase combinatorial strategy, *Tetrahedron*, 59, 2297, 2003.

287. VanWagenen, B.C. et al., Ulosantoin, a potent insecticide from the sponge *Ulosa ruetzleri*, *J. Org. Chem.*, 58, 335, 1993.

288. Pettit, G.R. et al., Antineoplastic agents. 168. Isolation and structure of axinohydantoin, *Can. J. Chem.*, 68, 1621, 1990.

289. Nefzi, A. et al., Combinatorial chemistry: from peptides and peptidomimetics to small organic and heterocyclic compounds, *Bioorg. Med. Chem. Lett.*, 8, 2273, 1998.

290. Bleicher, K.H. et al., Parallel solution- and solid-phase synthesis of spirohydantoin derivatives as neurokinin-1 receptor ligands, *Bioorg. Med. Chem. Lett.*, 12, 2519, 2002.

291. Boehm, H.-J. et al., Novel inhibitors of DNA gyrase: 3D structure based biased needle screening, hit validation by biophysical methods, and 3D guided optimization: a promising alternative to random screening, *J. Med. Chem.*, 43, 2664, 2000.

292. Brophy, G.C. et al., Novel lignans from a *Cinnamomum sp.* from bougainville, *Tetrahedron Lett.*, 10, 5159, 1969.
293. Fecik, R.A. et al., Use of combinatorial and multiple parallel synthesis methodologies for the development of anti-infective natural products, *Pure Appl. Chem.*, 71, 559, 1999.
294. Lindslay, C.W. et al., Solid-phase biomimetic synthesis of carpanone-like molecules, *J. Am. Chem. Soc.*, 122, 422, 2000.
295. Wipf, P., Reeves, J.T., and Day, B.W., Chemistry and biology of curacin A, *Curr. Pharm. Des.*, 10, 1417, 2004.
296. Wipf, P. et al., Synthesis and biological evaluation of a focused mixture library of analogs of the antimitotic marine natural product curacin A, *J. Am. Chem. Soc.*, 122, 9391, 2000.
297. Wipf, P. et al., Synthesis and biological evaluation of structurally highly modified analogs of the antimitotic natural product curacin A, *J. Med. Chem.*, 45, 1901, 2002.
298. Kulkarni, B.A. et al., Combinatorial synthesis of natural product-like molecules using a first-generation spiroketal scaffold, *J. Comb. Chem.*, 4, 56, 2002.
299. Green, J., Solid phase synthesis of lavendustin A and analogs, *J. Org. Chem.*, 60, 4287, 1995.
300. Onoda, T. et al., Isolation of a novel tyrosine kinase inhibitor, lavendustin A, from *Streptomyces griseolavendus*, *J. Nat. Prod.*, 52, 1252, 1989.
301. Nielsen, J. and Lyngsø, L.O., Combinatorial solid-phase synthesis of balanol analogs, *Tetrahedron Lett.*, 37, 8439, 1996 and references cited therein.
302. Toske, S.G. et al., Aspergillamides A and B: modified cytotoxic tripeptides produced by a marine fungus of the genus *Aspergillus*, *Tetrahedron*, 54, 13459, 1998.
303. Beck, B., Hess, S., and Dömling, A., One-pot synthesis and biological evaluation of aspergillamides and analogs, *Bioorg. Med. Chem. Lett.*, 10, 1701, 2000.
304. Quiñoà, E. and Crews, P., Phenolic constituents of *Psammaplysilla*, *Tetrahedron Lett.*, 28, 3229, 1987.
305. Nicolaou, K.C. et al., Combinatorial synthesis through disulfide exchange: discovery of potent psammaplin A type antibacterial agents active against methicillin-resistant *Staphylococcus aureus* (MRSA), *Chem. Eur. J.*, 7, 4280, 2001.
306. Nicolaou, K.C. et al., Optimization and mechanistic studies of psammaplin A type antibacterial agents active against methicillin-resistant *Staphylococcus aureus* (MRSA), *Chem. Eur. J.*, 7, 4296, 2001.
307. Lee, C. et al., Two new constituents of Isodon excisus and their evaluation in an apoptosis inhibition assay, *J. Nat. Prod.*, 64, 659, 2001.
308. Nesterenko, V., Putt, K.S., and Hergenrother, P.J., Identification from a combinatorial library of a small molecule that selectively induces apoptosis in cancer cells, *J. Am. Chem. Soc.*, 125, 14672, 2003.
309. Eldefrawi, A.T. et al., Structure and synthesis of a potent glutamate receptor antagonist in wasp venom, *Proc. Natl. Acad. Sci. U.S.A.*, 85, 4910, 1998.
310. Strømgaard, K. et al., Solid-phase synthesis and biological evaluation of a combinatorial library of philanthotoxin analogs, *J. Med. Chem.*, 43, 4526, 2000.
311. Kim, Y.J. et al., Isolation and structural elucidation of sekothrixide, a new macrolide effective to overcome drug-resistance of cancer cell, *J. Antibiot.*, 44, 553, 1991.
312. Andrus, M.B. et al., The synthesis and evaluation of a solution phase indexed combinatorial library of non-natural polyenes for reversal of P-glycoprotein mediated multidrug resistance, *J. Org. Chem.*, 65, 4973, 2000.
313. Perez, J.J., Managing molecular diversity, *Chem. Soc. Rev.*, 34, 143, 2004.
314. Webb, T.R., Current directions in the evolution of compound libraries, *Curr. Opin. Drug Discovery Dev.*, 8, 303, 2005.
315. Arya, P. and Roth, H.-J., Combinatorial chemistry, *Curr. Opin. Chem. Biol.*, 9, 229, 2005.
316. Reayi, A. and Arya, P., Natural product-like chemical space: search for chemical dissectors of macromolecular interactions, *Curr. Opin. Chem. Biol.*, 9, 240, 2005.
317. Shang, S. and Tan, D.S., Advancing chemistry and biology through diversity-oriented synthesis of natural product-like libraries, *Curr. Opin. Chem. Biol.*, 9, 248, 2005.
318. Dekker, F.J., Koch, M.A., and Waldmann, H., Protein structure similarity clustering (PSSC) and natural product structure as inspiration sources for drug development and chemical genomics, *Curr. Opin. Chem. Biol.*, 9, 232, 2005.
319. Xiao, X.-Y. et al., Solid-phase combinatorial synthesis using MicroKan reactors, Rf tagging, and directed sorting, *Biotechnol. Bioeng.*, 71, 44, 2000.

320. Jung, G., *Combinatorial Chemistry: Synthesis, Analysis, Screening*; ed., Wiley-VCH, Weinheim, 1999.
321. Shapiro, M.J. and Gounarides, J.S., NMR methods utilized in combinatorial chemistry research, *Prog. Nucl. Magn. Reson. Spectr.*, 35, 153, 1999.
322. Keifer, P.A., NMR tools for biotechnology, *Curr. Opin. Biotechnol.*, 10, 34, 1999.
323. Swali, V., Langley, G.J., and Bradley, M., Mass spectrometric analysis in combinatorial chemistry, *Curr. Opin. Chem. Biol.*, 3, 337, 1999.
324. Fergus, S., Bender, A., and Spring, D.R., Assessment of structural diversity in combinatorial synthesis, *Curr. Opin. Chem. Biol.*, 9, 304, 2005.
325. Langer, T. and Krovat, E.M., Chemical feature-based pharmacophores and virtual library screening for discovery of new leads, *Curr. Opin. Drug Discovery Dev.*, 6, 370, 2003.

Index

A

Printed and bound by CPI Group (UK) Ltd, Croydon, CR0 4YY

23/10/2024

01778250-0009